# STUDENT'S SOLUTIONS MANUAL

## DAVID LUND
*University of Wisconsin – Ea*

D0641897

# INTRODUCTORY STATISTICS

## EIGHTH EDITION

# Neil A. Weiss
*Arizona State University*

PEARSON

Addison
Wesley

Boston  San Francisco  New York
London  Toronto  Sydney  Tokyo  Singapore  Madrid
Mexico City  Munich  Paris  Cape Town  Hong Kong  Montreal

Reproduced by Pearson Addison-Wesley from electronic files supplied by the author.

Copyright © 2008 Pearson Education, Inc.
Publishing as Pearson Addison-Wesley, 75 Arlington Street, Boston, MA 02116.

ISBN-13:   978-0-321-43413-5
ISBN-10:   0-321-43413-7

2 3 4 5 6 BB 09 08 07

# PREFACE

*Student's Solutions Manual* is designed to be used with the text *Introductory Statistics, Eighth Edition* by Neil A. Weiss. It provides complete solutions to all (1003) of the odd-numbered exercises, and all (382) of the review problems in Chapters 1-16 of the text. The solutions are more than answers - intermediate steps in the process of solving the exercises are also provided.

**Note:** Many of the numerical answers presented here were obtained by using computer software and the original set of data. If you solve problems by hand and do some intermediate rounding or use the summary statistics that are provided in the text for many of the exercises, your answers may differ slightly from the ones given in this manual. In a few instances, the solutions provided here may also differ slightly from the answers provided in Appendix B of the text, but we are not aware of any instances in which the differences are significant enough to alter the conclusions reached. As noted by the author, you should retain intermediate results in your calculator or on your computer and delay any rounding until the end of your calculations.

My thanks go to a number of people. Completing a work as large as this has required the understanding, assistance, patience, stamina, and tolerance of Neil and the staff at Addison Wesley, especially Joanne Ha and Joe Vetere, who have been such a great help in completing this project. I would also like to thank Gary Williams, Professor of Mathematics at Trident Technical College of Charleston, South Carolina for his excellent work in checking all of the solutions in this manual. Finally, I want to express my thanks to my wife, Judy, for her support and understanding throughout this project.

D. R. L.
Eau Claire, WI

# CONTENTS

## Exercises 1.1

**1.1**   (a)  The *population* is the collection of all individuals or items under consideration in a statistical study.

(b)  A *sample* is that part of the population from which information is obtained.

**1.3**   Descriptive methods are used for organizing and summarizing information and include graphs, charts, tables, averages, measures of variation, and percentiles.

**1.5**   (a)  An *observational study* is a study in which researchers simply observe characteristics and take measurements.

(b)  A *designed experiment* is a study in which researchers impose treatments and controls and *then* observe characteristics and take measurements.

**1.7**   This study is inferential. Data from a sample of Americans are used to make an estimate of (or an inference about) average TV viewing time for all Americans.

**1.9**   This study is descriptive. It is a summary of the assessment results for level of performance of all senior geography majors in 2003 and 2004 at one institution.

**1.11**  This study is descriptive. It is a summary of the annual final closing values of the Dow Jones Industrial Average at the end of December for the years 1997-2004.

**1.13**  (a)  This study is inferential. It would have been impossible to survey all U.S. adults about their opinions on Darwinism. Therefore, the data must have come from a sample. Then inferences were made about the opinions of all U.S. adults.

(b)  The population consists of all U.S. adults. The sample consists only of those U.S. adults who took part in the survey.

**1.15**  (a)  This statement is descriptive since it only tells what was said by those who were surveyed.

(b)  Then the statement would be inferential since the data has been used to provide an estimate of what <u>all</u> Americans would choose.

**1.17**  Designed experiment. The researchers did not simply observe the two groups of children, but instead randomly assigned one group to receive the Salk vaccine and the other to get a placebo.

**1.19**  Observational study. The researchers had no control over who became an elite distance runner and who did not. They simply observed the skinfold thickness of a sample of people in each group.

**1.21**  Designed experiment. The researchers did not simply observe the three groups of patients, but instead randomly assigned some patients to receive optimal pharmacologic therapy, some to receive optimal pharmacologic therapy and a pacemaker, and some to receive optimal pharmacologic therapy and a pacemaker-defibrillator combination.

**1.23**  (a)  This statement is inferential since it is a statement about all Americans based on a poll. We can be reasonably sure that this is the case since the time and cost of questioning every single American on this issue would be prohibitive. Furthermore, by the time everyone could be questioned, many would have changed their minds.

(b)  To make it clear that this is a descriptive statement, the new statement could be, "Of 1032 American adults surveyed, 73% favored a law that would require every gun sold in the United States to be test-fired first, so law enforcement would have its fingerprint in case it were ever used in a crime." To rephrase it as an inferential statement, use "Based on a sample of 1032 American adults, it is estimated that 73% of American adults favor a law that would require every gun sold in the United States to be test-fired first, so law enforcement would have its fingerprint in case it were ever used in a crime."

**1.25** (a) The figure 42,800 is an inferential statistic since it is indicated in the statement that it is a projection (probably based on incomplete data for the year 2004). The data may be incomplete in part because in April, 2005, there might still be deaths to occur that are the result of traffic accidents in 2004.

    (b) The figure 42,643 is a descriptive statistic since it reflects the actual number of traffic deaths for the year 2003.

### Exercises 1.2

**1.27** A census is generally time consuming, costly, frequently impractical, and sometimes impossible.

**1.29** The sample should be representative so that it reflects as closely as possible the relevant characteristics of the population under consideration.

**1.31** Dentists form a high-income group whose incomes are not representative of the incomes of Seattle residents in general.

**1.33** (a) Probability sampling consists of using a randomizing device such as tossing a coin or consulting a random number table to decide which members of the population will constitute the sample.

    (b) No. It is possible for the randomizing device to randomly produce a sample that is not representative.

    (c) Probability sampling eliminates unintentional selection bias, permits the researcher to control the chance of obtaining a non-representative sample, and guarantees that the techniques of inferential statistics can be applied.

**1.35** Simple random sampling.

**1.37** (a) GLS, GLA, GLT, GSA, GST, GAT, LSA, LST, LAT, SAT.

    (b) There are 10 samples, each of size three. Each sample has a one in 10 chance of being selected. Thus, the probability that a sample of three officials is the first sample on the list presented in part (a) is 1/10. The same is true for the second sample and for the tenth sample.

**1.39** (a) E,M,P,L    E,M,L,B    E,P,A,B    M,P,A,B

        E,M,P,A    E,M,A,B    E,L,A,B    M,L,A,B

        E,M,P,B    E,P,L,A    M,P,L,A    P,L,A,B

        E,M,L,A    E,P,L,B    M,P,L,B

    (b) One procedure for taking a random sample of four representatives from the six is to write the initials of the representatives on six separate pieces of paper, place the six slips of paper into a box, and then, while blindfolded, pick four of the slips of paper. Or, number the representatives 1-6, and use a table of random numbers or a random-number generator to select four different numbers between 1 and 6.

    (c) 1/15; 1/15

**1.41** (a) C,W,H   C,W,V   C,W,A   C,H,V   C,H,A

        C,V,A   W,H,V   W,H,A   W,V,A   H,V,A

    (b) 1/10; 1/10

**1.43** (a) I am using Table I to obtain a list of 10 random numbers between 1 and 500 as follows. First, I pick a random starting point by closing my eyes and putting my finger down on the table.

My finger falls on the three digits located at the intersection of line number 00 with columns 34, 35, and 36.  The selected digits are 356.  This is my starting point.

I now go down the table and record the three-digit numbers appearing directly beneath 356.  Since I want numbers between 1 and 500 only, I throw out numbers between 501 and 999, inclusive.  I also discard the number 000.

After 356, I skip 876, record 351, skip 717, record 239, 455, 431, 008, skip 900, 721, record 259, 068, 156, skip 570, 540, 937, 989, and record 047.

I've finished recording the 10 random numbers.  In summary, these are:

$$356 \quad 239 \quad 431 \quad 259 \quad 156$$
$$351 \quad 455 \quad 008 \quad 068 \quad 047$$

(b)  We can use Excel to generate random numbers.  If we enter the expression =RAND() in a cell, Excel returns a random number between 0 and 1.  If we multiply this number by 500, we will get a number between 0 and 500, and if we take the integer part of that result, we will get one of the integers 0 through 499.  If we add 1 to this result, we will get a random number between 1 and 500.  For example, in cell A1, we enter the expression =INT(500*RAND())+1.  One of integers 1 through 500 will result.  We copy the content of A1 to the clipboard and paste it into cells A2 through A30.  Then we choose the first 10 unique digits for our sample.  Our result is 489, 451, 61, 114, 389, 381, 364, 166, 221, 437, 266, 46, 422, 388, 401, 387, 276, 248, 21, 198.  The first 10 unique digits are 489, 451, 61, 114, 389, 381, 364, 166, 221, and 437.  Your result may be different from ours.  [Note:  Each time the ENTER key is depressed, the random numbers will be recalculated, so make sure that you do not press the ENTER key until you have recorded all of your random numbers.]

1.45  (a)  The possible samples of size one are    G    L    S    A    T

      (b)  There is no difference between obtaining a sample of size one and selecting one official at random.

1.47  No.  Only adults with access to the Internet would have been able to respond to the survey.  Thus, not every adult had an equal chance of being chosen for the sample.

**Exercises 1.3**

1.49  (a)  Answers will vary, but here is the procedure:  (1) Divide the population size, 500, by the sample size, 10, and round down to the nearest whole number if necessary; this gives 50.  (2) Use a table of random numbers (or a similar device) to select a number between 1 and 50, call it $k$.  (3)  List every 50th number, starting with $k$, until 10 numbers are obtained; thus, the first number on the required list of 10 numbers is $k$, the second is $k+50$, the third is $k+100$, and so forth (e.g., if $k=6$, then the numbers on the list are 6, 56, 106, ...).

      (b)  Systematic random sampling is easier.

      (c)  The answer depends on the purpose of the sampling.  If the purpose of sampling is not related to the size of the sales outside the U.S., systematic sampling will work.  However, since the listing is a ranking by amount of sales, if k is low (say 2), then the sample will contain firms that, on the average, have higher sales outside the U.S. than the population as a whole.  If the k is high, (say 49) then the sample will contain firms that, on the average, have lower sales than the population as a whole.  In either of those cases, the sample would not be representative of the population in regard to the amount of sales outside the U.S.

**1.51**  (a)  Number the suites from 1 to 48, use a table of random numbers to randomly select three of the 48 suites, and take as the sample the 24 dormitory residents living in the three suites obtained.

(b)  Probably not, since friends are more likely to have similar opinions than are strangers.

(c)  There are 384 students in total.  Freshmen make up 1/3 of them. Sophomores make up 7/24 of them, Juniors 1/4, and Seniors 1/8. Multiplying each of these fractions by 24 yields the proportional allocation, which dictates that the number of freshmen, sophomores, juniors, and seniors selected should be, respectively, 8, 7, 6, and 3. Thus a stratified sample of 24 dormitory residents can be obtained as follows:  Number the freshmen dormitory residents from 1 to 128 and use a table of random numbers to randomly select 8 of the 128 freshman dormitory residents; number the sophomore dormitory residents from 1 to 112 and use a table of random numbers to randomly select 7 of the 112 sophomore dormitory residents; and so forth.

**1.53**  Stratified Sampling.  The entire population is naturally divided into subpopulations, one from each lake, and random sampling is done from each lake.  The stratified sampling is not with proportional allocation since that would require knowing how many fish were in each lake.

**1.55**  (a)  This is a poll taken by calling randomly selected U.S. adults.  Thus, the sampling design appears to be simple random sampling, although it is possible that a more complex design was used to ensure that various political, religious, educational, or other types of groups were proportionately represented in the sample.

(b)  The approximate sample size for the second question was 78% of 1010 or 788.

(c)  The approximate sample size for the third question was 28% of 788 or 221.

**1.57**  From the information about the sample, we can conclude that the population of interest consists of all adults in the continental U.S.  The sample size was 2010 except that for questions about politics, only registered voters were considered part of the sample.  The sample size for those questions was 1,637.

The overall procedure for drawing the sample was multistage (actually, three stages were used) sampling: the first stage was to randomly select 520 geographic points in the continental U.S.; then proportional sampling was used to randomly sample a number of households with telephones from each of the 520 regions in proportion to its population; finally, once each household was selected, a randomizing procedure was used to ensure that the correct numbers of adult male and female respondents were included in the sample.

The last paragraph indicates the confidence that the poll-takers had in the results of the survey, that is, that there is a 95% chance that the sample results will not differ by more than 2.2 percentage points in either direction from the true percentage that would have been obtained by surveying all adults in the actual population, or by more than 2.5 percentage points in either direction from the true percentage that would have been obtained by surveying all registered voters in the population. The last sentence says that smaller samples have a larger Amargin of error,@ an explanation for the difference in the maximum percentage points of error for all adults and for registered voters.

### Exercises 1.4

**1.59**  (a)  Experimental units are the individuals or items on which the experiment is performed.

(b)  When the experimental units are humans, we call them subjects.

**1.61**   (a)   There were three treatments.

        (b)   The first group, the one receiving only the pharmacologic therapy, would be considered the control group.

        (c)   There were three treatment groups. The first received only pharmacologic therapy, the second received pharmacologic therapy plus a pacemaker, and the third received pharmacologic therapy plus a pacemaker-defibrillator combination.

        (d)   The first group (control) contained 1/5 of the 1520 patients or 304. The other two groups each contained 2/5 of the 1520 patients or 608.

        (e)   Each patient could be randomly assigned a number from 1 to 1520. Any patient assigned a number between 1 and 304 would be assigned to the control group; any patient assigned to the next 608 numbers (305 to 912) would be assigned to receive the pharmacologic therapy plus a pacemaker; and any patient assigned a number between 913 and 1520 would receive pharmacologic therapy plus a pacemaker-defibrillator combination. Each random number would be used only once to ensure that the resulting treatment groups were of the intended sizes.

**1.63**   (a)   Experimental units: batches of the product being sold

        (b)   Response variable: the number of units of the product sold

        (c)   Factors: two factors - display type and pricing scheme

        (d)   Levels of each factor: three types of display of the product and three pricing schemes

        (e)   Treatments: the nine different combinations of display type and price resulting from testing each of the three pricing schemes with each of the three display types

**1.65**   (a)   Experimental units: female lions

        (b)   Response variable: whether or not the female lion approached the male lion dummy

        (c)   Factors: length and color of the mane on the male lion dummy

        (d)   Levels of each factor: two different mane lengths and two different mane colors

        (e)   Treatments: the four combinations of mane length and color

**1.67**   Double-blinding guards against bias, both in the evaluations and in the responses. In the Salk vaccine experiment, double-blinding prevented a doctor's evaluation from being influenced by knowing which treatment (vaccine or placebo) a patient received; it also prevented a patient's response to the treatment from being influenced by knowing which treatment he or she received.


**Review Problems for Chapter 1**

**1.**   Student exercise.

**2.**   Descriptive statistics are used to display and summarize the data to be used in an inferential study. Preliminary descriptive analysis of a sample often reveals features of the data that lead to the choice or reconsideration of the choice of the appropriate inferential analysis procedure.

**3.**   Descriptive study. The scores are merely reported.

**4.**   Descriptive study. The paragraph describes the results of a sample of U.S. residents.

**5.**   Inferential study. The results of a sample are used to make inferences about the age distribution of <u>all</u> British backpackers in South Africa.

**6.**   (a)   Since 18% <u>reported</u> that they had abused Vicodin, this figure applies to the sample and is therefore descriptive.

   (b)  The 4.3 million youths abusing Vicodin is clearly inferential since it applies to the entire population, not just the sample.

7.  (a)  An *observational study* is a study in which researchers simply observe characteristics and take measurements.

   (b)  A *designed experiment* is a study in which researchers impose treatments and controls and *then* observe characteristics and take measurements.

8.  This is an observational study. To be a designed experiment, the researchers would have to have the ability to assign some children at random to live in persistent poverty during the first 5 years of life or to not suffer any poverty during that period. Clearly that is not possible.

9.  This is a designed experiment since the researcher is imposing a treatment and then observing the results.

10.  A literature search should be made before planning and conducting a study.

11.  (a)  A representative sample is one that reflects as closely as possible the relevant characteristics of the population under consideration.

   (b)  Probability sampling involves the use of a randomizing device such as tossing a coin or die, using a random number table, or using computer software that generates random numbers to determine which members of the population will make up the sample.

   (c)  A sample is a simple random sample if all possible samples of a given size are equally likely to be the actual sample selected.

12.  Because Yale is a very expensive school, incomes of parents of Yale students will not be representative of the incomes of all college students' parents.

13.  (a)  This method does not involve probability sampling. No randomizing device is being used and people who do not visit the campus cafeteria have no chance of being included in the sample.

   (b)  The dart throwing is a randomizing device that makes all samples of size 20 equally likely. This is probability sampling.

14.  (a)  HA,3C,G4   HA,3C,ZV    HA,3C,F9    HA,G4,ZV    HA,G4,F9

          HA,ZV,F9   3C,G4,ZV    3C,G4,F9    3C,ZV,F9    G4,ZV,F9

   (b)  Since each of the 10 samples of size three is equally likely, there is a 1/10 chance that the sample chosen is the first sample in the list, 1/10 chance that it is the second sample in the list, and 1/10 chance that it is the tenth sample in the list.

   (c)  (i) Make five slips of paper with each airline on one slip. Draw three slips at random. (ii) Make 10 slips of paper, each having one of the combinations in part (a). Draw one slip at random. (iii) Number the five airlines from 1 to 5. Use a random number table or random number generator to obtain three distinct random numbers between 1 and 5 inclusive.

   (d)  Your method and result may differ from ours. We rolled a die (ignoring 6's) and got 2, 5, 2, 6, 4. So our sample consists of Corporate Express, Frontier Airlines, and Air Midwest.

15.  (a)  Table I can be employed to obtain a sample of 15 random numbers between 1 and 100 as follows. First, I pick a random starting point by closing my eyes and putting my finger down on the table.

        My finger falls on three digits located at the intersection of a line with three columns. (Notice that the first column of digits is labeled "00" rather than "01".) This is my starting point.

        I now go down the table and record all three-digit numbers appearing directly beneath the first three-digit number that are between 001 and

100 inclusive. I throw out numbers between 101 and 999, inclusive. I also discard the number 0000. When the bottom of the column is reached, I move over to the next sequence of three digits and work my way back up the table. Continue in this manner. When 10 distinct three-digit numbers have been recorded, the sample is complete.

(b) Starting in row 10, columns 7-9, we skip 484, 797, record 082, skip 586, 653, 452, 552, 155, record 008, skip 765, move to the right and record 016, skip 534, 593, 964, 667, 452, 432, 594, 950, 670, record 001, skip 581, 577, 408, 948, 807, 862, 407, record 047, skip 977, move to the right, skip 422 and all of the rest of the numbers in that column, move to the right, skip 732, 192, record 094, skip 615 and all of the rest of the numbers in that column, move to the right, record 097, skip 673, record 074, skip 469, 822, record 052, skip 397, 468, 741, 566, 470, record 076, 098, skip 883, 378, 154, 102, record 003, skip 802, 841, move to the right, skip 243, 198, 411, record 089, skip 701, 305, 638, 654, record 041, skip 753, 790, record 063.

The final list of numbers is 082, 008, 016, 001, 047, 094, 097, 074, 052, 076, 098, 003, 089, 041, 063.

(c) Using Excel, we enter the expression =INT(100*RAND())+1 in cell A1, copy the content of A1 to the clipboard and paste it into cells A2 to A20. Then we use the first 15 unique numbers as our random sample. Our results were the numbers 46, 99, 90, 31, 75, 98, 79, 14, 44, 13, 66, 49, 37, 87, 73, 26, 61, 71, 72, 2. Thus our sample consists of the first 15 numbers 46, 99, 90, 31, 75, 98, 79, 14, 44, 13, 66, 49, 37, 87, 73. Your sample may be different.

16. (a) Systematic random sampling is done by first dividing the population size by the sample size and rounding the result down to the next integer, say m. Then we select one random number, say k, between 1 and m inclusive. That number will be the first member of the sample. The remaining members of sample will be those numbered k+m, k+2m, k+3m, ... until a sample of size n has been chosen. Systematic sampling will yield results similar to simple random sampling as long as there is nothing systematic about the way the members of the population were assigned their numbers.

(b) In cluster sampling, clusters of the population (such as blocks, precincts, wards, etc.) are chosen at random from all such possible clusters. Then every member of the population lying within the chosen clusters is sampled. This method of sampling is particularly convenient when members of the population are widely scattered and is most appropriate when the members of each cluster are representative of the entire population. Cluster sampling can save both time and expense in doing the survey, but can yield misleading results if individual clusters are made up of subjects with very similar views on the topic being surveyed.

(c) In stratified random sampling with proportional allocation, the population is first divided into subpopulations, called strata, and simple random sampling is done within each stratum. Proportional allocation means that the size of the sample from each stratum is proportional to the size of the population in that stratum. This type of sampling may improve the accuracy of the survey by ensuring that those in each stratum are more proportionately represented than would be the case with cluster sampling or even simple random sampling. Ideally, the members of each stratum should be homogeneous relative to the characteristic under consideration. If they are not homogeneous within each stratum, simple random sampling would work just as well.

17. (a) Answers will vary, but here is the procedure: (1) Divide the population size, 100, by the sample size 15, and round down to the nearest whole number; this gives 6. (2) Use a table of random numbers

(or a similar device) to select a number between 1 and 6, call it *k*. (3) List every 6th number, starting with *k*, until 15 numbers are obtained; thus the first number on the required list of 15 numbers is *k*, the second is *k*+6, the third is *k*+12, and so forth (e.g., if *k*=4, then the numbers on the list are 4, 10, 16, ...).

(b)    Yes, unless for some reason there is some kind of trend or a cyclical pattern in the listing of the athletes.

18.    (a)    The number of full professors should be (205/820) x 40 = 10. Similarly, proportional allocation dictates that 16 associate professors, 12 assistant professors, and 2 instructors be selected.

(b)    The procedure is as follows: Number the full professors from 1 to 205, and use Table I to randomly select 10 of the 205 full professors; number the associate professors from 1 to 328, and use Table I to randomly select 16 of the 328 associate professors; and so on.

19.    The statement under the vote is a disclaimer as to the validity of the survey. Since the vote reflects only the responses of volunteers who chose to vote, it can not be regarded as representative of the public in general, some of whom do not use the Internet, nor as representative of Internet users since the sample was not chosen at random from either group.

20.    (a)    This is a designed experiment.

(b)    The treatment group consists of the 158 patients who took AVONEX. The control group consists of the 143 patients who were given a placebo. The treatments were the AVONEX and the placebo.

21.    The three basic principles of experimental design are control, randomization, and replication. Control refers to methods for controlling factors other than those of primary interest. Randomization means randomly dividing the subjects into groups in order to avoid unintentional selection bias in constituting the groups. Replication means using enough experimental units or subjects so that groups resemble each other closely and so that there is a good chance of detecting differences among the treatments when such differences actually exist.

22.    (a)    Experimental units: tomato plants

(b)    Response variable: yield of tomatoes

(c)    Factor(s): tomato variety and density of plants

(d)    Levels of each factor: These are not given, but tomato varieties tested would be the levels of variety and the different densities of plants would be the levels of density.

(e)    Treatments: Each treatment would be one of the combinations of a variety planted at a given plant density.

23.    (a)    Experimental Units: The children

(b)    Response variable: Whether or not the child was able to open the bottle

(c)    Factors: The container designs

(d)    Levels of each factor: Three (types of containers)

(e)    Treatments: The container designs

24.    This is a completely randomized design. All of the experimental units (batches of doughnuts) were assigned at random to the four treatments (four different fats).

25.    (a)    This is a completely randomized design since the 24 cars were randomly assigned to the 4 brands of gasoline.

(b)    This is a randomized block design. The four different gasoline brands are randomly assigned to the four cars in each of the six car model groups. The blocks are the six groups of four identical cars each.

(c) If the purpose is to learn about the mileage rating of one particular car model with each of the four gasoline brands, then the completely randomized design is appropriate. But if the purpose is to learn about the performance of the gasoline across a variety of cars (and this seems more reasonable), then the randomized block design is more appropriate and will allow the researcher to determine the effect of car model as well as of gasoline type on the mileage obtained.

26. The explanation informs the reader that only random sampling of a population can be used to draw valid inferences about the population as a whole.

27. The data in this study were clearly not collected via a controlled experiment in which some participants were forced to do crossword puzzles, practice musical instruments, play board games, or read while others were not allowed to do any of those activities. Therefore, any data relative to these activities and dementia arose as a result of observing whether or not the subjects in the study carried out any of those activities and whether or no they had some form of dementia. Since this would be an observational study, no statement of cause and effect can rightfully be made. It cannot be claimed as result of the study that "Crosswords Reduce Risk of Dementia."

**Exercises 2.1**

**2.1** (a) Hair color, model of car, and brand of popcorn are qualitative variables.

(b) Number of eggs in a nest, number of cases of flu, and number of employees are discrete, quantitative variables.

(c) Temperature, weight, and time are quantitative continuous variables.

**2.3** (a) Qualitative data result from observing and recording values of a qualitative variable, such as, color or shape.

(b) Discrete, quantitative data are values of a discrete quantitative variable. Values usually result from counting something.

(c) Continuous, quantitative data are values of a continuous variable. Values are usually the result of measuring something such as temperature that can take on any value in a given interval.

**2.5** Of qualitative and quantitative (discrete and continuous) types of data, only qualitative yields nonnumerical data.

**2.7** (a) The second column consists of *quantitative, discrete* data. This column provides the ranks of the cities with the highest temperatures.

(b) The third column consists of *quantitative, continuous* data since temperatures can take on any value from the interval of numbers found on the temperature scale. This column provides the highest temperature in each of the listed cities.

(c) The information that Phoenix is in Arizona is qualitative data since it is nonnumeric.

**2.9** (a) The first column consists of *quantitative, discrete* data. This column provides the ranks of the cities with the highest percentage of internet access connected via broadband in August, 2004. These are whole numbers.

(b) The cities listed in the second column are qualitative data since they are nonnumerical.

(c) The third column contains percentages, which are ratios of whole numbers. Ratios of whole numbers cannot be irrational numbers and therefore there are gaps in the number line that represent values that the ratios cannot assume. These data are therefore *quantitative, discrete*.

**2.11** The first two columns contain quantitative, discrete data in the form of ranks. These are whole numbers. The third and fourth columns contain qualitative data in the form of names. The last column contains the number of viewers of the programs. Total number of viewers is a whole number and therefore *quantitative, discrete* data.

**2.13** The first and fourth columns are nonnumerical and are therefore *qualitative* data. The second, third, and fifth columns are measures of size, time, and weight, all of which are *quantitative, continuous* data.

**Exercises 2.2**

**2.15** One important reason for grouping data is that grouping often makes a large and complicated set of data more compact and easier to understand.

**2.17** The three most important guidelines in choosing the classes for grouping a data set are: (1) the number of classes should be small enough to provide an effective summary, but large enough to display the relevant characteristics of the data; (2) each observation must belong to one, and only one, class; and (3) whenever feasible, all classes should have the same width.

**2.19** (a) True. Having identical frequency distributions implies that the total number of observations and the numbers of observations in each class are identical. Thus, the relative frequencies will also be identical.

(b) False. Having identical relative frequency distributions means that the ratio of the count in each class to the total is the same for both frequency distributions. However, one distribution may have twice (or some other multiple) the total number of observations as the other. For example, two distributions with counts of 5, 4, 1 and 10, 8, 2 would be different, but would have the same relative frequency distribution.

(c) If the two data sets have the same number of data values, either a frequency distribution or a relative-frequency distribution is suitable. If, however, the two data sets have different numbers of observations, using relative-frequency distributions is more appropriate because the total of each set of relative frequencies is 1, putting both distributions on the same basis for comparison.

**2.21** In the first method for depicting classes, we used the notation $a < b$ to mean values that are greater than or equal to $a$ and up to, but not including $b$, such as $30 < 40$ to mean a range of values greater than or equal to 30, but strictly less than 40. In the alternate method, we used the notation $a-b$ to indicate a class that extends from $a$ to $b$, including both. For example, 30-39 is a class that includes both 30 and 39. The alternate method is especially appropriate when all of the data values are integers. If the data include values like 39.7 or 39.93, the first method is more advantageous since the cutpoints remain integers; whereas, in the alternate method, the upper limits for each class would have to be expressed in decimal form such as 39.9 or 39.99.

**2.23** When grouping data using classes that each represents a single possible numerical value, the midpoint of each class would be the same as the value in that class. Thus, listing the midpoints would be redundant.

**2.25** The first class to construct is $52 < 54$. Since all classes are to be of equal width 2, the second class begins with 54. All of the classes are presented in column 1. The last class to construct is $74 < 76$, since the largest single data value is 75.3. Having established the classes, we tally the speed figures into their respective classes. These results are presented in column 2, which lists the frequencies. Dividing each frequency by the total number of observations, which is 35, results in each class's relative frequency. The relative frequencies for all classes are presented in column 3. By averaging the lower and upper class cutpoints for each class, we arrive at the class midpoint for each class. The class midpoints for all classes are presented in column 4.

| Speed (MPH) | Frequency | Relative Frequency | Midpoint |
|---|---|---|---|
| 52 < 54 | 2 | 0.057 | 53 |
| 54 < 56 | 5 | 0.143 | 55 |
| 56 < 58 | 6 | 0.171 | 57 |
| 58 < 60 | 8 | 0.229 | 59 |
| 60 < 62 | 7 | 0.200 | 61 |
| 62 < 64 | 3 | 0.086 | 63 |
| 64 < 66 | 2 | 0.057 | 65 |
| 66 < 68 | 1 | 0.029 | 67 |
| 68 < 70 | 0 | 0.000 | 69 |
| 70 < 72 | 0 | 0.000 | 71 |
| 72 < 74 | 0 | 0.000 | 73 |
| 74 < 76 | 1 | 0.029 | 75 |
|  | 35 | 1.001 |  |

Note that the relative frequencies sum to 1.001, not 1.00, due to round-off errors in the individual relative frequencies.

**2.27**  The first class to construct is 52-53.9.  Since all classes are to be of equal width, the second class has limits of 54 and 55.9.  All of the classes are presented in column 1.  The last class to construct is 74-75.9 since the largest single data value is 75.3.  Having established the classes, we tally the speed figures into their respective classes.  These results are presented in column 2, which lists the frequencies.  Dividing each frequency by the total number of observations, 35, results in each class's relative frequency, which is presented in column 3.  By averaging the lower limit for each class with the upper limit of the same class, we arrive at the class mark for each class.  The class marks for all classes are presented in column 4.

| Speed (MPH) | Frequency | Relative Frequency | Class Mark |
|---|---|---|---|
| 52-53.9 | 2 | 0.057 | 52.95 |
| 54-55.9 | 5 | 0.143 | 54.95 |
| 56-57.9 | 6 | 0.171 | 56.95 |
| 58-59.9 | 8 | 0.229 | 58.95 |
| 60-61.9 | 7 | 0.200 | 60.95 |
| 62-63.9 | 3 | 0.086 | 62.95 |
| 64-65.9 | 2 | 0.057 | 64.95 |
| 66-67.9 | 1 | 0.029 | 66.95 |
| 68-69.9 | 0 | 0.000 | 68.95 |
| 70-71.9 | 0 | 0.000 | 70.95 |
| 72-73.9 | 0 | 0.000 | 72.95 |
| 74-75.9 | 1 | 0.029 | 74.95 |
|  | 35 | 1.001 |  |

Note that the relative frequencies sum to 1.001, not 1.00, due to round-off errors in the individual relative frequencies.

**2.29**  The first class to construct is 11 < 14.  This has width 3 with its midpoint at 12.5.  Since all classes are to be of equal width 3, the second class begins with 14.  All of the classes are presented in column 1.  The last class to construct is 26 < 29, since the largest single data value is 27.0.  Having established the classes, we tally the audience sizes into their respective classes.  These results are presented in column 2, which lists the frequencies.  Dividing each frequency by the total number of observations, which is 35, results in each class's relative frequency.  The relative frequencies for all classes are presented in column 3.  By averaging the lower and upper class cutpoints for each class, we arrive at the class midpoint for each class.  The class midpoints for all classes are presented in column 4.

| Audience (Millions) | Frequency | Relative Frequency | Midpoint |
|---|---|---|---|
| 11 < 14 | 5 | 0.25 | 12.5 |
| 14 < 17 | 4 | 0.20 | 15.5 |
| 17 < 20 | 6 | 0.30 | 18.5 |
| 20 < 23 | 2 | 0.10 | 21.5 |
| 23 < 26 | 2 | 0.10 | 24.5 |
| 26 < 29 | 1 | 0.05 | 27.5 |
|  | 20 | 1.00 |  |

**2.31** Since the data values range from 0 to 4, we construct a table with classes based on a single value.  The resulting table follows.

| Number of Siblings | Frequency | Relative Frequency |
|---|---|---|
| 0 | 8 | 0.200 |
| 1 | 17 | 0.425 |
| 2 | 11 | 0.275 |
| 3 | 3 | 0.075 |
| 4 | 1 | 0.025 |
|  | 40 | 1.000 |

**2.33** The classes are the NCAA wrestling champions and are presented in column 1. The frequency distribution of the champions is presented in column 2. Dividing each frequency by the total number of champions, which is 25, results in each class's relative frequency.  The relative frequency distribution is presented in column 3.

| Champion | Frequency | Relative Frequency |
|---|---|---|
| Iowa | 15 | 0.60 |
| Iowa St. | 1 | 0.04 |
| Minnesota | 2 | 0.08 |
| Arizona St. | 1 | 0.04 |
| Oklahoma St. | 6 | 0.24 |
|  | 25 | 1.00 |

**2.35** (a)  Using Minitab, retrieve the data from the Weiss-Stats-CD.  Column 1 contains the numbers of pups borne in a lifetime for each of 80 female Great White Sharks.  From the tool bar, select **Stat ▶ Tables ▶ Tally Individual Variables**, double-click on PUPS in the first box so that PUPS appears in the **Variables** box, put a check mark next to Counts and Percents under Display, and click **OK**.  The result is

| PUPS | Count | Percent |
|---|---|---|
| 3 | 2 | 2.50 |
| 4 | 5 | 6.25 |
| 5 | 10 | 12.50 |
| 6 | 11 | 13.75 |
| 7 | 17 | 21.25 |
| 8 | 17 | 21.25 |
| 9 | 11 | 13.75 |
| 10 | 4 | 5.00 |
| 11 | 2 | 2.50 |
| 12 | 1 | 1.25 |

(b) As the number of pups increases from 3 to 12, the counts increase from 2 to a maximum of 17 at 7 and 8 pups, and then decrease back down to 1 again at 12 pups.

**2.37** (a) In Minitab, retrieve the data from the *WeissStats* CD. Then, assuming that the CD drive is drive D, type in Minitab's session window after the MTB> prompt [You may have to first choose **Enable Commands** from **Editor** on the Tool bar.] the command %D:\Minitab_Macros\Group.mac 'NAEP' and press the **ENTER** key. We are given three options for specifying the classes. Since we want the first class to have lower cutpoint 234.5 and a class width of 2, we select the third option (3) by entering 3 after the DATA> prompt, press the **ENTER** key, and then type 234.5 2 when prompted to enter the cutpoint and class width of the first class. Press the **ENTER** key again. The resulting output is

```
Grouped-data table for NAEP          N = 50
Row  LowerCut  UpperCut  Freq  RelFreq  Midpoint
  1    234.5    236.5     3    0.06     235.5
  2    236.5    238.5     3    0.06     237.5
  3    238.5    240.5     1    0.02     239.5
  4    240.5    242.5     3    0.06     241.5
  5    242.5    244.5     2    0.04     243.5
  6    244.5    246.5     5    0.10     245.5
  7    246.5    248.5     4    0.08     247.5
  8    248.5    250.5     6    0.12     249.5
  9    250.5    252.5     9    0.18     251.5
 10    252.5    254.5     9    0.18     253.5
 11    254.5    256.5     3    0.06     255.5
 12    256.5    258.5     2    0.04     257.5
```

Alternatively, enter the NAEP data in Excel with the variable name NAEP in the first row. Use the cursor to highlight the entire data set including the name. Then from the tool bar, select **DDXL ▶ Tables**; select **Frequency Table** from the **Function type** drop-down list box; specify NAEP in the **Categorical Variable** text box and click **OK**. The result is the following table:

```
Total Cases      50
Number of Categories 21
Group   Count  %
235       1    2
236       2    4
238       3    6
239       1    2
241       2    4
242       1    2
243       1    2
244       1    2
245       4    8
246       1    2
247       2    4
248       2    4
249       3    6
250       3    6
251       5   10
252       4    8
253       6   12
254       3    6
255       1    2
256       2    4
257       2    4
```

(b) Using Minitab, Column 3 contains the RESULT.  From the tool bar, select **Stat ▶ Tables ▶ Tally Individual Variables**, double-click on RESULT in the first box so that RESULT appears in the **Variables** box, put a check mark next to Counts and Percents under Display, and click **OK**.  The result is

```
RESULT  Count  Percent
Bush     30    60.00
Gore     20    40.00
N=       50
```

**2.39** (a) Since the volumes are given in hundreds, we must first convert the numbers to millions by dividing by 10,000.  The new data (in increasing order) are 3.960, 4.873, 5.199, 5.533, 6.105, 6.485, 7.625, 7.640, 7.925, 11.752, 11.761, 11.855, 12.995, 13.177, 13.635, 14.306, 15.095, 16.265, 17.337, 19.193, 19.557, 19.875, 24.198, 25.125, 30.379, 30.937, 40.119, 72.573, 96.678, 191.491.

The first class to construct is $0 < 5$.  Since all classes are to be of equal width 5, the second class is $5 < 10$.  All of the classes are presented in column 1. Having established the classes, we tally the volume figures into their respective classes.  These results are presented in column 2, which lists the frequencies.  Dividing each frequency by the total number of observations, which is 30, results in each class's relative frequency.  The relative frequencies for all classes are presented in column 3.  By averaging the cutpoints for each class, we arrive at the class midpoint for each class.  The class midpoints for all classes are presented in column 4.

| Volume | Frequency | Relative Frequency | Midpoint |
|---|---|---|---|
| $0 < 5$ | 2 | 0.07 | 2.5 |
| $5 < 10$ | 7 | 0.23 | 7.5 |
| $10 < 15$ | 7 | 0.23 | 12.5 |
| $15 < 20$ | 6 | 0.20 | 17.5 |
| $20 < 25$ | 1 | 0.03 | 22.5 |
| $25 < 30$ | 1 | 0.03 | 27.5 |
| $30 < 35$ | 2 | 0.07 | 32.5 |
| $35 < 40$ | 0 | 0.00 | 37.5 |
| $40 < 45$ | 1 | 0.03 | 42.5 |
| 45 & Over | 3 | 0.10 | |
| | 30 | 0.99 | |

(b) Since the last class has no upper cutpoint, the midpoint cannot be computed.

**2.41** (a) The classes are presented in column 1.  With the classes established, we then tally the exam scores into their respective classes.  These results are presented in column 2, which lists the frequencies.  Dividing each frequency by the total number of exam scores, which is 20, results in each class's relative frequency.  The relative frequencies for all classes are presented in column 3.  The class mark of each class is the average of the lower and upper limits.  The class marks for all classes are presented in column 4.

| Score | Frequency | Relative Frequency | Class Mark |
|-------|-----------|--------------------|-----------|
| 30-39 | 2 | 0.10 | 34.5 |
| 40-49 | 0 | 0.00 | 44.5 |
| 50-59 | 0 | 0.00 | 54.5 |
| 60-69 | 3 | 0.15 | 64.5 |
| 70-79 | 3 | 0.15 | 74.5 |
| 80-89 | 8 | 0.40 | 84.5 |
| 90-100 | 4 | 0.20 | 95.0 |
| | 20 | 1.00 | |

(b)   The first six classes have width 10; the seventh class had width 11.

(c)   Answers will vary, but one choice is to keep the first six classes the same and make the next two classes 90-99 and 100-109.  Another possibility is 31-40, 41-50, …, 91-100.

**Exercises 2.3**

2.43  A frequency histogram shows the actual frequencies on the vertical axis; whereas, the relative frequency histogram always shows proportions (between 0 and 1) or percentages (between 0 and 100) on the vertical axis.

2.45  By showing the cutpoints on the horizontal axis, the range of possible data values in each class is immediately known and the midpoint can be quickly determined.  This is particularly helpful if it is not convenient to make all classes the same width. The use of the midpoints is appropriate when each class consists of a single value (which is, of course, also the midpoint).  Use of the midpoints is not appropriate in other situations since it may be difficult to determine the location of the cutpoints from the values of the midpoints, particularly if the midpoints are not evenly spaced.  Midpoints cannot be used if there is an open class.

2.47  The histogram (especially one using relative frequencies) is generally preferable.  Data sets with a large number of observations may result in a stem of the stem-and-leaf diagram having more leaves than will fit on the line.  In that case, the histogram would be preferable.

2.49  Since a bar graph is used for qualitative data, we separate the bars from each other to emphasize that the graph is not a histogram. There is no special ordering of the classes.  If the bars were to touch, some viewers might infer an ordering representing quantitative data.

2.51  (a)   Each rectangle in the frequency histogram would be of a height equal to the corresponding number of dots in the dotplot.

(b)   If the classes for the histogram were based on multiple values, there would not be one rectangle for each column of dots (there would be fewer rectangles than columns of dots).  The height of each rectangle would be equal to the total number of dots between each adjacent pair of cutpoints.  If the classes were constructed so that only a few columns of dots corresponded to each rectangle, the general impression of the shape of the distribution should remain the same even though the details did not appear to be the same.

2.53  (a)   The frequency histogram in Figure (a) is constructed using the frequency distribution presented in this exercise; i.e., columns 1 and 2.  The lower class limits of column 1 are used to label the horizontal axis of the frequency histogram.  Suitable candidates for vertical axis units in the frequency histogram are the even integers 0 through 8, since these are representative of the magnitude and spread of the frequencies presented in column 2.  The height of each bar in the frequency histogram matches the respective frequency in column 2.

(b)    The relative-frequency histogram in Figure (b) is constructed using the relative-frequency distribution presented in this exercise; i.e., columns 1 and 3.  It has the same horizontal axis as the frequency histogram.  We notice that the relative frequencies presented in column 3 range in size from 0.000 to 0.229.  Thus, suitable candidates for vertical axis units in the relative-frequency histogram are increments of 0.05 (or 5%), from zero to 0.25 (or 25%).  The height of each bar in the relative-frequency histogram matches the respective relative frequency in column 3.

Figure (a)                                    Figure (b)

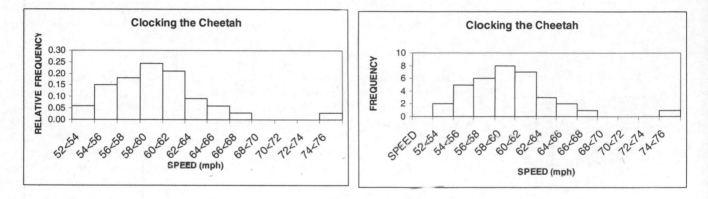

**2.55**  (a)    The frequency histogram in Figure (a) is constructed using the frequency distribution obtained in Exercise 2.29; i.e., columns 1 and 2.  Column 1 demonstrates that the data are grouped using classes with class widths of 3 starting at the midpoint 12.5.  Suitable candidates for vertical axis units in the frequency histogram are the integers within the range 0 through 6, since these are representative of the magnitude and spread of the frequencies presented in column 2.  Also, the height of each bar in the frequency histogram matches the respective frequency in column 2.

(b)    The relative-frequency histogram in Figure (b) is constructed using the relative-frequency distribution obtained in Exercise 2.29; i.e., columns 1 and 3.  It has the same horizontal axis as the frequency histogram.  We notice that the relative frequencies presented in column 3 range in size from 0.05 to 0.30.  Thus, suitable candidates for vertical axis units in the relative-frequency histogram are increments of 0.05 (5%), from zero to 0.30 (30%).  The middle of each histogram bar is placed directly over the class midpoint.  Also, the height of each bar in the relative-frequency histogram matches the respective relative frequency in column 3.

(c)     Figure (a)                          Figure (b)

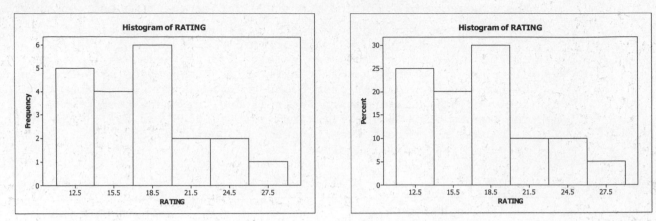

2.57    (a)     The frequency histogram in Figure (a) is constructed using the
                frequency distribution presented in Exercise 2.31; i.e., columns 1 and
                2.  Column 1 demonstrates that the data are grouped using classes
                based on a single value.  These single values in column 1 are used to
                label the horizontal axis of the frequency histogram.  Suitable
                candidates for vertical axis units in the frequency histogram are the
                integers within the range 0 through 7, since these are representative
                of the magnitude and spread of the frequencies presented in column 2.
                When classes are based on a single value, the middle of each histogram
                bar is placed directly over the single numerical value represented by
                the class.  Also, the height of each bar in the frequency histogram
                matches the respective frequency in column 2.

        (b)     The relative-frequency histogram in Figure (b) is constructed using
                the relative-frequency distribution presented in this exercise; i.e.,
                columns 1 and 3.  It has the same horizontal axis as the frequency
                histogram.  We notice that the relative frequencies presented in
                column 3 range in size from 0.025 to 0.425.  Thus, suitable candidates
                for vertical-axis units in the relative-frequency histogram are
                increments of 0.05 (or 5%), from zero to 0.45 (or 45%).  The middle of
                each histogram bar is placed directly over the single numerical value
                represented by the class.  Also, the height of each bar in the
                relative-frequency histogram matches the respective relative frequency
                in column 3.

                Figure (a)                          Figure (b)

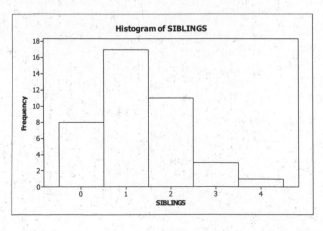
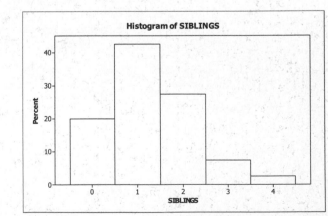

**2.59**  The horizontal axis of this dotplot displays a range of possible ages.  To complete the dotplot, we go through the data set and record each age by placing a dot over the appropriate value on the horizontal axis.

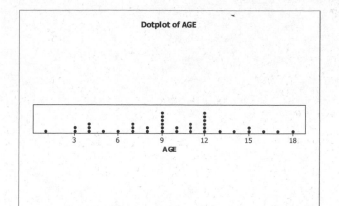

**2.61**  (a)   The data values range from 7 to 18, so the scale must accommodate those values.  We stack dots above each value on two different lines using the same scale for each line.  The result is

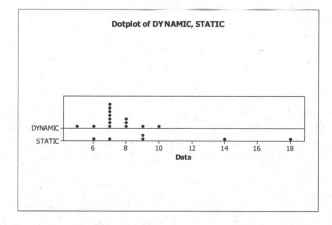

(b)   The dynamic system does seem to reduce acute postoperative days in the hospital on the average.  The Dynamic data are centered at about 7 days, whereas the Static data are centered at about 11 days and are much more spread out than the Dynamic data.

**2.63**  Since each data value consists of 2 digits, each beginning with 1, 2, 3, or 4, we will construct the stem-and-leaf diagram with two lines per stem.  The result is

```
1  23
1  8
2  1
2  678899
3  344
3  59
4  04
```

**2.65**  (a)   Since each data value lies between 26.6 and 34.0, we will construct the stem-and-leaf diagram with one line per stem.  The result is

```
26  6
27  7999
28  1456789
29  133447889
30  3579
31  008
32  7
33
34  0
```

(b)  Using two lines per stem, the same data result in the following diagram:

```
26  6
27
27  7999
28  14
28  56789
29  13344
29  7889
30  3
30  579
31  00
31  8
32
32  7
33
33
34  0
```

(c)  Both diagrams have an acceptable number of lines.  The second diagram makes it a little easier to see how extreme the youngest and oldest teams are, so we find that one slightly more useful.

**2.67**  (a)  We multiply each of the relative frequencies by 360 degrees to obtain the portion of the pie represented by each team.  The result is

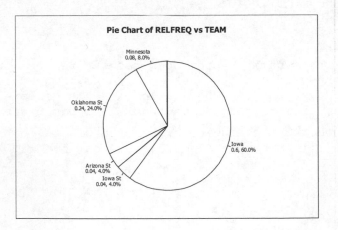

(b)  We use the bar chart to show the relative frequency with which each TEAM occurs.  The result is

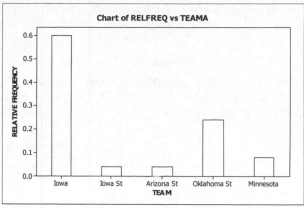

**2.69**  (a)  We first find each of the relative frequencies by dividing each of the frequencies by the total frequency of 509.  Then we multiply each of the relative frequencies by 360 degrees to obtain the portion of the pie represented by each color of M&M.  The result is

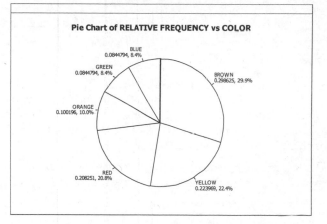

(b)  We use the bar chart to show the relative frequency with which each color occurs.  The result is

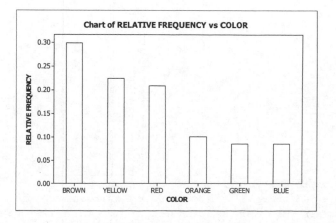

**2.71**  The graph indicates that:

(a)  20% of the patients have cholesterol levels between 205 and 209, inclusive.

(b)  20% are between 215 and 219; and 5% are between 220 and 224.  Thus, 25% (i.e., 20% + 5%) have cholesterol levels of 215 or higher.

(c)  35% of the patients have cholesterol levels between 210 and 214, inclusive.  With 20 patients in total, the number having cholesterol levels between 210 and 214 is 7 (i.e., 0.35 x 20).

**2.73**  (a)  After retrieving the data from the WeissStats CD, select **Graph ▶ Histogram,** choose **Simple** and click **OK**.  Double click on PUPS in the first box to enter PUPS in the **Graphs variables** box, and click **OK**.  The frequency histogram is (except for shading)

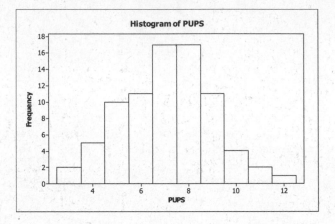

To change to a relative-frequency histogram, before clicking OK the second time, click on the **Scale** button and the **Y-Scale type** tab, and choose **Percent** and click **OK**.  The graph will look like the one in part (a), but will have relative frequencies on the vertical scale instead of counts.

(b)  The numbers of pups range from 1 to 12 per female with 7 and 8 pups occurring more frequently than any other values.

**2.75**  (a)  After entering the data from the WeissStats CD, in Minitab, select **Graph ▶ Histogram,** choose **Simple** and click **OK**.  Double click on NAEP to enter NAEP in the **Graph variables** box and click **OK**.  The result is

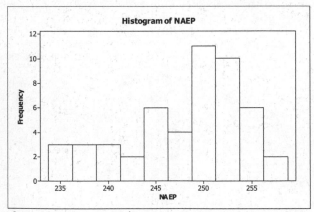

Minitab has chosen the class cutpoints.  We see that the scores tend to cluster near the high end of the range of scores with a considerable number of scores somewhat lower than the most common score.

(b)  To obtain the pie chart, select **Graph ▶ Pie Chart,** put the cursor in the **Categorical variables** box and double click on RESULT to enter

RESULT in the **Categorical variables** box.  Click on the **Labels** button and then on the **Slice Labels** tab.  Check all four boxes and click **OK** twice.  The result (except for colors and legend) is

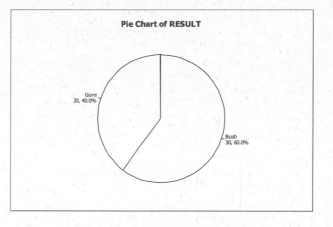

(c)  To obtain the bar graph, select **Graph ▶ Bar Chart,** select **Simple,** and click **OK.**  Double click on RESULT to enter RESULT in the **Categorical variables** box and   click **OK**.  The result (except for shading) is

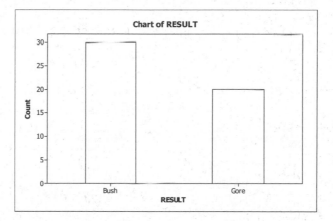

Both charts show that Bush won 30 states and Gore won 20.   These graphs do not reflect the proportion of votes cast for Bush and Gore.

**2.77**  (a)  After entering the data from the WeissStats CD, in Minitab, select

**Graph ▶ Stem-and-Leaf,** double click on PERCENT to enter PERCENT in the **Graph variables** box and enter a 1̲0̲ in the **Increment** box, and click **OK**.  The result is

```
Stem-and-leaf of PERCENT   N  = 51
Leaf Unit = 1.0


  2    7  79
 (41)  8  00111111123445566666667777888888999999999
  8    9  00011122
```

(b)  Repeat part (a), but this time enter a 5̲ in the **Increment** box.  The result is

```
Stem-and-leaf of PERCENT   N  = 51
Leaf Unit = 1.0

 2    7  79
15    8  0011111112344
(28)  8  5566666667777888888999999999
 8    9  00011122
```

(c) Repeat part (a) again, but this time enter a <u>2</u> in the **Increment** box.
   The result is

```
Stem-and-leaf of PERCENT   N  = 51
Leaf Unit = 1.0

 1    7  7
 2    7  9
11    8  001111111
13    8  23
17    8  4455
(11)  8  66666667777
23    8  888888999999999
 8    9  000111
 2    9  22
```

(d) The last graph is the most useful since it gives a better idea of the
   shape of the distribution.  Typically, we like to have five to fifteen
   classes and this is the only one of the three graphs that satisfies
   that condition.

**2.79** (a) After entering the data from the WeissStats CD, in Minitab, select

   **Graph ▶ Histogram,** select **Simple** and click **OK.**  double click on TEMP to
   enter TEMP in the **Graph variables** box and click **OK.**   The result is

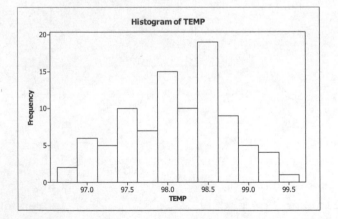

(b) Now select **Graph ▶ Dotplot,** select **Simple** in the **One Y** row, and click
   **OK.**  Double click on TEMP to enter TEMP in the **Graph variables** box and
   click **OK.**   The result is

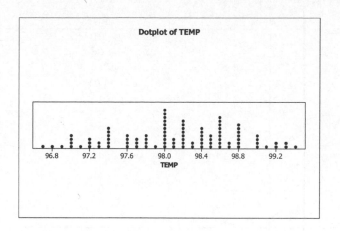

(c) Now select **Graph ▶ Stem-and-Leaf,** double click on TEMP to enter TEMP in the **Graph variables** box and click **OK.** Leave the **Increment** box blank to allow Minitab to choose the number of lines per stem. The result is

```
Stem-and-leaf of TEMP  N  = 93
Leaf Unit = 0.10

  1    96  7
  3    96  89
  8    97  00001
  13   97  22233
  19   97  444444
  26   97  6666777
  31   97  88889
  45   98  00000000000111
 (10)  98  2222222233
  38   98  4444445555
  28   98  66666666677
  17   98  8888888
  10   99  00001
  5    99  2233
  1    99  4
```

(d) The dotplot shows all of the individual values. The stem-and-leaf diagram used five lines per stem and therefore each line contains leaves with possibly two values. The histogram chose classes of width 0.25. This resulted in, for example, the class with midpoint 97.0 including all of the values 96.9, 97.0, and 97.1, while the class with midpoint 97.25 includes only the two values 97.2 and 97.3. Thus the 'smoothing' effect is not as good in the histogram as it is in the stem-and-leaf diagram. Overall, the dotplot gives the truest picture of the data and allows recovery of all of the data values.

**2.81** (a) Consider all three columns of the energy-consumption data given in Exercise 2.24. Column 1 is now reworked to present just the lower cutpoint of each class. Column 2 is reworked to sum the frequencies of all classes representing values less than the specified lower cutpoint. These successive sums are the cumulative frequencies. Column 3 is reworked to sum the relative frequencies of all classes representing values less than the specified cutpoints. These successive sums are the cumulative relative frequencies. (Note: The cumulative relative frequencies can also be found by dividing the each cumulative frequency by the total number of data values.)

| Less than | Cumulative Frequency | Cumulative Relative Frequency |
|---|---|---|
| 40 | 0 | 0.00 |
| 50 | 1 | 0.02 |
| 60 | 8 | 0.16 |
| 70 | 15 | 0.30 |
| 80 | 18 | 0.36 |
| 90 | 24 | 0.48 |
| 100 | 34 | 0.68 |
| 110 | 39 | 0.78 |
| 120 | 43 | 0.86 |
| 130 | 45 | 0.90 |
| 140 | 48 | 0.96 |
| 150 | 48 | 0.96 |
| 160 | 50 | 1.00 |

(b)    Pair each cutpoint in reworked column 1 with its corresponding cumulative relative frequency found in reworked column 3.  Construct a horizontal axis, where the units are in terms of the cutpoints and a vertical axis where the units are in terms of cumulative relative frequencies.  For each cutpoint on the horizontal axis, plot a point whose height is equal to the cumulative relative frequency.  Then join the points with connecting lines.  The result, presented in Figure (b), is an *ogive* using cumulative relative frequencies.  (Note: A similar procedure could be followed using cumulative frequencies.)

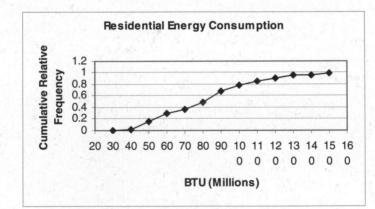

**2.83**  (a)  After rounding to the nearest 10 and then dropping the final zero, the stem-and-leaf diagram is

```
 9   1
 9
 9   5
 9   6667
 9   8888999999
10   0000011
10   223333
10
10   6
```

(b)  After truncating each observation to the 10s digit, the stem-and-leaf diagram is

```
 9   1
 9
 9   455
 9   67777
 9   88888999999
10   01111
10   2233
10
10   6
```

(c) The overall impression of the shape of the distribution is the same for the diagrams in parts (a) and (b) although there is a slight shift to lower values in part (b).  This is due to truncating instead of rounding.  The diagram in part (b) is the same as the one in Exercise 2.62.

## Section 2.4

**2.85** (a) The distribution of a data set is a table, graph, or formula that provides the values of the observations and how often they occur.

(b) Sample data are the values of a variable for a sample of the population.

(c) Population data are the values of a variable for the entire population.

(d) Census data are the same as population data, a complete listing of all data values for the entire population.

(e) A sample distribution is the distribution of sample data.

(f) A population distribution is the distribution of population data.

(g) A distribution of a variable is the same as a population distribution.

**2.87** A large simple random sample from a bell-shaped distribution would be expected to have roughly a bell-shaped distribution since more sample values should be obtained, on average, from the middle of the distribution.

**2.89** Three distribution shapes that are symmetric are bell-shaped, triangular, and rectangular, shown in that order below.  It should be noted that there are others as well.

Bell-shaped                    Triangular                    Uniform (or rectangular)

**2.91** (a) Except for the one data value between 74 and 76, this distribution is close to bell-shaped.  That one value makes the distribution slightly right skewed.

(b) The distribution is slightly right skewed.

**2.93** (a) The distribution of burrow depths is left skewed.

(b) The distribution is left skewed.

**2.95** (a) The distribution of shell thickness is approximately bell-shaped.

(b) The distribution is nearly symmetric.

**2.97** (a) The distribution of cholesterol levels appears to be slightly left skewed.

(b) This distribution is nearly symmetric, but is slightly left skewed.

Given that the data originated from patients who had high cholesterol levels, one would not expect symmetry. Individuals with low cholesterol levels were not patients and were not included in the testing.

**2.99** (a) The distribution of length of stay is nearly reverse J-shaped. It is certainly right skewed.

(b) The distribution is right skewed.

**2.101** (a) The distribution for 1993 is right skewed and the distribution for 1996 is reverse J shaped.

(b) Both distributions are right skewed.

**2.103** (a) After entering the data from the WeissStats CD, in Minitab, select

**Graph ▶ Histogram,** select **Simple** and click **OK.** double click on NAEP to enter NAEP in the **Graph variables** box and click **OK.** The result is

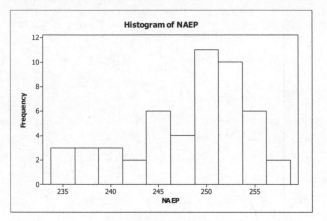

(b) The overall distribution of NAEP scores is left skewed.

(c) We classify this distribution as left skewed.

**2.105** (a) In Exercise 2.77, we used Minitab to obtain a stem-and-leaf diagram using 5 lines per stem. That diagram is shown below

```
Stem-and-leaf of PERCENT   N  = 51
Leaf Unit = 1.0

     1    7  7
     2    7  9
    11    8  001111111
    13    8  23
    17    8  4455
   (11)   8  66666667777
    23    8  888888999999999
     8    9  000111
     2    9  22
```

(b) The overall shape of this distribution is left skewed.

(c) We classify the distribution of PERCENT as left skewed.

**2.107** (a) After entering the data in Minitab, select **Graph ▶ Dotplot,** select **Simple** in the **One Y** row, and click **OK.** Double click on TEMP to enter TEMP in the **Graph variables** box and click **OK.** The result is

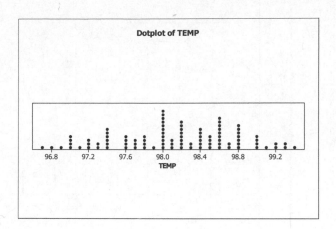

(b)  The overall distribution of temperatures is roughly triangular.

(c)  The distribution is fairly symmetric.

**2.109** The precise answers to this exercise will vary from class to class or individual to individual.  Thus your results will likely differ from our results shown below.

(a)  We obtained 50 random digits from a table of random numbers.  The digits were

4 5 4 6 8 9 9 7 7 2 2 2 9 3 0 3 4 0 0 8 8 4 4 5 3

9 2 4 8 9 6 3 0 1 1 0 9 2 8 1 3 9 2 5 8 1 8 9 2 2

(b)  Since each digit is equally likely in the random number table, we expect that the distribution would look roughly rectangular.

(c)  Using single value classes, the frequency distribution is given by the following table.  The histogram is shown below.

| Value | Frequency | Relative-Frequency |
|-------|-----------|--------------------|
| 0 | 5 | .10 |
| 1 | 4 | .08 |
| 2 | 8 | .16 |
| 3 | 5 | .10 |
| 4 | 6 | .12 |
| 5 | 3 | .06 |
| 6 | 2 | .04 |
| 7 | 2 | .04 |
| 8 | 7 | .14 |
| 9 | 8 | .16 |

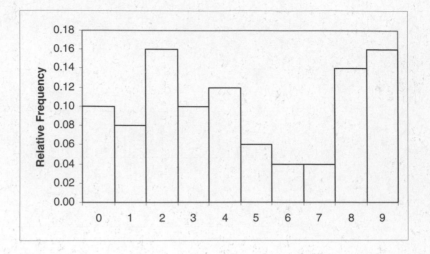

We did not expect to see this much variation.

(d)    We would have expected a histogram that was a little more 'even', more like a rectangular distribution, but the relatively small sample size can result in considerable variation from what is expected.

(e)    We should be able to get a more evenly distributed set of data if we choose a larger set of data.

(f)    Class project.

**2.111** (a)    Your result will differ from, but be similar to, the one below which was obtained using Minitab.  Choose **Calc ▶ Random Data ▶ Normal...**, type 3000 in the **Generate rows of data** text box, click in the **Store in column(s)** text box and type C1, and click **OK**.

(b)    Then choose **Graph ▶ Histogram**, choose the **Simple** version, click **OK**, enter C1 in the **Graph variables** text box, click on the **Scale** button and then on the **Y-Scale Type** tab.  Check the **Percent** box and click **OK** twice.

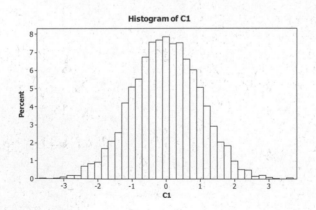

(c)    The histogram in part (b) has the shape of a standard normal distribution.  The sample of 3000 is representative of the population from which the sample was taken.

**Section 2.5**

**2.113** (a)  A truncated graph is one for which the vertical axis starts at a value other than its natural starting point, usually zero.

(b)  A legitimate motivation for truncating the axis of a graph is to place the emphasis on the ups and downs of the distribution rather than on the actual height of the graph.

(c)  To truncate a graph and avoid the possibility of misinterpretation, one should start the axis at zero and put slashes in the axis to indicate that part of the axis is missing.

**2.115** (a)  A large lower portion of the graph is eliminated. When this is done, differences between district and national averages appear greater than in the original figure.

(b)  Even more of the graph is eliminated. Differences between district and national averages appear even greater than in part (a).

(c)  The truncated graphs give the misleading impression that, in 2005, the district average is much greater relative to the national average than it actually is.

**2.117** (a)  The problem with the bar graph is that it is truncated. That is, the vertical axis, which should start at $0 (in trillions), starts with $3.05 (in trillions) instead. The part of the graph from $0 (in trillions) to $3.05 (in trillions) has been cut off. This truncation causes the bars to be out of correct proportion and hence creates the misleading impression that the money supply is changing more than it actually is.

(b)  A version of the bar graph with an untruncated and unmodified vertical axis is presented in Figure (a). Notice that the vertical axis starts at $0.00 (in trillions). Increments are in halves of trillion dollars. In contrast to the original bar graph, this one illustrates that the changes in money supply from week to week are not that different. However, the "ups" and "downs" are not as easy to spot as in the original, truncated bar graph.

(c)  A version of the bar graph in which the vertical axis is modified in an acceptable manner is presented in Figure (b). Notice that the special symbol "//" is used near the base of the vertical axis to signify that the vertical axis has been modified. Thus, with this version of the bar graph, not only are the "ups" and "downs" easy to spot but the reader is also aptly warned by the slashes that part of the vertical axis between $0.00 (in trillions) and $3.05 (in trillions) has been removed.

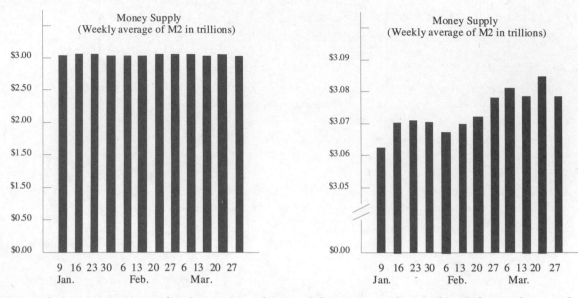

**2.119** (b) Without the vertical scale, it would appear that oil prices dropped about 75% from the peak price to the last day shown on the graph.

(c) The actual drop was from about $70 per barrel to $63.05 per barrel, a drop of about $7 dollars per barrel. This is a drop of 7/70 = 0.10 or about 10%.

(d) The graph is potentially misleading because if the reader doesn't pay attention to the vertical scale, he or she may be led to conclude that the oil price was much more volatile than it actually was.

(e) The graph could be made less potentially misleading by either making the vertical scale range from zero to 70 or by starting at zero and putting a break in the vertical axis to call attention to the fact that part of the vertical axis is missing.

**2.121** (a) The brochure shows a "new" ball with twice the radius of the "old" ball. The intent is to give the impression that the "new" ball lasts roughly twice as long as the "old" ball. However, if the "new" ball has twice the radius of the "old" ball, the "new" ball will have eight times the volume of the "old" ball (since the volume of a sphere is proportional to the cube of its radius, or the radius $2^3 = 8$). Thus, the scaling is improper because it gives the impression that the "new" ball lasts eight times as long as the "old" ball rather than merely two times as long.

Old Ball           New Ball

(b) One possible way for the manufacturer to illustrate the fact that the "new" ball lasts twice as long as the "old" ball is to present pictures of two balls, side by side, each of the same magnitude as the picture of the "old" ball and to label this set of two balls "new ball". This will illustrate the point that a purchaser will be getting twice as much for the money.

Old Ball                              New Ball

**Review Problems For Chapter 2**

1.   (a)   A variable is a characteristic that varies from one person or thing to another.

     (b)   Variables are quantitative or qualitative.

     (c)   Quantitative variables can be discrete or continuous.

     (d)   Data are values of a variable.

     (e)   The data type is determined by the type of variable being observed.

2.   It is important to group data in order to make large data sets more compact and easier to understand.

3.   The concepts of midpoints and cutpoints do not apply to qualitative data since no numerical values are involved in the data.

4.   (a)   The midpoint is halfway between the cutpoints.  Since the class width is 8, 10 is halfway between 6 and 14.

     (b)   The class width is also the distance between consecutive midpoints. Therefore, the second midpoint is at 10 + 8 = 18.

     (c)   The sequence of cutpoints is 6, 14, 22, 30, 38, ...  Therefore the lower and upper cutpoints of the third class are 22 and 30.

     (d)   An observation of 22 would go into the third class since that class contains data greater than or equal to 22 and strictly less than 30.

5.   (a)   The common class width is the distance between consecutive cutpoints, which is 15 - 5 = 10.

     (b)   The midpoint of the second class is halfway between the cutpoints 15 and 25, and is therefore 20.

     (c)   The sequence of cutpoints is 5, 15, 25, 35, 45, ...  Therefore, the lower and upper cutpoints of the third class are 25 and 35.

6.   Single value grouping is appropriate when the data are discrete with relatively few distinct observations.

7.   (a)   The vertical edges of the bars will be aligned with the cutpoints.

     (b)   The horizontal centers of the bars will be aligned with the midpoints.

8.   The two main types of graphical displays for qualitative data are the bar chart and the pie chart.

9.   A histogram is better than a stem-and-leaf diagram for displaying large quantitative data sets since it can always be scaled appropriately and the individual values are of less interest than the overall picture of the data.

**10.**    Bell-shaped        Right skewed        Reverse J shape        Uniform

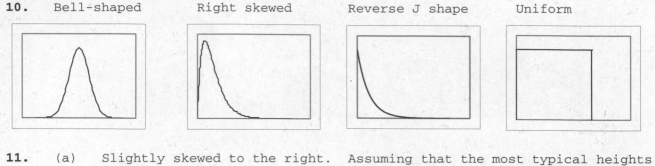

**11.** (a)    Slightly skewed to the right. Assuming that the most typical heights are around 5'10", most heights below that figure would still be above 5'4", whereas heights above 5'10" extend to around 7'.

(b)    Skewed to the right. High incomes extend much further above the mean income than low incomes extend below the mean.

(c)    Skewed to the right. While most full-time college students are in the 17-22 age range, there are very few below 17 while there are many above 22.

(d)    Skewed to the right. The main reason for the skewness to the right is that those students with GPAs below fixed cutoff points have been suspended by the time they would have been seniors.

**12.** (a)    The distribution of the large simple random sample will reflect the distribution of the population, so it would be left-skewed as well.

(b)    No. The randomness in the samples will certainly produce different sets of observations resulting in shapes that are not identical.

(c)    Yes. We would expect both of the simple random samples to reflect the shape of the population and be left-skewed.

**13.** (a)    The first column ranks the hydroelectric plants. Thus, it consists of *quantitative, discrete* data.

(b)    The fourth column provides measurements of capacity. Thus, it consists of *quantitative, continuous* data.

(c)    The third column provides nonnumerical information. Thus, it consists of *qualitative* data.

**14.** (a)    The first class to construct is 40-44. Since all classes are to be of equal width, and the second class begins with 45, we know that the width of all classes is 45 - 40 = 5. All of the classes are presented in column 1 of the grouped-data table in the figure below. The last class to construct does not go beyond 65-69, since the largest single data value is 69. Having established the classes, we tally the ages into their respective classes. These results are presented in column 2, which lists the frequencies. Dividing each frequency by the total number of observations, which is 43, results in each class's relative frequency. The relative frequencies for all classes are presented in column 3.

By averaging the lower and upper limits for each class, we arrive at the class mark for each class. The class marks for all classes are presented in column 4.

| Age at Inauguration | Frequency | Relative Frequency | Class Mark |
|---|---|---|---|
| 40-44 | 2 | 0.047 | 42 |
| 45-49 | 6 | 0.140 | 47 |
| 50-54 | 13 | 0.302 | 52 |
| 55-59 | 12 | 0.279 | 57 |
| 60-64 | 7 | 0.163 | 62 |
| 65-69 | 3 | 0.070 | 67 |
| | 43 | 1.001 | |

(b)   The lower cutpoint for the first class is 40.  The upper cutpoint for the first class is 45 since that is the smallest value that can go into the second class.

(c)   The common class width is 45 - 40 = 5.

(d)   The frequency histogram presented below is constructed using the frequency distribution presented above; i.e., columns 1 and 2.  Notice that the lower cutpoints of column 1 are used to label the horizontal axis of the frequency histogram.  Suitable candidates for vertical-axis units in the frequency histogram are the even integers within the range 0 through 14, since these are representative of the magnitude and spread of the frequencies presented in column 2.  The height of each bar in the frequency histogram matches the respective frequency in column 2.

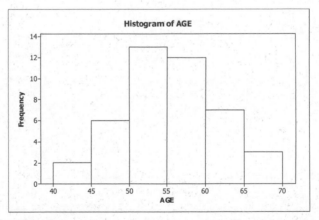

(e)  The overall shape of the inauguration ages is somewhere between triangular and bell-shaped.

(f)  The distribution is roughly symmetric.

15.   The horizontal axis of this dotplot displays a range of possible ages for the 43 Presidents of the United States.  To complete the dotplot, we go through the data set and record each age by placing a dot over the appropriate value on the horizontal axis.

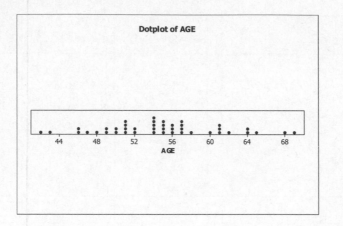

16.  (a)  Using *one* line per stem in constructing the ordered stem-and-leaf
          diagram means vertically listing the numbers comprising the stems
          *once*.  The leaves are then placed with their respective stems in
          order.  The ordered stem-and-leaf diagram using one line per stem is
          presented in Figure (a).

     (b)  Using *two* lines per stem in constructing the ordered stem-and-leaf
          diagram means vertically listing the numbers comprising the stems
          *twice*.  In turn, if the leaf is one of the digits 0 through 4, it is
          ordered and placed with the first of the two stem lines.  If the leaf
          is one of the digits 5 through 9, it is ordered and placed with the
          second of the two stem lines.  The ordered stem-and-leaf diagram using
          two lines per stem is presented in Figure (b).

(a)                                                        (b)

                                                           4 | 2 3

4 | 2 3 6 6 7 8 9 9                                         4 | 6 6 7 8 9 9

5 | 0 0 1 1 1 1 2 2 4 4 4 4 4 5 5 5 5 6 6 6 7 7 7 7 8       5 | 0 0 1 1 1 1 2 2 4 4 4 4 4 4

6 | 0 1 1 1 2 4 4 5 8 9                                     5 | 5 5 5 5 6 6 6 7 7 7 7 8

                                                           6 | 0 1 1 1 2 4 4

                                                           6 | 5 8 9

     (c)  Two lines per stem corresponds to the frequency distribution.

17.  (a)  The grouped-data table presented below is constructed using classes
          based on a single value.  Since each data value is one of the integers
          0 through 6, inclusive, the classes will be 0 through 6, inclusive.
          These are presented in column 1.  Having established the classes, we
          tally the number of busy tellers into their respective classes.  These
          results are presented in column 2, which lists the frequencies.
          Dividing each frequency by the total number of observations, which is
          25, results in each class's relative frequency.  The relative
          frequencies for all classes are presented in column 3.  Since each
          class is based on a single value, the class mark is the class itself.
          Thus, a column of midpoints is not given, since this same information
          is presented in column 1.

| Number Busy | Frequency | Relative Frequency |
|:---:|:---:|:---:|
| 0 | 1 | 0.04 |
| 1 | 2 | 0.08 |
| 2 | 2 | 0.08 |
| 3 | 4 | 0.16 |
| 4 | 5 | 0.20 |
| 5 | 7 | 0.28 |
| 6 | 4 | 0.16 |
|  | 25 | 1.00 |

(b)   The following relative-frequency histogram is constructed using the relative-frequency distribution presented in part (a); i.e., columns 1 and 3.   Column 1 demonstrates that the data are grouped using classes based on a single value.   These single values in column 1 are used to label the horizontal axis of the relative-frequency histogram.   We notice that the relative frequencies presented in column 3 range in size from 0.04 to 0.28 (4% to 28%).   Thus, suitable candidates for vertical axis units in the relative-frequency histogram are increments of 0.05, starting with zero and ending at 0.30.   The middle of each histogram bar is placed directly over the single numerical value represented by the class.   Also, the height of each bar in the relative-frequency histogram matches the respective relative frequency in column 3.

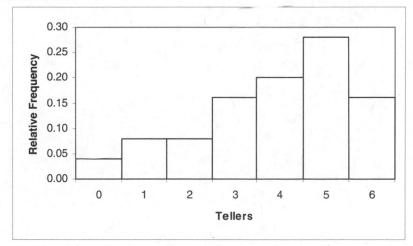

(c)   The overall shape of this distribution is left skewed.

(d)   The distribution is left skewed.

(e)

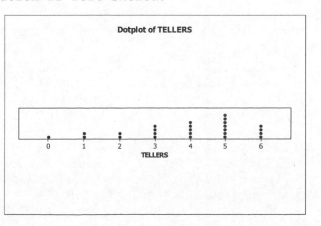

      (f)   Since both the histogram and the dotplot are based on single value grouping, they both convey exactly the same information.

**18.**    (a)   The dotplot of the ages of the oldest player on each major league baseball team is

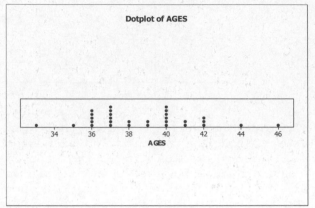

      (b)   The overall shape of the distribution of ages is bimodal.

      (c)   The distribution is roughly symmetric.

**19.**    (a)-(b)The two pie-charts are shown below.  In each case, the proportion of the circle for each category is found by multiplying 360 degrees by the category frequency and dividing by the total frequency (941 for Buybacks and 369 for Homicides).

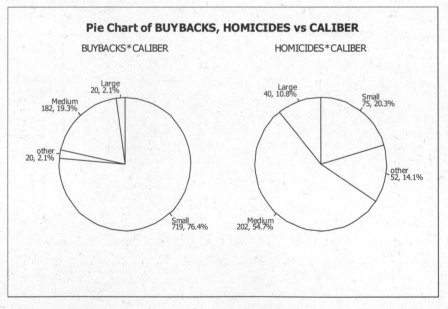

      (c)   The most striking characteristic of the two charts is that most of the buybacks are of small caliber guns while most of the homicides are committed with medium caliber guns.  It should also be noted that small and medium caliber guns are the two largest categories of both buybacks and homicides, accounting for 95.7% of the buybacks and 75.0% of the homicides.

**20.**    (a)   The table below shows both the frequency distribution and the relative frequency distribution.  If each frequency reported in the table is divided by the total number of students, which is 40, we arrive at the relative frequency (or percentage) of each class.

| Class | Frequency | Relative Frequency |
|-------|-----------|--------------------|
| **Fr** | 6 | 0.150 |
| **So** | 15 | 0.375 |
| **Jr** | 12 | 0.300 |
| **Sr** | 7 | 0.175 |

(b)  The following pie chart is used to display the percentage of students in each of the four classes.

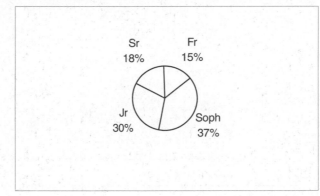

(c)  The following bar graph also displays the relative frequencies of each class.

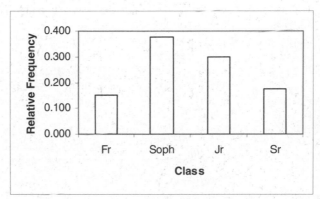

**21.**  (a)  The first class to construct is 0 ≤ 1000.  Since all classes are to be of equal width, we know that the width of all classes is 1000 - 0 = 1000.  All of the classes are presented in column 1 of Figure (a) below.  The last class to construct is 11000 ≤ 12000, since the largest single data value is 11722.98.  Having established the classes, we tally the highs into their respective classes.  These results are presented in column 2, which lists the frequencies.  Dividing each frequency by the total number of observations, which is 36, results in each class's relative frequency.  The relative frequencies for all classes are presented in column 3.  By averaging the numbers that appear as the lower and upper cutpoints for each class, we arrive at the midpoint for each class.  The midpoints for all classes are presented in column 4.

| High | Freq. | Relative Frequency | Midpoint |
|---|---|---|---|
| 0 < 1000 | 8 | 0.222 | 500 |
| 1000 < 2000 | 10 | 0.278 | 1500 |
| 2000 < 3000 | 4 | 0.111 | 2500 |
| 3000 < 4000 | 4 | 0.111 | 3500 |
| 4000 < 5000 | 0 | 0.000 | 4500 |
| 5000 < 6000 | 1 | 0.028 | 5500 |
| 6000 < 7000 | 1 | 0.028 | 6500 |
| 7000 < 8000 | 0 | 0.000 | 7500 |
| 8000 < 9000 | 1 | 0.028 | 8500 |
| 9000 < 10000 | 1 | 0.028 | 9500 |
| 10000 < 11000 | 3 | 0.083 | 10500 |
| 11000 < 12000 | 3 | 0.083 | 11500 |
|  | 36 | 1.000 |  |

(b)   The following relative-frequency histogram is constructed using the relative-frequency distribution presented above; i.e., columns 1 and 3.   The lower cutpoints of column 1 are at the left edges of each rectangle in the relative-frequency histogram.   We notice that the relative frequencies presented in column 3 range in size from 0.000 to 0.278.   Thus, suitable candidates for vertical axis units in the relative-frequency histogram are increments of 5%, from 0,00 (0%) to 0.30 (30%).   The height of each bar in the relative-frequency histogram matches the respective relative frequency in column 3.

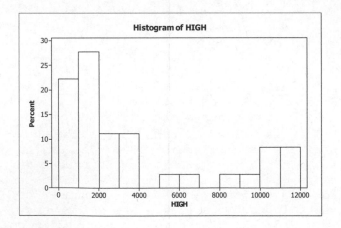

22.   Answers will vary, but here is one possibility:

23.    (a)    The break in the third bar is to emphasize that the bar as shown is
              not as tall as it should be.

       (b)    The bar for the space available in coal mines is at a height of about
              30 billion tonnes.  To accurately represent the space available in
              saline aquifers (10,000 billion tonnes), the third bar would have to
              be over 300 times as high as the first bar and more than 10 times as
              high as the second bar.  If the first two bars were kept at the sizes
              shown, there wouldn't be enough room on the page for the third bar to
              be shown at its correct height.  If a reasonable height were chosen
              for the third bar, the first bar wouldn't be visible.  The only
              apparent solution is to present the third bar as a broken bar.

24.    (a)    Covering up the numbers on the vertical axis totally obscures the
              percentages.

       (b)    Having followed the directions in part (a), we might conclude that the
              percentage of women in the labor force for 2000 is about three and
              one-half times that for 1960.

       (c)    Not covering up the vertical axis, we find that the percentage of
              women in the labor force for 2000 is about 1.8 times that for 1960.

       (d)    The graph is potentially misleading because it is truncated.  Notice
              that vertical axis units begin at 30 rather than at zero.

       (e)    To make the graph less potentially misleading, we can start it at zero
              instead of 30.

25.    (a)    The population consists of all of the states of the U.S.  The variable
              under consideration is the division to which the U.S. Census Bureau
              assigns each state.

       (b)    After entering the data in Minitab from the WeissStats CD, choose **Stat**

              ▶ **Tables** ▶ **Tally Individual Variables.**  Double-click on DIVISION to
              enter DIVISION in the **Variables** box.  Put check marks by **Counts** and
              **Percents** under **Display** and click **OK**. The result is shown in the Session
              Window as below:

              | DIVISION | Count | Percent |
              |---|---|---|
              | East North Central | 5 | 10.00 |
              | East South Central | 4 | 8.00 |
              | Middle Atlantic | 3 | 6.00 |
              | Mountain | 8 | 16.00 |
              | New England | 6 | 12.00 |
              | Pacific | 5 | 10.00 |
              | South Atlantic | 8 | 16.00 |
              | West North Central | 7 | 14.00 |
              | West South Central | 4 | 8.00 |
              | N= | 50 | |

       (c)    To obtain a pie chart, first enter the nine divisions in another column
              of Minitab, say C3, and enter the counts from the table in part (a) in
              the next column C4.  Name the columns C3 and C4 DIV and COUNT

              respectively.  Choose **Graph** ▶ **Pie Chart**, enter DIV in the **Categorical**

              **Variable** box and COUNT in the S**ummary variables** box.  Click on the
              **LABELS** button and then on the **Slice Labels** tab.  Put check marks in all
              four boxes and click **OK** twice.  The result is (except for colors and
              legend)

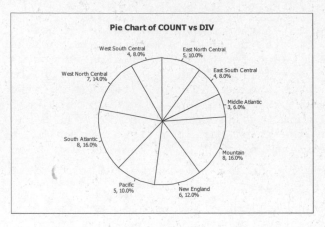

(d) Still in Minitab, choose **Graph ▶ Bar Chart,** make sure that Bars represents counts of unique values, click on **Simple** and on **OK.** Now enter DIVISION in the **Graph variables** box and click **OK.** The result is

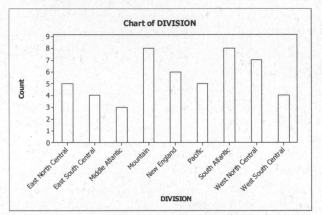

(e) Both charts indicate that Mountain and South Atlantic divisions have the most states (8) while the Middle Atlantic division has the fewest (3).

**26.** (a) The population consists of the states of the U.S. and the variable under consideration is the value of the exports of each state.

(b) Using Minitab, we enter the data from the WeissStats CD, choose **Graph ▶ Histogram**, click on **Simple** and click **OK.** Then double click on VALUE to enter it in the **Graph variables** box and click **OK.** The result is

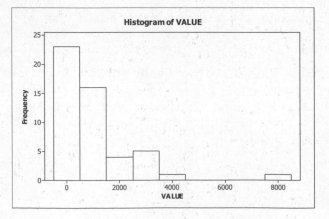

(c)  For the dotplot, we choose **Graph ▶ Dotplot**, click on **Simple** from the **One Y** row and click **OK**.  Then double click on VALUE to enter it in the **Graph variables** box and click **OK**.  The result is

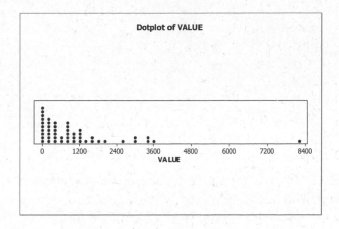

(d)  For the stem-and-leaf plot, we choose **Graph ▶ Stem-and-Leaf**, double click on VALUE to enter it in the **Graph variables** box and click **OK**. The result is

```
Stem-and-leaf of VALUE   N  = 50
Leaf Unit = 100

  23   0   0000000000011122223344444
 (10)  0   5677888899
  17   1   012224
  11   1   5679
   7   2
   7   2   69
   5   3   034
   2   3   6
   1   4
   1   4
   1   5
   1   5
   1   6
   1   6
   1   7
   1   7
   1   8   2
```

(e)  The overall shape of the distribution is reverse J shaped.

(f)  The distribution is right skewed.

27.  (a)  The population consists of countries of the world, and the variable under consideration is the expected life in years for people in those countries.

(b)  Using Minitab, we enter the data from the WeissStats CD, choose **Graph ▶ Histogram**, click on **Simple** and click **OK**.  Then double click on LIFE EXP to enter it in the **Graph variables** box and click **OK**.  The result is

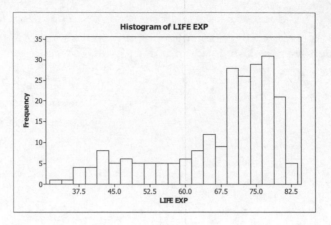

(c) For the dotplot, we choose **Graph ▶ Dotplot**, click on **Simple** from the **One Y** row and click **OK**. Then double click on LIFE EXP to enter it in the **Graph variables** box and click **OK**. The result is

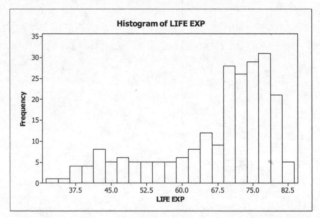

(d) For the stem-and-leaf plot, we choose **Graph ▶ Stem-and-Leaf**, double click on LIFE EXP to enter it in the **Graph variables** box and click **OK**. The result is

Stem-and-leaf of LIFE EXP  N = 224
Leaf Unit = 1.0

```
  1    3  2
  9    3  56788999
 21    4  112222233444
 31    4  5567788889
 42    5  01111333444
 50    5  56667779
 70    6  00111112233333444444
 99    6  5555666677788888899999999999
(52)   7  0000000001111111111111222222222233333333333444444444
 73    7  55555555555556666666666777777777778888888888888888889999999999
 11    8  00000011113
```

(e) The overall shape of the distribution is left skewed.
(f) This distribution is classified as left skewed.
28.  (a) The population consists of cities in the U.S., and the variables under consideration are their annual average maximum and minimum

temperatures.

(b)   Using Minitab, we enter the data from the WeissStats CD, choose **Graph ▶ Histogram**, click on **Simple** and click **OK**.  Double click on HIGH to enter it in the **Graph variables** box, and double click on LOW to enter it in the **Graph variables** box.  Now click on the **Multiple graphs** button and click to **Show Graph Variables on separate graphs** and also check both boxes under **Same Scales for Graphs**, and click **OK** twice.  The result is

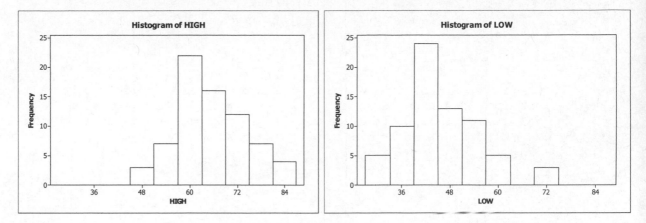

(c)   For the dotplot, we choose **Graph ▶ Dotplot**, click on **Simple** from the **Multiple Y's** row and click **OK**.  Then double click on HIGH and then LOW to enter them in the **Graph variables** box and click **OK**.  The result is

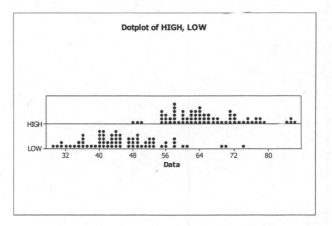

(d)   For the stem-and-leaf diagram, we choose **Graph ▶ Stem-and-Leaf**, double click on HIGH and then on LOW to enter then in the **Graph variables** box and click **OK**.  The result is

```
Stem-and-leaf of HIGH  N  = 71
Leaf Unit = 1.0

    3    4  789
    6    5  444
   21    5  555677777888999
  (17)   6  00001122222333444
   33    6  55556677779
   22    7  00001122234
   11    7  5577889
    4    8  444
    1    8  5
```

```
Stem-and-leaf of LOW   N  = 71
Leaf Unit = 1.0


  1    2  9
  7    3  001234
 19    3  555556779999
(20)   4  00011111223333344444
 32    4  5567777888899
 19    5  11122223
 11    5  5667889
  4    6  1
  3    6  9
  2    7  04
```

(e)  Both variables have distributions that are slightly right skewed.

(f)  Both distributions are close to symmetric, but are slightly right skewed.  It would take only a few changes in the data to make the distributions approximately symmetric.

## CHAPTER 3 ANSWERS

### Exercises 3.1

**3.1** The purpose of a measure of center is to indicate where the center or most typical value of a data set lies.

**3.3** Of the mean, median, and mode, only the mode is appropriate for use with qualitative data.

**3.5** (a) The mean is the sum of the values (45) divided by n (9). The result is 5. The median is the middle value in the ordered list and is also 5.

   (b) The mean is the sum of the values (135) divided by n (9). The result is 15. The median is the middle value in the ordered list and is 5, as previously. The median is more typical of most of the data than is the mean and works better here.

   (c) The mean does not have the property of being resistant to the influence of extreme observations.

**3.7** The median is more appropriate as a measure of central tendency than the mean because, unlike the mean, the median is not affected strongly by the relatively few homes that have extremely large or small areas.

**3.9** The mean number of values is calculated first by summing all seven values presented, which results in 51 days, and then dividing this by 7. Carrying this result to one more decimal place than the original data, the mean is 7.3 days.

   Calculating the median requires ordering the data from the smallest to the largest values: 5 5 6 **6** 7 11 11. Since the number of data values is odd, the median will be the observation exactly in the middle of this ordering, i.e., in the fourth ordered position. Thus, the median number of days is 6.

   The mode is the most frequently occurring data value or values. For these data, there are three modes because 5, 6, and 11 each occur twice.

**3.11** The mean number of tornado touchdowns is calculated by first summing all 12 data values presented, which results in 941, and then dividing this sum by 12. Thus, the mean is 78.4. This figure is already rounded to one more decimal place than the original data.

   Calculating the median requires ordering the data from the smallest to the largest values: 2, 3, 47, 57, 62, **68, 86**, 97, 98, 99, 118, 204. Since the number of data values is even, the median will be the mean of the two middle data values in the ordered list. The two middle values are the sixth and seventh ordered observations, or 68 and 86. Thus, the median age is found as (68 + 86)/2 = 77.0.

   The mode is the most frequently occurring data value or values. In these data, no value occurs more than once, so there is no mode.

**3.13** The mean wealth is calculated first by summing all 5 data values presented, which results in 148.5, and then dividing this sum by 5. Thus, the mean is 29.70. This figure is already rounded to one more decimal place than the original data.

   Calculating the median requires ordering the data from the smallest to the largest values: 17, 18, **22.5**, 40, 51. Since the number of data values is odd, the median will be the middle data value in the ordered list. The middle value is the third ordered observation. Thus, the median wealth is 22.5 billion dollars.

   The mode is the most frequently occurring data value or values. In these data, there is no mode since no value occurs more than once.

**3.15** The mean mpg is calculated first by summing all 14 data values presented, which results in 197, and then dividing this sum by 14.  Thus the mean is 14.07. Rounding to one decimal place yields 14.1 mpg.

Calculating the median requires ordering the data from the smallest to the largest values: 11, 12, 13, 13, 14, 14, **14**, **14**, 14, 15, 15, 15, 16, 17. Since the number of data values is even, the median will be the mean of the two middle data values in the ordered list.  The two middle values are the seventh and eighth ordered observations, which are both 14.  Thus, the median is 14 mpg.

The mode is the most frequently occurring data value or values.  In these data, the mode is 14 since 14 occurs 5 times, more than any other value.

3.17   (a)   The mean number of cremation burials is calculated first by summing all 17 data values presented, which results in 4977, and then dividing this sum by 17.  Thus, the mean is 292.8.  This figure is already rounded to one more decimal place than the original data.

Calculating the median requires ordering the data from the smallest to the largest values: 21, 34, 35, 46, 46, 48, 51, 64, **83**, 86, 119, 258, 265, 385, 429, 523, 2484.  Since the number of data values is odd, the median will be the middle data value in the ordered list.  The middle value is the ninth ordered observation.  Thus, the median is 83 cremation burials.

The mode is the most frequently occurring data value or values.  In these data, the mode is 46 since 46 occurs two times, more than any other value.

(b)   The median works best.  The mean is highly affected by the largest value and the mode is well to the left of the distribution's center.

3.19   (a)   The mean number of accidents is calculated first by summing all 7 data values presented, which results in 619, and then dividing this sum by 7.  Thus, the mean is 88.4.  This figure is already rounded to one more decimal place than the original data.

Calculating the median requires ordering the data from the smallest to the largest values: 69, 76, 85, **88**, 98, 100, 103.  Since the number of data values is odd, the median will be the middle data value in the ordered list. The middle value is the fourth ordered observation.  Thus the median is 88.

The mode is the most frequently occurring data value or values.  In these data, there is no mode since no number occurs more than once.

(b)   The mean number of accidents is calculated first by summing all 7 data values presented, which results in 492, and then dividing this sum by 7.  Thus, the mean is 70.3.  This figure is already rounded to one more decimal place than the original data.

Calculating the median requires ordering the data from the smallest to the largest values: 53, 56, 58, **59**, 70, 94, 102.  Since the number of data values is even, the median will be the middle data value in the ordered list.  The middle value is the fourth ordered observation. Thus the median is 59.

(c)   For built-up roads, the modal day is Friday since Friday would appear 103 times in the list, more than any other day.  For non-built-up roads the modal day is Sunday since Sunday would appear 102 times in the list, more than any other day.

(d)   Perhaps on Sunday, motorcycles are being used more for pleasure and sightseeing, possibly on unfamiliar roads.  This explanation seems reasonable since Saturday is the day with the second most number of accidents on non-built-up roads.  On the weekdays, which include the modal day Friday, motorcycles are perhaps being used more for transportation to and from work.  This traffic is more likely to be on built-up roads.

**3.21**   For a given population, the population mean is not a variable; it is a parameter, a constant.  The sample mean will vary from sample to sample and thus is a variable.

**3.23**   (a)  n = number of data values = 4            (b)  $\sum x_i$ = 12 + 8 + 9 + 17 = 46

   (c)  $\overline{x}$ = $\sum x_i/n$ = 46/4 = 11.5

**3.25**   (a)  n = number of data values = 10           (b)  $\sum x_i$ = 23.3

   (c)  $\overline{x}$ = $\sum x_i/n$ = 23.3/10 = 2.33 , which is given to two decimal places.

**3.27**   (a)  n = number of data values = 14           (b)  $\sum x_i$ = 16,458

   (c)  $\overline{x}$ = $\sum x_i$ /n = 16458/14 = 1175.57, which, rounded to one decimal place, is 1175.6

**3.29**   (a)  The mode is defined as the data value or values that occur most frequently. In this exercise, the data values are the NCAA wrestling champion universities.  During the period 1981 - 2005, Iowa was champion 15 times, Oklahoma State six times, Minnesota twice, and Arizona State and Iowa State each once.  Since Iowa is the data value that occurs most frequently, it is the mode.

   (b)  Again, the data values are the five champions.  There is no way to compute a mean or median for such nonnumerical data.  Thus, it would not be appropriate to use either the mean or median here.  The mode is the only measure of center that can be used for qualitative data.

**3.31**   (a)  The mode is defined as the data value or values that occur most frequently. In this exercise, the data values are the six colors of M&Ms.  In the three bag sample, brown occurred most often (152 times). Since brown is the data value that occurs most frequently, it is the mode.

   (b)  Again, the data values are the six M&M colors.  There is no way to compute a mean or median for such non-numerical data.  Thus, it would not be appropriate to use either the mean or median here.  The mode is the only measure of center that can be used for qualitative data.

**3.33**   Using Minitab, retrieve the data from the WeissStats CD.  Since the data are quantitative discrete, any of the three measures of center could be used.

To obtain the mode, select **Stat ▶ Tables ▶ Tally Individual Variables**, then enter PUPS in the **Variables** box and click **OK**. .  The result is

| PUPS | Count |
|------|-------|
| 3 | 2 |
| 4 | 5 |
| 5 | 10 |
| 6 | 11 |
| 7 | 17 |
| 8 | 17 |
| 9 | 11 |
| 10 | 4 |
| 11 | 2 |
| 12 | 1 |
| N= | 80 |

The mode is the most occurring value.  From the table, we see that 7 and 8 pups each occurred 17 times, more than any other number of pups.  Thus there are two modes - 7 and 8.

To obtain the mean and median, select **Stat ▶ Basic Statistics ▶ Display Descriptive Statistics**, then enter <u>PUPS</u> in the **Variables** box and click **OK**.  The result is

| Variable | N | N* | Mean | SE Mean | StDev | Minimum | Q1 | Median | Q3 |
|---|---|---|---|---|---|---|---|---|---|
| PUPS | 80 | 0 | 7.125 | 0.211 | 1.885 | 3.000 | 6.000 | 7.000 | 8.000 |
| Maximum | | | | | | | | | |
| 12.000 | | | | | | | | | |

From this display, we see that the mean is 7.125 and the median is 7.000.  Thus all three measures yield similar results.  If we didn't know the results, the one to be used might depend on the purpose of the investigation.  The median tells us that half of the sharks had 7 or fewer pups.  The mean would be more useful for estimating the total number of pups for all great white sharks if we knew how many great white shark females there were.  The modes tell us the most common numbers of pups for a female white shark.

**3.35  Using Minitab, r**etrieve the data from the WeissStats CD.  Since the data are quantitative discrete, any of the three measures of center could be used.

To obtain the mode, select **Stat ▶ Tables ▶ Tally Individual Variables**, then enter <u>NAEP</u> in the **Variables** box and click **OK**. .   The result is

| NAEP | Count |
|---|---|
| 235 | 1 |
| 236 | 2 |
| 238 | 3 |
| 239 | 1 |
| 241 | 2 |
| 242 | 1 |
| 243 | 1 |
| 244 | 1 |
| 245 | 4 |
| 246 | 1 |
| 247 | 2 |
| 248 | 2 |
| 249 | 3 |
| 250 | 3 |
| 251 | 5 |
| 252 | 4 |
| 253 | 6 |
| 254 | 3 |
| 255 | 1 |
| 256 | 2 |
| 257 | 2 |
| N= | 50 |

The mode is the most occurring value.  We see that an NAEP score of 253 occurred 6 times, more than any other score.  Thus, the mode is 253.

To obtain the mean and median, select **Stat ▶ Basic Statistics ▶ Display Descriptive Statistics**, then enter <u>NAEP</u> in the **Variables** box and click **OK**.  The result is

| Variable | N | N* | Mean | SE Mean | StDev | Minimum | Q1 | Median | Q3 | Maximum |
|---|---|---|---|---|---|---|---|---|---|---|
| NAEP | 50 | 0 | 248.16 | 0.851 | 6.02 | 235.00 | 244.75 | 250.00 | 253.00 | 257.00 |

From this display, we see that the mean NAEP is 248.16 and the median is 250.00. In this case, the median is the most useful statistic since it tells us that half of the states had mean NAEP scores less than or equal to 250. The mean is probably not a true mean of all scores since the states are unlikely to have had the same number of people taking the test.  Larger samples would have been taken from more populous states, and those states should count more heavily in estimating the mean score for all states.

**3.37**   Using Minitab, retrieve the data from the WeissStats CD.  Since the data are quantitative, any of the three measures of center could be used.  To obtain the mode, select **Stat ▶ Tables ▶ Tally Individual Variables**, then enter PERCENT in the **Variables** box and click **OK.** .  The result is

| PERCENT | Count |     | PERCENT | Count |
|---------|-------|-----|---------|-------|
| 77      | 1     |     | 86      | 7     |
| 79      | 1     |     | 87      | 4     |
| 80      | 2     |     | 88      | 6     |
| 81      | 7     |     | 89      | 9     |
| 82      | 1     |     | 90      | 3     |
| 83      | 1     |     | 91      | 3     |
| 84      | 2     |     | 92      | 2     |
| 85      | 2     |     | N=      | 51    |

The mode is the most occurring value, which, in this case, is 89 (occurring 9 times).

To obtain the mean and median, select **Stat ▶ Basic Statistics ▶ Display Descriptive Statistics**, then enter PERCENT in the **Variables** box and click **OK.**   The result is

| Variable | N  | N* | Mean   | SE Mean | StDev | Minimum | Q1     | Median | Q3     |
|----------|----|----|--------|---------|-------|---------|--------|--------|--------|
| PERCENT  | 51 | 0  | 86.118 | 0.530   | 3.782 | 77.000  | 83.000 | 87.000 | 89.000 |

| Variable | Maximum |
|----------|---------|
| PERCENT  | 92.000  |

From this display, we see that the mean is 86.118 and the median is 87.  In this case, the median is the most useful statistic since it tells us that half of the states had high school completion rates of 87 percent or less. The mean is not a true national completion rate since all states are counted equally in computing this mean.  A true national mean completion rate should be computed by dividing the total number of completions by the total numbers of high school students, not by averaging state completion rates.  By doing so, states with larger numbers of high school students will have a greater effect on the mean than states with smaller numbers of students.

**3.39**   Using Minitab, retrieve the data from the WeissStats CD.  Since the data are quantitative, any of the three measures of center could be used.  However, since the variable, temperature, is continuous, the data are temperatures rounded to 0.1 degree.  It is unlikely that two temperatures were exactly the same, so the use of the mode is not advised.

To obtain the mean and median, select **Stat ▶ Basic Statistics ▶ Display Descriptive Statistics**, then enter TEMP in the **Variables** box and click **OK.** The result is

| Variable | N  | N* | Mean   | SE Mean | StDev | Minimum | Q1     | Median | Q3     | Maximum |
|----------|----|----|--------|---------|-------|---------|--------|--------|--------|---------|
| TEMP     | 93 | 0  | 98.124 | 0.0671  | 0.647 | 96.700  | 97.650 | 98.200 | 98.600 | 99.400  |

From this display, we see that the mean temperature is 98.124 degrees and the median is 98.200.  In this case, the question was whether the **mean** body temperature of humans is 98.6 degrees Fahrenheit.  It makes more sense to use the sample mean to address this question even though the results of the study show the median and mean to be very close.

**3.41**   (a) The mean of the ratings is 2.5.  This is derived by summing the 14

individual ratings, which yields 35, and dividing by 14.

(b)  The median of the data is the average of the two observations appearing in the middle positions after ordering the data from the smallest to the largest values: 1, 1, 2, 2, 2, 2, **2, 2**, 3, 3, 3, 4, 4, 4.  The median is the average of the seventh and eighth ordered observations, or $(2 + 2)/2 = 2$.

(c)  In this case, the median provides a better descriptive summary of the data than does the mean.  Notice that six of the fourteen values are 2s and these are more in line with the median 2 than with the mean 2.5. Furthermore, the data are not actually measurements, but reflect the opinions of the participants, each of whom must decide what a '2' or a '3' really means.

**3.43**  The expression $(\sum x_i)^2$ represents the square of the sum of the data, whereas $\sum x_i^2$ represents the sum of the squares of the data.  To show that these two quantities are generally unequal, consider three observations on x as presented in column 1 of the following table.  Also, consider each observation's square, as presented in column 2.

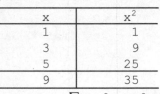

| x | $x^2$ |
|---|-------|
| 1 | 1 |
| 3 | 9 |
| 5 | 25 |
| 9 | 35 |

From the bottom row, we see that $(\sum x_i)^2 = 9^2 = 81$ and $\sum x_i^2 = 35$.  Thus, $(\sum x_i)^2 \neq \sum x_i^2$.

**Exercises 3.2**

**3.45**  The purpose of a measure of variation is to show the amount of variation or spread in a data set.  Any value of central tendency does not adequately characterize the elements of a data set.  Measures of central tendency merely describe the center of the observations, but do not show how observations differ from each other.  A measure of variation is intended to capture the degree to which observations differ among themselves.

**3.47**  When we use the standard deviation as a measure of variation, the reference point is the mean since all of the differences used in computing the standard deviation are between the data values and the mean.

**3.49**  (a)  The mean of this data set is 5.  Thus

$$\sum (x_i - \overline{x})^2 = (-4)^2 + (-3)^2 + (-2)^2 + (-1)^2 + 0^2 + 1^2 + 2^2 + 3^2 + 4^2 = 60$$

Dividing by $n - 1 = 8$, $s^2 = 60/8 = 7.5$.  The standard deviation, s, is the square root of 7.5, or 2.739.

(b)  The mean of this data set is 15.  Thus

$$\sum (x_i - \overline{x})^2 = (-14)^2 + (-13)^2 + (-12)^2 + (-11)^2 + 10^2 + (-9)^2 + (-8)^2 + (-7)^2 + 84^2 = 7980$$

Dividing by $n - 1 = 8$, $s^2 = 7980/8 = 997.5$.  The standard deviation, s, is the square root of 997.5, or 31.583.

(c)  Changing the 9 to 99 greatly increases the standard deviation, illustrating that it lacks the property of resistance to the influence of extreme data values.

**3.51**  (a)  The range is $54 - 9 = 45$.

(b)  The defining formula for $s$ is $s = \sqrt{\dfrac{\sum(x_i - \bar{x})^2}{n-1}}$ .

The first three columns of the following table present the calculations that are needed to compute $s$ by the defining formula:

| x | x−x̄ | (x−x̄)² | x² |
|---|---|---|---|
| 21 | −15 | 225 | 441 |
| 54 | 18 | 324 | 2916 |
| 9 | −27 | 729 | 81 |
| 45 | 9 | 81 | 2025 |
| 51 | 15 | 225 | 2601 |
| 180 | 0 | 1584 | 8064 |

With $n = 5$ and using the bottom figure of column 1, we find that $\bar{x} = \sum x_i/n = 180/5 = 36$.  This value is needed to construct the differences presented in column 2.  Column 3 squares these differences. The figure presented at the bottom of column 3 is the computation needed for the numerator of the defining formula for $s$.  Thus,

$$s = \sqrt{\frac{1584}{5-1}} = 19.900$$

(c)  The computing formula for s is

$$s = \sqrt{\frac{\sum x_i^2 - (\sum x_i)^2/n}{n-1}}$$

The computing formula for s with figures used from the bottom of Columns 1 and 4 of the previous table yields

$$s = \sqrt{\frac{8064 - 180^2/5}{5-1}} = 19.900$$

(d)  The computing formula was a time-saver since only one column of intermediate numbers needed to be calculated, whereas the defining formula required three additional columns.

**3.53**  The range is 11 - 5 = 6.

The defining formula for $s$ is $s = \sqrt{\dfrac{\sum(x_i - \bar{x})^2}{n-1}}$ .

The following table presents the calculations that are needed to compute $s$ by the defining formula:

| x | x-x̄ | (x-x̄)² |
|---|---|---|
| 6 | −1.3 | 1.69 |
| 7 | −0.3 | 0.09 |
| 11 | 3.7 | 13.69 |
| 6 | −1.3 | 1.69 |
| 5 | −2.3 | 5.29 |
| 5 | −2.3 | 5.29 |
| 11 | 3.7 | 13.69 |
| 51 | −0.1 | 41.43 |

With n = 7 and using the bottom figure of column 1, we find that $\bar{x} = \sum x_i/n$ = 51/7 = 7.3. This value is needed to construct the differences presented in column 2. Column 3 squares these differences. The figure presented at the bottom of column 3 is the computation needed for the numerator of the ↗ defining formula for s. Thus,

$$s = \sqrt{\frac{41.43}{7-1}} = 2.63$$

**3.55** The range of a data set is defined to be the difference between the largest and smallest data values in the data set. The largest number of tornado touchdowns is 204. The smallest number is 2. The range is 204 - 2 = 202 tornadoes.

The defining formula for s is $s = \sqrt{\dfrac{\sum (x_i - \bar{x})^2}{n-1}}$ .

The first three columns of the following table present the calculations that are needed to compute s by the defining formula:

| x | x-x̄ | (x-x̄)² | x² |
|---|---|---|---|
| 3 | −75.4 | 5685.16 | 9 |
| 2 | −76.4 | 5836.96 | 4 |
| 47 | −31.4 | 985.96 | 2204 |
| 118 | 39.6 | 1568.16 | 13924 |
| 204 | 125.6 | 15775.36 | 41616 |
| 97 | 18.6 | 345.96 | 9409 |
| 68 | −10.4 | 108.16 | 4624 |
| 86 | 7.6 | 57.76 | 7396 |
| 62 | −16.4 | 268.96 | 3844 |
| 57 | −21.4 | 457.96 | 3249 |
| 98 | 19.6 | 384.16 | 9604 |
| 99 | 20.6 | 424.36 | 9801 |
| 941 | 0.2 | 31898.92 | 105689 |

$$s = \sqrt{\frac{31898.92}{12-1}} = 53.85$$

The computing formula for s is

$$s = \sqrt{\frac{\sum x_i^2 - (\sum x_i)^2/n}{n-1}}$$

Columns 1 and 4 of the previous table present the calculations needed to compute s by the computing formula. Using the figures presented at the bottom of each of these columns and substituting into the formula itself, we get:

$$s = \sqrt{\frac{105689 - 941^2/12}{12-1}} = 53.85$$

This is, of course, the same result that we obtained by the defining formula.

**3.57**   The range of a data set is defined to be the difference between the largest and smallest data values in the data set.  The highest value is 51 billion and the lowest value is 17 billion.  The range is 51 - 17 = 34 billion.

The defining formula for $s$ is  $s = \sqrt{\frac{\sum(x_i - \overline{x})^2}{n-1}}$ .

The following table presents the calculations that are needed to compute $s$ by the defining formula:

| X | $x - \overline{x}$ | $(x-\overline{x})^2$ |
|---|---|---|
| 51 | 21.3 | 453.69 |
| 40 | 10.3 | 106.09 |
| 22.5 | -7.2 | 51.84 |
| 18 | -11.7 | 136.89 |
| 17 | -12.7 | 161.29 |
| 148.5 | 0.0 | 909.80 |

With n = 5 and using the bottom figure of column 1, we find that $\overline{x} = x_i/n = 148.5/5 = 29.7$.  This value is needed to construct the differences in column 2.  Column 3 squares the differences presented in column 2.  The figure presented at the bottom of column 3 is the computation needed for the numerator of the defining formula $s$.  Thus,

$$s = \sqrt{\frac{909.80}{5-1}} = 15.08$$

**3.59**   The range of a data set is defined to be the difference between the largest and smallest data values in the data set.  The highest overall mpg is 17 and the lowest overall mpg is 11.  The range is 17 - 11 = 6.

The defining formula for $s$ is  $s = \sqrt{\frac{\sum(x_i - \overline{x})^2}{n-1}}$ .

The following table presents the calculations that are needed to compute $s$ by the defining formula:

| x | $x-\overline{x}$ | $(x-\overline{x})^2$ |
|---|---|---|
| 14 | -0.071 | 0.00504 |
| 12 | -2.071 | 4.28904 |
| 13 | -1.071 | 1.14704 |
| 15 | 0.929 | 0.86304 |
| 14 | -0.071 | 0.00504 |
| 15 | 0.929 | 0.86304 |
| 13 | -1.071 | 1.14704 |
| 17 | 2.929 | 8.57904 |
| 14 | -0.071 | 0.00504 |
| 14 | -0.071 | 0.00504 |
| 14 | -0.071 | 0.00504 |
| 15 | 0.929 | 0.86304 |
| 11 | -3.071 | 9.43104 |
| 16 | 1.929 | 3.72104 |
| 197 | 0.000 | 30.9286 |

With n = 14 and using the bottom figure of column 1, we find that $\overline{x}$ = .$x_i$/n = 197/14 = 14.071.  This value is needed to construct the differences in column 2. Column 3 squares the differences presented in column 2.  The figure presented at the bottom of column 3 is the computation needed for the numerator of the defining formula $s$.  Thus,

$$s = \sqrt{\frac{30.9286}{14-1}} = 1.542$$

Rounding to one decimal place, s = 1.5.

**3.61**   (a)  We will use the computing formula to find s.  The following table presents the calculations that are needed to compute $s$ by the computing formula:

| x | $x^2$ | x | $x^2$ |
|---|---|---|---|
| 83 | 6889 | 429 | 184041 |
| 46 | 2116 | 35 | 1225 |
| 64 | 4096 | 51 | 2601 |
| 385 | 148225 | 34 | 1156 |
| 46 | 2116 | 258 | 66564 |
| 21 | 441 | 265 | 70225 |
| 48 | 2304 | 119 | 14161 |
| 86 | 7396 | 2484 | 6170256 |
| 523 | 273529 | | |
| | | 4977 | 6957341 |

The computing formula for $s$ is

$$s = \sqrt{\frac{\sum x_i^2 - (\sum x_i)^2 / n}{n-1}}$$

The previous table presents the calculations needed to compute $s$ by the computing formula. Using the figures presented at the bottom of the right two columns and substituting into the formula itself, we get:

$$s = \sqrt{\frac{6957341 - 4977^2 / 17}{17 - 1}} = 586.3$$

(b) No. The standard deviation is based on deviations from the mean. However, these data are right skewed and contain a very large outlier (2484) that greatly increases the value of the mean. The deviation of 2484 from the mean accounts for about 87% of the total of the squared deviations from the mean, so the influence of that data value on both the mean and the standard deviation is very great.

**3.63** (a) We would guess that the non-built up roads would have the greater variation due to more variation in the recreational use of those roads at various times of the week.

(b) The range for built-up roads is 103 - 69 = 34, and the range for non-built-up roads is 102 - 53 = 49.

For built-up roads, n = 17, $\sum x_i = 619$, $\sum x_i^2 = 55719$, and

$$s = \sqrt{\frac{55719 - 619^2 / 7}{7 - 1}} = 12.79$$

For non-built-up-roads, n = 7, $\sum x_i = 492$, $\sum x_i^2 = 36930$, and

$$s = \sqrt{\frac{36930 - 492^2 / 7}{7 - 1}} = 19.79$$

Thus both measures of variation confirm that our guess is correct. The range and standard deviation are greater for non-built-up roads than for built-up roads.

**3.65** Using Minitab, retrieve the data from the WeissStats CD. Choose **Stat ▶**

**Basic Statistics ▶ Display Descriptive Statistics**, enter PUPS in the

**Variables** text box. Now click on the **Statistics** button and check the range to add it to the list of statistics to be computed. Click **OK** twice. The output appears in the Session Window as

| Variable | N | N* | Mean | SE Mean | StDev | Minimum | Q1 | Median | Q3 | Maximum |
|----------|-----|----|-------|---------|-------|---------|-------|--------|-------|---------|
| PUPS | 80 | 0 | 7.125 | 0.211 | 1.885 | 3.000 | 6.000 | 7.000 | 8.000 | 12.000 |

| Variable | Range |
|----------|-------|
| PUPS | 9.000 |

The range (the difference between the maximum and the minimum data values) is 9 and the standard deviation is 1.885. Roughly speaking, on the average, the number of pups differs from the mean number of pups by 1.885.

**3.67** Using Minitab, retrieve the data from the WeissStats CD. . Select **Stat ▶**

**Basic Statistics ▶ Display Descriptive Statistics**, enter <u>NAEP</u> in the **Variables** text box.  Now click on the **Statistics** button and check the range to add it to the list of statistics to be computed.  Click **OK** twice.  The output appears in the Session Window as

| Variable | N | N* | Mean | SE Mean | StDev | Minimum | Q1 | Median | Q3 |
|---|---|---|---|---|---|---|---|---|---|
| NAEP | 50 | 0 | 248.16 | 0.851 | 6.02 | 235.00 | 244.75 | 250.00 | 253.00 |

| Variable | Maximum | Range |
|---|---|---|
| NAEP | 257.00 | 22.00 |

The range (the difference between the maximum and the minimum data values) is 22 and the standard deviation is 6.02.  Roughly speaking, on the average, the mean scores differ from the overall mean score by 6.02.  One should keep in mind, however, that the overall mean computed by Minitab assumes that the mean scores for each state count equally.  In fact, the true overall mean score would take into account that there are different numbers of people in each state

**3.69**  Using Minitab, retrieve the data from the WeissStats CD.  .  Select **Stat ▶**

**Basic Statistics ▶ Display Descriptive Statistics**, enter <u>PERCENT</u> in the **Variables** text box. Now click on the **Statistics** button and check the range to add it to the list of statistics to be computed.  Click **OK** twice.  The output appears in the Session Window as

| Variable | N | N* | Mean | SE Mean | StDev | Minimum | Q1 | Median | Q3 |
|---|---|---|---|---|---|---|---|---|---|
| PERCENT | 51 | 0 | 85.490 | 0.551 | 3.936 | 77.000 | 83.000 | 86.000 | 88.000 |

| Variable | Maximum | Range |
|---|---|---|
| PERCENT | 92.000 | 15.000 |

The range (the difference between the maximum and the minimum data values) is 15 and the standard deviation is 3.936.  Roughly speaking, on the average, the completion rates differ from the overall completion rate by 3.936.  One should keep in mind, however, that the overall mean computed by Minitab assumes that the completion rates for each state count equally.  In fact, the true overall mean completion rate would take into account that there are different numbers of students in each state.

**3.71**  Using Excel, retrieve the data from the WeissStats CD.  .  Select **DDXL ▶ Summaries**, select **Summary of one Variable** from the **Function type** drop-down box, specify <u>TEMP</u> in the **Quantitative Variable** text box and click **OK**.  The output appears as

| | |
|---|---|
| Count | 93 |
| Mean | 98.124 |
| Median | 98.2 |
| Std Dev | 0.647 |
| Variance | 0.418 |
| Range | 2.7 |
| Min | 96.7 |
| Max | 99.4 |
| IQR | 0.925 |
| 25th% | 97.675 |
| 75th% | 98.6 |

The range (the difference between the maximum and the minimum data values) is 2.7 and the standard deviation is 0.647.  Roughly speaking, on the average, the crime rates differ from the overall mean by 0.647.

**3.73** (a) We will compute $s$ for each data set using the computing formula.  The computing formula requires the following column manipulations.

<table>
<tr><th colspan="2">Data Set I</th><th colspan="2">Data Set II</th></tr>
<tr><th>x</th><th>$x^2$</th><th>x</th><th>$x^2$</th></tr>
<tr><td>0</td><td>0</td><td>10</td><td>100</td></tr>
<tr><td>0</td><td>0</td><td>12</td><td>144</td></tr>
<tr><td>10</td><td>100</td><td>14</td><td>196</td></tr>
<tr><td>12</td><td>144</td><td>14</td><td>196</td></tr>
<tr><td>14</td><td>196</td><td>14</td><td>196</td></tr>
<tr><td>14</td><td>196</td><td>15</td><td>225</td></tr>
<tr><td>14</td><td>196</td><td>15</td><td>225</td></tr>
<tr><td>15</td><td>225</td><td>15</td><td>225</td></tr>
<tr><td>15</td><td>225</td><td>16</td><td>256</td></tr>
<tr><td>15</td><td>225</td><td>17</td><td>289</td></tr>
<tr><td>16</td><td>256</td><td>142</td><td>2,052</td></tr>
<tr><td>17</td><td>289</td><td></td><td></td></tr>
<tr><td>23</td><td>529</td><td></td><td></td></tr>
<tr><td>24</td><td>576</td><td></td><td></td></tr>
<tr><td>189</td><td>3,157</td><td></td><td></td></tr>
</table>

For each data set, the calculations for $s$ are presented in column 2 of the following table.

| Data Set | $s$ | Range |
|----------|-----|-------|
| I | $s = \sqrt{\dfrac{3157 - 189^2/14}{14-1}} = 6.8$ | 24 |
| II | $s = \sqrt{\dfrac{2052 - 142^2/10}{10-1}} = 0.20.$ | 7 |

(b) The range for Data Set I is 24 - 0 = 24.  The range for Data Set II is 17 - 10 = 7.  These are recorded in column 3 of the previous table.

(c) Outliers increase the variation in a data set; in other words, removing the outliers from a data set results in a decrease in the variation.

**3.75** (a) We create a table with columns headed by x, f, xf, $(x-\overline{x})$, $(x-\overline{x})^2$, and $(x-\overline{x})^2 f$ to estimate the sample mean and standard deviation of the days-to-maturity data.

| Midpoint | Frequency | | | | |
|----------|-----------|-----|--------------------|----------------------|------------------------|
| x | f | xf | $x-\overline{x}$ | $(x-\overline{x})^2$ | $f(x-\overline{x})^2$ |
| 35 | 3 | 105 | -33.5 | 122.25 | 366.75 |
| 45 | 1 | 45 | -23.5 | 552.25 | 552.25 |
| 55 | 8 | 440 | -13.5 | 182.25 | 458.00 |
| 65 | 10 | 650 | -3.5 | 12.25 | 122.50 |
| 75 | 7 | 525 | 6.5 | 42.25 | 295.75 |
| 85 | 7 | 595 | 16.5 | 272.25 | 1905.75 |
| 95 | 4 | 380 | 26.5 | 702.25 | 2809.00 |
| | 40 | 2740 | | | 7510.00 |

The mean $\overline{x}$ = 2740/40 =68.5.  This is subtracted from each of the midpoints in column 1 to get the entries in column 4.  The grouped data formula yields the following estimate of the standard deviation:

$$s = \sqrt{\frac{\sum (x - \bar{x})^2 f}{n-1}} = \sqrt{\frac{10510}{40-1}} = 16.4$$

(b) The estimated mean is very close to the actual mean.  There is a greater discrepancy between the estimated standard deviation and the actual value. These discrepancies occur because in the grouped data formulas, every actual data value in a given class is replaced by the class midpoint even though most values in the class are not equal to the midpoint.  For example, every data value in the first class is represented by the midpoint 35, but the three data values are 36, 38, and 39, all of them larger than the midpoint.  This introduces a small error in each difference between an observation and the mean.  It is unlikely that all of these small errors will cancel each other out causing both the mean and the standard deviation to exhibit small errors.

**3.77** (a) In Figure 3.6, nine of the ten data points (90%) lie within two standard deviations to either side of the mean.

(b) In Figure 3.6, ten of the ten data points (100%) lie within three standard deviations to either side of the mean.

(c) In Figure 3.7, nine of the ten data points (90%) lie within two standard deviations to either side of the mean and ten of the ten data points (100%) lie within three standard deviations to either side of the mean.

(d) Chebychev's Rule clarifies that the phrase "almost all" in Key Fact 3.2 means "at least 89%".  Another reason for the importance of Chebychev's Rule is that is applies to all sets of data, regardless of the sample size or the shape of the distribution of the data, whether it be a sample or an entire population.  This rule also makes it possible to make statements about a data set even if only the mean and standard deviation of the data are known.

**3.79** (a) Since 30 of the 40 equates to 75%, k must be 2.  Thus at least 30 of the books' costs lie within two standard deviations of the mean, or between $76.75 - 2(10.42) = and $76.75 + 2(10.42) or between $55.16 and $96.84.

(b) The difference between $108.81 and the mean $76.75 is $32.26 or 3($10.42).  The difference between $76.75 and $45.49 is the same.  Chebychev's Rule, using k = 3, says that at least 89% of the 40 books must have costs in this range.  Since 0.89(40) = 35.6, at least 35.6 of the 40 sociology books cost between $45.49 and $108.01.  Since the number of books must be a whole number, this really means that at least 36 of the 40 books' costs must fall in this range.

**3.81** (a) Assuming that these weights have roughly a bell-shaped distribution, approximately 68% of the observations should lie within 4.16kg of the mean 45.30 or within the interval from 41.14 to 49.46.  Approximately 95% of the observations should lie within 2(4.16)kg of the mean 45.30 or within the interval from 36.98 to 53.62.  Approximately 99.7% of the observations should lie within 3(4.16)kg of the mean 45.30 or within the interval from 32.82 to 57.78.

(b) Forty-four of the 60 observations (73.3%) lie between 41.14 and 49.46.0.  Fifty-three of the 60 observations (88.3%) lie between 36.98 and 53.62.  All 60 of the observations (100%)lie between 32.82 and 57.78.

(c)  The actual percentage within one standard deviation to either side of the mean is close but a little higher than expected from the Empirical Rule.  The actual percentage within two standard deviations to either side of the mean is close but a little lower than expected from the Empirical Rule.  The actual percentage within three standard deviations to either side of the mean is very close to the expected from the Empirical Rule.

(d)  The histogram shown in Exercise 2.92 is a little right-skewed, but not far from bell-shaped.  Thus one should expect that the Empirical Rule percentages would not be exact for these data.

(e)  The differences between the actual percentages and the Empirical Rule percentages may be the result of this slight right skewness.  In each case, the difference in percentages is the result of only three or four observations out of 60 (41 instead of 44 and 57 instead of 53).  These could easily be the result of random variation in the sampling process.  Thus, use of the Empirical rule is reasonable.

**3.83**  (a)  The mean cost of the books was $76.75 and the standard deviation was $10.42.  Thirty-eight out of 40 is 95%.  If the costs have a bell-shaped distribution, we would expect the 38 costs to fall within two standard deviations ($20.84) to either side of the mean, or between $76.75 - $20.84 = $\underline{55.91}$ and $76.75 + $20.84 = $\underline{96.59}$.

(b)  The interval from $45.49 to 108.01 is $31.26 to either side of $76.75, or 3($10.42).  We would expect that about 99.7% (or all $\underline{40}$) of the costs would fall in this range if the costs had a bell-shaped distribution.

## Section 3.3

**3.85**  The median and the interquartile range have the advantage over the mean and standard deviation that they are not sensitive to extreme values; they are said to be resistant.

**3.87**  No.  An extreme observation may be an outlier, but it may also be an indication of skewness.

**3.89**  (a)  The interquartile range is a descriptive measure of variation.

(b)  It measures the spread of the middle two quarters (50%) of the data.

**3.91**  The adjacent values are just the minimum and maximum observations when there are no potential outliers.

**3.93**  (a)  First arrange the 20 data values in increasing order and note the two middle values:

45 48 64 70 73 74 74 78 78 **79 79** 80 80 80 80 80 80 81 82 82

The median (second quartile) is the average of the two middle values, 79 and 79, and thus equals (79 + 79)/2 = 79.

The first quartile $Q_1$ is the median of the lower half of the ordered data set 45 48 64 70 **73 74** 74 78 78 79 and thus is the average of the fifth and sixth values, 73 and 74.  $Q_1$ = (73 + 74)/2 = 73.5

The third quartile $Q_3$ is the median of the upper half of the ordered data set 79 80 80 80 **80 80** 80 81 82 82 and thus is the average of the fifth and sixth values, 80 and 80.  $Q_3$ = (80 + 80)/2 = 80.0

Thus 25% of the data values are less than or equal to 73.5, the next 25% are between 73.5 and the median 79, the next 25% are between 79 and 80, and the upper 25% are greater than or equal to 80.

(b)  The interquartile range is $Q_3 - Q_1 = 80.0 - 73.5 = 6.5$.

(c)  The five-number summary consists of the Minimum, $Q_1$, Median, $Q_3$, and the Maximum.  Respectively, these are 45, 73.5, 79.0, 80.0, and 82.  In addition to the explanation of the quartiles in part (a), all of the values are greater than or equal to the Minimum, and all of the values are less than or equal to the Maximum.

(d)  To identify any potential outliers, we first find the lower and upper limits. These are found by subtracting 1.5 x IQR from $Q_1$ and adding 1.5 x IQR to $Q_3$.  The lower limit is $73.5 - 1.5(6.5) = 63.75$ and the upper limit is $80.0 + 1.5(6.5) = 89.75$.  Potential outliers consist of any values that are less than 63.75 or greater than 89.75.  We see that 45 and 48 are both less than 63.75, so these two values are potential outliers.

(e)  We will construct a modified boxplot so as to show the potential outliers.  The boxplot consists of a rectangle with its ends at the first and third quartiles, a line through it at the median, and two "whiskers" that extend from the quartiles outward to the adjacent values.  The adjacent values are the most extreme data values that do not lie outside the lower and upper limits.  For these data, the adjacent values are 64 and 82.  The potential outliers, 45 and 48, are plotted individually.  The modified boxplot, prepared using Minitab, is

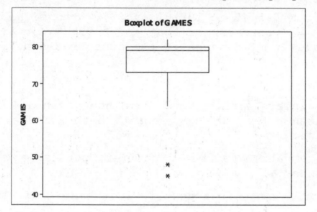

We see that the data is very left skewed, even without considering the two potential outliers.  This is because, in most seasons, Gretzky played almost every game, but in a few seasons, he played considerably fewer than the maximum number of games, probably due to injuries.

**3.95**  (a)  First arrange the data in increasing order:

1  1  3  3  4  4  5  6  6  7  **7**  9  9  10  12  12  13  15  18  23  55

Since this data set has on odd number of values (21), the median is the middle value (11$^{th}$) in the ordered list.  The median, or second quartile, is 7.

The first quartile $Q_1$ is the median of the lower half of the ordered data set (including the median) 1  1  3  3  4  **4**  5  6  6  7  7.  Since this data set has an odd number of values (11), $Q_1$ is the sixth value, so $Q_1 = 4$.

The third quartile $Q_3$ is the median of the upper half of the ordered data set (including the median) 7  9  9  10  12  **12**  13  15  18  23  55. Since this data set has an odd number of values (11), $Q_3$ is the sixth value, so $Q_3 = 12$.

Thus 25% of the data values are less than or equal to 4, the next 25% are between 4 and the median 7, the next 25% are between 7 and 12, and the upper 25% are greater than or equal to 12.

(b)   The interquartile range is $Q_3 - Q_1 = 12 - 4 = 8$.

(c)   The five-number summary consists of the Minimum, $Q_1$, Median, $Q_3$, and the Maximum. Respectively, these are 1, 4, 7, 12, and 55. In addition to the explanation of the quartiles in part (a), all of the values are greater than or equal to the Minimum, and all of the values are less than or equal to the Maximum.

(d)   To identify any potential outliers, we first find the lower and upper limits. These are found by subtracting 1.5 x IQR from $Q_1$ and adding 1.5 x IQR to $Q_3$. Thus the lower limit is $4 - 1.5(8) = -8$ and the upper limit is $12 + 1.5(8) = 24$. Potential outliers consist of any values that are less than -8 or greater than 24. We see that 55 is the only potential outlier.

(e)   We will construct a modified boxplot, which consists of a rectangle with its ends at the first and third quartiles, a line through it at the median, and two "whiskers" that extend from the quartiles outward to the adjacent values. The adjacent values are the most extreme data values that do not lie outside the lower and upper limits. For these data, the adjacent values are 1 and 23. The potential outlier, 55, is plotted individually. Thus the modified boxplot, prepared using Minitab, is

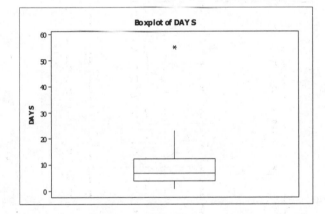

We see that the data are slightly right skewed if the potential outlier is not considered. The potential outlier is far from the other values, making it important to investigate the observation further.

3.97   (a)   First arrange the data in increasing order:

57   66   88   96   **116**   **147**   147   154   154   175

Since this data set has on even number of values (10), the median is the mean of the middle two values (5th and 6th) in the ordered list. Thus the median, or second quartile, is $(116 + 147)/2 = 131.5$.

The first quartile $Q_1$ is the median of the lower half of the ordered data set 57   66   **88**   96   116. Since this data set has an odd number of values (5), $Q_1$ is the middle value (3rd) in the ordered list = 88.

The third quartile $Q_3$ is the median of the upper half of the ordered data set 147   147   **154**   154   175. Since this data set has an odd number of values (5), $Q_3$ is the middle value in the ordered list = 154.

Thus 25% of the data values are less than or equal to 88, the next 25% are between 88 and the median 131.5, the next 25% are between 131.5 and 154, and the upper 25% are greater than or equal to 154.

(b)   The interquartile range is $Q_3 - Q_1 = 154 - 88 = 66$.

(c)   The five-number summary consists of the Minimum, $Q_1$, Median, $Q_3$, and the Maximum.  Respectively, these are 57, 88, 131.5, 154, and 175.  In addition to the explanation of the quartiles in part (a), all of the values are greater than or equal to the Minimum, and all of the values are less than or equal to the Maximum.

(d)   To identify any potential outliers, we first find the lower and upper limits. These are found by subtracting 1.5 x IQR from $Q_1$ and adding 1.5 x IQR to $Q_3$.  Thus the lower limit is $88 - 1.5(66) = -11$ and the upper limit is $154 + 1.5(66) = 253$.  Potential outliers consist of any values that are less than -11 or greater than 253.  We see that there are no potential outliers.

(e)   We will construct a boxplot, which consists of a rectangle with its ends at the first and third quartiles, a line through it at the median, and two "whiskers" that extend from the quartiles outward to the minimum and maximum values.  Thus the boxplot, prepared using Minitab, is

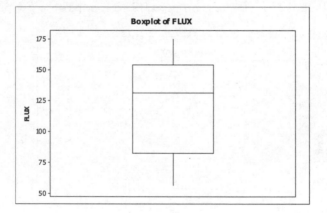

We see that the data are roughly symmetrical with no potential outliers.

**3.99**   (a)   First arrange the data in increasing order:

21   70   125   195   389   649   656   664   682   1006   1300   1403   1433   **1800**
1982   2205   2515   3027   3634   4200   5299   5947   7886   8543   9310   11189
17341

Since this data set has on odd number of values (27), the median is the middle value (14th) in the ordered list.  The median, or second quartile, is 1800.

The first quartile $Q_1$ is the median of the lower half of the ordered data set (including the median) 21   70   125   195   389   649   **656   664** 682   1006   1300   1403   1433   1800. Since this data set has an even number of values (14), $Q_1$ is the average of the seventh and eighth values, so $Q_1 = (656 + 664)/2 = 660.0$.

The third quartile $Q_3$ is the median of the upper half of the ordered data set (including the median) 1800   1982   2205   2515   3027   3634 **4200   5299**   5947   7886   8543   9310   11189   17341.  Since this data set has an even number of values (14), $Q_3$ is the average of the seventh and eighth values, so $Q_3$ = (4200 + 5299)/2 = 4749.5.

Thus 25% of the data values are less than or equal to 660.0, the next 25% are between 660.0 and the median 1800, the next 25% are between 1800 and 1719.5, and the upper 25% are greater than or equal to 4749.5.

(b)  The interquartile range is $Q_3$ − $Q_1$ = 4749.5 − 660.0 = 4089.5

(c)  The five-number summary consists of the Minimum, $Q_1$, Median, $Q_3$, and the Maximum.  Respectively, these are 21, 660, 1800, 4749.5, and 17341.  In addition to the explanation of the quartiles in part (a), all of the values are greater than or equal to the Minimum, and all of the values are less than or equal to the Maximum.

(d)  To identify any potential outliers, we first find the lower and upper limits. These are found by subtracting 1.5 x IQR from $Q_1$ and adding 1.5 x IQR to $Q_3$.  Thus the lower limit is 660.0 − 1.5(4089.5) = −5474.25 and the upper limit is 4749.5 + 1.5(4089.5) = 10883.75.  Potential outliers consist of any values that are less than −5474.25 or greater than 10883.75.  We see that 11189 and 17341 are potential outliers.

(e)  We will construct a modified boxplot that consists of a rectangle with its ends at the first and third quartiles, a line through it at the median, and two "whiskers" that extend from the quartiles outward to the adjacent values.  The adjacent values are the most extreme data values that do not lie outside the lower and upper limits.  For these data, the adjacent values are 21 and 9310.  The potential outliers, 11189 and 17341, are plotted individually.  Thus the boxplot, prepared using Minitab, is

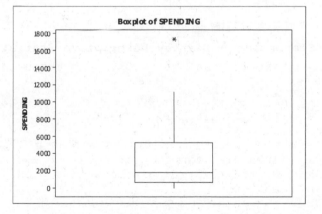

We have shown you the modified boxplot from Minitab to illustrate that software may produce a slightly different boxplot than we expected from our own computations.  Minitab uses a different definition of the quartiles, which leads to a larger value of the IQR (4643).  This, in turn, results in upper and lower limits that are farther from the median than ours are.  Consequently, 11189 shows up as an adjacent value, not an outlier, in Minitab's modified boxplot.  The overall picture of the data being right skewed with outlier(s) remains much the same.

**3.101** (a)  First, arrange the data in increasing order

6 7 8 8 8 8 9 **9** 9 10 10 10 10 10 10

Since there is an odd number of data values (15), the median is the middle (8th) value.  Thus the median is 9.

The first quartile $Q_1$ is the median of the lower half of the ordered data set (including the median) 6 7 8 **8 8** 8 9 9. Since this data set has an even number of values (8), $Q_1$ is the average of the fourth and fifth values, so $Q_1 = (8 + 8)/2 = 8$.

The third quartile $Q_3$ is the median of the upper half of the ordered data set (including the median) 9 9 10 **10 10** 10 10 10. Since this data set has an even number of values (8), $Q_3$ is the average of the fourth and fifth values, so $Q_3 = (10 + 10)/2 = 10$.

(b) The quartiles are not particularly useful for this set of data due to the small range of the data and the large numbers of identical values at 8, 9, and 10. For example, $Q_3$ and the maximum are equal as a result of the highest six values all being 10.

**3.103** The center of the distribution of weight loss, as measured by the median, is very similar for the two groups. However, group 1 has less variation in weight loss, both in the middle 50% and overall. Group 2 has a larger range than group 1. The distribution of weight loss for both groups is approximately symmetric.

**3.105** The median hemoglobin level is much lower in the HB SS patients than in the other two types of patients (which were roughly the same). The least variation in hemoglobin levels occurred in the HB SC patients, while the greatest variation occurred in the HB ST patients All three distributions were roughly symmetric about their medians.

**3.107** The central box is divided into equal parts by the median. The two whiskers will be the same length, although they may not necessarily be the same length as each part of the central box.

**3.109** (a) Using Minitab, retrieve the data from the WeissStats CD. Then choose

**Stat ▶ Basic Statistics ▶ Display Descriptive Statistics**, enter <u>PUPS</u> in the **Variables** text box, and click **OK**. The output appears in the Session Window as

| Variable | N | N* | Mean | SE Mean | StDev | Minimum | Q1 | Median | Q3 | Maximum |
|----------|---|----|------|---------|-------|---------|-----|--------|-----|---------|
| PUPS | 80 | 0 | 7.125 | 0.211 | 1.885 | 3.000 | 6.000 | 7.000 | 8.000 | 12.000 |

We see from the output that $Q_1 = 6.0$, the median = 7.0, and $Q_3 = 8.0$. Thus 25% of the values are less than or equal to 6.0, another 25% are between 6.0 and 7.0, the next 25% are between 7.0 and 8.0, and the top 25% are greater than 8.0.

(b) The interquartile range is $Q_3 - Q_1 = 8.0 - 6.0 = 2.0$. The middle 50% of the data values span an interval of 2.0.

(c) The five-number summary consists of the Minimum, $Q_1$, median, $Q_3$, and the Maximum. These are, respectively, 3, 6, 7, 8, and 12. In addition to the interpretation in part (a), all of the values are greater than or equal to the minimum, 3, and all are less than or equal to the maximum, 12.

(d) To identify any potential outliers, we first find the lower and upper limits. These are found by subtracting 1.5 x IQR from $Q_1$ and adding 1.5 x IQR to $Q_3$. Thus the lower limit is $6 - 1.5(2) = 3$, and the upper limit is $8 + 1.5(2) = 11$. Potential outliers consist of any values that are less than 3 or greater than 11. We see that 12 is a potential outlier. The maximum, 128, is greater than the upper limit, and it occurs once, so there is one potential outlier.

(e) To obtain a boxplot, we choose **Graph ▶ Boxplot**, select **Simple** in the **One Y** row and click **OK**. Enter <u>PUPS</u> in the **Graph Variables** text box and

click **OK.**   The result is

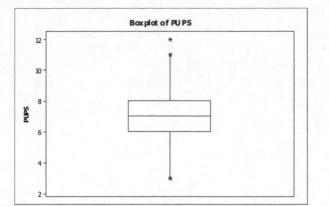

With the exception of the one potential outlier, the distribution of the data is very symmetric, approximately bell-shaped.

**3.111** (a)  Using Minitab, retrieve the data from the WeissStats CD.   Then choose

**Stat ▶ Basic Statistics ▶ Display Descriptive Statistics,** enter

PERCENT in the **Variables** text box, and click **OK.**   The output appears in the Session Window as

```
Variable   N  N*    Mean   SE Mean   StDev   Minimum      Q1  Median      Q3
PERCENT   51  0   85.490    0.551   3.936    77.000  83.000  86.000  88.000


Variable  Maximum
PERCENT    92.000
```

We see from the output that $Q_1$ = 83.0, the median = 86.0, and $Q_3$ = 88.0. Thus 25% of the values are less than or equal to 83.0, another 25% are between 83.0 and 86.0, the next 25% are between 86.0 and 88.0, and the top 25% are greater than 88.0.

(b)  The interquartile range is $Q_3 - Q_1$ = 88.0 - 83.0 = 5.0.   The middle 50% of the data values span an interval of 5.0.

(c)  The five-number summary consists of the Minimum, $Q_1$, median, $Q_3$, and the Maximum.   These are, respectively, 77.0, 83.0, 86.0, 88.0, and 92.0. In addition to the interpretation in part (a), all of the values are greater than or equal to the minimum, 77.0, and all are less than or equal to the maximum, 92.0.

(d)  To identify any potential outliers, we first find the lower and upper limits. These are found by subtracting 1.5 x IQR from $Q_1$ and adding 1.5 x IQR to $Q_3$.   Thus the lower limit is 83.0 - 1.5(5.0) = 75.5, and the upper limit is 88.0 + 1.5(5.0) = 95.5.   Potential outliers consist of any values that are less than 75.5 or greater than 95.5. We see that there are no potential outliers.

(e)  To obtain a boxplot, we choose **Graph ▶ Boxplot,** select **Simple** in the **One Y** row and click **OK.**   Enter PERCENT in the **Graph Variables** text box and click **OK.**   The result is

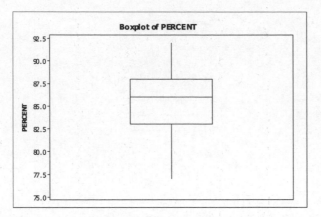

The distribution of the data roughly symmetrical, maybe slightly left skewed.  The whiskers on the graph extend out to the minimum and maximum values.

**3.113** (a) Using Minitab, retrieve the data from the WeissStats CD.  Then choose

**Stat ▶ Basic Statistics ▶ Display Descriptive Statistics**, enter

PERCENT in the **Variables** text box, and click **OK**.  The output appears in the Session Window as

```
Variable   N  N*    Mean  SE Mean  StDev  Minimum      Q1  Median      Q3
BODYTEMP  93  0   98.124   0.0671  0.647   96.700  97.650  98.200  98.600


Variable  Maximum
BODYTEMP   99.400
```
We see from the output that $Q_1$ = 97.65, the median = 98.20, and $Q_3$ = 98.60. Thus 25% of the values are less than or equal to 97.65, another 25% are between 97.65 and 98.20, the next 25% are between 98.20 and 98.60, and the top 25% are greater than 98.60.

(b) The interquartile range is $Q_3 - Q_1$ = 98.60 - 97.65 = 0.95  The middle 50% of the data values span an interval of 0.95.

(c) The five-number summary consists of the Minimum, $Q_1$, median, $Q_3$, and the Maximum.  These are, respectively, 96.70, 97.65, 98.20, 98.60, and 99.40.  In addition to the interpretation in part (a), all of the values are greater than or equal to the minimum, 96.70, and all are less than or equal to the maximum, 99.40.

(d) To identify any potential outliers, we first find the lower and upper limits. These are found by subtracting 1.5 x IQR from $Q_1$ and adding 1.5 x IQR to $Q_3$.  Thus the lower limit is 97.65 - 1.5(.095) = 96.225, and the upper limit is 98.60 + 1.5(0.95) = 100.025.  Potential outliers consist of any values that are less than 96.225 or greater than 100.025. We see that there are no potential outliers.

(e) To obtain a boxplot, we choose **Graph ▶ Boxplot**, select **Simple** in the

**One Y** row and click **OK**.  Enter RATE in the **Graph Variables** text box and click **OK**.  The result is

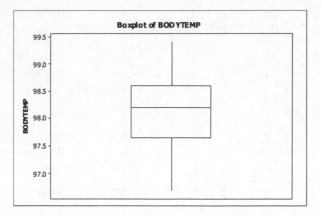

We see that the distribution of the data is roughly symmetrical with no outliers.

**3.115** (a) Using Minitab, retrieve the data from the WeissStats CD.  Then choose

**Graph ▶ Boxplot,** select **Simple** in the **Multiple Y's** row and click **OK.**
Enter VEGETARIANS and OMNIVORES in the **Graph Variables** text box and click **OK.**   The result is

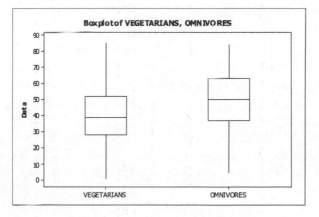

(b) The daily median amount of protein consumed by vegetarians is about 10 grams less than for omnivores.  The same is true for the first and third quartiles.  However, the maximums and minimums are about the same for the two groups.  The variation is about the same in the two groups. The IQR is approximately 25 and the range is about 80.

**3.117** (a) Using Minitab, retrieve the data from the WeissStats CD.  Then choose

**Graph ▶ Boxplot,** select **Simple** in the **Multiple Y's** row and click **OK.**
Enter STOMACH, BRONCHUS, COLON, OVARY, and BREAST in the **Graph Variables** text box and click **OK.**   The result is

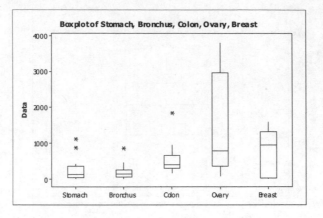

(b)   The median survival times were greatest for breast cancer patients, followed by those with cancer of the ovary, colon, bronchus, and stomach.  The variation in survival times for colon, bronchus, and stomach cancer patients, as measured by the interquartile range, were considerably smaller than the variation experienced by the ovary and breast cancer patients.  The first three also contained outliers in the data, whereas the last two did not.  All of the data distributions except for breast cancer are right skewed, whereas the breast cancer survival times are slightly left skewed.

### Section 3.4

**3.119** The ultimate objective of dealing with sample data in inferential studies is to describe the entire population.

**3.121** (a)  A standardized variable always has mean <u>0</u> and standard deviation <u>1</u>.

(b)  The z-score corresponding to an observed value of a variable tells you <u>how many standard deviations the observation is from the mean and, by its sign, what direction it is from the mean.</u>

(c)  A positive z-score indicates that the observation is <u>greater than</u> the mean, whereas a negative z-score indicates that the observation is <u>less than</u> the mean.

**3.123** The number 62.55 is a parameter since it is the mean of the entire population of players on the 2003 U.S. Women's World Cup soccer team.

**3.125** (a)  The population consists of all people living in the United States.  The variable of interest is the age of each person.

(b)  The mean is the total of the six ages divided by 6, i.e. 195/6 = 32.5.  The median is the mean of the two middle ages when they are listed in increasing order as 7, 9, **29, 45**, 51, 54.  Thus the median is (29 + 45)/2 = 37.0.  These are statistics since they are derived from a sample.  We write our results as  $\bar{x}$ = 32.5 and M = 37.0.

(c)  This mean and median are parameters since they are derived from the entire population.  Thus we write μ = 35.8 and η = 35.3.

**3.127** (a)  The population mean is  $\mu = \dfrac{\sum x_i}{N} = \dfrac{1385}{15} = 92.33$

(b)  We will use the computing formula to find the standard deviation.  For that we need  $\sum x_i^2 = 120^2 + 65^2 + \ldots + 50^2 = 153,825$ .   The standard deviation is then $\sigma = \sqrt{\dfrac{\sum x_i^2}{N} - \mu^2} = \sqrt{\dfrac{153825}{15} - 92.33^2} = 41.59$ .

(c) The median is the middle observation in the following ordered list of 15 observations: 45, 45, 50, 50, 50, 65, 70, **75**, 105, 115, 120, 140, 145, 145, 165.   The middle observation is the eighth value, so η = 75.

(d) The mode is the value that occurs more often than any other value, so the mode is 50.

(e) The first quartile is the median of the first eight values in the ordered list of part (c), so it is the mean of the fourth and fifth values.   Thus $Q_1$ = (50 + 50)/2 = 50.   The third quartile is the median of the last eight values in the ordered list of part (c), so $Q_3$ = (120 + 140)/2 = 130.   The IQR is the difference between $Q_1$ and $Q_3$.   Thus IQR = $Q_3$ - $Q_1$ = 130 - 50 = 80.

**3.129** (a) For Phoenix, the population mean is $\mu = \dfrac{\sum x_i}{N} = \dfrac{4100}{5} = 820.0$ .   For Los

Angeles, the population mean is $\mu = \dfrac{\sum x_i}{N} = \dfrac{7763}{5} = 1552.6$ .

(b) The range of values for Los Angeles is almost 700, while it is only a little over 200 for Phoenix, so Phoenix is likely to have the smaller standard deviation.

(c) We will use the computing formula to find the standard deviation.   For Phoenix, we need $\sum x_i^2 = 722^2 + 957^2 + ... + 829^2 = 3,398,568$ .   The standard

deviation is then $\sigma = \sqrt{\dfrac{\sum x_i^2}{N} - \mu^2} = \sqrt{\dfrac{3398568}{5} - 820.0^2} = 85.52$ .   For Los

Angeles, we need $\sum x_i^2 = 1189^2 + 1857... + 1752^2 = 12,368,471$ .   The standard

deviation is then $\sigma = \sqrt{\dfrac{\sum x_i^2}{N} - \mu^2} = \sqrt{\dfrac{12368471}{5} - 1552.6^2} = 251.25$ .

(d) Yes.   We expected the standard deviation to be smaller for Phoenix than for Los Angeles, and that turned out to be the case.

**3.131** (a)   The standardized version of x is z = (x - 16.3)/17.9.

(b)   The mean and standard deviation of z are 0 and 1 respectively.

(c)   For x = 64.7, the z score is (64.7 - 16.3)/17.9 = 2.70

For x = 4.2, the z score is (4.2 - 16.3)/17.9 = -0.68

(d)   The value 64.7 is 2.70 standard deviations above the mean 16.3.

The value 4.2 is 0.68 standard deviations below the mean 16.3.

(e)

| $\bar{x}$-3s | $\bar{x}$-2s | $\bar{x}$-s | $\bar{x}$ | $\bar{x}$+s | $\bar{x}$+2s | $\bar{x}$+3s |
|---|---|---|---|---|---|---|
| -37.4 | -19.5 | -1.6 | 16.3 | 34.2 | 52.1 | 70.0 | x |
| -3 | -2 | -1 | 0 | 1 | 2 | 3 | z |

**3.133** (a)  The standardized version of x is z = (x − 6.71)/0.67.

(b)  For x = 5.2, the z score is (5.2 − 6.71)/0.67 = −2.25

For x = 8.1, the z score is (8.1 − 6.71)/0.67 = 2.07

The value 5.2 is 2.25 standard deviations below the mean 6.71.

The value 8.1 is 2.07 standard deviations above the mean 6.71.

**3.135** (a)  z = (21.4 − 25)/1.15 = −3.13.

(b)  Yes.  If the gas mileages have a distribution that is roughly bell-shaped, we would expect more than 99% of cars to get mileage within three standard deviations from the mean of 25 mpg.  Your car's mpg is more than three standard deviations below the mean.  Even if the distribution is not bell-shaped, Chebychev's Rule says that at least 89% of the mileages must be within three standard deviations.

**3.137** The formulas used here to compute s and σ are

$$s = \sqrt{\frac{\sum (x_i - \bar{x})^2}{n-1}} \qquad \text{and} \qquad \sigma = \sqrt{\frac{\sum (x_i - \mu)^2}{N}}$$

For each data set, the relevant calculations are:

| Data Set | Number of Observations | $\sum (x_i - \text{mean})^2$ | s | σ |
|----------|------------------------|------------------------------|------|------|
| 1 | 4 | 14.00 | 2.16 | 1.87 |
| 2 | 7 | 24.86 | 2.04 | 1.88 |
| 3 | 10 | 36.40 | 2.01 | 1.91 |

(a)  The sample standard deviations are found in the fourth column of the previous table.

(b)  The population standard deviations are found in the fifth column of the previous table.

(c)  Comparing s and σ for a given data set, the two measures will tend to be closer together if the data set is large.

**3.139** (a)  Using Excel, we enter HEIGHT in A1 and the 20 data values in cells A2 through A21.  The data values in inches are

175    165    165    170    161    171    173    164    163    165

168    180    163    170    170    157    168    173    168    180

To obtain the mean, in cell A22, we type =AVERAGE(A2:A21) and the result is 168.45 cm.

To obtain the population standard deviation directly, in cell A23 we type =STDEVP(A2:A21).  This yields the value 5.77 cm.

You could also obtain the first result in Excel by selecting **DDXL ▶ Summaries**.  Select **Summary of One Variable** from the Function type box, select cells A1 to A21 with the cursor, enter HEIGHT in the **Quantitative Variable** box by dragging it from the **Names and Columns** box and click **OK**.  Unfortunately, this procedure produces a sample standard deviation instead of a population standard deviation, so you than have to convert s to σ using the formula derived in Exercise 3.138.

**3.141** z = (205500 − 220258)/5237 = −2.82.  Applying Chebychev's rule to that z-score with k = 82.82, we conclude that at least $100(1 - 1/2.82^2)\% = 87.4\%$ of the sale prices lie within 2.82 standard deviations to either side of the mean of $220,258.  Therefore, the $205,500 price of the home you are contemplating buying is lower than at least 87.4% of the prices of comparable homes. It appears that this home is bargain.

### Review Problems For Chapter 3

1.   (a)   Descriptive measures are numbers used to describe data sets.

     (b)   Measures of center indicate where the center or most typical value of a data set lies.

     (c)   Measures of variation indicate how much variation or spread the data set has.

2.   The two most commonly used measures of center for quantitative data are the mean and the median.  The mean uses all of the data, but can be influenced by the presence of a few outliers.  The median is computed from the one or two center values in an ordered list of the data.  It does not make use of all of the data, but it has the advantage that it is not influenced by the presence of a few outliers.

3.   The only measure of center, among those we discussed, that is appropriate for qualitative data is the mode.

4.   (a)   standard deviation

     (b)   interquartile range

5.   (a)  $\bar{x}$          (b)   s          (c)   $\mu$          (d)   $\sigma$

6.   (a)   This is not necessarily true.

     (b)   This is necessarily true.

7.   Almost all of the observations in any data set lie within $\underline{3}$ standard deviations to either side of the mean.

8.   (a)   The components of the five-number summary are the minimum, $Q_1$, median, $Q_3$, and the maximum.

     (b)   The median is a measure of the center.  The interquartile range, which is found as $Q_3 - Q_1$, and the range, which is the difference between the maximum and the minimum, are measures of variation.

     (c)   The boxplot is based on the five-number summary.

9.   (a)   An outlier is an observation that falls well outside the overall pattern of the data.

     (b)   First compute the interquartile range IQR = $Q_3 - Q_1$.  Then compute two limits as $Q_1 - 1.5IQR$ and $Q_3 + 1.5IQR$.  Observations that are lower than the lower limit or higher than the upper limit are potential outliers and require further study.

10.   (a)   A z-score for an observation x is obtained by subtracting the mean of the data set from x and then dividing the result by the standard deviation; i.e., $z = (x - \mu)/\sigma$ for population data or $z = (x - \bar{x})/s$ for sample data.

      (b)   A z-score indicates how many standard deviations an observation is below or above the mean of the data set.

      (c)   An observation with a z-score of 2.9 is likely to be greater than all or almost all of the other data values.

**11.** (a) The mean is calculated as $\bar{x} = \Sigma x_i/n$. For the sample of 20 guests, $\bar{x}$ = 47/20 = 2.35 alcoholic drinks.

The median is found by ordering the observations from lowest to highest and finding the average of the two observations in the middle. The list is

0 0 1 1 1 1 1 2 2 **2 2** 2 3 3 4 4 4 4 5 5

The median is the average of the 10th and 11th values, and is therefore (2 + 2)/2 = 2.

The mode is the value that occurs the most times. In this data set, 1 and 2 both occur 5 times, and so the modes are 1 and 2.

(b) If the purpose is to help in estimating the cost of the party, the mean is the best since it takes into account the drinking practices of all of the guests. If the purpose is to describe the average guest, the median is more typical of the list of values. The mode is less useful for this data since there are two modes.

**12.** Many more marriages are characterized as being of short duration than other durations. Since the median is ordinarily preferred for data sets that have exceptional (very large or small) values, the median is more appropriate than the mean as a measure of central tendency for data on the duration of marriages.

**13.** Regarding death certificates, we are interested in the most frequent cause of death. Causes of death are qualitative. There is no way to compute a mean or median for such data. The mode is the only measure of central tendency that can be used for qualitative data.

**14.** The mean is $\bar{x} = \dfrac{\sum x_i}{n} = \dfrac{305.3}{10} = 30.53$; since n is even (10), the median is the mean of the two middle observations in the ordered list of the data values: 17.4 21.0 17.4 31.5 **32.0 33.0** 33.0 34.5 37.5 38.0. Thus the median is (32 + 33)/2 = 32.5. The mode is the most frequently occurring value. Since 33.0 occurs twice and no other value occurs more than once, 33.0 is the mode.

**15.** (a) $\bar{x} = \dfrac{\sum x_i}{n} = \dfrac{457}{10} = 45.7$ kilograms

(b) range = 54 - 37 = 17 kilograms

(c) $s = \sqrt{\dfrac{\sum x^2 - (\sum x)^2/n}{n-1}} = \sqrt{\dfrac{21109 - 457^2/10}{9}} = 5.0$ kg

**16.** (a)

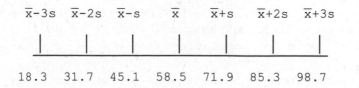

$\bar{x}-3s$ $\bar{x}-2s$ $\bar{x}-s$ $\bar{x}$ $\bar{x}+s$ $\bar{x}+2s$ $\bar{x}+3s$

18.3 31.7 45.1 58.5 71.9 85.3 98.7

(b) Almost all of the ages of the 36 millionaires are between <u>18.3</u> and <u>98.7</u> years old.

17. (a) The first quartile is the median of the lower half of the ordered list. Since there is a total of 36 ages, the first 18 are in the lower half and the median of these is the average of the middle two, the $9^{th}$ and $10^{th}$ values.   Thus $Q_1 = (48 + 48)/2 = 48$.

    The second quartile is the median of the entire data set.   The number of pieces of data is 36, and so the position of the median is $(36 + 1)/2 = 18.5$, halfway between the eighteenth and nineteenth data values. Thus the median of the entire data set is $(59 + 60)/2 = 59.5$. That is, $Q_2 = 59.5$.

    The third quartile is the median of the upper half of the ordered list. Since there is a total of 36 ages, the last 18 are in the upper half and the median of these is the average of the middle two, the $9^{th}$ and $10^{th}$ values.   Thus $Q_3 = (68 + 69)/2 = 68.5$.

    Interpreting our results, we conclude that 25% of the ages are less than 48 years; 25% of the ages are between 48 and 59.5 years; 25% of the ages are between 59.5 and 68.5 years; and 25% of the ages are greater than 68.5 years.

    (b) The IQR $= Q_3 - Q_1 = 68.5 - 48 = 20.5$.   Thus, the middle 50% of the ages has a range of 20.5 years.

    (c) Min $= 31$, $Q_1 = 48$, $Q_2 = 59.5$, $Q_3 = 68.5$, Max $= 79$

    (d) The limits are given by:

    Lower limit $= Q_1 - 1.5(IQR) = 48.0 - 1.5(20.5) = 17.25$

    Upper limit $= Q_3 + 1.5(IQR) = 68.5 + 1.5(20.5) = 99.25$

    (e) There are no values below 17.25 or above 99.25, so there are no potential outliers.

    (f) A boxplot is constructed easily using the information in part (a).

    (i)     low data value = 31 years

    (ii)    high data value = 79 years

    (iii)   $Q_1 = 48.0$ years

    (iv)    $Q_3 = 68.5$ years

    (v)     median $= Q_2 = 59.5$ years.

    These values are used to construct the boxplot as follows:

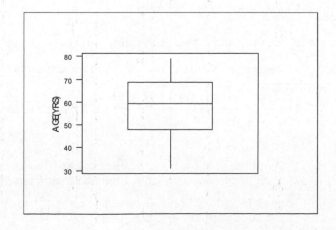

**18.**   (a)   The order list of the data is 0.7   1.0   1.1   1.1   1.2   1.5   1.8   1.8
1.8   1.8   **1.9   2.0**   2.0   2.0   2.3   2.7   3.3   3.4   3.6   3.8   6.7   7.6.
Since n is even (22), the median is mean of the two middle values, the
$11^{th}$ and $12^{th}$.   Thus the median is (1.9 + 2.0)/2 = 1.95.

The first quartile is the median of the first 11 values in the list, so
$Q_1$ = the $6^{th}$ value = 1.5.   The third quartile is the median of the last
11 values in the list, so $Q_3$ = the $6^{th}$ value from the upper end = 3.3.
Thus, the five-number summary is Minimum = 0.7, $Q_1$ = 1.5, Median = 1.95,
$Q_3$ = 3.3, and Maximum = 7.6.

(b)   The IQR = 3.3 – 1.95 = 1.35.   The lower limit = Q1 – 1.5 x IQR = 1.5 –
1.5(1.35) = -0.525, while the upper limit = Q3 + 1.5 x IQR = 3.3 +
1.5(1.35) = 5.325.   There are no values below the lower limit, but 6.7
and 7.6 lie above the upper limit and are therefore potential outliers.

(c)   The boxplot consists of a rectangle with its ends at 1.5 and 3.3 and a
line across the middle at 1.95.   Whiskers extend from the ends of the
box out to the adjacent values (the most extreme values on each end of
the box that are not beyond the limits).   Thus the lower whisker
extends to 0.7, and the upper whisker extends to 3.8.   The two
potential outliers are plotted individually.   The result is

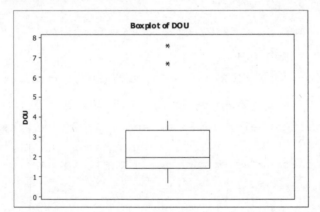

**19.**   The median number of traffic fatalities in Wisconsin is about 770 for the
years 1982-2003 while the median number in New Mexico is about 490.   In
fact, the maximum number in New Mexico (about 590) is less than the minimum
number in Wisconsin (about 650) for those years.   The IQR and the range of
traffic fatalities are both slightly larger for Wisconsin than for New
Mexico.

**20.**   (a)   $\mu = \dfrac{\sum x_i}{n} = \dfrac{158.3}{8} = 19.79$ thousand students

(b)   Squaring each data value and then summing, we find that

$\sum x_i^2 = 3241.09$.   Then the population standard deviation is

$\sigma = \sqrt{\dfrac{\sum x_i^2 - (\sum x_i)^2 / N}{N}} = \sqrt{\dfrac{3241.09 - 158.3^2 / 8}{8}} = 3.69$ thousand students

(c)   z = (x – μ)/σ = (x – 19.79)/3.69

(d)   The mean of z is 0, and the standard deviation of z is 1.   All
standardized variables have mean 0 and standard deviation 1.

(e)   Converting each x value (dotplot on the left) to its corresponding
z-score results in the dotplot on the right.

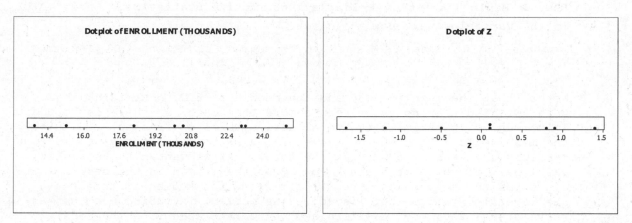

The relative positions of the points in the two plots are the same, but the scales are different.

(f)

| x | z = (x - μ)/σ |
|---|---|
| 24.9 | (24.9 - 19.79)/3.69 = 1.38 |
| 15.2 | (15.2 - 19.79)/3.69 = -1.24 |

The enrollment at Los Angeles is 1.38 standard deviations above the mean and the enrollment at Riverside is 1.24 standard deviation below the mean.

21.  (a)  The mean price given here is a sample mean.  This is because it is the mean price per gallon for the sample of 800 gasoline service stations.

(b)  The letter used to designate the mean of $2.603 is $\overline{x}$.

(c)  The mean price given here is a statistic.  It is a descriptive measure.

22.  Using Minitab, retrieve the data from the WeissStats CD.  Then choose **Stat**

▶ **Tables** ▶ **Tally Individual Variables**.  Enter <u>REGION</u> and <u>DIVISION</u> in the **Variables** text box and click **OK**.  The result is the table below.

| REGION | Count | DIVISION | Count |
|---|---|---|---|
| Midwest | 12 | East North Central | 5 |
| Northeast | 9 | East South Central | 4 |
| South | 16 | Middle Atlantic | 3 |
| West | 13 | Mountain | 8 |
| N= | 50 | New England | 6 |
| | | Pacific | 5 |
| | | South Atlantic | 8 |
| | | West North Central | 7 |
| | | West South Central | 4 |
| | | N= | 50 |

(a)  We see from the first column of the above table that the South region occurs 16 times, more than any other region, so South is the mode of the regions.

(b)  We see from the second column of the above table that the Mountain and the South Atlantic divisions each occur 8 times, more than any other region, so Mountain and South Atlantic are both modal divisions.

**23.**    (a)  Using Minitab, retrieve the data from the WeissStats CD.  Then choose

**Stat ▶ Basic Statistics ▶ Display Descriptive Statistics.**  Enter <u>VALUE</u>
in the **Variables** text box and click **OK.**   The result is

| Variable | N | N* | Mean | SE Mean | StDev | Minimum | Q1 | Median | Q3 | Maximum |
|---|---|---|---|---|---|---|---|---|---|---|
| VALUE | 50 | 0 | 1080 | 202 | 1427 | 0.900 | 155 | 677 | 1329 | 8210 |

We see that the mean is 1080 and the Median is 677.  Examining the
data, we find that no value occurs more than once, so there is no mode.
If we wanted to estimate the total agricultural exports for the nation,
the mean would be the best measure of center to use since we could just
multiply the mean by 50 to get the total.  If we wanted to know what
value half of the states are below, then the median is the best measure
of center to use.  Another reason for using the median is that there
are several outliers that lie above the median, including one that is
very large (8210.0 for California).  Since the mean is quite sensitive
to outliers such as this one, the median is the better choice for
describing these data.

(b)  From the table above, Range = 8210 - 0.9 = 8209.1, and the sample
standard deviation s = 1427.

(c)  The five-number summary consists of the minimum, $Q_1$, the median, $Q_3$, and
the maximum.  These are, respectively, 0.900, 155, 677, 1329, and 8210.
 The interquartile range IQR = $Q_3$ - $Q_1$ = 1329 - 155= 1174.

(d)  We first find the lower limit as $Q_1$ - 1.5 x IQR = 155 - 1.5(1174) =    -
1606.  The upper limit is $Q_3$ + 1.5 x IQR = 1329 + 1.5(1174) = 3090.  Any
data values below the lower limit or above the upper limit are
potential outliers.  Thus 3363.5, 3427.8, 3637.8, and 8210 are
potential outliers.

(e)  To obtain a boxplot of the data, choose **Graph ▶ Boxplot**, select **Simple**
from the **One Y** row and click **OK.**  Enter <u>VALUE</u> in the **Variables** text box
and click **OK.**   The result is

The data distribution is quite right skewed with a median of 155.  The
plot shows four potential outliers, all on the high side of the data.

**24.**    (a)  Using Minitab, retrieve the data from the WeissStats CD.  Then choose

**Stat ▶ Basic Statistics ▶ Display Descriptive Statistics.**  Enter '<u>LIFE</u>
<u>EXP'</u> in the **Variables** text box and click **OK.**   The result is

| Variable | N | N* | Mean | SE Mean | StDev | Minimum | Q1 | Median | Q3 |
|----------|---|-----|------|---------|-------|---------|-----|--------|-----|
| LIFE EXP | 224 | 0 | 67.188 | 0.813 | 12.175 | 32.300 | 61.400 | 71.250 | 76.625 |

| Variable | Maximum |
|----------|---------|
| LIFE EXP | 83.500 |

We see that the mean is 67.188 and the median is 71.250.  Finding the mode is easiest if we choose **Stat ▶ Tables ▶ Tally Individual Variables** and enter `LIFE EXP` in the **Variables** text box.  Examining the resulting table (not shown here), we find that three values (68.9, 75.4, and 75.5) each occur four times, more than any other values, so there are three modes.  The mean is not appropriate for these data since each value is a mean life expectancy based on population sizes that are different for different countries.  The median is the best measure of center to use since it tells us that half of the listed countries have life expectancies below this number.  Another reason for using the median is that the distribution is left skewed with several potential outliers at the lower end, as will be seen in parts (d) and (e).

(b) From the table above, Range = 83.5 – 32.3 = 51.2, and the sample standard deviation s = 12.175.  [Note:  The sample standard deviation is based on a calculation that uses the sample mean.  Since the sample mean being used does not account for the fact that the different countries have different populations, the true sample standard deviation may differ somewhat from this value.]

(c) The five-number summary consists of the minimum, $Q_1$, the median, $Q_3$, and the maximum.  These are, respectively, 32.3, 61.4, 71.250, 76.625, and 83.500.  The interquartile range IQR = $Q_3 - Q_1$ = 76.625 – 61.4= 15.225.

(d) We first find the lower limit as $Q_1$ – 1.5 x IQR = 61.4 – 1.5(15.225) = 38.5625.  The upper limit is $Q_3$ + 1.5 x IQR = 76.625 + 1.5(15.225) = 99.4625.  Any data values below the lower limit or above the upper limit are potential outliers.  Thus 32.3, 35.3, 36.9, 37.0, and 38.0 are potential outliers.

(e) To obtain a boxplot of the data, choose **Graph ▶ Boxplot**, select **Simple** from the **One Y** row and click **OK**.  Enter `LIFE EXP` in the **Variables** text box and click **OK**.  The result is

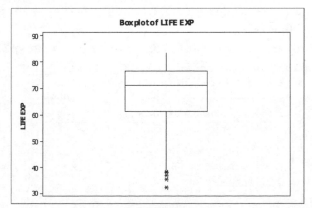

We see that the data are left skewed with the median at about 71.  There are several potential outliers at the lower end of the data.

25.  (a) Using Minitab, retrieve the data from the WeissStats CD.  Then choose **Stat ▶ Basic Statistics ▶ Display Descriptive Statistics**.  Enter <u>HIGH</u> and <u>LOW</u> in the **Variables** text box and click **OK**.  The result is

| Variable | N | N* | Mean | SE Mean | StDev | Minimum | Q1 | Median | Q3 | Maximum |
|---|---|---|---|---|---|---|---|---|---|---|
| HIGH | 71 | 0 | 65.33 | 1.04 | 8.72 | 47.60 | 58.40 | 64.00 | 71.10 | 85.50 |
| LOW | 71 | 0 | 45.52 | 1.11 | 9.34 | 29.30 | 39.80 | 44.20 | 51.40 | 74.20 |

We see that the mean HIGH temperature is 65.33 and the median HIGH is 64.00. The mean LOW temperature is 45.52 and the median LOW is 44.20.

Finding the mode is easiest if we choose **Stat ▶ Tables ▶ Tally Individual Variables** and enter HIGH and LOW in the **Variables** text box. Examining the resulting table (not shown here), we find for HIGH that 54.5, 55.9, 57.6, 59.8, 62.6, 63.6, 65.1, 67.4, 67.8, and 70.6 all occur twice, more than any other values, so there are ten modes. For LOW, the values 35.8, 39.8, 41.2, 43.2, 44.8, 47.4, 48.6, and 52.5 all occur twice, more than any other values, so there are eight modes. The median is the best measure of center to use since it tells us that half of the listed cities have minimums (or maximums) below this number. Another reason for using the median is that the distribution for LOW has several potential outliers at the upper end, as will be seen in parts (d) and (e).

(b)    From the table above, for HIGH, Range = 85.5 - 47.6 = 37.9, and the sample standard deviation s = 8.72. For LOW, Range = 74.20 - 29.30 = 44.90 and the sample standard deviation s = 9.34.

(c)    The five-number summary consists of the minimum, $Q_1$, the median, $Q_3$, and the maximum. For HIGH, these are, respectively, 47.60, 58.40, 64.00, 71.10, and 85.50. The interquartile range for HIGH is IQR = $Q_3 - Q_1$ = 71.10 - 58.40= 12.70. For LOW, the five-number summary is 29.30, 39.80, 44.20, 51.40, and 74.20. IQR = $Q_3 - Q_1$ = 51.40 - 39.80 = 11.60.

(d)    For HIGH, we first find the lower limit as $Q_1$ - 1.5 x IQR = 58.40 - 1.5(12.70) = 39.35. The upper limit is $Q_3$ + 1.5 x IQR = 71.10 + 1.5(12.70) = 90.15. Any data values below the lower limit or above the upper limit are potential outliers. Examining the data, we find that there are no potential outliers for HIGH.

For LOW, we first find the lower limit as $Q_1$ - 1.5 x IQR = 39.80 - 1.5(11.60) = 22.40. The upper limit is $Q_3$ + 1.5 x IQR = 51.40 + 1.5(11.60) = 68.80. Any data values below the lower limit or above the upper limit are potential outliers. Examining the data, we see that 69.1, 70.2, and 74.2 are potential outliers. If you are looking for someplace that is warm year around, these values are for Miami, Honolulu, and San Juan.

(e)    To obtain a boxplot of the data, choose **Graph ▶ Boxplot**, select **Simple** from the **Multiple Y's** row and click **OK**. Enter HIGH and LOW in the **Variables** text box and click **OK**. The result is

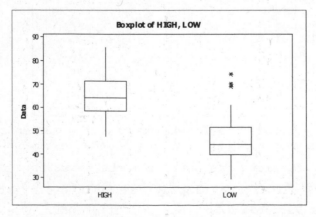

The distribution of HIGH temperatures is quite symmetrical, while the distribution of LOW temperatures is also quite symmetrical except for three outliers.  The three cities with high LOW temperatures are all affected by warm southern ocean waters that keep the temperature quite moderate year around.

26. **(a)** To obtain a boxplot of the height data, choose **Graph ▶ Boxplot**, select **Simple** from the **Multiple Y's** row and click **OK**.  Enter 'HT_C' and 'HT_H' in the **Variables** text box and click **OK**.  The result is

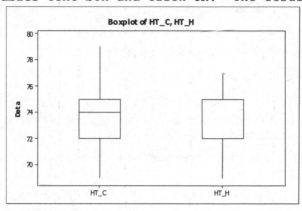

**(b)** Before commenting on center, note that the median line for Houston seems to be missing.  On examining the data, we see that the median and the first quartile are both the same.  Thus, the median height for Chicago is 74" while for Houston, the median is 72".  The IQR is the same for both teams, but Chicago has a larger range by 2".

**(c)** To obtain a boxplot of the weight data, choose **Graph ▶ Boxplot**, select **Simple** from the **Multiple Y's** row and click **OK**.  Enter 'WT_C' and 'WT_H' in the **Variables** text box and click **OK**.  The result is

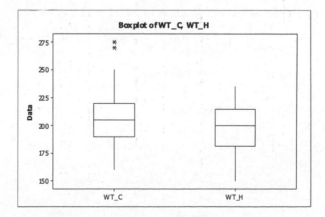

**(d)** The medians are about the same for both groups.  The variation as measured by the IQR is slightly greater for Houston, but, if measured by the range, it is greater for Chicago.  Chicago has two players whose weights are potential outliers on the high side.  Otherwise, the distributions have very similar shapes.

# CHAPTER 4 ANSWERS

## Exercises 4.1

**4.1** An experiment is an action the result of which cannot be predicted with certainty. An event is a specified result that may or may not occur when the experiment is performed.

**4.3** There is no difference.

**4.5** The frequentist interpretation of probability is that the probability of an event is the proportion of times the event occurs in a large number of repetitions of the experiment.

**4.7** The following could not possibly be probabilities:

(b) -0.201: A probability cannot be negative.

(e)  3.5: A probability cannot exceed 1.

**4.9** (a) G,L,S,A   G,L,S,T   G,L,A,T   G,S,A,T   L,S,A,T

(b) Two samples include the governor, attorney general, and treasurer. Therefore, the probability is $f/N = 2/5$.

(c) Three samples include the governor and treasurer. Therefore, the probability is $f/N = 3/5$.

(d) Four samples include the governor. Thus, the probability is $f/N = 4/5$.

**4.11** (a) There was a total of 66,529,566 votes cast. Therefore, the probability that a randomly selected voter voted for Vladimir Putin is $f/n = 49,565,238/66,529,566 = 0.745$.

(b) There were 1,405,315 + 524,324 = 1,929,639 votes cast for Malyshkin or Mironov. . Therefore, the probability that a randomly selected voter voted for Malyshkin or Mironov is $f/n = 1,929,639/66,529,566 = 0.029$.

(c) The number of votes not for Putin is 66,529,566 - 49,565,238 = 16,964,328. Thus, the probability that a randomly selected voter did not vote for Putin is 16,964,328/66,529,566 = 0.255.

**4.13** The total number of units (in thousands) is N = 120,783.

(a) The probability that the unit has 4 rooms is $f/N = 23,363/120,783 = 0.193$ (to three decimal places).

(b) The probability that the unit has more than 4 rooms is $f/N = (27976 + 24646 + 14676 + 17234)/120,783 = 84532/120,783 = 0.700$.

(c) The probability that the unit has 1 or 2 rooms is $f/N = (520 + 1425)/120,783 = 19450/120,783 = 0.016$.

(d) The probability that the unit has fewer than one room is $f/N = 0/120,783 = 0.000$.

(e) The probability that the unit has one or more rooms is $f/N = 120,783/120,783 = 1.000$.

**4.15** The total number of graduate students is 49,088.

(a) The probability that his occupation is service is $f/N = 9274/49088 = 0.189$.

(b) The probability that his occupation is administrative is $f/N = (2197 + 6450)/49088 = 8647/49088 = 0.176$.

(c) The probability that his occupation is manufacturing is $f/N = (2197 + 2166 + 1640 + 5721)/49088 = 11724/49088 = 0.239$

(d) The number whose occupation is not manufacturing is 49088 - 11724 = 37364. The probability that his occupation is not manufacturing is $f/N$

= 37364/49088 = 0.761.

**4.17**  (a)  The total number of graduate science students in doctorate-granting institutions is 297.0 thousand.  Of these, 40.8 thousand are in psychology.  Thus, the probability that a randomly selected graduated student in science is in the field of psychology is f/N = 40800/297000 = 0.137.

(b)  There are 31.3 + 78.8 = 110.1 thousand = 110100 graduate students in physical or social science.  Thus, the probability that a randomly selected graduated student in science is in physical or social science is f/N = 110100/297000 = 0.371.

(c)  There are 47.6 thousand graduate students in computer science and 297.0 − 47.6 = 249.4 thousand graduate students not in computer science.  Thus, the probability that a randomly selected graduated student in science is not in computer science is f/N = 249400/297000 = 0.840.

**4.19**  (a)  There are five ways [(1,5), (2,4), (3,3), (4,2), (5,1)] among the 36 possibilities that sum to 6.  Thus, f/N = 5/36 = 0.139.

(b)  There are 18 ways among the 36 possibilities that provide a sum that is even.  Thus, f/N = 18/36 = 0.500.

(c)  There are six ways among the 36 possibilities that sum to 7 and two ways among the 36 possibilities that sum to 11.  Thus, f/N = (6 + 2)/36 = 8/36 = 0.222.

(d)  There is one way among the 36 possibilities that sums to 2, two ways in which the sum is 3, and one way in which the sum is 12.  Thus, f/N = (1 + 2 + 1)/36 = 4/36 = 0.111.

**4.21**  The event in part (e), that the housing unit has one or more rooms, is a certainty.  The event in part (d), that the housing unit has fewer than one room, is impossible.

**4.23**  When rolling two balanced dice and observing the sum, there are 36 possible equally likely outcomes, not 11.  The probability that the sum is 12 equals 1/36, not 1/11.

**4.25**  Since there are 18 red numbers and 20 that are not red, for the bet to be fair, the odds should be 20 to 18.

**4.27**  (a)  The odds against Funny Cide winning were 1 to 1.  These are in the ratio of 1-p to p.  Thus 1/1 = (1-p)/p.  Multiplying both sides of the equation by p, we have p = (1-p).  Adding p to both sides of the equation yields 2p = 1, from which we conclude that p = 1/2.

(b)  Similarly, the odds against Empire Maker winning were 2 to 1. These are in the ratio of 1-p to p. Thus 2/1 = (1-p)/p.  Multiplying both sides of the equation by p, we have 2p = (1-p).  Adding p to both sides of the equation yields 3p = 1, from which we conclude that p = 1/3.  Thus the probability that Funny Cide would win the race was 1/2 and the probability that Empire Maker would win was 1/3.

**Exercises 4.2**

**4.29**  Venn diagrams are useful for portraying events and relationships between events.

**4.31**  Two events are mutually exclusive if they cannot occur at the same time, i.e., they have no outcomes in common.  Three events are mutually exclusive if no two of them can occur at the same time, i.e., no pair of the events has any outcomes in common.

**4.33**  A = {2, 4, 6}; B = {4, 5, 6}; C = {1, 2}; D = {3}

**4.35**  A contains the outcomes JM, JS, JH, JB, WM, WS, WH, WJ

B contains the outcomes HM, HS, HJ, HW

C contains the outcomes MW, SW, HW, JW

D contains the outcomes MS, MH, SM, SH, HM, HS

**4.37**  (a)  (not A) = {1, 3, 5} = the event the die comes up odd.

(b)  (A & B) = {4, 6} = the event the die comes up four or six.

(c)  (B or C) = {1, 2, 4, 5, 6} = the event the die does *not* come up three.

**4.39**  (a)  (not A) = (MS, MH, MJ, MW, SM, SH, SJ, SW, HM, HS, HJ, HW) = the event that a female is appointed chairperson

(b)  (B & D) = (HM, HS) = the event that Holly is appointed chairperson and a woman is appointed secretary

(c)  (B or C) = (HM, HS, HJ, HW, MW, SW, JW) = the event that Holly is appointed chairperson or Will is appointed secretary.

**4.41**  (a)  (not C) = the event that a state has a diabetes prevalence percentage less than 5% or is 10% or more.  There are 8 + 1 = 9 states in (not C).

(b)  (A&B) = the event that a state has a diabetes prevalence percentage at least 8% and is less than 7%.  There are no states in this event.

(c)  (C or D) = the event that a state has a diabetes prevalence percentage that is less than 10%.  There are 49 states in this event.

(d)  (C&B) = the event that a state has a diabetes prevalence percentage of at least 5% and less than 7%.  There are 10 + 15 = 25 states in this event.

**4.43**  (a)  (A or D) = the event that the bill was paid by Medicare or by the patient or a charity.  There are 9983 + 5512 = 15,495 bills in this event.

(b)  (not C) = the event that private insurance **did** pay the bill.  There are 26,825 bills in this event.

(c)  (B & (not A)) = the event that some government agency other than Medicare paid the bill.  There are 8142 + 1777 = 9919 bills in this event.

(d)  Since D is included in C, (C or D) is just C = the event that private insurance did not pay the bill.  Therefore (not(C or D)) = (not C) = the event that private insurance did pay the bill.  There are 26,825 bills in this event.

**4.45**  (a)  (not A) is the event that the unit has more than 4 rooms. There are 27,976 + 24,646 + 14,670 + 17,234 = 84,526 (thousand) such units.

(b)  (A&B) is the event that the unit has at most 4 rooms and at least 2 rooms, i.e., has 2 or 3 or 4 rooms.  There are 1,425 + 10,943 + 23,363 = 35,731 (thousand) such units.

(c)  The event (C or D) is the event that the unit has between 5 and 7 rooms inclusive or more than seven rooms, i.e., has 5 or more rooms.  This is the same as the event (not A) and therefore there are 84,526 (thousand) such units.

**4.47**  (a)  Events A and B are not mutually exclusive.  They have a four and a six in common.

(b)  Events B and C are mutually exclusive.  They have no outcomes in common.

(c)  Events A, C, and D are not mutually exclusive.  The outcome two is
common to A and C.

(d)  Among A, B, C, and D, there are three mutually exclusive events.  These
are B, C, and D.  Among B, C, and D, there are no outcomes in common.
There are not, however, four mutually exclusive events.  The outcome
two is common to A and C, four is common to A and B, and six is common
to A and B.

**4.49**  The groups that are mutually exclusive are A and C; A and D; C and D; and A,
C, and D.

**4.51**  Below is a Venn diagram portraying four mutually exclusive events.

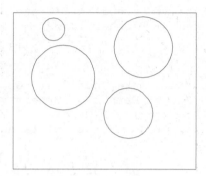

**4.53**  To say that A, B, and C do not all occur simultaneously does not necessarily
imply that A, B, and C are mutually exclusive.  In the following diagram, C does
not touch A or B.  That is, we have a situation where A, B, and C do not occur
simultaneously.  However, these three events clearly are not mutually exclusive.

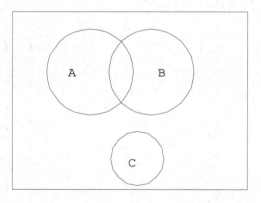

### Exercises 4.3

**4.55**  (a)  P(S) = f/N = (12 + 33 + 32)/100 = 0.77.

(b)  S = (A or B or C)

(c)  P(A) = 12/100 = 0.12;      P(B) = 33/100 = 0.33;
P(C) = 32/100 = 0.32.

(d)  P(S) = P(A or B or C) = P(A) + P(B) + P(C) = 0.12 + 0.33 + 0.32 = 0.77.
This result is identical to the one found in part (a).

**4.57**  (a)  The percentage who smoked within the last 30 days is 9.8 + 7.8 + 5.3 +
2.8 + 0.7 + 0.3 = 26.7.  Therefore, the probability that a randomly
selected twelfth grader smoked is 0.267.

(b)   The percentage who smoked at least one cigarette per day within the last 30 days is 7.8 + 5.3 + 2.8 + 0.7 + 0.3 = 16.9.   Therefore, the probability that a randomly selected twelfth grader smoked at least one cigarette per day is 0.169.

(d)   The percentage who smoked between 6 and 34 cigarettes per day, inclusive, within the last 30 days is 5.3 + 2.8 + 0.7 = 8.8.   Therefore, the probability that a randomly selected twelfth grader smoked between 6 and 34 cigarettes per day is 0.088.

**4.59**   (a)   P(Spill occurs in an ocean) = P(Atlantic Ocean or Pacific Ocean) = P(Atlantic Ocean) + P(Pacific Ocean) = 0.011 + 0.064 = 0.075.   That is, 7.5% of the spills occur in an ocean.

(b)   P(Spill occurs in a lake or harbor) = P(Great Lakes or Other lakes or Harbors) = P(Great Lakes) + P(Other lakes) + P(Harbors)

= 0. 014 + 0.005 + 0.118 = 0.137.   That is, 13.7% of the spills occur in a lake or harbor.

(c)   P(Spill does not occur in a lake, ocean, river, or canal) = P(Gulf of Mexico or Bays and Sounds or Harbor or Other) = P(Gulf of Mexico) + P(Bays and Sounds) + P(Harbor) + P(Other) = 0.229 + 0.151 + 0.118 + 0.185 = 0.683.   That is, 68.3% of the spills do not occur in a lake, ocean, river, or canal.

**4.61**   (a)   P(senator is at least 50) = $\dfrac{33}{100}+\dfrac{32}{100}+\dfrac{18}{100}+\dfrac{5}{100}=0.88$ .

This is accomplished more easily using the complementation rule:

P(senator is at least 50) = 1 - P(senator is under 50) = $1-\dfrac{12}{100}=0.88$ .

(b)   P(senator is under 70) = $\dfrac{12}{100}+\dfrac{33}{100}+\dfrac{32}{100}=\dfrac{77}{100}=0.77$ .

This is accomplished more easily using the complementation rule:

P(senator is under 70) = 1 - P(senator is at least 70) =

$1-\left(\dfrac{18}{100}+\dfrac{5}{100}\right)=1-\dfrac{23}{100}=\dfrac{77}{100}=0.77$ .

**4.63**   (a)   The percentage of day laborers who have lived in the United States either between 1 and 20 years, inclusive, or less than 11 years is just the sum of the percentages in the first five categories, or 93%.   Thus the probability is 0.93.

(b)   P(1 to 20) = (30 + 21 + 12 + 13)/100 = 0.76

P(Less than 11) = (17 + 30+ 21 + 12)/100 = 0.80

P(1 to 20 & Less than 11) = (30 + 21 + 12)/100 = 0.63

P(1 to 20 or Less than 11) = P(1 to 20) + P(Less than 11) - P(1 to 20 & Less than 11) = 0.76 + 0.80 - 0.63 = 0.93

(c)   In this case, it was easier in part (a).   If the three probabilities needed in part (b) had been provided, the general addition rule would have been the easier method.

**4.65**   (a)   $P(A)=\dfrac{6}{36}=0.167$   $P(B)=\dfrac{2}{36}=0.056$   $P(C)=\dfrac{1}{36}=0.028$   $P(D)=\dfrac{2}{36}=0.056$

$$P(E) = \frac{1}{36} = 0.028 \quad P(F) = \frac{5}{36} = 0.139 \quad P(G) = \frac{6}{36} = 0.167$$

(b)  P(7 or 11)= P(A) + P(B) = 0.167 + 0.056 = 0.223.

(c)  P(2 or 3 or 12) = P(C) or P(D) or P(E).

   = 0.028 + 0.056 + 0.028 = 0.112

(d)  P(8 or doubles):   Using Figure 4.1:  $\frac{10}{36} = 0.278$

(e)  P(8) + P(doubles) - P(8 & doubles) = 0.139 + 0.167 - 0.028 = 0.278.

**4.67**  P(Public school or college) = P(Public school) + P(College) - P(Public school & college) = 0.855 + 0.223 - 0.171 = 0. 907

**4.69**  (a) For A and B to be mutually exclusive, P(A & B) must equal zero. Recalling the general addition rule and making the appropriate substitutions, we have:

P(A or B) = P(A) + P(B) - P(A & B)

or

1/2  =  1/4  +  1/3  - P(A & B).

Rearranging terms and solving for P(A & B), we find:

P(A & B) = 1/4 + 1/3 - 1/2 = 3/12 + 4/12 - 6/12 = 1/12.

Thus, A and B are not mutually exclusive because

P(A & B) = 1/12 ≠ 0.

(b) From part (a), P(A & B) = 1/12 = 0.083.

**4.71**  Let:

J = person enjoys job          L = person enjoys personal life

$\overline{J}$ = person doesn't enjoy job    $\overline{L}$ = person doesn't enjoy personal life

We are given: $P(\overline{J} \& \overline{L}) = 0.15$, $P(J \& \overline{L}) = 0.80$, and $P(J \& L) = 0.04$ .

(a) We must find P(J or L), which can be expressed in terms of the general addition rule as:

P(J or L) = P(J) + P(L) - P(J & L)   .

Notice that we are given P(J & L) = 0.04.

We find P(J) and P(L) as follows:

(i)  $P(J) = P(J \& \overline{L}) + P(J \& L) = 0.80 + 0.04 = 0.84$

(ii)  $P(\overline{L}) = P(\overline{J} \& \overline{L}) + P(J \& \overline{L}) = 0.15 + 0.80 = 0.95$

Using $P(\overline{L}) = 0.95$, we know $P(L) = 1 - P(\overline{L}) = 1 - 0.95 = 0.05$

Thus, P(J or L) = 0.84 + 0.05 - 0.04 = 0.85.

(b) We must find $P(L \& \overline{J}) = P(\overline{J} \& L)$. Using $P(L) = P(J \& L) + P(\overline{J} \& L)$,

and making the appropriate substitutions for items already solved,

$0.05 = 0.04 + P(\overline{J} \& L)$ .

Rearranging terms, $P(\overline{J} \& L) = 0.05 - 0.04 = 0.01$.

### Exercises 4.4

**4.73** The total number of observations of bivariate data can be obtained from the frequencies in a contingency table by summing the counts in the cells, summing the row totals, or summing the column totals.

**4.75** (a) Data obtained by observing values of one variable of a population are called <u>univariate</u> data.

   (b) Data obtained by observing values of two variables of a population are called <u>bivariate</u> data.

**4.77** (a) This contingency table has 12 cells.

   (b) The total number of players on the New England Patriots as of November 7, 2005 is 67.

   (c) There are 13 rookies.

   (d) There are 40 players who weigh between 200 and 300 pounds.

   (e) There are 7 rookies who weigh between 200 and 300 pounds.

**4.79** (a) To fill in the five empty cells, we first determine the $A_2$ total as $107,414 - 36,013 - 34,314 = 37,087$ and the $S_3$ total as $107,414 - 21,949 - 38,806 - 31,829 = 14,770$. Then the number in the $A_2S_2$ cell is $37,087 - 7,880 - 5,146 - 10,440 = 13,621$ and the number in the $A_3S_2$ cell is $38,806 - 13,194 - 13,621 = 11,991$. The last number in $A_3S_3$ can be determined from the S3 total as $14,830 - 4,805 - 5,146 = 4,879$. The final result can be checked by showing that the $A_3$ total is $34,314$. Thus the table is now

|  |  | Under 35 $A_1$ | 35-44 $A_2$ | 45 and over $A_3$ | Total |
|---|---|---|---|---|---|
| Family practice | $S_1$ | 7,638 | 7,880 | 6,431 | 21,949 |
| Internal medicine | $S_2$ | 13,194 | 13,621 | 11,991 | 38,806 |
| Obstetrics/gynecology | $S_3$ | 4,805 | 5,146 | 4,879 | 14,830 |
| Pediatrics | $S_4$ | 10,376 | 10,440 | 11,013 | 31,829 |
| Total |  | 36,013 | 37,087 | 34,314 | 107,414 |

   (b) The number between 35 and 44 years old is 37,087.

   (c) The number of pediatricians under 35 is 10,376.

   (d) The number who are either pediatricians or who are under 35 is $31,829 + 36,013 - 10,376 = 57,466$.

   (e) The event of being neither a pediatrician nor under 35 is the complement of the event in part (d). Thus the number of such female physicians can be found as $107,414 - 57,466 = 49,948$.

   (f) The number not in family practice is $107,414 - 21,949 = 85,465$.

**4.81** (a) Thirty-two teachers offered field trips.

   (b) Twenty-three of the teachers have master's degrees.

   (c) Fourteen of the bachelor's degree teachers offered field trips.

   (d) $D_1$ is the event that one of these teachers has a bachelor's degree.

   ($D_2$ and $F_2$) is the event that one of these teachers has a master's degree and does not offer field trips.

**4.83** (a) $Y_3$ is the event that the player has 6-10 years of experience. $W_2$ is the event that the player weighs between 200 and 300 pounds. $W_1\&Y_2$ is the event that the player has between 1 and 5 years of experience and weighs under 200 pounds.

(b) $P(Y_3) = 17/67 = 0.254$; $P(W_2) = 40/67 = 0.597$; $P(W_1\&Y_2) = 7/67 = 0.104$. Thus 25.4% of the players have 6-10 years of experience; 59.7% of the players weigh between 200 and 300 pounds; and 10.4% of the players have 1-5 years of experience and weigh less than 200 pounds.

(c) We divide each cell frequency by the total 67 to obtain the following table.

|  |  | Rookie $Y_1$ | 1-5 $Y_2$ | 6-10 $Y_3$ | 10+ $Y_4$ | Total |
|---|---|---|---|---|---|---|
| Under 200 | $W_1$ | 0.030 | 0.104 | 0.045 | 0.030 | 0.209 |
| 200-300 | $W_2$ | 0.104 | 0.224 | 0.209 | 0.060 | 0.597 |
| Over 300 | $W_3$ | 0.060 | 0.134 | 0.000 | 0.000 | 0.194 |
| Total |  | 0.194 | 0.463 | 0.254 | 0.090 | 1.000 |

(b) The sum of each row and column of joint probabilities equals the marginal probability to within 0.001.

**4.85** (a) $S_2$ is the event the physician is an internist; $A_3$ is the event the physician is 45 or over; and $S_1\&A_1$ is the event that the physician is in family practice and under 35.

(b) $P(S_2) = 38806/107414 = 0.361$; $P(A_3) = 34314/107414 = 0.319$; $P(S_1\&A_1) = 7638/107414 = 0.071$.

(c) Percentage Distribution

|  |  | Under 35 $A_1$ | 35-44 $A_2$ | 45 and over $A_3$ | Total |
|---|---|---|---|---|---|
| Family practice | $S_1$ | 7.1 | 7.3 | 6.0 | 20.4 |
| Internal medicine | $S_2$ | 12.3 | 12.7 | 11.2 | 36.1 |
| Obstetrics/gynecology | $S_3$ | 4.5 | 4.8 | 4.5 | 13.8 |
| Pediatrics | $S_4$ | 9.7 | 9.7 | 10.3 | 29.6 |
| Total |  | 33.5 | 34.5 | 31.9 | 100.0 |

**4.87** The categories for each of the characteristics in the cross classification are selected so that no member of the population (or sample) can belong to more than one of the categories (cells).

**4.89** (a) A member of the population (or sample) that belongs to category $R_1$ must also belong to exactly one of the categories $C_1$, $C_2$, ..., $C_n$.

(b) The events $(R_1 \& C_1)$, $(R_1 \& C_2)$, ..., $(R_1 \& C_n)$ are mutually exclusive since no member of the population (or sample) can belong to more than one of the categories $C_1$, $C_2$, ..., $C_n$.

(c) We can conclude that this equation holds because of the special addition rule that applies to mutually exclusive events.

### Exercises 4.5

**4.91** The conditional probability of tossing a head on the second toss given that a head occurred on the first toss is the same as the unconditional probability of tossing a head on the second toss without knowing the result of the first toss.

**4.93** (a)  $P(B) = 4/52 = 0.077$; the probability of randomly selecting a king from an ordinary deck of 52 playing cards is 0.077.

(b)  $P(B|A) = 4/12 = 0.333$; given that the selection is made from among (the 12) face cards, the probability of selecting a king is 0.333.

(c)  $P(B|C) = 1/13 = 0.077$; given that the selection is made from among (the 13) hearts, the probability of selecting a king is 0.077.

(d)  $P(B|(\text{not } A)) = 0/40 = 0$; given that the selection is made from among (the 40) non-face cards, the probability of selecting a king is 0.

(e)  $P(A) = 12/52 = 0.231$; the probability of selecting a face card from an ordinary deck of 52 playing cards is 0.231.

(f)  $P(A|B) = 4/4 = 1$; given that the selection is made from among (the four) kings, the probability of selecting a face card is 1.

(g)  $P(A|C) = 3/13 = 0.231$; since the selection is made from among (the 13) hearts, the probability of selecting a face card is 0.231.

(h)  $P(A|(\text{not } B)) = 8/48 = 0.167$; given that the selection is made from among (the 48) non-kings, the probability of selecting a face card is 0.167.

**4.95** (a)  P(exactly 4 rooms) = 23363/120777 = 0.193

(b)  P(exactly 4 rooms | at least 2 rooms) = f/N

= 23363/(120777 - 520) = 23363/120257 = 0.194

(c)  P(at most 4 rooms | at least 2 rooms) = f/N

= (1425 + 10943 + 23363)/120257 = 35731/120257 = 0.297

(d)  (i)   19.3% of the units have exactly 4 rooms;

(ii)  Of those with at least 2 rooms, 19.4% have exactly 4 rooms;

(iii) Of those with at least 2 rooms, 29.7% have at most 4 rooms.

**4.97** (a)  P(Rookie) = 13/67 = 0.194

(b)  P(Weighs under 200 pounds) = 14/67 = 0.209

(c)  P(Rookie | Weighs under 200 Pounds) = 2/14 = 0.143

(d)  P(Weighs under 200 pounds | Rookie) = 2/13 = 0.154

(e)  Of the players, 19.4% are rookies and 20.9% weigh under 200 pounds; 14.3% of those under 200 pounds are rookies; and 15.4% of the rookies weigh under 200 pounds.

**4.99** (a)  $P(L_2) = P(\text{lives with spouse}) = 0.529$

(b)  $P(A_4) = P(\text{over } 64) = 0.153$

(c)  $P(L_2 \& A_4) = P(\text{lives with spouse and is over } 64) = 0.084$

(d)  $P(L_2|A_4) = P(\text{lives with spouse | over } 64) = P(L_2 \& A_4)/P(A_4) = 0.084/0.153 = 0.549.$

(e)  $P(A_4|L_2) = P(\text{over } 64 \mid \text{lives with spouse}) = P(L_2 \& A_4)/P(L_2) = 0.084/0.529 = 0.159$

(f)  Parts (a) - (e) are interpreted, respectively, as follows:

(a)   The probability is 0.529 that the person selected lives with a spouse;

(b)   The probability is 0.153 that the person selected is over 64;

(c)   The probability is 0.084 that the person selected lives with a spouse and is over 64;

(d)   The probability is 0.549 that the person selected lives with a spouse  if the person is over 64;

(e)   The probability is 0.159 that the person selected is over 64 if the person lives with a spouse.

**4.101** (a)   $P(T_1) = 0.441$

(b)   $P(\text{not } A_2) = 1 - P(\text{not } A_2) = 1 - 0.314 = 0.686$

(c)   $P(A_4 \& T_5) = 0.022$

(d)   $P(T_4|A_1) = P(T_4 \& A_1)/P(A_1) = 0.070/0.526 = 0.133$

(e)   $P(A_1|T_4) = P(A_1 \& T_4)/P(T_4) = 0.070/0.122 = 0.574$

Of U.S. adults, for 44.1%, it has been less than half a year since their last visit to a dentist.  69.9% of U.S. adults are not between 45 and 64 years old.  2.2% are over 75 and have not seen a dentist for 5 or more years.  Of those 18-44 years old, for 13.3%, it has been between 2 and 5 years since they last have seen a dentist.  Of those for whom it has been between 2 and five years since their last dental visit, 57.4% are between 18 and 44 years old.

**4.103** Let: R = property crime that is committed in a rural area;

B = property crime that is a burglary.

We are given:  P(R) = 0.049 and P(B & R) = 0.019.  The percentage of property crimes committed in rural areas that are burglaries is

P(B|R) = P(B & R)/ P(R) = 0.019/0.049 = 0.388 or 38.8%.

**4.105** (a)   Event B is positively correlated with Event A if the probability of B increases when it becomes known that A has occurred.  Event B is negatively correlated with Event A if the probability of B decreases when it becomes known that A has occurred.  Event B is independent of Event A if the probability of B does not change when it becomes known that A has occurred.

(b)   We need to show that P(A|B) > P(A) if and only if P(B|A) > P(B). Formula 4.4 implies that P(A & B) = P(A|B)P(B) = P(A & B) = P(B|A)P(A).

$$P(A|B) = \frac{P(A \& B)}{P(B)} = \frac{P(B|A)P(A)}{P(B)} > \frac{P(B)P(A)}{P(B)} = P(A)$$ where the ">" sign

occurs because P(B|A) > P(B).  This proves that P(A|B) > P(A) $\underline{if}$ P(B|A) > P(B).  By interchanging A & B in the above proof, we can show that P(B|A) > P(B) $\underline{if}$ P(A|B) > P(A).  This is the proof of the $\underline{only\ if}$ part of the statement.

(c)   We need to show that P(A|B) < P(A) if and only if P(B|A) < P(B).

$$P(A|B) = \frac{P(A \& B)}{P(B)} = \frac{P(B|A)P(A)}{P(B)} < \frac{P(B)P(A)}{P(B)} = P(A)$$ where the "<" sign

occurs because P(B|A) < P(B).  This proves that P(A|B) < P(A) $\underline{if}$ P(B|A) < P(B).  By interchanging A & B in the above proof, we can show that P(B|A) < P(B) $\underline{if}$ P(A|B) < P(A).  This is the proof of the $\underline{only\ if}$ part of the statement.

(d) We need to show that P(A|B) = P(A) if and only if P(B|A) = P(B).

$$P(A \mid B) = \frac{P(A \& B)}{P(B)} = \frac{P(B \mid A)P(A)}{P(B)} = \frac{P(B)P(A)}{P(B)} = P(A)$$ where the third "="

sign occurs because P(B|A) = P(B). This proves that P(A|B) = P(A) <u>if</u> P(B|A) = P(B). By interchanging A & B in the above proof, we can show that P(B|A) = P(B) <u>if</u> P(A|B) = P(A). This is the proof of the <u>only if</u> part of the statement.

## <u>Exercises 4.6</u>

**4.107** (a) General Multiplication rule:  P(A&B) = P(A)P(B|A)

   Conditional probability rule:  P(B|A) = P(A&B)/P(A)

(b) The general probability can be obtained from the conditional probability rule by multiplying both sides of the equation by P(A).

(c) We can use the general multiplication rule when we know the marginal and conditional probabilities to find a joint probability.  We can use the conditional probability rule when we know the joint and marginal probabilities to find a conditional probability.

**4.109** We have P(Holiday Depression | Woman) = 0.44 and P(Woman) = 0.52.  The probability that a randomly selected U.S. adult is a woman who suffers from holiday depression is P(Woman & Holiday Depression) = P(Woman)P(Holdiay Depression | Woman) = (0.52)(0.44) = 0.2288.  Thus, about 23% of U.S. adults are women who suffer from holiday depression.

**4.111** (a) P(H) = 1/6 = 0.167.

(b) After selecting a 3, there are five numbers remaining.  With two numbers exceeding 4, P(K|H) = 2/5 = 0.4.

(c) P(H & K) = P(H)·P(K|H) = (0.167)(0.4) = 0.067.

(d) The probability that both numbers picked are less than 3 is (2/6)(1/5) = 2/30 = 0.067.

(e) The probability that both numbers picked are greater than 3 is (3/6)(2/5) = 6/30 = 0.2.

**4.113** (a) P(First is Jr & Second is Sr) = P(First is Jr)P(Second is Sr | First is Jr) = (12/40)(7/39) = 84/1560 = 0.054.

(b) P(First is So & Second is So) = P(First is So)P(Second is So | First is So) = (15/40)(14/39) = 210/1560 = 0.135.

(c)

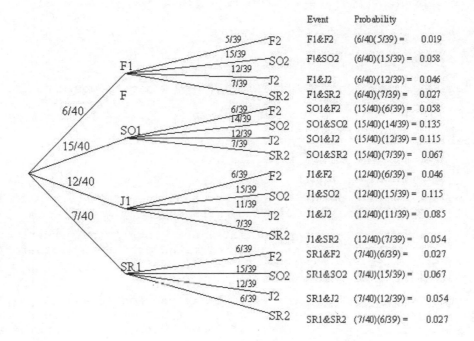

| Event | Probability | | |
|-------|-------------|---|---|
| F1&F2 | (6/40)(5/39) = | 0.019 |
| F1&SO2 | (6/40)(15/39) = | 0.058 |
| F1&J2 | (6/40)(12/39) = | 0.046 |
| F1&SR2 | (6/40)(7/39) = | 0.027 |
| SO1&F2 | (15/40)(6/39) = | 0.058 |
| SO1&SO2 | (15/40)(14/39) = | 0.135 |
| SO1&J2 | (15/40)(12/39) = | 0.115 |
| SO1&SR2 | (15/40)(7/39) = | 0.067 |
| J1&F2 | (12/40)(6/39) = | 0.046 |
| J1&SO2 | (12/40)(15/39) = | 0.115 |
| J1&J2 | (12/40)(11/39) = | 0.085 |
| J1&SR2 | (12/40)(7/39) = | 0.054 |
| SR1&F2 | (7/40)(6/39) = | 0.027 |
| SR1&SO2 | (7/40)(15/39) = | 0.067 |
| SR1&J2 | (7/40)(12/39) = | 0.054 |
| SR1&SR2 | (7/40)(6/39) = | 0.027 |

(d) Using the tree diagram, P(One is F and one is SO) = P(F1 & SO2) + P(SO1 & F2) = 0.058 + 0.058 = 0.116

**4.115** (a) $P(R_3)$ = 40379/98993 = 0.408

(b) $P(R_3|G_1)$ = (25888/70000) = 0.370

(c) No.  The probability of $R_3$ changes when it becomes know that $G_1$ has occurred.  To be independent, the probability must not change.

(c) Since $P(R_2)$ = 21732/98993 = 0.215 and $P(R_2|G_2)$ = 5400/28993 = 0.186, the events of being female and being associate professor are not independent.  To be independent, the probability must not change.

**4.117** (a) $P(P_1)$ = 0.459; $P(C_2)$ = 0.187; $P(P_1 \& C_2)$ = 0.082

(b) $P(P_1)$ $P(C_2)$ = (0.459)(0.187) = 0.085.  Since this is not equal to $P(P_1 \& C_2)$, these events are not independent.

**4.119** (a) P(A) = 4/8 = 1/2; P(B) = 4/8 =1/2; P(C) = 3/8.

(b) P(B|A) =2/4 = 1/2

(c) Yes.  P(B|A) is the same as P(B).

(d) P(C|A) = 1/4

(e) No.  P(C|A) and P(C) are not equal.

**4.121** Let $A_i$ = draw an ace on selection i, where i = 1,2.

(a) If the first card is replaced before the second card is drawn:
$$P(A_1 \& A_2) = (4/52)(4/52) = 0.006 .$$

(b) If the first card is not replaced before the second card is drawn:
$$P(A_1 \& A_2) = (4/52)(3/51) = 0.005 .$$

**4.123** Let:

$\overline{F_i}$ = non-failure of "criticality 1" item i, where i = 1, 2, …, 748

$F_i$ = failure of "criticality 1" item i, where i = 1, 2, …, 748

Also, $P(\overline{F_i})$ = 0.9999 and $P(F_i)$ = 0.0001

(a) The probability that none of the "criticality 1" items would fail is

$$P(\overline{F_1}) \cdot P(\overline{F_2}) \cdot \ldots \cdot P(\overline{F_{748}}) = (0.9999)^{748} = 0.928$$

(b) The probability that at least one "criticality 1" item would fail is

1 − 0.928 = 0.072.

(c) There was a 7.2% chance that at least one "criticality 1" item would fail. In other words, on the average at least one "criticality 1" item will fail in 7.2 out of every 100 such missions.

**4.125** (a) The age of a driver at fault in one fatal accident is independent of the age of a driver at fault in any other fatal accident. Therefore, P(16-24 & 25-34 & 34-64) = P(16-24)P(25-34)P(34-64) = (0.255)(0.238)(0.393) = 0.024.

(b) Let A be the event that a driver at fault is between 16 and 24 years old and let B be the event that a driver at fault is 65 or older. Having two drivers at fault between 16 and 24 years old and one being 65 or older can occur as A&A&B, A&B&A, or B&A&A. These three outcomes are mutually exclusive, so we can find the probabilities of each one and then add the three results. P(A&A&B) = P(A)P(A)P(B) = (0.255)(0.255)(0.114) = 0.0074. The other two products have the same factors, but written in a different order. Therefore all three probabilities are the same, so the probability of having two of the drivers being between 16 and 24 and one being 65 or older is 0.0074 + 0.0074 + 0.0074 = 0.0222.

**4.127** Let:

$A_i$ = Age Group i, where i = 1, 2, or 3; i.e. $A_1$.= 65-74, $A_2$.= 75-79, $A_3$.= 80 or older

(a) $P(A_3.\& A_3.\& A_3.)$ = $P(A_3)P(A_3)P(A_3)$ = (0.033)(0.033)(0.033) = 0.0000359

(b) $P(A_3.\& A_3.\& A_3.)$ = $P(A_3)P(A_3)P(A_3)$ = (0.008)(0.008)(0.008) = 0.000000512

(c) $P(A_1.\& A_2.\& A_3.)$ = $P(A_1)P(A_2)P(A_3)$ = (0.066)(0.027)(0.033) = 0.0000588

(d) We are assuming sampling with replacement. This assumption is not critical since the populations are so large that the percentages would change very little when one person is chosen from the population.

**4.129** Let: M = male;     F = female;     L = activity limitation.

Also, P(L|M) = 0.136 and P(L|F) = 0.144. Note that these probabilities are different from each other. This means that activity limitation is influenced by a person's gender.

If gender and activity limitation were statistically independent, the following would have to hold:

P(L|M) = P(L) and P(L|F) = P(L).

But P(L) is different depending upon a person's gender. Thus, gender and activity limitation are not statistically independent.

**4.131** (a) If two events A and B are mutually exclusive, they cannot both happen, so P(A & B) = 0.

(b)   If neither A nor B is impossible, then P(A) > 0 and P(B) > 0.   Since A
and B are independent, P(A & B) = P(A) P(B), but this cannot be zero
since neither A nor B has a probability of 0.

(c)   Answers will vary.   If A and B are two events such that P(A) = 0.4,
P(B) = 0.3, and P(A & B) = 0.2, then A and B are not mutually exclusive
since they can happen simultaneously, and they are not independent
since P(A & B) ≠ P(A) P(B).

**4.133** (a)   We have

$$P(A) = P(B) = P(C) = \frac{18}{36} = \frac{1}{2}$$

$$P(A \& B) = P(A \& C) = P(B \& C) = \frac{9}{36} = \frac{1}{4}$$

$$P(A \& B \& C) = \frac{9}{36} = \frac{1}{4}$$

$$P(A \& B) = P(A)P(B)$$

Consequently,    $$P(A \& C) = P(A)P(C)$$

$$P(B \& C) = P(B)P(C)$$

However,    $$P(A \& B \& C) = \frac{1}{4} \neq \frac{1}{2} \cdot \frac{1}{2} \cdot \frac{1}{2} = P(A) \cdot P(B) \cdot P(C)$$

Thus, the events A, B, and C are *not* independent.

(b)   We have

$$P(D) = \frac{18}{36} = \frac{1}{2} \qquad P(E) = \frac{18}{36} = \frac{1}{2} \qquad P(F) = \frac{4}{36} = \frac{1}{9}$$

$$P(D \& E) = \frac{6}{36} = \frac{1}{6} \qquad P(D \& F) = \frac{3}{36} = \frac{1}{12} \qquad P(E \& F) = \frac{2}{36} = \frac{1}{18}$$

$$P(D \& E \& F) = \frac{1}{36}$$

Consequently,    $$P(D \& E \& F) = \frac{1}{36} = \frac{1}{2} \cdot \frac{1}{2} \cdot \frac{1}{9} = P(D)P(E)P(F)$$

However, since    $$P(D \& E) = \frac{1}{6} \neq \frac{1}{2} \cdot \frac{1}{2} = P(D)P(E),$$

the events D, E, and F are *not* independent.

## Exercises 4.7

**4.135** (a)   Four events are exhaustive if at least one of them must occur.

(b)   Four events are mutually exclusive if no two of them can occur at the
same time.

(c)   No.   Consider an example in which a die is rolled.   Let A be the event
that the number showing is even, let B be the event that the number is
odd, let C be the event that the number is a 1, and let D be the event
that the number is 4 or greater.   These four events contain all of the
possible outcomes from 1 to 6 and are thus exhaustive.   However, B and
C both contain 1, A and D both contain 4 and 6, and B and D both
contain 5.   Therefore the events are not mutually exclusive.

(d)   No.  Consider an example in which two dice are tossed.  Let A be the event the total is 2, let B be the event that the total is between 3 and 5 inclusive, let C be the event that the total is between 7 and 9 inclusive, and let D be the event that the total is 12.  These four events are mutually exclusive, but if the total is 6, 10, or 11, none of these events will occur.  Thus, these events are not exhaustive.

**4.137** (a)   $P(R_3)$

(b)   $P(S|R_3)$

(c)   $P(R_3|S)$

**4.139** (a)   P(believe in aliens) = $0.54 \cdot 0.48 + 0.33 \cdot 0.52 = 0.431$, so 43.1% of U.S. adults believe in aliens.

(b)   P(believe in aliens|woman) = 0.33; so 33% of women believe in aliens.

(c)   P(woman|believe in aliens) = (0.33)(0.52)/.431 = 0.398; so 39.8% of adults who believe in aliens are women.

**4.141** (a)   Let A be the event that the selected person reads an astrology report every day, let L be the event that the person has less than high school education, let H be the event that the person is a high school graduate, and let B be the event that the person has Baccalaureate or higher.  $P(A) = P(A|L)P(L) + P(A|H)P(H) + P(A|B)P(B) = (0.09)(0.074) + (0.07)(0.533) + (0.04)(0.393) = 0.05969$

(b)   $P(H|A) = P(H \& A)/P(A) = P(A|H)P(H)/P(A) = (0.09)(0.074)/0.05969 = 0.1116$.  From this, $P((\text{not } H)|A) = 1 - P(H|A) = 1 - 0.1116 = 0.8884$.

(c)   $P(B|A) = P(B \& A)/P(A) = P(A|B)P(B)/P(A) = (0.04)(0.393)/0.05969 = 0.26336$.

**4.143** (a)   Let A1 to A4 represent the four age categories in the order shown in the table and let F by the event that a sElected person is obese or overweight.  $P(F) = P(A1)P(F|A1) + P(A2)P(F|A2) + P(A3)P(F|A3) + P(A4)P(F|A4) = (0.425)(0.411) + (0.285)(0.579) + (0.164)(0.682) + (0.126)(0.553) = 0.5212$.  Thus 52.12% of Utah adults are overweight or obese.

(b)   57.9% of Utah adults between 35 and 49 years old are overweight or obese.

(c)   $P(A2|D) = P(A2 \& D)/P(D) = P(D|A2)P(A2)/P(D) = (0.579)(0.285)/0.5212 = 0.3166$.  Thus 31.66% of overweight or obese Utah adults are between 35 and 49 years old, inclusive.

(d)   52.12% of Utah adults are overweight or obese.  57.9% of Utah adults between 35 and 49 years old are overweight or obese.  31.66% of overweight or obese Utah adults are between 35 and 49 years old, inclusive.

**4.145** Let: M = sell more than projected P(M) = 0.10 P(R|M) = 0.70
        C = sell close to projected P(C) = 0.30 P(R|C) = 0.50
        L = sell less than projected P(L) = 0.60 P(R|L) = 0.20
        R = revised for a second edition

(a)   $P(R) = P(M) \cdot P(R|M) + P(C) \cdot P(R|C) + P(L) \cdot P(R|L)$

      $= (0.10 \cdot 0.70) + (0.30 \cdot 0.50) + (0.60 \cdot 0.20) = 0.34$ or 34%

(b)

$$P(L|R) = \frac{P(L \& R)}{P(R)} = \frac{P(L)P(R|L)}{P(R)} = \frac{(0.60)\ (0.20)}{0.34} = 0.353 \text{ or } 35.3\%$$

**4.147** (a)  Of those individuals who have the disease, 93.4% will test positive. Also, of those individuals who do not have the disease, 96.8% will test negative.

(b)  Let: P= test is positive                    $P(P|D) = 0.934$

not P= test is negative                 $P(\text{not } P|\text{not } D) = 0.968$

   D= person has the disease          $P(D) = 0.002$

not  D= person does not have the disease

$$P(D|P) = \frac{P(D) \cdot P(P|D)}{P(D) \cdot P(P|D) + P(\text{not } D) \cdot P(P|\text{not } D)}$$

Two of the items on the right-hand side of the equation above are not provided in the statement of the problem.  These are P(not D) and

P(P|not D).  Because D and not D are complements, we know that P(not D) = 1 - P(D) = 1 - 0.002 = 0.998.  Also, P(P|not D) and P(not P|not D) are complements.  (Although not so readily apparent, this can be demonstrated with the assistance of a tree diagram.)  Thus, P(P|not D) =  1 - P(not P|not D) = 1 - 0.968 = 0.032.

Substituting all of these items into the equation above results in

$$P(D|P) = \frac{(0.002)(0.934)}{(0.002)(0.934) + (0.998)(0.032)} = 0.055$$

(c)  In terms of a percentage, the result from part (b) means that 5.5% of those who initially test positive for the disease actually have the disease.

**4.149** Let: N = enjoy neither their job nor their personal life  $P(N) = 0.15$

O = enjoy their job but not their personal life       $P(O) = 0.80$

B = enjoy both their job and their personal life      $P(B) = 0.04$

J = people interviewed that enjoy their job

I = people interviewed that enjoy their job who also enjoy their personal life

(a)  $P(J) = P(O) + P(B) = 0.80 + 0.04 = 0.84$

84% of the people interviewed enjoy their jobs.

(b)  $P(I) = P(B)/P(J) = 0.04/0.84 = 0.048$

4.8% of the people interviewed who enjoy their jobs also enjoy their personal lives.

**Exercises 4.8**

**4.151** Counting rules are techniques for determining the number of ways something can happen without directly listing all of the possibilities.  They are important because sometimes the number of possibilities is so large that a direct listing is impractical.

**4.153** (a)  A permutation of *r* objects from a collection of *m* objects is any *ordered* rearrangement of *r* of the *m* objects.

(b)  A combination of *r* objects from a collection of *m* objects is any *unordered* set of *r* of the *m* objects.

(c)  Permutations are ordered rearrangements while combinations are sets in which the order of listing is not important.

**4.155** (a)

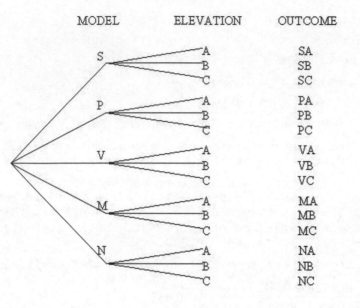

(b) There are 15 choices for the selection of a home, including both model and elevation.

(c) There are 5 ways to choose the model and for each of these, there are 3 ways to choose the elevation. By the BCR, there are (5)(3) = 15 ways to choose both model and elevation.

**4.157** There are $6\cdot8\cdot5\cdot19\cdot16\cdot14 = 1{,}021{,}440$ possibilities for answering all six questions.

**4.159** There are $(8)(9)(7)(8)(6) = 24{,}192$ possibilities for the five-question quiz.

**4.161** There are $(39)(19)(8)(6)(24)(5)(5)(6)(5) = 640{,}224{,}000$ possibilities to create an i Doll.

**4.163** (a) $_7P_3 = 7\cdot6\cdot5 = 210$      (d) $_6P_0 = 1$

(b) $_5P_2 = 5\cdot4 = 20$      (e) $_9P_9 = 9\cdot8\cdot7\cdot6\cdot5\cdot4\cdot3\cdot2\cdot1 = 362{,}880$

(c) $_8P_4 = 8\cdot7\cdot6\cdot5 = 1{,}680$

**4.165** (a) There are $10\cdot9\cdot8\cdot7\cdot6\cdot5\cdot4\cdot3\cdot2\cdot1 = 3{,}628{,}800$ possibilities for the order in which the subject writes down the numbers.

(b) With 3,628,800 possibilities in total, the subject has a one in 3,628,800 chance of guessing the numbers in the correct order.

(c) We would conclude that the subject really does possess ESP because obtaining these results by chance is extremely improbable.

**4.167** There are $_{18}P_3 = 18\cdot17\cdot16 = 4896$ ways to pick the first, second, and third place winners.

**4.169** (a) Since the order in which the cards are dealt in Stud Poker is important, there are $_{52}P_5 = 52!/47! = 52\cdot51\cdot50\cdot49\cdot48 = 311{,}875{,}200$ different five-card hands.

(b)   There are $_4C_3 = 4!/3!1! = 4$ ways of getting three kings and $_4C_2 = 4!/2!2!$ $= 4\cdot3/2\cdot1 = 6$ ways of getting two queens.  Once the suits have been chosen, there are $_5P_5 = 5!/0! = 120$ ways of ordering the 5 cards. Altogether, there are $4\cdot6\cdot120 = 2880$ ways of getting three kings and two queens.  Another way to do this is to say that there are $_5C_3 = 5!/3!2! = 10$ ways to pick the positions in the deal for the three kings.  Then there are $_4P_3 = 4!/1! = 24$ and $_4P_2 = 4!/2! = 12$ ways to order the three kings within their positions and to order the two queens within their positions.  So there are $(10)(24)(12) = 2880$ ways to deal three kings and two queens.

(c)   There are $_{13}C_1 = 13!/1!12! = 13$ ways to pick the three-card denomination and then $_{12}C_1 = 12!/1!11! = 12$ ways to pick the two-card denomination. Then as in part (b), there are 2880 ways to pick the three cards of one denomination and two cards of the other denomination.  Altogether, there are $13\cdot12\cdot2880 = 449,280$ different full houses possible.

(d)   P(full house) = $449,280/311,875,200 = 0.00144$

**4.171** (a)

$$_7C_3 = \frac{7!}{3!(7-3)!} = \frac{7!}{3!4!} = \frac{7\cdot6\cdot5\cdot4!}{3!4!} = \frac{7\cdot6\cdot5}{3\cdot2\cdot1} = 35$$

(b)

$$_5C_2 = \frac{5!}{2!(5-2)!} = \frac{5!}{2!3!} = \frac{5\cdot4\cdot3!}{2!3!} = \frac{5\cdot4}{2\cdot1} = 10$$

(c)

$$_8C_4 = \frac{8!}{4!(8-4)!} = \frac{8!}{4!4!} = \frac{8\cdot7\cdot6\cdot5\cdot4!}{4!4!} = \frac{8\cdot7\cdot6\cdot5}{4\cdot3\cdot2\cdot1} = 70$$

(d)

$$_6C_0 = \frac{6!}{0!(6-0)!} = \frac{6!}{0!6!} = 1$$

(e)

$$_9C_9 = \frac{9!}{9!(9-9)!} = \frac{9!}{9!0!} = 1$$

**4.173** (a)   If the three prizes are equivalent, the number of possible outcomes is

$$_{100}C_3 = \frac{100!}{3!(100-3)!} = \frac{100!}{3!97!} = \frac{100\cdot99\cdot98\cdot97!}{3!97!} = \frac{100\cdot99\cdot98}{3\cdot2\cdot1} = 161,700$$

(b)   If there are $1^{st}$, $2^{nd}$, and $3^{rd}$ prizes, the number of possible outcomes is

$$_{100}P_3 = \frac{100!}{(100-3)!} = \frac{100!}{97!} = \frac{100\cdot99\cdot98\cdot97!}{97!} = \frac{100\cdot99\cdot98}{1} = 970,200$$

**4.175** (a)   Suppose the two teams are A and B.  There are 2 sequences, AAAA and BBBB, in which the series ends in 4 games.  This could be expressed as

$$2\cdot{_4C_4} = 2\cdot\frac{4!}{4!(4-4)!} = 2\cdot\frac{4!}{4!0!} = 2\cdot1 = 2$$

(b)   If A wins the series, A must win the fifth game and any 3 of the first 4.  For each series that A wins there is a similar one that B wins, the number of series ending in 5 games is

$$2 \cdot {}_4C_3 = 2 \cdot \frac{4!}{3!(4-3)!} = 2 \cdot \frac{4!}{3!1!} = 2 \cdot 4 = 8$$

(c)  If A wins the series, A must win the sixth game and any 3 of the first 5.  For each series that A wins there is a similar one that B wins, the number of series ending in 6 games is

$$2 \cdot {}_5C_3 = 2 \cdot \frac{5!}{3!(5-3)!} = 2 \cdot \frac{5!}{3!2!} = 2 \cdot 10 = 20$$

(d)  If A wins the series, A must win the seventh game and any 3 of the first 6. For each series that A wins there is a similar one that B wins, the number of series ending in 7 games is

$$2 \cdot {}_6C_3 = 2 \cdot \frac{6!}{3!(6-3)!} = 2 \cdot \frac{6!}{3!3!} = 2 \cdot 20 = 40$$

(e)  There is a total of 70 different equally likely ways the series can end if the teams are evenly matched.  The probability that it ends in 4 games is P(4) = 2/70 = 1/35.  Similarly, P(5) = 8/70 = 4/35; P(6) = 20/70 = 10/35; and P(7) = 40/70 = 20/35.

**4.177** (a)

$$_{100}C_5 = \frac{100!}{5!(100-5)!} = \frac{100!}{5!95!} = \frac{100 \cdot 99 \cdot 98 \cdot 97 \cdot 96 \cdot 95!}{5!95!} = \frac{100 \cdot 99 \cdot 98 \cdot 97 \cdot 96}{5 \cdot 4 \cdot 3 \cdot 2 \cdot 1} = 75,287,520$$

(b)

$$_{50}C_5 \cdot 2^5 = \frac{50!}{5!(50-5)!} \cdot 32 = 67,800,320$$

(c)  The probability that no state will have both of its senators on the committee is the answer in part (b) divided by the answer in part (a), which is 67,800,320/75,287,520 = 0.901.

**4.179** You have eight tries to find the right key.  Each try has the same probability of occurrence.

(a)  The probability that you get the right key on the first try is 1/8.

(b)  The probability that you get the key right on the last try is 1/8.

$$P(F_1F_2F_3F_4F_5F_6F_7S_8) = \frac{7 \cdot 6 \cdot 5 \cdot 4 \cdot 3 \cdot 2 \cdot 1 \cdot 1}{8 \cdot 7 \cdot 6 \cdot 5 \cdot 4 \cdot 3 \cdot 2 \cdot 1} = \frac{1}{8}$$

(c)  The probability that you get the right key on or before the fifth try is 5(1/8) = 5/8.

P(S) + P(FS) + P(FFS) + P(FFFS) + P(FFFFS) = 5(1/8) = 5/8.

**4.181** Given 365 days in a year and 38 students, the probability that no two students have the same birthday is

$$\frac{365!}{(365-38)!}/365^{38} = \frac{3.165\,5642\,x10^{96}}{23.287\,820\,x10^{96}} = 0.1359$$

The probability that at least two students in the class have the same birthday is then 1- 0.1359 = 0.8641.

**4.183** There are 15 true-false questions.

(a)  The probability of getting at least one question correct is

$$1 - \frac{{}_{15}C_0}{2^{15}} = 1 - \frac{1}{32768} = 0.999969$$

(b) Getting a 60% or better on the·exam means getting nine or more questions correct.   The required calculations are

$$_{15}C_9 / 32768 = 0.1527 \qquad _{15}C_{13}/32768 = 0.0032$$

$$_{15}C_{10}/32768 = 0.0916 \qquad _{15}C_{14}/32768 = 0.0005$$

$$_{15}C_{11}/32768 = 0.0417 \qquad _{15}C_{15}/32768 = 0.0000$$

$$_{15}C_{12}/32768 = 0.0139$$

The probability of a 60% or better is the sum of the seven items above, or 0.3036.

**4.185** (a) The number of ways that n items can be selected from a collection of N items is

$$_{N}C_n = \frac{N!}{n!\,(N-n)!}$$

To determine f, defined as the number of ways that any particular sample of size n is the one selected, we begin by partitioning the observations comprising the sample of size n into 2 groups; namely, a subset containing the desired n observations and a subset containing none of the remaining N-n observations.

There are n items in the subset containing the desired n observations. Of these n items, n are to be selected.   This can be done in

$$_{n}C_n = \frac{n!}{n!(n-n)!} = 1 \text{ way}$$

There are N-n items in the subset containing none of the desired n observations.   Of these N-n items, n-n (or zero) are to be selected. This can be done in

$$_{N-n}C_{n-n} = \frac{(N-n)!}{(n-n)![(N-n)-(n-n)]!} = \frac{(N-n)!}{0!(N-n)!} = 1 \text{ way}$$

Using the two results above in the context of the fundamental counting rule, there is a total of

$$1 \cdot 1 = 1$$

outcome in which exactly n of the n items are selected.   Thus, f = 1. Now, by the f/N-rule, the probability that any particular sample of size n is selected from a population of size N equals

$$\frac{f}{N} = \frac{1}{\dfrac{N!}{n!(N-n)!}} = \frac{1}{_{N}C_n}$$

(b) The probability that any specified member of the population is included in the sample is a simple extension of part (a) above.  First, the number of ways that a sample of n items can be selected from a collection of N items is again

$$_N C_n = \frac{N!}{n!(N-n)!}$$

To determine f, defined as the number of ways that any (one) specified member of the population is included in the sample, we begin by partitioning the observations comprising the sample of size n into 2 groups; namely, a subset containing the desired single observation and a subset containing the remaining N-1 observations.

There is 1 item in the subset containing the desired 1 observation.  Of this 1 item, 1 is to be selected.  This can be done in

$$_1 C_1 = \frac{1!}{1!(1-1)!} = 1 \text{ way}$$

There are N-1 items in the subset containing the remaining observations.  Of these N-1 items, n-1 are to be selected.  This can be done in

$$_{N-n}C_{n-n} = \frac{(N-1)!}{(n-1)![(N-1)-(n-1)]!} = \frac{(N-1)!}{(n-1)!(N-n)!} \text{ ways}$$

Using the two results above in the context of the fundamental counting rule, there is a total of

$$1 \cdot \frac{(N-1)!}{(n-1)!(N-n)!}$$

outcomes in which exactly 1 of the n items is selected.  This is also the value of f.  Now, by the f/N-rule, the probability that any (one) specified member of the population is included in the sample is f/N =

$$\frac{(N-1)!}{(n-1)!(N-n)!} \div \frac{N!}{n!(N-n)!} = \frac{(N-1)!}{(n-1)!(N-n)!} \cdot \frac{n!(N-n)!}{N!}$$

$$= \frac{n!}{(n-1)!} \cdot \frac{(N-1)!}{N!} \cdot \frac{(N-n)!}{(N-n)!} = \frac{n}{N}$$

Notice that the expression before the last equality sign simplifies to n/N because n!/(n-1)! = n and (N-1)!/N! = 1/N.

(c)  The probability f that any k specified members of the population are included in the sample is

$$f = {}_k C_k \cdot {}_{N-k} C_{n-k} = 1 \cdot \frac{(N-k)!}{(n-k)![(N-k)-(n-k)]!} = \frac{(N-k)!}{(n-k)!(N-n)!}$$

So, the probability that any k specified members of the population are included in the sample is

$$\frac{f}{N} = \frac{\dfrac{(N-k)!}{(n-k)!(N-n)!}}{\dfrac{N!}{n!(N-n)!}} = \frac{(N-k)!}{(n-k)!(N-n)!} \cdot \frac{n!(N-n)!}{N!} = \frac{(N-k)!}{(n-k)!} \cdot \frac{n!}{N!}$$

Review Problems for Chapter 4

1.    Probability theory enables us to control and evaluate the likelihood that a statistical inference is correct, and it provides the mathematical basis for statistical inference.

2.    (a)  The equal-likelihood model is used for computing probabilities when an experiment has N possible outcomes, all equally likely.

      (b)  If an experiment has N equally likely outcomes and an event can occur in f ways, then the probability of the event is f/N.

3.    In the frequentist interpretation of probability, the probability of an event is the relative frequency of the event in a large number of experiments.

4.    The numbers (b) -0.047 and (c) 3.5 cannot be probabilities because probabilities must always be between 0 and 1.

5.    The Venn diagram is a common graphical technique for portraying events and relationships among events.

6.    Two or more events are mutually exclusive if no two of the events have any outcomes in common, i.e., no two of the events can occur at the same time.

7.    (a)  P(E)

      (b)  P(E) = 0.436

8.    (a)  False.  For any two events, the probability that one or the other occurs equals the sum of the two individual probabilities minus the probability that both events occur.

      (b)  True.  Either the event occurs or it does not.  Therefore, the probability that the event occurs plus the probability that it does not occur equals 1. Thus the probability that it occurs is 1 minus the probability that it does not occur.

9.    Frequently, it is quicker to compute the probability of the complement of an event than it is to compute the probability of the event itself.  The complement rule allows one to use the probability of the complement to obtain the probability of the event itself.

10.   (a)  Data obtained by observing values of one variable of a population are called <u>univariate</u> data.

      (b)  Data obtained by observing values of two variables of a population are called <u>bivariate</u> data.

      (c)  A frequency distribution for bivariate data is called a <u>contingency table</u>.

11.   The sum of the joint probabilities in a row or column of a joint probability distribution equals the <u>marginal</u> probability of that row or column.

12.   (a)  P(B|A)

      (b)  A

13.   Conditional probabilities can sometimes be computed directly as f/N or by using the formula P(B|A) = P(A&B)/P(A).

14.   The joint probability of two independent events is the product of their marginal probabilities.

15.   If two or more events have the property that at least one of them must occur when the experiment is performed, then the events are said to be <u>exhaustive</u>.

**16.** If r actions are to be performed in a definite order and there are $m_1$ possibilities for the first action, $m_2$ possibilities for the second action, ..., and $m_r$ possibilities for the $r^{th}$ action, then there are   $m_1\ m_2\ m_3...m_r$ possibilities altogether for the r actions.

**17.** (a)  ABC  ACB   BAC   BCA   CAB   CBA

ABD  ADB   BAD   BDA   DAB   DBA

ACD  ADC   CAD   CDA   DAC   DCA

BCD  BDC   CBD   CDB   DBC   DCB

(b)  ABC  ABD   ACD   BCD

(c)  $_4P_3 = 24$       $_4C_3 = 4$

(d)  $_4P_3 = 4!/(4-3)! = 24$        $_4C_3 = 4!/3!1! = 4$

**18.** (a)  $P(A) = \dfrac{26015}{130423} = 0.199$

(b)  $P(D\ or\ E\ or\ F) = P(D) + P(E) + P(F) = \dfrac{13957 + 10452 + 26915}{130423} = \dfrac{51324}{130423} = 0.394$

(c)

| Event | Probability |
|-------|-------------|
| A | 0.199 |
| B | 0.179 |
| C | 0.141 |
| D | 0.107 |
| E | 0.080 |
| F | 0.206 |
| G | 0.088 |

**19.** (a)  (not J) is the event that the return selected shows an adjusted gross income of at least $100,000.  There are 11,415 thousand such returns.

(b)  (H & I) is the event that the return selected shows an adjusted gross income of between $20,000 and $50,000.  There are 42,782 thousand such returns.

(c)  (H or K) is the event that the return selected shows an adjusted gross income of at least $20,000.  There are 81,112 thousand such returns.

(d)  (H & K) is the event that the return selected shows an adjusted gross income of between $50,000 and $100,000.  There are 26,915 thousand such returns.

**20.** (a)  Events H and I are not mutually exclusive.  Both have events C, D, and E in common.

(b)  Events I and K are mutually exclusive.  They have no events in common.

(c)  Events H and (not J) are mutually exclusive.  H = (C or D or E or F) while (not J) = G.  H and (not J) have no events in common.

(d)  Events H, (not J), and K are not mutually exclusive.  While H and (not J) have no events in common, a part of K (i.e., event F) is common to event H, and the other part of K (i.e., event G) is common to event (not J).

**21.**   (a)

$$P(H) = \frac{18373 + 13957 + 10452 + 26915}{130423} = \frac{69697}{130423} = 0.534$$

$$P(I) = \frac{26015 + 23296 + 18373 + 13957 + 10452}{130423} = \frac{93602}{130423} = 0.706$$

$$P(J) = \frac{130423 - 11415}{130423} = \frac{119008}{130423} = 0.912$$

$$P(K) = \frac{26915 + 11415}{130423} = \frac{38330}{130423} = 0.294$$

(b)   H =   (C or D or E or F)

   I =   (A or B or C or D or E)

   J =   (A or B or C or D or E or F)

   K =   (F or G)

(c)   P(H) = P(C) + P(D) + P(E) + P(F) = 0.141 + 0.107 + 0.080 + 0.206

   = 0.534

   P(I) = P(A) + P(B) + P(C) + P(D) + P(E)

   = 0.199 + 0.179 + 0.141 + 0.107 + 0.080 = 0.706

   P(J) = P(A) + P(B) + P(C) + P(D) + P(E) + P(F)

   = 0.199 + 0.179 + 0.141 + 0.107 + 0.080 + 0.206 = 0.912

   P(K) = P(F) + P(G) = 0.206 + 0.088 = 0.294

**22.**   (a)

$$P(not\ J) = \frac{11415}{130423} = 0.088$$

$$P(H\ \&\ I) = \frac{42782}{130423} = 0.328$$

$$P(H\ or\ K) = \frac{81112}{130423} = 0.622$$

$$P(H\ \&\ K) = \frac{26915}{130423} = 0.206$$

(b)   P(J) = 1 - P(not J) = 1 - 0.059 = 0.941

(c)   P(H or K) = P(H) + P(K) - P(H & K) = 0.534 + 0.294 - 0.206

   = 0.622.

(d)   The answer in (c) agrees with that in (a).

**23.**   (a)   The contingency table has six cells.

(b)   There are 15,058 thousand students in high school.

(c)   There are 59,921 thousand students in public schools.

(d)   There are 3,695 thousand students in private colleges.

**24.**   (a)   (i)      $L_3$ = the student selected is in college;

   (ii)      $T_1$ = the student selected attends a public school;

   (iii)      ($T_1$ & $L_3$) = the student selected attends a public college.

(b)

$$P(L_3) = \frac{15\,928}{69\,818} = 0.228$$

$$P(T_1) = \frac{59\,921}{69\,818} = 0.858$$

$$P(T_1 \,\&\, L_3) = \frac{12\,233}{69\,818} = 0.175$$

The interpretation of each result above as a percentage is as follows: 22.8% of students attend college, 85.8% attend public schools, and 17.5% attend public colleges.

(c)

|  | Public $T_1$ | Private $T_2$ | $P(L_i)$ |
|---|---|---|---|
| Elementary   $L_1$ | 0.486 | 0.070 | 0.556 |
| High School $L_2$ | 0.197 | 0.019 | 0.216 |
| College      $L_3$ | 0.175 | 02.053 | 0.228 |
| $P(T_j)$ | 0.858 | 0.142 | 1.000 |

(d)  P(T₁ or L₃) = (33952 + 13730 + 12233 + 3695)/69818 = 0.911

(e)  P(T₁ or L₃) = P(T₁) + P(L₃) - P(T₁ & L₃) = 0.858 + 0.228 - 0.175 = 0.911.

25.    (a)  $P(L_3 \mid T_1) = \frac{12233}{59921} = 0.204$

20.4% of students attending public schools are in college.

(b)  $P(L_3 \mid T_1) = \frac{P(L_3 \,\&\, T_1)}{P(T_1)} = \frac{0.175}{0.858} = 0.204$

26.    (a)  $P(T_2) = \frac{9897}{69818} = 0.142; P(T_2 \mid L_2) = \frac{1322}{15058} = 0.088$

(b)  No.  For L₂ and T₂ to be independent, P(T₂|L₂) must equal P(T₂).  From part (a), however, we see that this is not the case.  In terms of percentages, each of these probabilities, respectively, means that 8.8% of high school students attend private schools, whereas 14.2% of all students attend private schools.

(c)  Events L₂ and T₂ are not mutually exclusive.  From Table 4.17, we see that both events can occur simultaneously.  There are 1,322 thousand students who attend private high schools.

(d)  No.  Let: L₁ = a student is in elementary school;

T₁ = a student attends public school.

For L₁ and T₁ to be independent, P(L₁|T₁) must equal P(L₁).  But, P(L₁) = 0.556 and P(L₁|T₁) = 0.486/0.858 = 0.567.  Since P(L₁|T₁) ≠ P(L₁), the event that a student is in elementary school is *not* independent of the event that a student attends public school.

27.   Let MA1, MP1, and MS1 denote, respectively, the events that the first student selected received a master of arts, a master of public administration, and a master of science; and let MA2, MP2, and MS2 denote, respectively, the events that the second student selected received a master of arts, a master of public administration, and a master of science.

(a)   $P(MA1 \,\&\, MS2) = \dfrac{3}{50} \cdot \dfrac{19}{49} = 0.023$

(b)   $P(MP1 \,\&\, MP2) = \dfrac{28}{50} \cdot \dfrac{27}{49} = 0.309$

(c)

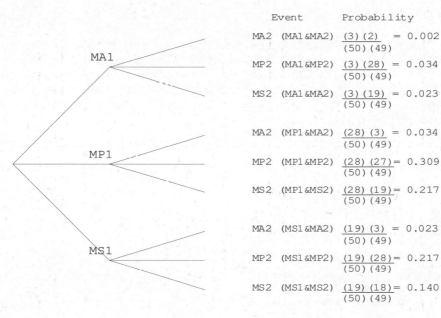

| Event | | Probability | |
|---|---|---|---|
| MA2 | (MA1&MA2) | $\dfrac{(3)(2)}{(50)(49)}$ | = 0.002 |
| MP2 | (MA1&MP2) | $\dfrac{(3)(28)}{(50)(49)}$ | = 0.034 |
| MS2 | (MA1&MA2) | $\dfrac{(3)(19)}{(50)(49)}$ | = 0.023 |
| MA2 | (MP1&MA2) | $\dfrac{(28)(3)}{(50)(49)}$ | = 0.034 |
| MP2 | (MP1&MP2) | $\dfrac{(28)(27)}{(50)(49)}$ | = 0.309 |
| MS2 | (MP1&MS2) | $\dfrac{(28)(19)}{(50)(49)}$ | = 0.217 |
| MA2 | (MS1&MA2) | $\dfrac{(19)(3)}{(50)(49)}$ | = 0.023 |
| MP2 | (MS1&MP2) | $\dfrac{(19)(28)}{(50)(49)}$ | = 0.217 |
| MS2 | (MS1&MS2) | $\dfrac{(19)(18)}{(50)(49)}$ | = 0.140 |

(d)   P(MA1 & MA2) + P(MP1 & MP2) + P(MS1 & MS2)

   = 0.002 + 0.309 + 0.140 = 0.451.

28.   Let F = event that pairs raised their offspring to the <u>F</u>ledgling stage, and D = event that pairs divorced.  The given information is

P(D) = 0.63, P(D|F) = 0.81, P(D|not F) = 0.43.

(a)   We are trying to find P(Offspring died) = P(not F). The total probability rule states that

   P(D) = P(D|F)P(F) + P(D|not F)P(not F) or

   P(D) = P(D|F)P(F) + P(D|not F)[1 - P(F)].  Substituting known values,

   0.63 = (0.81)P(F) + (0.43)[1 - P(F)] or

   0.20 = 0.38 P(F).  Thus

   P(F) = 0.20/0.38 = 0.526.  Therefore, P(not F) = 1 - 0.526 = 0.474 or 47.4%.

(b)  We can use the multiplication rule to find

P(D & not F) = P(D|not F)P(not F) = (0.43)(0.474) = 0.204 or 20.4%

Thus 20.4% divorced and had offspring that died.

(c)  P(not F|D) = P(D & not F)/P(D) = (0.204)/(0.63) = 0.324

Thus, among those that divorced, 32.4% had offspring that died.

29.   (a)  Let $C_i$ be the event that the $i^{th}$ man selected is color blind.

P(none color blind) = P{(not $C_1$)& (not $C_2$)& (not $C_3$)& (not $C_4$)} =

P(not $C_1$)P(not $C_2$)P(not $C_3$)P(not $C_4$) = $0.91^4$ = 0.686

(b)  P{(not $C_1$) & (not $C_2$) & (not $C_3$) & ($C_4$)} =

P(not $C_1$)P(not $C_2$)P(not $C_3$)P($C_4$) = 0.91· 0.91· 0.91· 0.09 = 0.068

(c)  There are four sequences like the one in part (b) in which the fourth
man is color blind.  The others have the first man color blind and the
others not; the second color blind and the others not; and the third
color blind and the others not.  Each sequence has the same probability
as the one in part (b).  Thus, P(exactly one man is color blind) is
4 $(0.91)^3 \cdot (0.09)$ = 0.271

30.   (a)  No.  A&B has a probability of 0.2.  Thus A and B can occur together.

(b)  Yes.   P(A&B) = P(A)P(B).

31.   Let: A1 = Driver is 16-20 years old

A2 = Driver is 21-24 years old

A3 = Driver is 21-24 years old

A4 = Driver is 21-24 years old

A5 = Driver is 21-24 years old

A6 = Driver is 21-24 years old

B  = Driver has BAC of 0.10% or greater

(a)  P(B|A2) = 0.278

(b)

$$P(B) = \sum P(B \mid A_i) P(A_i)$$

=(0.127)(0.141)+(0.278)(0.1+(0.050)(0.114)=0.192

(c)  P(A2|B) = P(A2 & B)/P(B) = P(B|A2)P(A2)/P(B) = (0.278)(0.114)/0.192 =
0.165

(d)  Thus 27.8% of drivers aged 21-24 at fault in fatal crashes had a BAC of
0.10% or greater; 19.2% of all drivers at fault in fatal crashes had a
BAC of 0.10% or greater; and of those drivers at fault in fatal crashes
with a BAC of 0.10% or greater, 16.5% were in the 21-24 age group.

(e)  The probabilities in parts (a) and (b) are prior; the probability in
part (c) is posterior because it represents the probability that the
person selected was in the 21-24 age group **after** knowing that the
person has a BAC over 0.10%.

32.   (a)  $_{12}C_2 = \dfrac{12!}{2!10!} = \dfrac{12 \cdot 11}{2 \cdot 1} = 66$

(b) $\quad _{12}P_2 = \dfrac{12!}{(12-3)!} = 12 \cdot 11 \cdot 10 = 1320$

(c) (i) $\quad _8C_2 = \dfrac{8!}{2!6!} = \dfrac{8 \cdot 7}{2 \cdot 1} = 28$

     (ii) $\quad _8P_2 = \dfrac{12!}{(8-3)!} = 8 \cdot 7 \cdot 6 = 336$

**33.** (a) $\quad _{52}C_{13} = \dfrac{52!}{13!\,39!} = 635{,}013{,}559{,}600$

(b) $\quad \dfrac{_4C_2 \; _{48}C_{11}}{_{52}C_{13}} = \dfrac{\dfrac{4!}{2!(4-2)!} \cdot \dfrac{48!}{11!(48-11)!}}{\dfrac{52!}{13!(52-13)!}} = 0.213$

(c) With four choices for the eight-card suit, three choices for the four-card suit, and two choices for the one-card suit, the probability of being dealt an 8-4-1 distribution is

$$4 \cdot 3 \cdot 2 \cdot \frac{_{13}C_8 \; _{13}C_4 \; _{13}C_1 \; _{13}C_0}{_{52}C_{13}} = 24 \cdot \frac{\dfrac{13!}{8!(13-8)!} \cdot \dfrac{13!}{4!(13-4)!} \cdot \dfrac{13!}{1!(13-1)!} \cdot \dfrac{13!}{0!(13-13)!}}{635{,}013{,}559{,}600}$$

$$= 24 \cdot \frac{1287 \cdot 715 \cdot 13 \cdot 1}{635{,}013{,}559{,}600} = 24 \cdot \frac{11{,}962{,}665}{635{,}013{,}559{,}600} = 0.00045$$

(d) Initially, two suits from among the four are to be selected, from which five cards from each suit are drawn. From the remaining two suits, one is to be selected from which two cards are drawn. Finally, only one suit remains from which to draw the final card. The probability of being dealt a 5-5-2-1 distribution is

$$_4C_2 \cdot _2C_1 \cdot _1C_1 \cdot \frac{_{13}C_5 \; _{13}C_5 \; _{13}C_2 \; _{13}C_1}{_{52}C_{13}} = 12 \cdot \frac{\dfrac{13!}{5!(13-5)!} \cdot \dfrac{13!}{5!(13-5)!} \cdot \dfrac{13!}{2!(13-2)!} \cdot \dfrac{13!}{1!(13-1)!}}{635{,}013{,}559{,}600}$$

$$= 12 \cdot \frac{1287 \cdot 1287 \cdot 78 \cdot 13}{635{,}013{,}559{,}600} = 12 \cdot \frac{1{,}679{,}558{,}166}{635{,}013{,}559{,}600} = 0.032$$

The probability of being dealt a hand void in a specified suit is

$$\frac{_{39}C_{13} \; _{13}C_0}{_{52}C_{13}} = \frac{\dfrac{39!}{13!(39-13)!} \cdot \dfrac{138!}{0!(39-0)!}}{\dfrac{52!}{13!(52-13)!}} = 0.013$$

**34.** $\quad _{64}C_8 = \dfrac{64!}{8!\,56!} = 4{,}426{,}165{,}368$

**35.** (a) We would need to assume that people don't own a VCR unless they also have a TV set. This is a reasonable assumption since the VCR is almost useless without a TV set (It could be used in conjunction with another VCR to duplicate tapes.).

(b)   P(VCR) = P(VCR & TV) = P(TV)P(VCR|TV) = (0.982)(0.912) = 0.896 or 89.6%.

(c)   We would need to know how many (or the percentage of) households have a VCR, but don't have a TV.

**CHAPTER 5 ANSWERS**

**Exercises 5.1**

**5.1**  (a)  probability

(b)  probability

**5.3**  The notation {X=3} denotes an event that occurs when X=3, whereas P(X=3) denotes the probability that the event {X=3} will occur. Another way of thinking about the difference is that the first is an event and the second is a number between zero and one.

**5.5**  This table will resemble the probability distribution of the random variable.

**5.7**  (a)  X = 2, 3, 4, 5, 6, 7, 8

(b)  {X = 7}

(c)  The total number of shuttle missions is 96.  P(X = 4) = 2/96 = 0.021, so 2.1% of the shuttle crews consist of exactly four people.

(d)

| Size of Crew | Probability |
|---|---|
| 2 | 0.042 |
| 3 | 0.010 |
| 4 | 0.021 |
| 5 | 0.375 |
| 6 | 0.188 |
| 7 | 0.344 |
| 8 | 0.021 |

(e)

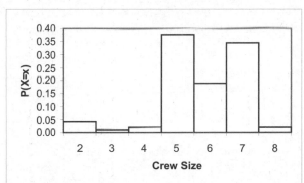

**5.9**  (a)  {Y ≥ 1}    (b)   {Y = 2}       (c)    {1 ≤ Y ≤ 3}

(d)  {Y = 1} or {Y = 3} or {Y = 5}

(e)  P(Y ≥ 1) = P(Y = 1) + P(Y = 2) + P(Y = 3) + P(Y = 4) = P(Y = 5)
= 0.376 + 0.371 + 0.167 + 0.061 + 0.016 = 0.991

(f)  P(Y = 2) = 0.371

(g)  P(1 ≤ y ≤ 3) = P(Y = 1) + P(Y = 2) + P(Y = 3) = 0.376 + 0.371 + 0.167
= 0.914

(h)  P(Y = 1 or 2 or 3) = P(Y = 1) + P(Y = 3) + P(Y = 5) = 0.376 + 0.167 + 0.016 = 0.559

**5.11**  (a)  Y = 2,3,4,5,6,7,8,9,10,11,12

(b)  {Y = 7}

(c)  P(Y = 7) = 6/36 = 1/6 = 0.167

(d)

| Y | P(Y) |
|----|------|
| 2 | 1/36 |
| 3 | 2/36 |
| 4 | 3/36 |
| 5 | 4/36 |
| 6 | 5/36 |
| 7 | 6/36 |
| 8 | 5/36 |
| 9 | 4/36 |
| 10 | 3/36 |
| 11 | 2/36 |
| 12 | 1/36 |

(e)

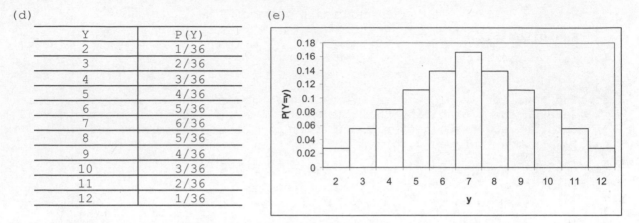

**5.13**  (a)  The area of the square is $6^2$ = 36 square feet.  The area of the bull's eye is $\pi(1)^2$ = 3.1416 square feet.  The area of the ring is the area of a circle with radius 2 less the area of the bull's eye, or $\pi(2)^2 - \pi(1)^2$ = 12.5664 – 3.1416 = 9.4248.  The remaining area of the square outside the ring has area 36 – 12.5664 = 23.4336.  Thus, P(S = 10) = 3.1416/36 = 0.087; P(S = 5) = 9.4248/36 = 0.262; and P(S = 0) = 23.3226/36 = 0.651.  These are shown in the table below.

| S | P(S = s) |
|----|----------|
| 0 | 0.651 |
| 5 | 0.262 |
| 10 | 0.087 |

Assuming that all points in the six foot square are equally likely, the archer has almost a 2/3 chance of scoring 0 points and about a 1/12 chance of scoring 10 points.

(b)  P(S = 5) = 0.262

P(S > 0) = P(S = 5 or 10) = 0.262 + 0.087 = 0.349

P(S $\leq$ 7) = P(S = 0 or 5) = 0.651 + 0.262 = 0.913

P(5 < S $\leq$ 15) = P(S = 10) = 0.087

P(S < 15) = P(S = 0 or 5 or 10) = 1; This event always occurs.

P(S < 0) = 0.000;  This event cannot occur.

**5.15**  P(Z $\leq$ 1.96) + P(Z > 1.96) = 1 .

Since P(Z > 1.96) = 0.025,

P(Z $\leq$ 1.96) + 0.025 = 1, or

P(Z $\leq$ 1.96) = 1 – 0.025 = 0.975.

**5.17**  (a)  P(X $\leq$ c) + P(X > c) = 1.

Since P(X > c) = $\alpha$,

P(X $\leq$ c) + $\alpha$ = 1, or

P(X $\leq$ c) = 1 – $\alpha$.

(b)  P(Y < –c) + P(–c $\leq$ Y $\leq$ c) + P(Y > c) = 1.

Since  P(Y < –c) = P(Y > c) = $\alpha/2$,

$\alpha/2$ + P(–c $\leq$ Y $\leq$ c) + $\alpha/2$ = 1, or

P(–c $\leq$ Y $\leq$ c) = 1 – $\alpha/2$ – $\alpha/2$ = 1 – $\alpha$.

(c)   P(T < -c) + P(-c $\leq$ T $\leq$ c) + P(T > c) = 1.

Since   P(-c $\leq$ T $\leq$ c) = 1 - $\alpha$,

P(T < -c) + (1 - $\alpha$) + P(T > c) = 1, or

P(T < -c) + P(T > c) = 1 - 1 + $\alpha$ = $\alpha$.

Since   P(T < -c) = P(T > c),

2· [P(T > c)] = $\alpha$      or      P(T > c) = $\alpha$/2.

**Exercises 5.2**

**5.19**   The mean of a discrete random variable generalizes the concept of a population mean.

**5.21**   The required calculations are

| x | P(X=x) | xP(X=x) | $x^2$ | $x^2$ P(X=x) |
|---|--------|---------|-------|--------------|
| 2 | 0.042 | 0.084 | 4.000 | 0.168 |
| 3 | 0.010 | 0.030 | 9.000 | 0.090 |
| 4 | 0.021 | 0.084 | 16.000 | 0.336 |
| 5 | 0.375 | 1.875 | 25.000 | 9.375 |
| 6 | 0.188 | 1.128 | 36.000 | 6.768 |
| 7 | 0.344 | 2.408 | 49.000 | 16.856 |
| 8 | 0.021 | 0.168 | 64.000 | 1.344 |
|   |        | 5.777 |       | 34.937 |

(a)   $\mu_x = \sum xP(X=x) = 5.777$.   The average number of persons in a shuttle crew is about 5.8.

(b)   $\sigma = \sqrt{\sum x^2 P(X=x) - \mu_x^2} = \sqrt{34.937 - 5.777^2} = 1.250$

(c)

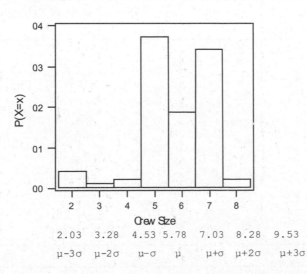

| | | | | | | |
|---|---|---|---|---|---|---|
| 2.03 | 3.28 | 4.53 | 5.78 | 7.03 | 8.28 | 9.53 |
| $\mu-3\sigma$ | $\mu-2\sigma$ | $\mu-\sigma$ | $\mu$ | $\mu+\sigma$ | $\mu+2\sigma$ | $\mu+3\sigma$ |

**5.23**  The required calculations are

| y | P(Y=y) | yP(Y=y) | $y^2$ | $y^2$P(Y=y) |
|---|--------|---------|-------|-------------|
| 0 | 0.009 | 0.000 | 0 | 0.000 |
| 1 | 0.376 | 0.376 | 1 | 0.376 |
| 2 | 0.371 | 0.742 | 4 | 1.484 |
| 3 | 0.167 | 0.501 | 9 | 1.503 |
| 4 | 0.061 | 0.244 | 16 | 0.976 |
| 5 | 0.016 | 0.080 | 25 | 0.400 |
|   | 1 | 1.943 |   | 4.739 |

$\mu_y = \sum yP(Y=y) = 1.943$.  The mean number of color TVs owned by households with annual incomes between \$15,000 and \$29,999 is 1.943.

(b)  $\sigma = \sqrt{\sum y^2 P(Y = y) - \mu_y^2} = \sqrt{4.739 - 1.943^2} = 0.982$.  Roughly speaking, the number of color TVs per one of these households averages about 1 away from the mean of 1.943.

(c)

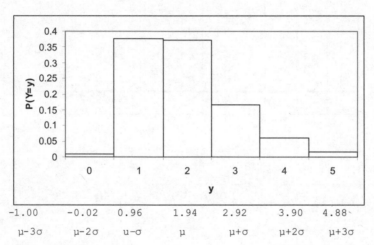

| -1.00 | -0.02 | 0.96 | 1.94 | 2.92 | 3.90 | 4.88 |
|-------|-------|------|------|------|------|------|
| μ-3σ | μ-2σ | u-σ | μ | μ+σ | μ+2σ | μ+3σ |

**5.25**  The required calculations are

| $y$ | $P(Y = y)$ | $yP(Y = y)$ | $y - \mu_y$ | $(y - \mu_y)^2$ | $(y - \mu_y)^2 P(Y = y)$ | $y^2$ | $y^2 P(Y = y)$ |
|-----|-----------|-------------|-------------|------------------|----------------------------|-------|-----------------|
| 2 | 1/36 | 2/36 | -5 | 25 | 25/36 | 4 | 4/36 |
| 3 | 2/36 | 6/36 | -4 | 16 | 32/36 | 9 | 18/36 |
| 4 | 3/36 | 12/36 | -3 | 9 | 27/36 | 16 | 48/36 |
| 5 | 4/36 | 20/36 | -2 | 4 | 16/36 | 25 | 100/36 |
| 6 | 5/36 | 30/36 | -1 | 1 | 5/36 | 36 | 180/36 |
| 7 | 6/36 | 42/36 | 0 | 0 | 0/36 | 49 | 294/36 |
| 8 | 5/36 | 40/36 | 1 | 1 | 5/36 | 64 | 320/36 |
| 9 | 4/36 | 36/36 | 2 | 4 | 16/36 | 81 | 324/36 |
| 10 | 3/36 | 30/36 | 3 | 9 | 27/36 | 100 | 300/36 |
| 11 | 2/36 | 22/36 | 4 | 16 | 32/36 | 121 | 242/36 |
| 12 | 1/36 | 12/36 | 5 | 25 | 25/36 | 144 | 144/36 |
|   |   | 252/36=7 |   |   | 210/36 |   | 1,974/36 |

(a)   $\mu_y = \sum yP(Y=y) = 252/36 = 7$.   When rolling two balanced dice, the mean of their sum is 7.

(b)   $\sigma = \sqrt{\sum (y-\mu_y)^2 P(Y=y)} = \sqrt{210/36} = 2.415$

or

$$\sigma = \sqrt{\sum y^2 P(Y=y) - \mu_y^2} = \sqrt{\dfrac{1974}{36} - 7^2} = 2.415$$

(c)

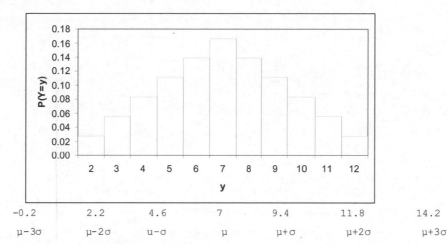

| -0.2 | 2.2 | 4.6 | 7 | 9.4 | 11.8 | 14.2 |
|------|-----|-----|---|-----|------|------|
| $\mu-3\sigma$ | $\mu-2\sigma$ | $u-\sigma$ | $\mu$ | $\mu+\sigma$ | $\mu+2\sigma$ | $\mu+3\sigma$ |

**5.27**   (a)   The necessary calculations are

| s | P(S=s) | sP(S=s) | $s^2$ | $s^2$P(S=s) |
|---|--------|---------|-------|-------------|
| 0 | 0.651 | 0.000 | 0 | 0.000 |
| 5 | 0.262 | 1.310 | 25 | 6.550 |
| 10 | 0.087 | 0.870 | 100 | 8.700 |
|   |       | 2.180 |   | 15.250 |

On average, the archer will score 2.180 points per arrow shot.

(b)   $\sigma = \sqrt{\sum s^2 P(S=s) - \mu_s^2} = \sqrt{15.250 - 2.180^2} = 3.24$.   Roughly speaking, on average, the number of points on a shot will differ from the mean by 3.24 points.

**5.29**   (a)

$$P(X=1) = \frac{18}{38} = 0.474$$

$$P(X=-1) = \frac{20}{38} = 0.526; \quad 0.474 + 0.526 = 1.000$$

(b)   $\mu = \sum x \cdot P(X=x) = (1)(0.474) + (-1)(0.526) = -0.052$

(c)   On the average, you will lose 5.2¢ per play.

(d)   If you bet $1 on red 100 times, you can expect to lose $100 \cdot 0.052 =$ $5.20.   If you bet $1 on red 1000 times, you can expect to lose $52.00.

(e)   Roulette is not a profitable game for a person to play.   Parts (c) and

(d) demonstrate that, no matter how much you play, you can expect to lose.  Also, part (a) shows that a higher probability is associated with losing rather than winning.

**5.31**

| w | P(W=w) | w·P(w) | $w^2$ | $w^2$·P(W=w) |
|---|---|---|---|---|
| 0 | 0.80 | 0.00 | 0 | 0.00 |
| 1 | 0.15 | 0.15 | 1 | 0.15 |
| 2 | 0.05 | 0.10 | 4 | 0.20 |
|   |   | 0.25 |   | 0.35 |

(a)

$$\mu_w = \sum wP(W=w) = 0.25$$

$$\sigma_w = \sqrt{\sum w^2 P(W=w) - \mu_w^2} = \sqrt{0.35 - 0.25^2} = \sqrt{0.2875} = 0.536$$

(b)  On the average, there are 0.25 breakdowns per day.

(c)  Assuming 250 workdays per year, the number of breakdowns expected per year is (0.25)(250) = 62.5.

**5.33**  (a)  The necessary calculations are

| y | P(Y=y) | yP(Y=y) | $y^2$ | $y^2$P(Y=y) |
|---|---|---|---|---|
| 0 | 0.424 | 0.000 | 0 | 0.000 |
| 1 | 0.161 | 0.161 | 1 | 0.161 |
| 2 | 0.134 | 0.268 | 4 | 0.536 |
| 3 | 0.111 | 0.333 | 9 | 0.999 |
| 4 | 0.093 | 0.372 | 16 | 1.488 |
| 5 | 0.077 | 0.385 | 25 | 1.925 |
|   |   | 1.519 |   | 5.109 |

The mean of Y, the mean number of customers waiting is 1.519.  On the average, there are about 1.5 customers waiting at any given time.

(b)  In a large number of independent observations, 1.519 customers will be waiting, on average.

(c)  We will use Minitab to simulate these observations.  Enter the values of Y in column C1 and their probabilities in column C2.  Then select

**Calc ▶ Random data ▶ Discrete**, enter 500 in the **Generate Rows of data** text box, enter Customers in the **Store in Column(s)** text box, then enter C1 in the **Values in** text box and C2 in the **Probabilities in** text box and click **OK**.  This will produce 500 random values in a column

named Customeers.  Now select **Stat ▶ Basic Statistics ▶ Display Descriptive Statistics**, enter Customers in the **Variables** text box and click **OK**.  The result appears in the Sessions Window.  Our result is

```
Variable    N  N*    Mean  SE Mean   StDev     Minimum                Q1  Median
Customers  500  0  1.5800   0.0767  1.7151  0.000000000  0.000000000  1.0000


Variable       Q3  Maximum
Customers  3.0000   5.0000
```

(d)  The mean for our observations was 1.58, a little bit higher than the exact mean of 1.519.  Your data and mean will likely be different.

(e)  Part (d) illustrates that simulations yield sample means that are close
to the population mean.

**5.35** (a)

|   |   | y | | | |
|---|---|---|---|---|---|
|   |   | 0 | 1 | 2 | P(X=x) |
|   | 0 | 0.6400 | 0.1200 | 0.0400 | 0.8000 |
| x | 1 | 0.1200 | 0.0225 | 0.0075 | 0.1500 |
|   | 2 | 0.0400 | 0.0075 | 0.0025 | 0.0500 |
|   | P(Y=y) | 0.8000 | 0.4500 | 0.0500 | 1.0000 |

(b)

| x + y | P(x+y) | (x+y)·P(x+y) | (x + y)² | (x+y)²·P(x+y) |
|---|---|---|---|---|
| 0 | 0.6400 | 0.000 | 0 | 0.000 |
| 1 | 0.2400 | 0.240 | 1 | 0.240 |
| 2 | 0.1025 | 0.205 | 4 | 0.410 |
| 3 | 0.0150 | 0.045 | 9 | 0.135 |
| 4 | 0.0025 | 0.010 | 16 | 0.040 |
|   |   | 0.500 |   | 0.825 |

Columns 1 and 2 of the preceding table comprise the probability
distribution of the random variable X + Y.  We have used P(x+y) for
P(X+Y=x+y) in the table headings.

Thus the table to be completed is

| u | 0 | 1 | 2 | 3 | 4 |
|---|---|---|---|---|---|
| P(X + Y) = u | 0.6400 | 0.2400 | 0.1025 | 0.0150 | 0.0025 |

(c)  Column 3 of the table above provides the calculations for the mean:
$\mu_{x+y}$ = 0.500.  Columns 4 and 5 provide the calculations for the
variance:

$$\sigma^2_{x+y} = \sum (x+y)^2 P(X+Y=x+y) - \mu^2_{x+y} = 0.825 - (0.50)^2 = 0.5750$$

(d)  Using the result from Exercise 5.31 and the information about X and Y

in this part of the exercise, we have

$$\mu_w = \mu_x = \mu_y = 0.25 \ ; \quad \sigma^2_w = \sigma^2_x = \sigma^2_y = 0.2875 \ .$$

Now,  $\mu_{x+y} = \mu_x + \mu_y = 0.25 + -.25 = 0.50$

and  $\sigma^2_{x+y} = \sigma^2_x + \sigma^2_y = 0.2875 + 0.2875 = 0.5750$

Notice that these two results match those obtained in part (c).

(e)  The mean of the sum of two random variables equals the sum of their
means.  The variance of the sum of two *independent* random variables
equals the sum of their variances.

### Exercises 5.3

**5.37**  (1) Randomly selected pieces of identical rope are tested by subjecting each to a 1000 pound force. Each rope either breaks or it doesn't. (2) People are selected at random by ticket numbers at a large convention and their gender is noted. Each is either male or female.

**5.39**  $3! = 3 \cdot 2 \cdot 1 = 6$;  $7! = 7 \cdot 6 \cdot 5 \cdot 4 \cdot 3 \cdot 2 \cdot 1 = 5,040$;

$8! = 8 \cdot 7 \cdot 6 \cdot 5 \cdot 4 \cdot 3 \cdot 2 \cdot 1 = 40,320$;  $9! = 9 \cdot 8 \cdot 7 \cdot 6 \cdot 5 \cdot 4 \cdot 3 \cdot 2 \cdot 1 = 362,880$

**5.41**  (a) $\dbinom{5}{2} = \dfrac{5!}{2!(5-2)!} = \dfrac{5 \cdot 4 \cdot 3 \cdot 2 \cdot 1}{2 \cdot 1 \cdot 3 \cdot 2 \cdot 1} = 10$

(b) $\dbinom{7}{4} = \dfrac{7!}{4!(7-4)!} = \dfrac{7 \cdot 6 \cdot 5 \cdot 4 \cdot 3 \cdot 2 \cdot 1}{4 \cdot 3 \cdot 2 \cdot 1 \cdot 3 \cdot 2 \cdot 1} = 35$

(c) $\dbinom{10}{3} = \dfrac{10!}{3!(10-3)!} = \dfrac{10 \cdot 9 \cdot 8 \cdot 7 \cdot 6 \cdot 5 \cdot 4 \cdot 3 \cdot 2 \cdot 1}{3 \cdot 2 \cdot 1 \cdot 7 \cdot 6 \cdot 5 \cdot 4 \cdot 3 \cdot 2 \cdot 1} = 120$

(d) $\dbinom{12}{5} = \dfrac{12!}{5!(12-5)!} = \dfrac{12 \cdot 11 \cdot 10 \cdot 9 \cdot 8 \cdot 7 \cdot 6 \cdot 5 \cdot 4 \cdot 3 \cdot 2 \cdot 1}{5 \cdot 4 \cdot 3 \cdot 2 \cdot 1 \cdot 7 \cdot 6 \cdot 5 \cdot 4 \cdot 3 \cdot 2 \cdot 1} = 792$

**5.43**  (a) $\dbinom{4}{1} = \dfrac{4!}{1!(4-1)!} = \dfrac{4 \cdot 3 \cdot 2 \cdot 1}{1 \cdot 3 \cdot 2 \cdot 1} = 4$

(b) $\dbinom{6}{2} = \dfrac{6!}{2!(6-2)!} = \dfrac{6 \cdot 5 \cdot 4 \cdot 3 \cdot 2 \cdot 1}{2 \cdot 1 \cdot 4 \cdot 3 \cdot 2 \cdot 1} = 15$

(c) $\dbinom{8}{3} = \dfrac{8!}{3!(8-3)!} = \dfrac{8 \cdot 7 \cdot 6 \cdot 5 \cdot 4 \cdot 3 \cdot 2 \cdot 1}{3 \cdot 2 \cdot 1 \cdot 5 \cdot 4 \cdot 3 \cdot 2 \cdot 1} = 56$

(d) $\dbinom{9}{6} = \dfrac{9!}{6!(9-6)!} = \dfrac{9 \cdot 8 \cdot 7 \cdot 6 \cdot 5 \cdot 4 \cdot 3 \cdot 2 \cdot 1}{6 \cdot 5 \cdot 4 \cdot 3 \cdot 2 \cdot 1 \cdot 3 \cdot 2 \cdot 1} = 84$

**5.45**  (a) Each trial consists of observing whether the child is cured of the pinworm infestation by pyrantel pamoate. Each is either cured or not cured. The results are independent assuming that the children are not in the same families. Since a success $s$ is that the child is cured, the success probability p is 0.90.

(b)

| Outcome | Probability |
|---------|-------------|
| sss | $(0.90)(0.90)(0.90) = 0.729$ |
| ssf | $(0.90)(0.90)(0.10) = 0.081$ |
| sfs | $(0.90)(0.10)(0.90) = 0.081$ |
| sff | $(0.90)(0.10)(0.10) = 0.009$ |
| fss | $(0.10)(0.90)(0.90) = 0.081$ |
| fsf | $(0.10)(0.90)(0.10) = 0.009$ |
| ffs | $(0.10)(0.10)(0.90) = 0.009$ |
| fff | $(0.10)(0.10)(0.10) = 0.001$ |

(c)

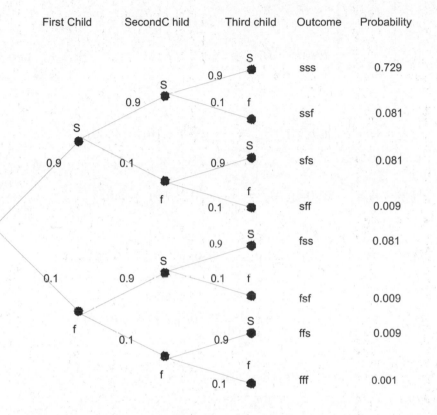

| First Child | SecondC hild | Third child | Outcome | Probability |

(d)   The outcomes in which exactly two of the three children are cured are ssf, sfs, and fss.

(e)   Each outcome in part (d) has the probability 0.081.  This probability is the same for each outcome because each probability is obtained by multiplying two success probabilities of 0.90 and one failure probability of 0.10.

(f)   P(exactly two children are cured) = P(ssf) + P(sfs) + P(fss)

$$= 0.081 + 0.081 + 0.081 = 0.243$$

(g)   P(exactly one child is cured) = P(sff) + P(fsf) + P(ffs)

$$= 0.009 + 0.009 + 0.009 = 0.027$$

P(exactly three children are cured) = P(sss) = 0.729, and

P(exactly zero children are cured) = P(fff) = 0.001.

Thus the complete probability distribution of X, the number of children out of three that are cured is

| X | 0 | 1 | 2 | 3 |
|---|---|---|---|---|
| P(X=x) | 0.001 | 0.027 | 0.243 | 0.729 |

**5.47**   Step 1: A success is that a treated child is cured.

Step 2: The success probability is p = 0.90.

Step 3: The number of trials is n = 3.

Step 4: The formula for x successes is

$$P(X = x) = \binom{3}{x}(.90)^x(.10)^{3-x}.$$   For x = 0, 1, 2, and 3, the probabilities

are

$$P(X = 0) = \binom{3}{0}(.90)^0(.10)^3 = \frac{3!}{0!\,3!}(.90)^0(.10)^3 = 0.001$$

$$P(X = 1) = \binom{3}{1}(.90)^1(.10)^2 = \frac{3!}{1!\,2!}(.90)^1(.10)^2 = 0.027$$

$$P(X = 2) = \binom{3}{2}(.90)^2(.10)^1 = \frac{3!}{2!\,1!}(.90)^2(.10)^1 = 0.243$$

$$P(X = 3) = \binom{3}{3}(.90)^3(.10)^0 = \frac{3!}{0!\,3!}(.90)^3(.10)^0 = 0.729$$

**5.49**   (a)   Since the graph is symmetric, p = 0.5.

(b)   Since the graph is right skewed, p is less than 0.5.

**5.51**   The calculations required to answer all parts of this exercise are

$$P(0) = \binom{5}{0}\cdot(.67)^0(.33)^5 = 0.004 \quad P(3) = \binom{5}{3}\cdot(.67)^3(.33)^2 = 0.328$$

$$P(1) = \binom{5}{1}\cdot(.67)^1(.33)^4 = 0.040 \quad P(4) = \binom{5}{4}\cdot(.67)^4(.33)^1 = 0.332$$

$$P(2) = \binom{5}{2}\cdot(.67)^2(.33)^3 = 0.161 \quad P(5) = \binom{5}{5}\cdot(.67)^5(.33)^0 = 0.135$$

(a)   P(2) = 0.161                    (b)   P(4) = 0.332

(c)   P(X ≥ 4) = P(4) + P(5) = 0.332 + 0.135 = 0.467 (actually 0.468 if rounding is done after the addition).

(d)   P(2 ≤ X ≤ 4) = P(2) + P(3) + P(4) = 0.161 + 0.328 + 0.332 = 0.821

(e)

| x | P(X=x) |
|---|--------|
| 0 | 0.004 |
| 1 | 0.040 |
| 2 | 0.161 |
| 3 | 0.328 |
| 4 | 0.332 |
| 5 | 0.135 |

(f)   Left skewed

(g)

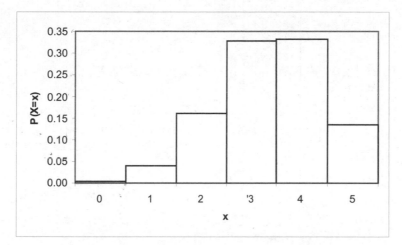

(h)

| X | P(X=x) | xP(X=x) | $x^2$ | $x^2$P(X=x) |
|---|--------|---------|-------|-------------|
| 0 | 0.004 | 0.000 | 0 | 0.000 |
| 1 | 0.040 | 0.040 | 1 | 0.040 |
| 2 | 0.161 | 0.322 | 4 | 0.644 |
| 3 | 0.328 | 0.984 | 9 | 2.952 |
| 4 | 0.332 | 1.328 | 16 | 5.312 |
| 5 | 0.135 | 0.675 | 25 | 3.375 |
|   |   | 3.349 |   | 12.323 |

$\mu$ = 3.349

$\sigma^2$ = 12.323 - 3.349$^2$ = 1.107;  $\sigma = \sqrt{1.107} = 1.052$

(i)   $\mu$ = np = 5(0.67) = 3.35

$\sigma^2$ = np(1-p) = 5(0.67)(0.33) = 1.1055

$\sigma = \sqrt{1.1055} = 1.051$

(j)   Out of any five races, the average number of favorites that finish in the money is 3.35.

**5.53**   (a)   n = 8; p = 0.40

$$P(\text{Exactly3}) = \binom{8}{3}(0.4)^3(0.6)^5 = 0.0241$$

$$P(\text{3or more}) = 1 - P(\text{0or 1or 2}) = 1 - \binom{8}{0}(0.4)^0(0.6)^8 - \binom{8}{1}(0.4)(0.6)^7 - \binom{8}{2}(0.4)^2(0.6)^6$$

$$= 1. - 0.0168 - 0.0896 - 0.2090 = 0.6846$$

P(atmost 3)= P(0)+ P(1) +P(2) +P(3)

$$= \binom{8}{0}(0.4)^0(0.6)^8 + \binom{8}{1}(0.4)(0.6)^7 + \binom{8}{2}(0.4)^2(0.6)^6 + \binom{8}{3}(0.4)^3(0.6)^5$$

$$= 0.0168 + 0.0896 + 0.2090 + 0.2787 = 0.5941$$

(b)
$$P(2 \le Y \le 4) = P(2) + P(3) + P(4)$$

$$= \binom{8}{2}(0.4)^2(0.6)^6 + \binom{8}{3}(0.4)^3(0.6)^5 + \binom{8}{4}(0.4)^4(0.6)^4 = 0.2092 + 0.2787 + 0.2322 = 0.7201$$

(If rounding is done after the addition, you will get 0.7200.)

(c) The mean of Y is $\mu$ = np = 8(0.4) = 3.2. On average, of eight traffic fatalities, 3.2 will involve an intoxicated or alcohol-impaired driver or non-occupant.

(d) $\sigma^2$ =np(1 - p) = 8(0.4)(0.6) = 19.2, so $\sigma$ = 4.3418.

**5.55** (a) n = 9, p = 0.35, so

$$P(\text{Exactly5}) = \binom{9}{5}(0.35)^5(0.65)^4 = 0.1181$$

$$P(\text{Atleast 5}) = \binom{9}{5}(0.35)^5(0.65)^4 + \binom{9}{6}(0.35)^6(0.65)^3 + \binom{9}{7}(0.35)^7(0.65)^2$$

$$+ \binom{9}{8}(0.35)^8(0.65)^1 + \binom{9}{9}(0.35)^9(0.65)^1$$

$$= 0.1181 + 0.0424 + 0.0098 + 0.0013 + 0.0001 = 0.1717$$

$$P(At \text{ most5}) = \binom{9}{0}(0.35)^0(0.65)^9 + \binom{9}{1}(0.35)(0.65)^8 + \binom{9}{2}(0.35)^2(0.65)^7$$

$$+ \binom{9}{3}(0.35)^3(0.65)^6 + \binom{9}{4}(0.35)^4(0.65)^5 \binom{9}{5}(0.35)^5(0.65)^4$$

$$= 0.0207 + 0.1004 + 0.2162 + 0.2716 + 0.2194 + 0.1181 = 0.9464$$

$$P(\text{Atleast 1correct}) = 1 - P(\text{0Correct}) = 1 - \binom{9}{0}(0.35)^0(0.65)^9 = 1 - 0.0207 = 0.9793$$

(b) P(Atmost 1correct) =P(0) +P(1) = $\binom{9}{0}(0.35)^0(0.65)^9 + \binom{9}{1}(0.35)(0.65)^8$

$$= 0.0207 + 0.1004 = 0.1211$$

(c)   $P(6 \text{ or } 7 \text{ or } 8) = \binom{9}{6}(0.35)^6(0.65)^3 + \binom{9}{7}(0.35)^7(0.65)^2 + \binom{9}{8}(0.35)^8(0.65)^1$

$= 0.0424 + 0.0098 + 0.0013 = 0.0535$

(d)   The formula is $P(X = x) = \binom{9}{x}(0.35)^x(0.65)^{9-x}$ for x = 0, 1, 2, ..., 9

Applying this formula for each value of x results in the table below.

| x | P(X=x) |
|---|--------|
| 0 | 0.0207 |
| 1 | 0.1004 |
| 2 | 0.2162 |
| 3 | 0.2716 |
| 4 | 0.2194 |
| 5 | 0.1181 |
| 6 | 0.0424 |
| 7 | 0.0098 |
| 8 | 0.0013 |
| 9 | 0.0001 |

(e)   The sampling was actually done without replacement, so the trials are not independent and the success probability changes very slightly from trial to trial.  The exact probability distribution is called a hypogeometric distribution.

**5.57** (a)   n = 4 and p = 0.165, so the formula for the probability distribution is

$$P(X = x) = \binom{4}{x}(0.165)^x(0.835)^{4-x} \text{ for x} = 0, 1, 2, 3, 4$$

Applying this formula for each value of x results in the table below

| x | P(X=x) |
|---|--------|
| 0 | 0.4059 |
| 1 | 0.4011 |
| 2 | 0.1585 |
| 3 | 0.0313 |
| 4 | 0.0031 |

(b)   The mean $\mu = np = 4(0.165) = 0.66$.  Thus, on the average, we would expect 0.66 people out of four under the age of 65 to have no health insurance.

(c)   Yes.  If, in fact, 16.5% of people under 65 have no health insurance, then finding that 3 or more out of 4 had no health insurance would mean that an event with probability 0.0313 + 0.0031 = 0.0344 had just occurred.  This is a fairly rare event if p is really 0.165, but would be less rare if p were greater than 0.165.  Rather than believe that we had just observed a rare event, we would be inclined to conclude that our assumption (p = 0.165) was incorrect and that p has increased from the 16.5% rate of 2002.

(d)   No.  If, in fact, 16.5% of people under 65 have no health insurance, then finding that two or more out of four had no health insurance would mean that an event with probability 0.1585 + 0.0313 + 0.0031 = 0.1929 had just occurred.  This is a fairly common event if p is really 0.165, so we would not be inclined to take 2 out of 4 as evidence that the uninsured rate has increased from the 16.5% rate in 2002.

**5.59**  (a)  n = 4, p = 18/38 = 0.4737

$$P(2) = \binom{4}{2}(0.4737)^2 (0.5263)^2 = 0.3729$$

(b)  $$P(\text{Atleast } 1) = 1 - P(0) = 1 - \binom{4}{0}(0.4737)^0 (0.5263)^4 = 1 - 0.0767 = 0.9233$$

**5.61**  (a)  P(Both parents pass hemoglobin S to an offspring) = (0.5)(0.5) = 0.25

(b)  n = 5, p = 0.25; X = number of children with sickle cell anemia.

$$P(X \geq 1) = 1 - P(X = 0) = 1 - \binom{5}{0}(0.25)^0 (0.75)^5 = 1 - 0.2373 = 0.7627$$

(c)

| x | P(X=x) |
|---|--------|
| 0 | 0.2373 |
| 1 | 0.3955 |
| 2 | 0.2637 |
| 3 | 0.0879 |
| 4 | 0.0146 |
| 5 | 0.0010 |

(d)

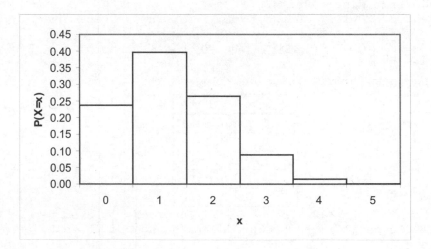

(e)  On average, they can expect μ = np = 5(.25) = 1.25 children to have sickle cell anemia.

**5.63**  The probability that a given party will show is 0.96.  If n parties make a reservation, the probability that all n parties will show is

$$P(n) = \binom{n}{n}(0.96)^n (0.04)^0 = (0.96)^n.$$  We want to find the largest n for which this

quantity is still 0.80 or larger.

| n | P(n) |
|---|------|
| 1 | 0.9600 |
| 2 | 0.9216 |
| 3 | 0.8847 |
| 4 | 0.8493 |
| 5 | 0.8154 |
| 6 | 0.7827 |

Thus, we see from the table that 5 is the largest value of n for which the owner can be at least 80% certain that all parties that make a reservation will show.

**5.65**  (a)  P(both are male) = (17/40)(17/40) = 289/1600 = 0.180625 if sampling is done with replacement.

(b)  P(both are male) = (17/40)(16/39) = 272/1560 = 0.174359 if sampling is done without replacement.

(c)  The probability without replacement is slightly smaller than the probability with replacement.  There is about a 3% difference in the probabilities.

(d)  P(both are male) = (170/400)(170/400) = 28900/160000 = 0.180625 if sampling is done with replacement.  P(both are male) = (170/400)(169/399) = 28730/159600 = 0.180013 if sampling is done without replacement.  The probability without replacement is just barely smaller than the probability with replacement.  There is only about a 0.3% difference in the probabilities.

(e)  In the second case, there is less difference between sampling without and with replacement.  This is because removing the first male from the population makes less difference in the probability of the second person being a male when there are 400 people in the class than when there are only 40 people in the class.  In other words, 169/399 is closer to 170/400 than 16/39 is to 17/40.

**5.67**  (a)  p = 0.0290647 or 0.029 to three decimals; $P(X = x) = 0.029(0.971)^{x-1}$

This formula gives the probability that a person will first win a prize in week x.

(b)  $P(3) = 0.029(0.971)^2 = 0.0273$

$P(X \le 3) = P(1) + P(2) + P(3) = 0.0290 + 0.0282 + 0.0273 = 0.0845$

$P(X \ge 3) = 1 - P(X \le 2) = 1 - [0.0290 + 0.0282] = 1 - 0.0572$

$= 0.9428$

(c)  On the average, it will take $\mu = 1/p = 1/0.029 = 34.5$ weeks until you win a prize.

## Exercises 5.4

**5.69**  Two uses of Poisson distributions would be (1) to model the frequency with which a specified event occurs during a particular period of time, and (2) to approximate binomial distribution probabilities when n is at least 100 and np is no more than 10..

**5.71** (a)  $P(X=5)=e^{-4.7}\cdot\dfrac{4.7^5}{5!}=0.1738$

(b)  $P(X<2)=P(0)+P(1)=e^{-4.7}\cdot\dfrac{4.7^0}{0!}+e^{-4.7}\cdot\dfrac{4.7^1}{1!}=0.0091+0.0427=0.0518$

(c)  $P(X\ge3)=1-[P(0)+P(1)+P(2)]==1-\left[e^{-4.7}\cdot\dfrac{4.7^0}{0!}+e^{-4.7}\cdot\dfrac{4.7^1}{1!}+e^{-4.7}\cdot\dfrac{4.7^2}{2!}\right]$

$=1-[0.0091+0.0427+0.1005]=0.8477$

(d)  $\mu=\lambda=4.7$

(e)  $\sigma=\sqrt{\lambda}=\sqrt{4.7}=2.168$

**5.73** (a)  $P(Y=4)=e^{-3.87}\cdot\dfrac{3.87^4}{4!}=0.1949$

(b)  $P(\text{Atmost }1)=P(Y=0\text{ or }1)=e^{-3.87}\cdot\dfrac{3.87^0}{0!}+e^{-3.87}\cdot\dfrac{3.87^1}{1!}=0.0209+0.0807=0.1016$

(c)

$P(2\le Y\le5)=e^{-3.87}\cdot\dfrac{3.87^2}{2!}+e^{-3.87}\cdot\dfrac{3.87^3}{3!}+e^{-3.87}\cdot\dfrac{3.87^4}{4!}+e^{-3.87}\cdot\dfrac{3.87^5}{5!}$

$=0.1561+0.2015+0.1949+1509=0.7034$

(d)                                                (e)

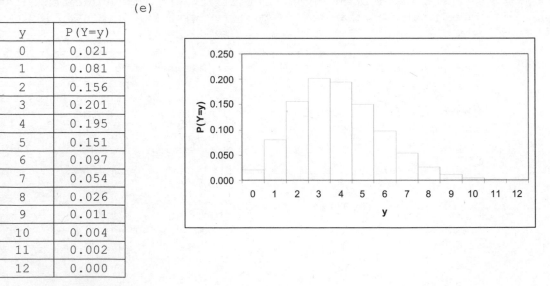

| y | P(Y=y) |
|---|--------|
| 0 | 0.021 |
| 1 | 0.081 |
| 2 | 0.156 |
| 3 | 0.201 |
| 4 | 0.195 |
| 5 | 0.151 |
| 6 | 0.097 |
| 7 | 0.054 |
| 8 | 0.026 |
| 9 | 0.011 |
| 10 | 0.004 |
| 11 | 0.002 |
| 12 | 0.000 |

(f)  On average, the number of alpha particles reaching the screen during an 8-minute interval is $\mu=\lambda=3.87$.

**5.75** (a)  $P(X=0)=e^{-0.7}\cdot\dfrac{0.7^0}{0!}=0.4966$

(b)
$$P(\text{Atmost } 2)= P(X = 0\text{ or } 1\text{ or } 2) = e^{-0.7} \cdot \frac{0.7^0}{0!} + e^{-0.7} \cdot \frac{0.7^1}{1!} + e^{-0.7} \cdot \frac{0.7^2}{2!}$$
$$= 0.4966 + 0.3476 + 0.1217 = 0.9659$$

(c)
$$P(1 \le X \le 3)= P(X = 1\text{ or } 2\text{ or } 3) = e^{-0.7} \cdot \frac{0.7^1}{1!} + e^{-0.7} \cdot \frac{0.7^2}{2!} + e^{-0.7} \cdot \frac{0.7^3}{3!}$$
$$= 0.3476 + 0.1217 + 0.0284 = 0.4977$$

(d) The mean number of wars that begin during a given year is the mean of X, $\mu = \lambda = 0.7$.

(e) $\sigma_x = \sqrt{\lambda} = \sqrt{0.7} = 0.8367$

**5.77** (a) The mean number of cherries per pie is 42/35 = 1.2

(b)

| x | f | f/n |
|---|---|---|
| 0 | 11 | 0.3143 |
| 1 | 12 | 0.3429 |
| 2 | 8 | 0.2286 |
| 3 | 2 | 0.0571 |
| 4 | 2 | 0.0571 |
| Total | 35 | 1.0000 |

(c) Using $\lambda = 1.2$, we obtain

| x | P(X=x) |
|---|---|
| 0 | 0.301 |
| 1 | 0.361 |
| 2 | 0.217 |
| 3 | 0.087 |
| 4 | 0.026 |
| 5 | 0.006 |
| 6 | 0.001 |
| 7 | 0.000 |

(d) All of the actual relative frequencies are quite close to the Poisson probabilities using a mean of 1.2. They would be even closer if the counts for 3 and 4 cherries were 3 and 1 instead of 2 and 2, respectively. We conclude that the Poisson distribution with a mean of 1.2 does a very good job of modeling the distribution of the number of cherries in the pies.

**5.79** We use $\lambda = \mu = np = 500(1/146) = 3.425$. First, we note that $n \ge 100$ and $np \le 10$. Then

$$P(\text{Atmost } 3) = P(X = 0 \text{ or } 1 \text{ or } 2 \text{ or } 3) = e^{-3.425} \cdot \frac{3.425^0}{0!} + e^{-3.425} \cdot \frac{3.425^1}{1!}$$

$$+ e^{-3.425} \cdot \frac{3.425^2}{2!} + e^{-3.425} \cdot \frac{3.425^3}{3!}$$

$$= 0.0325 + 0.1115 + 0.1909 + 0.2180 = 0.5529$$

**5.81** (a)  We would expect $\mu$ = np = (10000)(1/1500) = 6.667 to have Fragile X Syndrome.

(b)  First, we note that n ≥ 100 and np ≤ 10.  Then we use the Poisson probability formula to obtain the probabilities for x = 0 through 10. The required calculations are

$$P(0) = \frac{e^{-6.67} 6.67^0}{0!} = 0.0013 \qquad P(1) = \frac{e^{-6.67} 6.67^1}{1!} = 0.0085$$

$$P(2) = \frac{e^{-6.67} 6.67^2}{2!} = 0.0282 \qquad P(3) = \frac{e^{-6.67} 6.67^3}{3!} = 0.0627$$

$$P(4) = \frac{e^{-6.67} 6.67^4}{4!} = 0.1046 \qquad P(5) = \frac{e^{-6.67} 6.67^5}{5!} = 0.1395$$

$$P(6) = \frac{e^{-6.67} 6.67^6}{6!} = 0.1551 \qquad P(7) = \frac{e^{-6.67} 6.67^7}{7!} = 0.1478$$

$$P(8) = \frac{e^{-6.67} 6.67^8}{8!} = 0.1232 \qquad P(9) = \frac{e^{-6.67} 6.67^9}{9!} = 0.0913$$

$$P(10) = \frac{e^{-6.67} 6.67^{10}}{10!} = 0.0609$$

The probabilities that more than 7 of the males have Fragile X Syndrome and that at most 10 of the males have Fragile X Syndrome are (from the preceding calculations)

$$P(X > 7) = 1 - P(X \le 7) = 1 - (P(0) + P(1) + \cdots + P(7))$$

$$= 1 - 0.0013 - 0.0085 - 0.0282 - 0.0627 - 0.1046 - 0.1395 - 0.1551 - 0.1478$$

$$= 1 - 0.6477 = 0.3523$$

$$P(X \le 10) = 0.0013 + 0.0085 + \cdots + 0.0609 = 0.9231$$

**5.83** $\mu = np = (100,000,000) \cdot \dfrac{1}{30,000,000} = \dfrac{10}{3} = 3.33$

(a)  $P(3 \le X \le 5) = \dfrac{e^{-3.33} 3.33^3}{3!} + \dfrac{e^{-3.33} 3.33^4}{4!} + \dfrac{e^{-3.33} 3.33^5}{5!} = 0.5258$

(b)  Assuming that n lobsters must be hatched, the mean will be n/30,000,000.

Then $P(X \geq 1) = 1 - P(X = 0) = 1 - \dfrac{e^{-n/30000000}(n/30000000)^0}{0!} = 1 - e^{-(n/30000000)} \geq 0.9$

This implies that $e^{-(n/30000000)} \leq 0.1$. Taking the natural log of both sides of this inequality, we find that $-n/30000000 \leq \ln(0.1)$. Multiplying both sides of this inequality by -30,000,000 and reversing the direction of the inequality, we have

$n \geq (-30000000)\ln(0.1) = 69,077,552.79 \; or \; 69,077,553$.

**5.85** When n is large and the success probability p is near 1, 1 - p will be near zero. If we let q = 1 - p, then we can approximate the probabilities of the number of failures Y = n - X by letting $\lambda$ = nq. The probability of x successes in n trials is the same as the probability of n - x failures in n trials. Thus

$$P(X = x) = P(Y = n - x) = e^{-\lambda}\frac{\lambda^{n-x}}{(n-x)!} \; where \; \lambda = nq \; .$$

### Review Problems for Chapter 5

1. (a) random variable

   (b) finite (or countably infinite)

2. A probability distribution of a discrete distribution of a discrete random variable gives us a listing of the possible values of the random variable and their probabilities; or a formula for the probabilities.

3. Probability histogram

4. 1

5. (a) P(X = 2) = 0.386

   (b) 38.6%

   (c) 50(0.386) = 19.3 or about 19; 500(0.386) = 193

6. 3.6

7. X is more likely to take a value close to its mean because it has a smaller standard deviation and therefore less variation.

8. Each trial must have the same two possible outcomes (success and failure), the trials must be independent, and they must have a probability of success p that remains constant for all trials.

9. The binomial distribution is a probability distribution for the number of successes in a sequence of n Bernoulli trials.

10. $\dbinom{10}{3} = \dfrac{10!}{3!\,7!} = \dfrac{10(9)\,(8)}{3(2)\,(1)} = 120$

11. Definition 5.4 for the mean and definition 5.5 for the standard deviation are applied to the binomial and Poisson distribution. For example, in Definition 5.4, the formula for the binomial probability function is substituted for P(X = x) to get the mean.

12. (a) Binomial distribution

    (b) Hypergeometric distribution

    (c) The hypergeometric distribution may be approximated by the binomial distribution if the population size N is much greater than the sample size n.   When n is no more than 5% of N, the probability of a success does not change much from trial to trial.

**13.**   (a)   X = 1, 2, 3, 4                    (b)          {X = 3}

(c)   P(X = 3) = 10,028/39,649 = 0.2529.

25.3% of the undergraduates at this university are juniors.

(d)                                                  (e)

| Class level x | Probability P(X=x) |
|---|---|
| 1 | 0.2200 |
| 2 | 0.2153 |
| 3 | 0.2529 |
| 4 | 0.3118 |

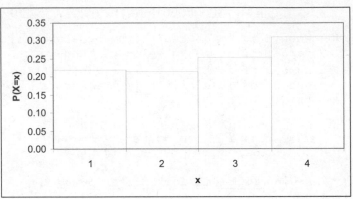

**14.**   (a)   {Y = 4}

(b)   {Y ≥ 4}

(c)   {2 ≤ Y ≤ 4}

(d)   {Y ≥ 1}

(e)   P(Y = 4) = 0.174

(f)   P(Y ≥ 4) = P(4) + P(5) + P(6) = 0.174 + 0.105 + 0.043 = 0.322

(g)   P(2 ≤ Y ≤ 4) = P(2) + P(3) + P(4) = 0.232 + 0.240 + 0.174 = 0.646

(h)   P(Y ≥ 1) = 1 - P(Y ≤ 0) = 1 - 0.052 = 0.948

**15.**   The required calculations are

| y | P(Y=y) | yP(Y=y) | $y^2$ | $y^2$P(Y=y) |
|---|---|---|---|---|
| 0 | 0.052 | 0.000 | 0 | 0.000 |
| 1 | 0.154 | 0.154 | 1 | 0.154 |
| 2 | 0.232 | 0.464 | 4 | 0.928 |
| 3 | 0.240 | 0.720 | 9 | 2.160 |
| 4 | 0.174 | 0.696 | 16 | 2.784 |
| 5 | 0.105 | 0.525 | 25 | 2.625 |
| 6 | 0.043 | 0.258 | 36 | 1.548 |
| | | 2.817 | | 10.199 |

(a)   (a)   $\mu_Y = \sum yP(Y=y) = 2.817$.

(b)   On the average, the number of busy lines is about 2.817.

(c)

$$\sigma_y = \sqrt{\sum (Y - \mu_y)^2 P(Y = y)} = \sqrt{2.2635} = 1.504, \text{or}$$

$$\sigma_y = \sqrt{\sum Y^2 P(Y = y) - \mu_y^2} = \sqrt{10.199 - 2.817^2} = 1.504$$

(d)

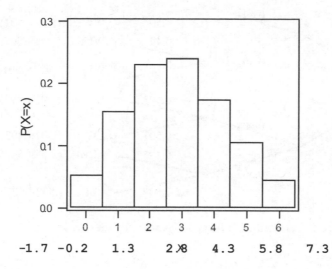

$$\mu-3\sigma \quad \mu-2\sigma \quad \mu-\sigma \quad \mu \quad \mu+\sigma \quad \mu+2\sigma \quad \mu+3\sigma$$

**16.**   $0! = 1$       $3! = 3\cdot2\cdot1 = 6$     $4! = 4\cdot3\cdot2\cdot1 = 24$

$7! = 7\cdot6\cdot5\cdot4\cdot3\cdot2\cdot1 = 5040$

**17.**

$(a)\ \dbinom{8}{3} = \dfrac{8!}{3!5!} = 56$        $(b)\ \dbinom{8}{5} = \dfrac{8!}{5!3!} = 56$        $(c)\ \dbinom{6}{6} = \dfrac{6!}{6!6!} = 1$

$(d)\ \dbinom{10}{2} = \dfrac{10!}{2!8!} = 45$        $(e)\ \dbinom{40}{4} = \dfrac{40!}{4!36!} = 91390$        $(f)\ \dbinom{100}{0} = \dfrac{100!}{0!100!} = 1$

**18.**   (a)  The success probability is p = 0.493.

(b)  The outcomes and their probabilities are shown in the table.

| Outcome | Probability |
|---------|-------------|
| WWW | 0.120 |
| WWL | 0.123 |
| WLW | 0.123 |
| WLL | 0.127 |
| LWW | 0.123 |
| LWL | 0.127 |
| LLW | 0.127 |
| LLL | 0.130 |

See the following tree diagram for the details of each probability.

(c)

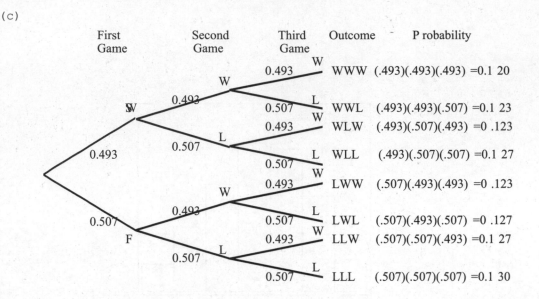

| | First<br>Game | Second<br>Game | Third<br>Game | Outcome | Probability |

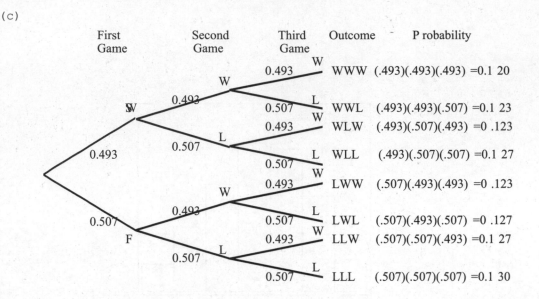

| Outcome | Probability |
|---|---|
| WWW | (.493)(.493)(.493) =0.1 20 |
| WWL | (.493)(.493)(.507) =0.1 23 |
| WLW | (.493)(.507)(.493) =0 .123 |
| WLL | (.493)(.507)(.507) =0.1 27 |
| LWW | (.507)(.493)(.493) =0 .123 |
| LWL | (.507)(.493)(.507) =0 .127 |
| LLW | (.507)(.507)(.493) =0.1 27 |
| LLL | (.507)(.507)(.507) =0.1 30 |

(d)  The outcomes in which the player wins two out of the three times are WWL, WLW, and LWW.

(e)  Each outcome in part (d) has probability 0.123.  The probabilities are equal for each outcome because each probability is obtained by multiplying two WIN probabilities of 0.493 and one LOSS probability of 0.507.

(f)  P(2 wins) = P(WWL) + P(WLW) + P(LWW) = 0.123 + 0.123 + 0.123
$$= 0.369$$

(g)  P(0 wins) = P(LLL) = 0.130

P(1 win) = P(WLL) + P(LWL) + P(LLW) = 0.123 + 0.123 + 0.123 = 0.369

P(3 wins) = P(WWW) = 0.120

| Y | P(Y=y) |
|---|---|
| 0 | 0.120 |
| 1 | 0.381 |
| 2 | 0.369 |
| 3 | 0.130 |

(h)  Binomial distribution with parameters n = 3 and p = 0.493.

19.  The calculations required to answer all parts of this exercise are

$$P(0) = \binom{4}{0} \cdot (0.6)^0 (0.4)^4 = 0.0256 \qquad P(1) = \binom{4}{1} \cdot (0.6)^1 (0.4)^3 = 0.1536$$

$$P(2) = \binom{4}{2} \cdot (0.6)^2 (0.4)^2 = 0.3456 \qquad P(3) = \binom{4}{3} \cdot (0.6)^3 (0.4)^1 = 0.3456$$

$$P(4) = \binom{4}{4} \cdot (0.6)^4 (0.4)^0 = 0.1296$$

(a)  P(3) = 0.3456

(b)  P(X ≥ 3) = P(3) + P(4) = 0.3456 + 0.1296 = 0.4752

(c)  P(X ≤ 3) = P(0) + P(1) + P(2) + P(3)

                = 0.0256 + 0.1536 + 0.3456 + 0.3456 = 0.8704

(d)

| Number with Pets x | Probability P(X=x) |
|:---:|:---:|
| 0 | 0.0256 |
| 1 | 0.1536 |
| 2 | 0.3456 |
| 3 | 0.3456 |
| 4 | 0.1296 |

(e)  Left skewed, since p > 0.5.

(f)

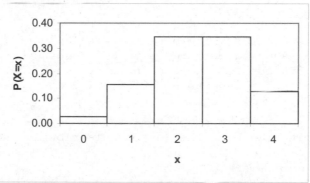

(g)  The distribution is only approximate for several reasons: the actual distribution is hypergeometric, based on sampling without replacement; and the success probability p = 0.60 is probably based on a sample.

(h)  $\mu$ = np = 4(.6) = 2.4; On the average, 2.4 out of 4 households have one or more pets.

(i)  $\sigma^2$ = np(1-p) = 4(.6)(.4) = 0.96;  $\sigma$ = $\sqrt{0.96}$ = 0.98

**20.**  (a)  p < 0.5 since the distribution is left-skewed.

      (b)  p = 0.5 since the distribution is symmetric.

**21.**  (a)  $P(X = 2) = e^{-1.75}\dfrac{1.75^2}{2!} = 0.266$

      (b)

$$P(4 \le X \le 6) = e^{-1.75}\frac{1.75^4}{4!} + e^{-1.75}\frac{1.75^5}{5!} + e^{-1.75}\frac{1.75^6}{6!} = 0.068 + 0.024 + 0.007$$

$$= 0.099$$

      (c)  $P(X \ge 1) = 1 - P(0) = 1 - e^{-1.75}\dfrac{1.75^0}{0!} = 1 - 0.174 = 0.826$

(d)                                              (e)

| x | P (X=x) |
|---|---------|
| 0 | 0.174 |
| 1 | 0.304 |
| 2 | 0.266 |
| 3 | 0.155 |
| 4 | 0.068 |
| 5 | 0.024 |
| 6 | 0.007 |
| 7 | 0.002 |
| 8 | 0.000 |

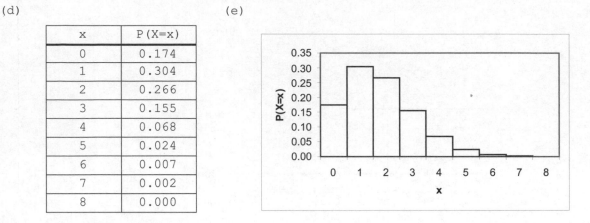

(f)   The distribution is right skewed.  Yes, this is typical of Poisson distributions.

(g)   $\mu = \lambda = 1.75$; on the average, there are 1.75 calls to a wrong number per minute.

(h)   $\sigma = \sqrt{\lambda} = \sqrt{1.75} = 1.323$

22.  (a)  We would expect about 3 meteoroids to be alien matter from outside our solar system.  Assuming meteoroids to be independent of each other, the number out of 300 consisting of such matter would have a binomial distribution with n = 300 and $\mu$ = 0.01.  The mean, or expected number, is np = 300(0.01) = 3.

(b)  Since n is at least 100 and np is less than 10, we can approximate the binomial distribution with the Poisson distribution having parameter $\lambda$ = 3.  Then

$$P(2 \le Y \le 4) = e^{-3} \cdot \frac{3^2}{2!} + e^{-3} \cdot \frac{3^3}{3!} + e^{-3} \cdot \frac{3^4}{4!} = 0.2240 + 0.2240 + 0.1680 = 0.6160 .$$

(c)  $P(Y \ge 1) = 1 - P(Y = 0) = 1 - e^{-3} \cdot \frac{3^0}{0!} = 1 - 0.0498 = 0.9502 .$

23.  (a)  n = 100 and p = 0.015.

(b)  $\lambda$ = np = 1.5

(c)  The binomial probability function is

$$P(Y = y) = \binom{100}{y}(0.015)^y (0.985)^{100-y} \text{ for} y = 0, \ 1, 2, \ ...,1\,00$$

Starting at y = 0 and evaluating this function until it is zero to four decimal places, we obtain the following table.

| y | P(Y = y) |
|---|---|
| 0 | 0.2206 |
| 1 | 0.3360 |
| 2 | 0.2532 |
| 3 | 0.1260 |
| 4 | 0.0465 |
| 5 | 0.0136 |
| 6 | 0.0033 |
| 7 | 0.0007 |
| 8 | 0.0001 |
| 9 | 0.0000 |

(d)   The Poisson probability function is

$$P(Y = y) = e^{-1.5} \cdot \frac{1.5^y}{y!} \text{ for } y = 0, \ 1, 2, \ ,,,$$

Starting at 7 = 0 and evaluating this function until it is zero to four decimal places, we obtain the following table.

| Y | P(Y = y) |
|---|---|
| 0 | 0.2231 |
| 1 | 0.3347 |
| 2 | 0.2510 |
| 3 | 0.1255 |
| 4 | 0.0471 |
| 5 | 0.0141 |
| 6 | 0.0035 |
| 7 | 0.0008 |
| 8 | 0.0001 |
| 9 | 0.0000 |

(e)   All of the probabilities for y = 0 to 9 agree to two decimal places and are usually at most a couple of thousandths apart.  The agreement is very good.

(f)

| Event | Binomial Probability | Poisson Probability |
|---|---|---|
| Y = 3 | 0.1260 | 0.1255 |
| 2 ≤ Y ≤ 5 | 0.4393 | 0.4377 |
| Y < 4 | 0.9358 | 0.9343 |
| Y > 2 | 0.1902 | 0.1912 |

Each pair of probabilities agree to within 0.0016 or less.

**Exercises 6.1**

**6.1**   The histogram will be roughly bell-shaped.

**6.3**   Their distributions are identical.  The mean and standard deviation completely determine the shape of a normal distribution.  Thus if two normally distributed variables have the same mean and standard deviation, they also have the same distribution.

**6.5**   (a)  True.  Both normal curves have the same shape.  Spread (or shape) is represented by σ.  For each normal curve, σ = 3.

   (b)  False.  The parameter μ affects where the normal curve is centered.  Since this parameter is different for each normal curve, each normal curve is centered at a different place.

**6.7**   True.  The value of the parameter μ has no effect on the shape of a normal curve.  The parameter μ affects only where the normal curve is centered.  The shape of the normal curve is determined by the parameter σ.

**6.9**

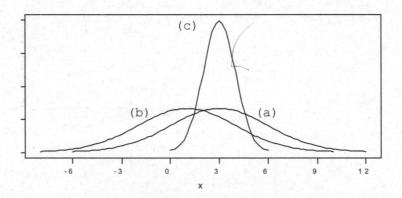

**6.11**  The percentage of all possible observations of a normally distributed variable that lie between 2 and 3 is the same as the area under the associated normal curve between 2 and 3.  If the variable is only approximately normally distributed, the percentage of all possible observations between 2 and 3 is only approximately the area under the associated normal curve between 2 and 3.

**6.13**  (a)  The percentage of female students who are between 60 and 65 inches tall is 100(0.0450 + 0.0757 + 0.1170 + 0.1480 + 0.1713) = 55.70%.

   (b)  The area under the normal curve with parameters μ = 64.4 and σ = 2.4 between 60 and 65 is approximately 0.5570.  This is only an estimate because the distribution of heights is only approximately normally distributed.

**6.15**   (a)

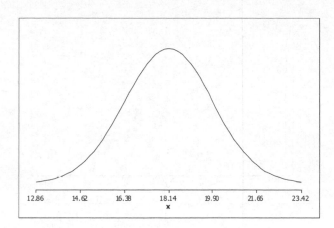

(b)   z = (x − 18.14)/1.76

(c)   z has a standard normal distribution (μ = 0 and σ = 1).

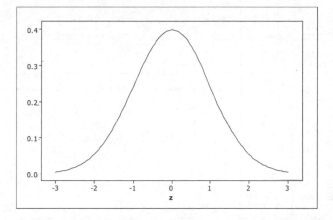

(d)   The percentage of adult males *G. mollicoma* that have carapace lengths between 16 mm and 17 mm is equal to the area under the standard normal curve between <u>-1.22</u> and <u>-0.65</u>.

(e)   The percentage of adult males *G. mollicoma* that have carapace lengths exceeding 19 mm is equal to the area under the standard normal curve that lies to the <u>right</u> of <u>0.49</u>.

**6.17**   (a)

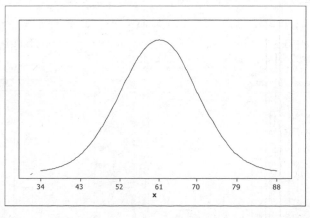

(b)   z = (x − 61)/9

(c)   z has a standard normal distribution (μ = 0 and σ = 1).

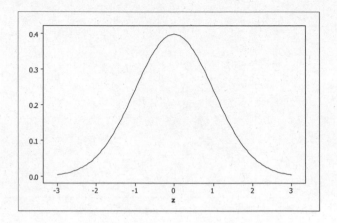

(d) The percentage of finishers with times between 50 and 70 minutes is equal to the area under the standard normal curve between -1.22 and 1.00.

(e) The percentage of finishers with times less than 75 minutes is equal to the area under the standard normal curve that lies to the left of 1.56.

**6.19** (a)

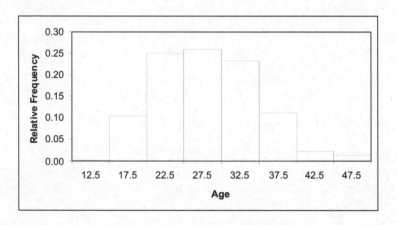

(b) Yes.  The distribution appears to be slightly right skewed, but could be approximated quite well by a normal distribution.

**6.21** (a) Using Minitab, we obtain the data from the WeissStats CD and choose

**Graph ▶ Histogram**, select **Simple** from the first row, and click **OK.**

Then enter VERBAL and MATH in the Graphs variables text box and click OK.  The resulting graphs follow.

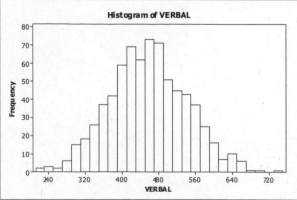
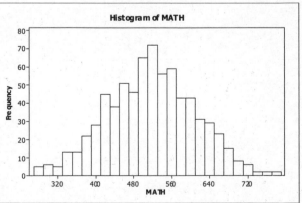

    (a)  Yes.  The SAT verbal scores appearing in the left hand histogram have roughly a bell shaped distribution.

    (b)  Yes.  The SAT verbal scores appearing in the right hand histogram have roughly a bell shaped distribution.

**6.23**  The number of chips per bag could <u>not</u> be exactly normally distributed because that number is a discrete random variable, whereas any variable having a normal distribution is a continuous random variable.

**6.25**  (a)  Using Minitab we choose **Calc ▶ Make patterned data ▶ Simple set of numbers...**, enter x in the **Store patterned data** in text box, type <u>218</u> in the **From first value** text box, type <u>314</u> in the **To last value** text box, type <u>.5</u> in the **In steps of** text box, and click **OK**.  This will provide x values within 3 standard deviations on both sides of the

mean.  Now choose **Calc ▶ Probability distribution ▶ Normal...**, click on **Probability density**, type <u>266</u> in the **Mean** text box and <u>16</u> in the **Standard deviation** text box, click in the **Input column** text box and select x, click in the **Optional storage** text box and type <u>P(x)</u>, and

click **OK**.  Now choose **Graph ▶ Scatterplot...**, select the **With connect line** version, select P(X) in the **Y** column in row **1** and X for the **X** column. Click on the **Data view** button and check only the **Connect line** box.  Click **OK** twice.  The result is

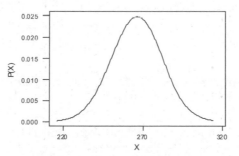

    (b)  Following the procedure in Example 6.2, we choose **Calc ▶ Random Data ▶ Normal...**, type <u>1000</u> in the **Generate rows of data** text box, click in the **Store in column(s)** text box and type <u>DAYS</u>, click in the **Mean** text box and type <u>266</u>, click in the **Standard deviation** text box and type <u>16</u>, and click **OK**.

    (c)  We would expect the sample mean and sample deviation to be approximately equal to the mean and standard deviation of the population, 266 and 16 respectively.  This is because we expect the sample to reflect approximately the characteristics of the population.

    (d)  We choose **Calc ▶ Column Statistics...**, click on the **Mean** button, click in the **Input variable** text box and select DAYS, and click **OK**.  Then repeat the process, selecting the **Standard deviation** button.  The results are shown in the Session Window.  The results we obtained were Mean of DAYS = 266.88 and  Standard deviation of DAYS  = 16.048.  Your results will vary from ours.

(e)  We would expect a histogram of the 1000 observations to look roughly
     like a normal curve with mean 266 and standard deviation 16.

(f)  Choose **Graph ▶ Histogram...**, select
     the **Simple** version, select DAYS in
     the **Graph variables** text box and
     click **OK**. The result shown at the
     right is as expected.

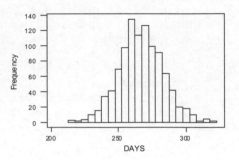

### Exercises 6.2

**6.27**  Finding areas under the standard normal curve is important because for <u>any</u>
          normally distributed variable, we can obtain the percentage of all possible
          observations that lie within any specified range by first converting x
          values to z-scores and then finding the corresponding area under the
          standard normal curve.

**6.29**  The total area under the curve is 1, and the standard normal curve is
          symmetric about 0.  Therefore, the area to the left of 0 is 0.5, and the
          area to the right of 0 is 0.5.

**6.31**  The area under the standard normal curve to the right of 0.43 is 1 - the
          area to the left of 0.43.  The area to the left of 0.43 is 0.6664.
          Therefore, the area to the right of 0.43 is 1 - 0.6664 = 0.3336.

**6.33**  The area to the left of z = 3.00 is 0.9987 and the area to the left of -3.00
          is 0.0013.  Therefore the area between -3.00 and 3.00 is 0.9987 - 0.0013 =
          0.9974 (99.74%).

**6.35**  (a)  Locate the row (tenths digit) and column (hundredths digit) of the
               specified z-score.  The corresponding table entry is the area under the
               standard normal curve that lies to the left of the z-score.

          (b)  The area that lies under the standard normal curve to the right of a
               specified z-score is 1 - (area to the left of the z-score).

          (c)  The area that lies under the standard normal score between two
               specified z-scores, say *a* and *b*, where a < b, is found by subtracting
               the area to the left of *a* from the area to the left of *b*.

**6.37**  (a)                                          (b)

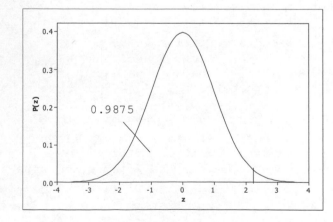

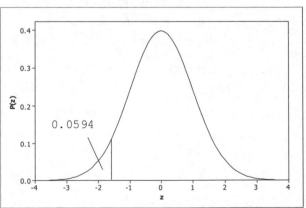

Area = 0.9875                          Area = 0.0594

(c)

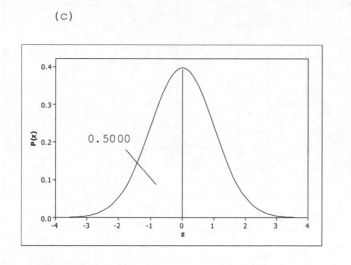

Area = 0.5

(d)

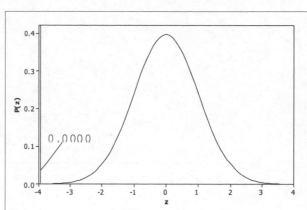

Area = 0.0000 (to 4 dp)

**6.39**  (a)

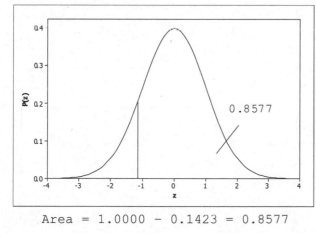

Area = 1.0000 − 0.1423 = 0.8577

(b)

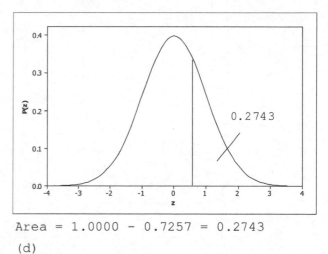

Area = 1.0000 − 0.7257 = 0.2743

(c)

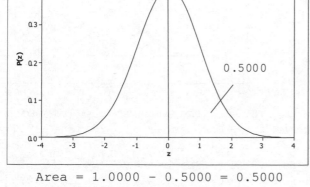

Area = 1.0000 − 0.5000 = 0.5000

(d)

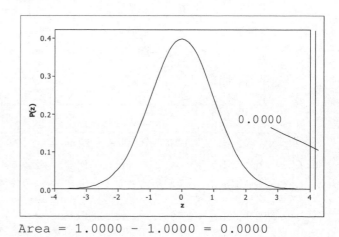

Area = 1.0000 − 1.0000 = 0.0000

**6.41** (a)                                   (b)

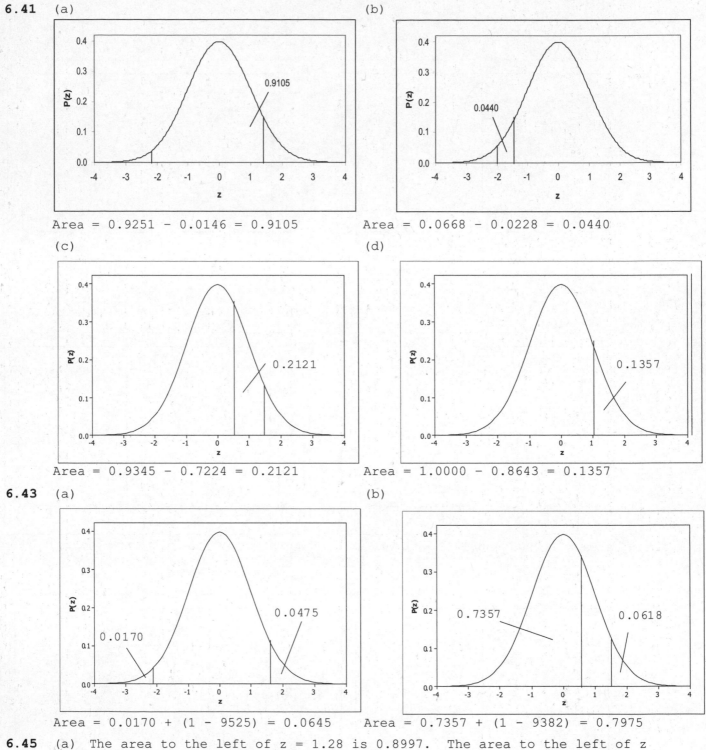

Area = 0.9251 - 0.0146 = 0.9105    Area = 0.0668 - 0.0228 = 0.0440

(c)                                   (d)

Area = 0.9345 - 0.7224 = 0.2121    Area = 1.0000 - 0.8643 = 0.1357

**6.43** (a)                                   (b)

Area = 0.0170 + (1 - 9525) = 0.0645    Area = 0.7357 + (1 - 9382) = 0.7975

**6.45** (a) The area to the left of z = 1.28 is 0.8997. The area to the left of z = -1.28 is 0.1003. The area between z = -1.28 and z = 1.28 is 0.8997 - 0.1003 = 0.7994.

(b) The area to the left of z = 1.64 is 0.9495. The area to the left of z = -1.64 is 0.0505. The area between z = -1.64 and z = 1.64 is 0.9495 - 0.0505 = 0.8990.

(c) The area to the left of z = -1.96 is 0.0250. The area to the right of z = 1.96 is 1.0000 - 0.9750 = 0.0250. The area either to the left of z

= -1.96 or to the right of z = 1.96 is 0.0250 + 0.0250 = 0.0500.

(d)   The area to the left of z = -2.33 is 0.0099.   The area to the right of
z = 2.33 is 1.0000 - 0.9901 = 0.0099.   The area either to the left of z
= -2.33 or to the right of z = 2.33 is 0.0099 + 0.0099 = 0.0198.

**6.47**   (a)                                                        (b)

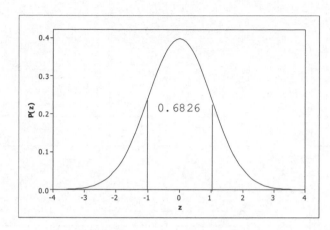

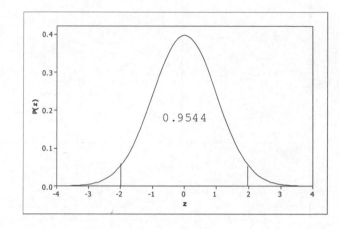

Area = 0.8413 - 0.1587                    Area = 0.9772 - 0.0228

      = 0.6826                                     = 0.9544

  (c)

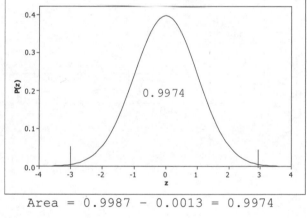

Area = 0.9987 - 0.0013 = 0.9974

**6.49**                                              **6.51**

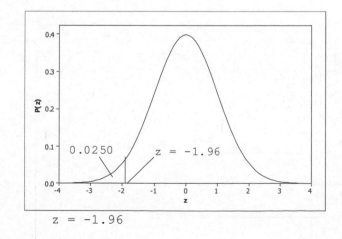

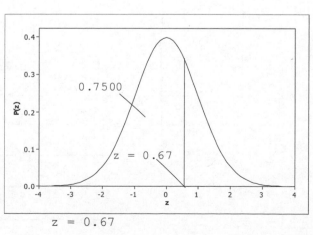

z = -1.96                                          z = 0.67

**6.53**

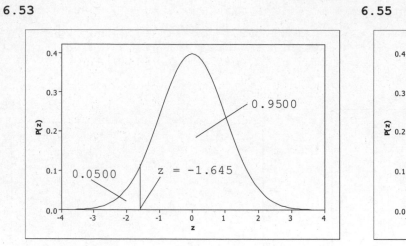

z = -1.645

**6.55**

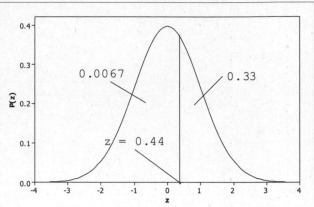

z = 0.44

**6.57**  (a)

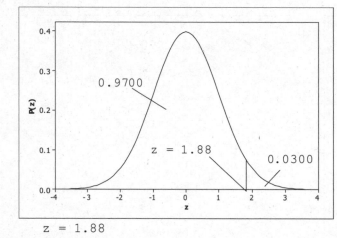

z = 1.88

(b)

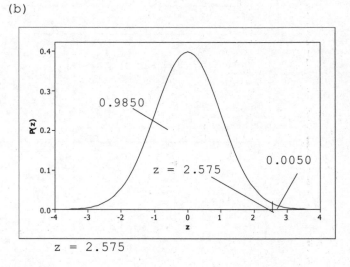

z = 2.575

**6.59**

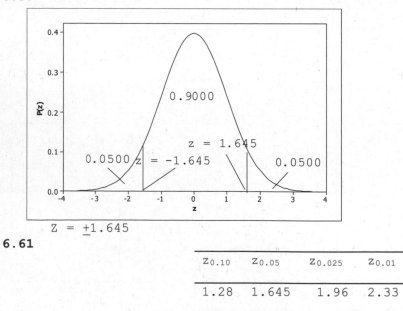

Z = ±1.645

**6.61**

| $z_{0.10}$ | $z_{0.05}$ | $z_{0.025}$ | $z_{0.01}$ | $z_{0.005}$ |
|---|---|---|---|---|
| 1.28 | 1.645 | 1.96 | 2.33 | 2.575 |

**6.63**   (a)   $z_\alpha$                        (b)   $-z_\alpha$                    (c)   $\pm\, z_{\alpha/2}$
(d)

For part (a):                                      For part (b):

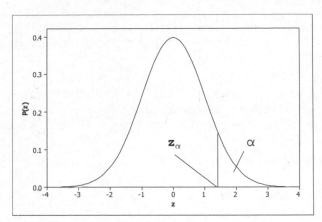     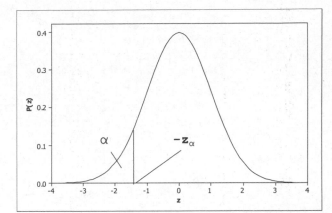

For part (c):

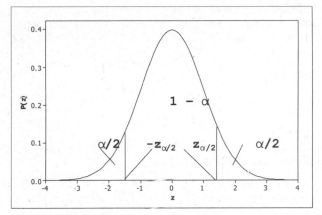

## Exercises 6.3

**6.65**   The x values delimiting the interval within two standard deviations either side of the mean are $\mu - 2\sigma$ and $\mu + 2\sigma$. When these are standardized using $z = (x - \mu)/\sigma$, the former becomes $z = [(\mu - 2\sigma) - \mu]/\sigma = -2$ and the latter becomes $z = [(\mu + 2\sigma) - \mu]/\sigma = +2$.

**6.67**   (a)   For carapace lengths 16 mm and 17 mm, the z-values are

$$z = \frac{16-18.14}{1.76} = -1.22 \text{ and } z = \frac{17-18.14}{1.76} = -0.65$$

The area to the left of z = -1.22 is 0.1112 and the area to the left of z = -0.65 is 0.2578. Therefore the area between z = -1.22 and z = -0.65 is 0.2578 - 0.1112 = 0.1466. Thus the percentage of carapace lengths that are between 16 mm and 17 mm is 14.66%.

(b)   For a carapace length of 19 mm, the z value is

$$z = \frac{19-18.14}{1.76} = 0.49$$

The area to the left of z = 0.49 is 0.6879. The area to right of 0.49 is  1 - 0.6879 = 0.3121. Thus the percentage of adult *G. mollicoma* with carapace lengths exceeding 19 mm is 31.21%.

(c)   Using Table II, we find that an area of 0.2500 lies to left of z = -0.67, an area of .5000 lies to the left of z = 0.00 and an area of 0.7500 lies to the left of 0.67. We convert each of these z-values to x-values using $x = \mu + z\sigma$. Thus 25% of the carapace lengths of adult *G. mollicoma* lie to the left of the first quartile, x = 18.14 + (-0.67)(1.76) = 16.96 mm. Half (50%) of the carapace lengths of adult *G. mollicoma* lie to the left of the second quartile (median), x = 18.14 + (0.00)(1.76) = 18.14 mm. Finally. 75% of the carapace lengths of adult *G. mollicoma* lie to the left of the third quartile, x = 18.14 + (0.67)(1.76) = 19.32 mm.

(d) Using Table II, we find that an area of 0.9500 lies to the left of z = 1.645. We convert this z-value to an x-value using $x = \mu + z\sigma$. Thus, the 95th percentile of the carapace lengths of adult *G. mollicoma* is x = 18.14 + (1.645)(1.76) = 21.04 mm.

**6.69** (a) For finishers with times of 50 and 70 minutes, the z-values are

$$z = \frac{50 - 61}{9} = -1.22 \text{ and } z = \frac{70 - 61}{9} = 1.00$$

The area to the left of z = -1.225 is 0.1112 and the area to the left of z = 1.00 is 0.8413. Therefore the area between z = -1.22 and z = 1.00 is 0.8413 - 0.1112 = 0.7301. Thus the percentage of finishers with times between 50 minutes and 70 minutes in the New York City 10 km run is 73.01%.

(b) For a finishing time of 75 minutes, the z value is

$$z = \frac{75 - 61}{9} = 1.56$$

The area to the left of z = 1.56 is 0.9406. Thus the percentage of finishers with times less than 765 minutes is 94.06%.

(c) Using Table II, we find that an area of 0.4000 lies to left of z = -0.25. We convert this z-value to an x-value using $x = \mu + z\sigma$. Thus 40% of the finishers had times less than the 40th percentile, x = 61 + (-0.25)(9) = 58.75 minutes.

(d) The eighth decile is the same as the 80th percentile. Using Table II, we find that an area of 0.8000 lies to the left of z = 0.84. We convert this z-value to an x-value using $x = \mu + z\sigma$. Thus, 80% of the finishing times were less than x = 61 + (0.84)(9) = 68.56 minutes.

**6.71** (a) For x = 260 and 280, the z-values are

$$z = \frac{260 - 272.2}{8.12} = -1.50 \text{ and } z = \frac{280 - 272.2}{8.12} = 0.96$$

The area to the left of z = -1.50 is 0.0668 and the area to the left of z = 0.96 is 0.8315. Therefore the area between z = -1.50 and z = 0.96 is 0.8315 - 0.0668 = 0.7647. Thus the percentage of tee shots that went between 260 and 280 yards is 76.47%.

(b) For x = 300, the z-value is

$$z = \frac{300 - 272.2}{8.12} = 3.42 \text{ .}$$ The area to the left of z = 3.42 is 0.9997.

Thus the area to the right of z = 3.42 is 1 - 0.9997 = 0.0003, implying that only 0.03% of tee shots went more than 300 yards.

**6.73**

| Part | Standard deviations to either side of the mean | Area under normal curve | Percent |
|------|------|------|------|
| (a) | 1 | 0.3413 x 2 = 0.6826 | 68.26 |
| (b) | 2 | 0.4772 x 2 = 0.9544 | 95.44 |
| (c) | 3 | 0.4987 x 2 = 0.9974 | 99.74 |

**6.75** (a)  68.26% of Swedish men have brain weights between <u>1.29</u> and <u>1.51</u> kg.

(b)  95.44% of Swedish men have brain weights between <u>1.18</u> and <u>1.62</u> kg.

(c)  99.74% of Swedish men have brain weights between <u>1.07</u> and <u>1.73</u> kg.

(d)

**6.77** (a)  From Table 6.1, exactly 11.70% of the heights of female students are between 62 and 63.  For heights of 62 and 63 inches, the z values are

$$z = \frac{62 - 64.4}{2.4} = -1.00 \text{ and } z = \frac{63 - 64.4}{2.4} = -0.58$$

The area to the left of z = -1.00 is 0.1587 and the area to the left of z = -0.58 is 0.2810.  Thus, the area between z = -1.00 and z = -0.58 is 0.2810 - 0.1587 = 0.1223 (12.23%).  The two percentages are quite close.

(b)  From Table 6.1, exactly = 0.1575 + 0.1100 + 0.0735 + 0.0374 + 0.0199 = 0.3983 (39.83%) of the heights of female students are between 65 and 70 inches.  For heights of 65 and 70 inches, the z values are

$$z = \frac{65 - 64.4}{2.4} = 0.25 \text{ and } z = \frac{70 - 64.4}{2.4} = 2.33$$

The area to the left of z = 0.25 is 0.5987 and the area to the left of z = 2.33 is 0.9901.  Thus, the area between z = 0.25 and a = 2.33 is 0.9901 - 0.5987 = 0.3914 (39.14%).  The two percentages are quite close.

**6.79** (a)  $P(Y > 5) = P(Z > \frac{5 - 4.66}{0.75}) = P(Z > 0.45) = 1 - 0.6736 = 0.3264$ .  We interpret this to mean that about 32.64% of the nests of the booted eagle of western Europe are more than 5 km to the nearest marshland.

(b)  $P(3 \leq Y \leq 6) = P(\frac{3 - 4.66}{0.75} \leq Z \leq \frac{6 - 4.66}{0.75}) = P(-2.21 \leq Z \leq 1.79)$

$$= 0.9633 - 0.0136 = 0.9497$$

We interpret this to mean that about 95% of the nests of the booted eagle of western Europe are between 3 and 6 km from the nearest marshland.

**6.81** (a) 99% of all possible observations lie within <u>2.575</u> standard deviations to either side of the mean.

(b) 80% of all possible observations lie within <u>1.28</u> standard deviations to either side of the mean.

**6.83** $P(\mu - z_{\alpha/2}\,\sigma < x < \mu + z_{\alpha/2}) = P(-z_{\alpha/2}\,\sigma < x - \mu < z_{\alpha/2}\,\sigma)$

$$= P(-z_{\alpha/2} < (x - \mu)/\sigma < z_{\alpha/2})$$

$$= P(-z_{\alpha/2} < z < z_{\alpha/2}) = (1 - \alpha/2) - \alpha/2$$

$$= 1 - \alpha$$

### Exercises 6.4

**6.85** Decisions about whether a variable is normally distributed often are important in subsequent analyses such as percentile calculations or in determining the type of statistical inference procedure to be used.

**6.87** (a) A normal probability plot is a plot of the observed values of a variable versus the normal scores - the observations expected for a variable having a standard normal distribution. If the variable is normally distributed, then the normal probability plot should yield roughly a straight line. If the plot is roughly linear, then it is accepted as reasonable that the variable is approximately normally distributed. If it is not roughly linear, then the variable is probably not normally distributed.

(b) If the normal probability plot is roughly linear except for a small number of points that lie well outside the overall pattern of the plot, it is possible that some or all of those points are outliers. If there is sufficient reason to remove the potential outliers from the sample and doing so results in a plot that is linear without outliers, the analysis can often be carried out with the remaining observations.

**6.89** (a)

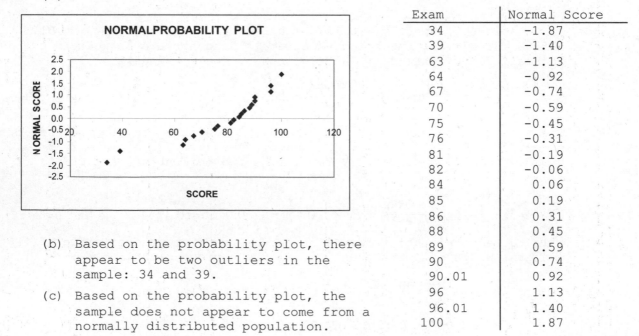

| Exam | Normal Score |
|---|---|
| 34 | -1.87 |
| 39 | -1.40 |
| 63 | -1.13 |
| 64 | -0.92 |
| 67 | -0.74 |
| 70 | -0.59 |
| 75 | -0.45 |
| 76 | -0.31 |
| 81 | -0.19 |
| 82 | -0.06 |
| 84 | 0.06 |
| 85 | 0.19 |
| 86 | 0.31 |
| 88 | 0.45 |
| 89 | 0.59 |
| 90 | 0.74 |
| 90.01 | 0.92 |
| 96 | 1.13 |
| 96.01 | 1.40 |
| 100 | 1.87 |

(b) Based on the probability plot, there appear to be two outliers in the sample: 34 and 39.

(c) Based on the probability plot, the sample does not appear to come from a normally distributed population.

**6.91**  (a)

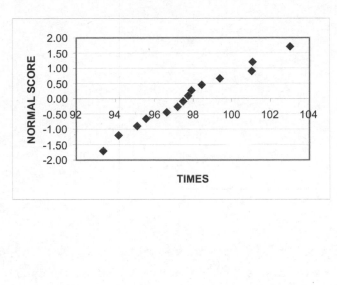

| Times<br>X | Normal Score<br>y |
|---|---|
| 93.37 | −1.71 |
| 94.15 | −1.20 |
| 95.10 | −0.90 |
| 95.57 | −0.66 |
| 96.63 | −0.45 |
| 97.19 | −0.27 |
| 97.47 | −0.09 |
| 97.73 | 0.09 |
| 97.91 | 0.27 |
| 98.44 | 0.45 |
| 99.38 | 0.66 |
| 101.05 | 0.90 |
| 101.09 | 1.20 |
| 103.02 | 1.71 |

(b)  Based on the probability plot, there do not appear to be any outliers in the sample.

(c)  Based on the probability plot, the sample appears to be from an approximately normally distributed population.

**6.93**  (a)  We enter the data in Excel in a column which we named "x" at the top.

Highlight the name and the data with your mouse and then choose **DDXL** ▶ **Charts and Plots**.  Select **Normal Probability Plot** from the drop down **Function type** box and enter x in both the **Quantitative variable** and **Label variable** boxes.  Click **OK**.  The resulting plot is

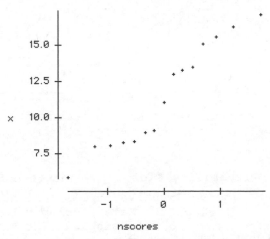

(b)  From the plot, there do not appear to be any outliers.

(c)  From the plot, normality seems to be a reasonable assumption.

**6.95**  (a)  We enter the data in Excel in a column which we named "DOU" at the top.

Highlight the name and the data with your mouse and then choose **DDXL** ▶ **Charts and Plots**.  Select **Normal Probability Plot** from the drop down

**Function type** box and enter <u>DOU</u> in both the **Quantitative variable** and **Label variable** boxes.  Click **OK**.  The resulting plot is

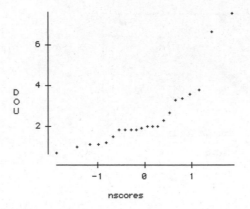

(b)  From the plot, it appears that there may be two outliers, 7.6 and 6.7.

(c)  From the plot, the data do not appear to be normally distributed. However, if the two outliers were excluded from the data, the remaining data may be normally distributed.

**6.97**  (a)  Using Minitab and assuming that the data are in a column named 'TEMP', choose **Graph ▶ Histogram...**, select the **Simple** plot and click **OK**. Specify TEMP in the **Graph Variables** text box and click **OK**.  The resulting plot is

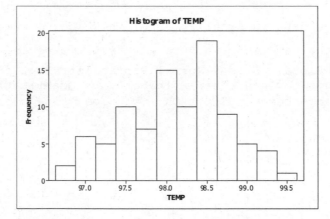

Aside from the unusual 'saw tooth' pattern of the left side of the graph, an assumption of normality for this data appears to be reasonable.

(b)  Choose **Graph ▶ Probability Plot...**, select the **Single** plot and click **OK**.  Specify TEMP in the Graph Variables text box.  Click on the **Distribution** button, click the **Data Display** tab, select the **Symbols only** option button from the **Data Display** list and click **OK**.  Now click the **Scale** button, click on the **Y-Scale Type** tab, select the **Score** option button from the **Y-Scale Type** list, and click **OK** twice.  The resulting plot is

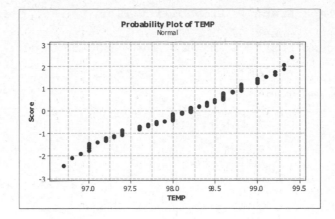

(b)  There are no apparent outliers and the graph is roughly linear, so it seems reasonable to assume normality for these data.

(c)  We arrived at the same conclusion from both graphs.

**6.99**  Choose **Graph ▶ Probability Plot...**, select the **Single** plot and click **OK**. Specify CHIPS in the **Graph Variables** text box. Click on the **Distribution** button, click the **Data Display** tab, select the **Symbols only** option button from the **Data Display** list and click **OK**. Now click the **Scale** button, click on the **Y-Scale Type** tab, select the **Score** option button from the **Y-Scale Type** list, and click **OK** twice. The resulting plot is

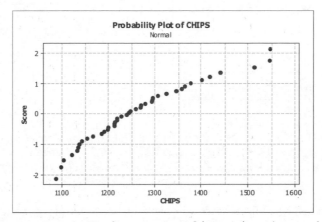

Yes. There do not appear to be any outliers in these data. Since the plot follows a fairly straight line, normality of the chips data seems to be a reasonable assumption.

**6.101** (a)  Using Minitab, we will generate four columns of fifty observations each by choosing **Calc ▶ Random data ▶ Normal...**, entering 50 in the **Generate rows of data** text box, clicking in the **Store in column(s)** text box and typing C1-C4, clicking in the **Mean** text box and entering 266, clicking in the **Standard deviation** text box and entering 16, and clicking **OK**. Now click in the worksheet column title row and name the four columns GEST1, GEST2, GEST3, and GEST4.

Next we select **Graph** ▶ **Probability Plot**, select the **Single** version and click **OK**, then enter GEST1, GEST2, GEST3, and GEST4 in the **Graph Variables** text box.  Click **OK**.  The resulting four graphs are shown below.

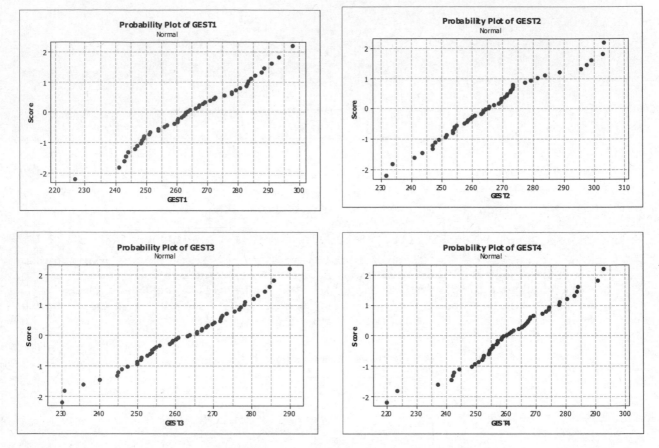

(c)  Yes.  Since the data were generated from a normal distribution, we would expect the four plots to be roughly linear, as they are.  Your simulation will likely result in data and graphs that differ from ours.

### Exercises 6.5

**6.103** It is not practical to use the binomial probability formula when the number of trials, *n*, is large.

**6.105** (a)  (i)  P(x = 4 or 5) = 0.2051 + 0.2461 = 0.4512

(ii) P(3 $\leq$ x $\leq$ 7) = 0.1172 + 0.2051 + 0.2461 + 0.2051 + 0.1172 = 0.8907

(b)  (i)  P(x = 4 or 5):

Step 1:  n = 10; p = 0.5

Step 2:  np = 5; n(1 - p) = 5.  Since both np and n(1 - p) are at least 5, the normal approximation can be used.

Step 3:

$$\mu_x = np = 10(0.5) = 5$$

$$\sigma_x = \sqrt{np(1 - p)} = \sqrt{10(0.5)\,(0.5)} = 1.58$$

Step 4:     Making the continuity correction, we find the area under the normal curve with parameters μ = 5 and σ = 1.58 that lies between x = 3.5 and x = 5.5.   The z-values for these x values are

$$z = \frac{3.5-5}{1.58} = -0.95 \ \text{ and } \ z = \frac{5.5-5}{1.58} = 0.32$$

The area to the left of z = -0.95 is 0.1711 and the area to the left of z = 0.32 is 0.6255.  Therefore, the approximate probability that x equals 4 or 5 is 0.6255 - 0.1711 = 0.4544.  This is very close the actual probability of 0.4512.

(ii) P(3 ≤ x ≤ 7):

Steps 1, 2 and 3 are the same as above.

Step 4:     Making the continuity correction, we find the area under the normal curve with parameters μ = 5 and σ = 1.58 that lies between x = 2.5 and x = 7.5.  The z-values for these x values are

$$z = \frac{2.5-5}{1.58} = -1.58 \ \text{ and } \ z = \frac{7.5-5}{1.58} = 1.58$$

The area to the left of z = -1.58 is 0.0571 and the area to the left of z = 1.58 is 0.9429.  Therefore, the approximate probability that x lies between 3 and 7, inclusive, is 0.9429 - 0.0571 = 0.8858.  This is very close the actual probability of 0.8907.

**6.107** Since both np and n(1 - p) are at least 5, we would use the normal curve with

$$\mu = np = 25(0.5) = 12.5 \ \text{ and } \ \sigma = \sqrt{np(1\text{-}p)} = \sqrt{25(0.5)(1\text{-}0.5)} = \sqrt{6.25} = 2.5.$$

**6.109** (a)  The mean is $\mu = np = 200(0.65) = 130$ and $\sigma = \sqrt{np(1\text{-}p)} = \sqrt{200(0.65)(0.35)} = 6.75.$

Note that np = 130 and n(1 - p) = 70, so it is reasonable to use the normal approximation for binomial probabilities.  If X represents the number of preschool children living in poverty that have been exposed to cigarette smoke at home, we want P(X ≥ 125).  With the continuity correction, this becomes P(X > 124.5).  For X = 124.5, we have z = (124.5 - 130)/6.75 = -0.81.  The area to the left of z = -0.81 is 0.2090.  Therefore P(X > 124.5) = 1 - 0.2090 = 0.7910 (approximately).

(b)  The mean is $\mu = np = 200(0.45) = 90$ and $\sigma = \sqrt{np(1\text{-}p)} = \sqrt{200(0.45)(0.65)} = 7.04.$

If X represents the number of preschool children not living in poverty that have been exposed to cigarette smoke at home, we want P(X ≥ 125).  With the continuity correction, this becomes P(X > 124.5).  For X = 124.5, we have z = (124.5 - 90)/7.04 = 4.90.  The area to the left of z = 4.90 is 1.0000.  Therefore P(X > 124.5) = 1 - 1.0000 = 0.0000.

**6.111** For parts (a), (b),and (c), steps 1-3 are as follows:

Step 1:   n = 100;  p = 0.058

Step 2:   np = 5.8; n(1 - p) = 94.2.  Since both np and n(1 - p) are at least

5, the normal approximation can be used.

Step 3:

$$\mu_x = np = 100(0.058) = 5.8$$

$$\sigma_x = \sqrt{np(1-p)} = \sqrt{100(0.058)(0.942)} = 2.34$$

(a) P(X = 6)

Step 4: For x = 5.5 and x = 6.5, the z-scores are

$$z = \frac{5.5 - 5.8}{2.34} = -0.13 \text{ and } z = \frac{6.5 - 5.8}{2.34} = 0.30.$$

The area to the left of z = -0.13 is 0.4483 and the area to the left of z = 0.30 is 0.6179. Therefore, the approximate probability that X = 6 is 0.6179 - 0.4483 = 0.1696.

(b) P(5 ≤ X ≤ 10)

Step 4: For X = 4.5 and X = 10.5, the z-scores are

$$z = \frac{4.5 - 5.8}{2.34} = -0.56 \text{ and } z = \frac{10.5 - 5.8}{2.34} = 2.01.$$

The area to the left of z = -0.56 is 0.2877 and the area to the left of z = 2.01 is 0.9778. Therefore, the approximate probability that X is between 5 and 10, inclusive, is 0.9778 - 0.2877 = 0.6901.

(c) P(X ≥ 8)

Step 4: For X = 7.5, the z-score is

$$z = \frac{7.5 - 5.8}{2.34} = 0.73.$$

The area to the left of z = 0.73 is 0.7673. Therefore, the approximate probability that X is at least 8 is 1.0000 - 0.7673 = 0.2327.

**6.113** For parts (a), (b), (c), and (d), steps 1-3 are as follows:

Step 1: n = 42; p = 0.16

Step 2: np = 6.72; n(1 - p) = 35.28. Since both np and n(1 - p) are at least 5, the normal approximation can be used.

Step 3:

$$\mu_x = np = 42(0.16) = 6.72$$

$$\sigma_x = \sqrt{np(1-p)} = \sqrt{42(0.16)(0.84)} = 2.38$$

(a) P(x = 5):

Step 4: For x = 4.5 and x = 5.5, the z-scores are

$$z = \frac{4.5 - 6.72}{2.38} = -0.93 \text{ and } z = \frac{5.5 - 6.72}{2.38} = -0.51.$$

The area to the left of z = -0.93 is 0.1762 and the area to the left of z = -0.51 is 0.3050. Therefore, the approximate probability that X is five is 0.3050 - 0.1762 = 0.1288.

(b) P(9 ≤ x ≤ 12):

Step 4: For x = 8.5 and x = 12.5, the z-scores are

$$z = \frac{8.5 - 6.72}{2.38} = 0.75 \text{ and } z = \frac{12.5 - 6.72}{2.38} = 2.43$$

The area to the left of z = 0.75 is 0.7734 and the area to the left of z = 2.43 is 0.9925.  Therefore, the approximate probability that X is between 9 and 12, inclusive, is 0.9925 − 0.7734 = 0.2191.

(c)  P(x ≥ 1):

   Step 4:    For x = 0.5, the z-score is

$$z = \frac{0.5 - 6.72}{2.38} = -2.61.$$

The area to the left of z = -2.61 is 0.0045.  Therefore, the approximate probability that X is at least one is 1.0000 − 0.0045  = 0.9955.

(d)  P(x ≤ 2):

   Step 4:    For x = 2.5, the z-score is

$$z = \frac{2.5 - 6.72}{2.38} = -1.77.$$

The area to the left of z = -1.77 is 0.0384.  Therefore, the approximate probability that X is at most two is .0384.

**6.115** For parts (a), (b), and (c), steps 1-3 are as follows:

   Step 1:    n = 250;  p = 0.527

   Step 2:    np = 131.75; n(1 - p) = 118.25.  Since both n(1 -p) and np are at least 5, the normal approximation can be used.

   Step 3:

$$\mu_x = np = 250(0.527) = 131.75$$

$$\sigma_x = \sqrt{np(1 - p)} = \sqrt{250(0.527)(0.473)} = 7.89$$

(a)  One-half of 250 is 125.  P(x = 125):

   Step 4:    For x = 124.5 and x = 125.5, the z scores are

$$z = \frac{124.5 - 131.75}{7.89} = -0.92 \text{ and } z = \frac{125.5 - 131.75}{7.89} = -0.79.$$

The area to the left of z = -0.92 is 0.1788 and the area to the left of z = -0.79 is 0.2148.  Therefore, the approximate probability that X is 125 is 0.2148 − 0.1788 = 0.0360.

(b)  Step 4:    For x = 124.5, the z-score is

$$z = \frac{124.5 - 131.75}{7.89} = -0.92.$$

The area to the left of z = -0.92 is 0.1788.  Therefore, the approximate probability that X is at least 125 is 1.0000 − 0.1788 = 0.8212.

(c)  Step 4:    For x = 114.5 and x = 130.5, the z-scores are

$$z = \frac{114.5 - 131.75}{7.89} = -2.19 \text{ and } z = \frac{114.5 - 131.75}{7.89} = -0.16.$$

The area to the left of z = -2.19 is 0.0143 and the area to the left of z = -0.16 is 0.4364.  Therefore, the approximate

probability that X is between 115 and 130, inclusive, is
0.4364 - 0.0143 = 0.4221

**6.117** (a) 100 bets are made. The gambler is ahead if she has won more than 50 bets. So, we are concerned with finding $P(x > 50)$ or $P(x \geq 51)$.

Step 1: $n = 100$; $p = 18/38 = 0.47368$

Step 2: $np = 47.368$; $n(1 - p) = 52.632$. Since both $np$ and $n(1-p)$ are at least 5, the normal approximation can be used.

Step 3:

$$\mu_x = np = 100(0.47368) = 47.368$$

$$\sigma_x = \sqrt{np(1 - p)} = \sqrt{100(0.47368)(0.52632)} = 4.993$$

Step 4: For $x = 50.5$, the z-score is

$$z = \frac{50.5 - 47.368}{4.993} = 0.63$$

The area to the left of $z = 0.63$ is 0.7357. Therefore, the approximate probability that X is at least 51 is 1.0000 – 0.7357 = 0.2643.

(b) 1000 bets are made. The gambler is ahead if she has won more than 500 bets. So, we are concerned with finding $P(x > 500)$ or $P(x \geq 501)$.

Step 1: $n = 1000$; $p = 18/38 = 0.47368$

Step 2: $np = 473.68$; $n(1 - p) = 526.32$. Since both $np$ and $n(1-p)$ are at least 5, the normal approximation can be used.

Step 3:

$$\mu_x = np = 1000(0.47368) = 473.68$$

$$\sigma_x = \sqrt{np(1 - p)} = \sqrt{1000(0.47368)(0.52632)} = 15.789$$

Step 4: For $x = 500.5$, the z-score is

$$z = \frac{500.5 - 473.68}{15.789} = 1.70$$

The area to the left of $z = 1.70$ is 0.9554. Therefore, the approximate probability that X is at least 501 is 1.0000 – 0.9554 = 0.0446.

(c) 5000 bets are made. The gambler is ahead if she has won more than 2500 bets. So, we are concerned with finding $P(x > 2500)$ or $P(x \geq 2501)$.

Step 1: $n = 5000$; $p = 18/38 = 0.47368$

Step 2: $np = 2368.4$; $n(1 - p) = 2631.6$. Since both $np$ and $n(1-p)$ are at least 5, the normal approximation can be used.

Step 3:

$$\mu_x = np = 5000(0.47368) = 2368.4$$

$$\sigma_x = \sqrt{np(1 - p)} = \sqrt{300(0.47368)(0.52632)} = 35.306$$

Step 4: For $x = 2500.5$, the z-score is

$$z = \frac{2500.5 - 2368.4}{35.306} = 3.74.$$

The area to the left of $z = 3.74$ is 0.9999. The approximate probability that X is at least 2501 is 1.0000 – 0.9999 = 0.0001.

**6.119** (a)  The expected number of males with Fragile X Syndrome is np = 10000(1/1500) = 6.67.

(b)  To use the normal approximation to the binomial distribution, we first need the standard deviation, which is

$$\sigma_x = \sqrt{np(1-p)} = \sqrt{10000(\frac{1}{1500})(\frac{1499}{1500})} = 2.581$$

To find the approximate probability that more than 7 males have Fragile X syndrome, we want P(X > 7) or P(X ≥ 8) or, approximating, we use P(X > 7.5). For x = 7.5, the z-score is

$$z = \frac{7.5 - 6.67}{2.581} = 0.32.$$

The area to the left of z = 0.32 is 0.6255. Thus, the probability that more than 7 males have Fragile X Syndrome is 1.0000 - 0.6255 = 3745.

To find the approximate probability that at most 10 males have Fragile X syndrome, we want P(X ≤ 10) or, approximating, we use P(X < 10.5). For x = 10.5, the z=score is

$$z = \frac{10.5 - 6.67}{2.581} = 1.48.$$

The area to the left of z = 1.48 is 0.9306. Thus, the probability that at most 10 males have Fragile X Syndrome is 0.9306.

(c)  Although both np and n(1 - p) are at least 5, a binomial distribution with p = 1/1500 is highly skewed to the right whereas the normal distribution being used to approximate it is a symmetric distribution. The Poisson distribution used in Exercise 5.53 reflects the skewness of the binomial distribution and thus we would expect it to provide the better approximation.

## Review Problems for Chapter 6

1.  Two primary reasons for studying the normal distribution are that:

(a)  It is often appropriate to use the normal distribution as the distribution of a population or random variable.

(b)  The normal distribution is frequently employed in inferential statistics.

2.  (a)  A variable is normally distributed if its distribution has the shape of a normal curve.

(b)  A population is normally distributed if a variable of the population is normally distributed and it is the only variable under consideration.

(c)  The parameters for a normal curve are the mean μ and the standard deviation σ.

3.  (a)  False. There are many types of distributions that could have the same mean and standard deviation.

(b)  True. The mean and standard deviation completely determine a normal distribution, so if two normal distributions have the same mean and standard deviation, then those two distributions are identical.

4.  The percentages for a normally distributed variable and areas under the corresponding normal curve (expressed as a percentage) are identical.

5.  The distribution of the standardized version of a normally distributed

variable is the standard normal distribution, that is, a normal distribution with a mean of 0 and standard deviation of 1.

6.    (a)  True.  The mean completely determines the center of the distribution.

      (b)  True.  The standard deviation completely determines the shape of the distribution.

7.    (a)  The (second) curve with $\sigma$ = 6.2 has the largest spread.

      (b)  The first and second curves are centered at $\mu$ = 1.5.

      (c)  The first and third curves have the same shape because $\sigma$ = 3 for both.

      (d)  The third curve is centered farthest to the left because it has the smallest value of $\mu$.

      (e)  The fourth curve is the standard normal curve because $\mu$ = 0 and $\sigma$ = 1.

8.    Key Fact 6.2.

9.    (a)  The table entry corresponding to the specified z-score is the area to the left of that z-score.

      (b)  The area to the right of a specified z-score is found by subtracting the table entry from 1.

      (c)  The area between two specified z-scores is found by subtracting the table entry for the smaller z-score from the table entry for the larger z-score.

10.   (a)  Find the table entry that is closest to the specified area.  The z-score determined by locating the corresponding marginal values is the z-score that has the specified area to its left.

      (b)  Subtract the specified area from 1.  Find the entry in the table that is closest to the result of the subtraction.  The z-score determined by locating the corresponding marginal values is the z-score that has the specified area to its right.

11.   The value $z_\alpha$ is the z-score that has area $\alpha$ to its right under the standard normal curve.

12.   The 68.26-95.44-99.74 rule states that for a normally distributed variable: 68.26% of all possible observations lie within one standard deviation to either side of the mean; 95.44% of all possible observations lie within two standard deviations to either side of the mean; 99.74% of all possible observations lie within three standard deviation to either side of the mean.

13.   The normal scores for a sample of observations are the observations we would expect to get for a sample of the same size for a variable having the standard normal distribution.

14.   If we observe the values of a normally distributed variable for a sample, then a normal probability plot should be roughly <u>linear</u>.

15.

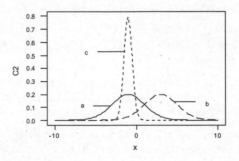

**16.**   (a)

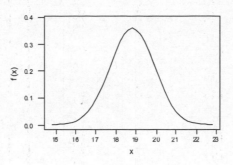

(b)   z = (x - 18.8)/1.1

(c)

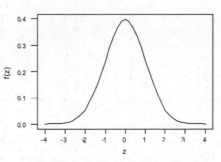

(d)   P(17 ≤ x ≤ 20) = 0.8115

(e)   The percentage of men who have forearm lengths less than 16 inches
equals the area under the standard normal curve that lies to the <u>left</u>
of <u>-2.55</u>.

**17.**   (a)   The area to the right of 1.05 is 1.0000 - 0.8531 = 0.1469.
(b)   The area to the left of -1.05 is 0.1469 (by symmetry).
(c)   The area between -1.05 and 1.05 is 0.8531 - 0.1469 = 0.7062.

**18.**   (a)   Area = 0.0013                          (b)   Area = 1 - 0.7291 = 0.2709

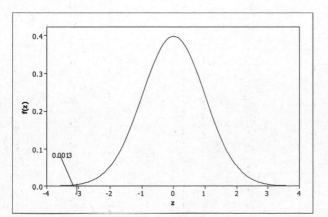

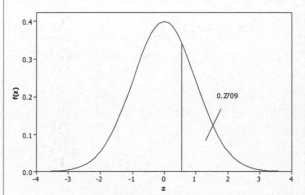

(c)  Area = 0.9970 - 0.8665 = 0.1305      (d)    Area = 1.000 - 0.0197 = 0.9803

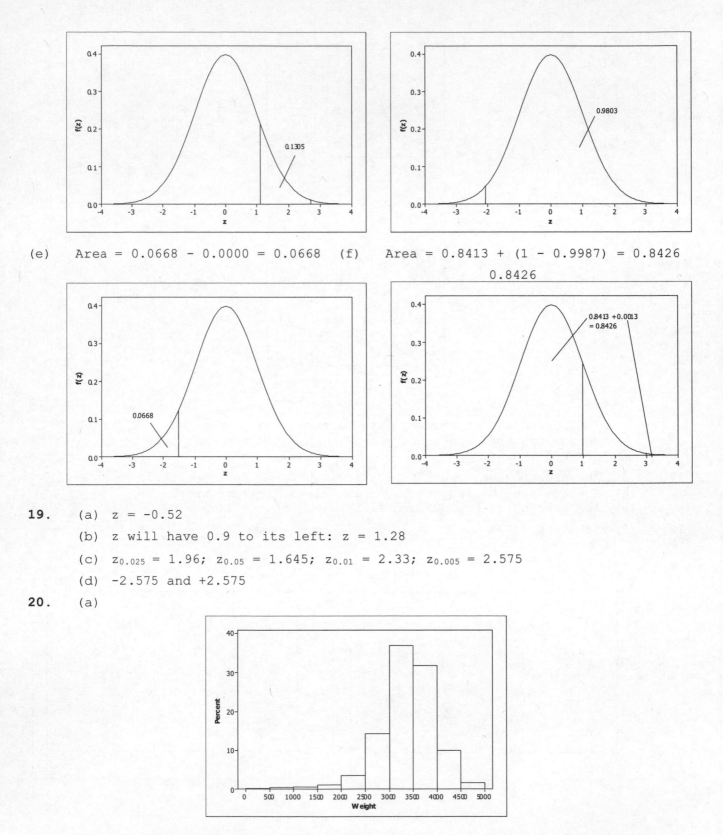

(e)   Area = 0.0668 - 0.0000 = 0.0668   (f)   Area = 0.8413 + (1 - 0.9987) = 0.8426

                                              0.8426

**19.**   (a)  z = -0.52

      (b)  z will have 0.9 to its left: z = 1.28

      (c)  $z_{0.025}$ = 1.96; $z_{0.05}$ = 1.645; $z_{0.01}$ = 2.33; $z_{0.005}$ = 2.575

      (d)  -2.575 and +2.575

**20.**   (a)

      (b)  No.  The histogram indicates that the data are left skewed.

21. (a) The z-score for x = 1.4 is (1.4 - 1.3)/0.4 = 0.25. The area to the left of z = 0.25 is 0.5987, so the percentage of patients receiving total knee arthroplasty who have a knee hyaluronic acid concentration below 1.4 mg/ml is 0.5987.

    (b) The z-score for x = 1.0 is (1.0 - 1.3)/0.4 = -0.75 and for x = 2.0, z = (2.0 - 1.3)/0.4 = 1.75. The area to the left of z = -0.75 is 0.2266 and the area to the left of z = 1.75 is 0.9599. Therefore, the probability that the percentage of patients receiving total knee arthroplasty who have a knee hyaluronic acid concentration between 1 and 2 mg/ml is 0.9599 - 0.2266 = 0.7333.

    (c) The z-score for x = 2.1 is (2.1 - 1.3)/0.4 = 2.00. The area to the left of z = 2.00 is 0.9772, so the percentage of patients receiving total knee arthroplasty who have a knee hyaluronic acid concentration above 2.1 mg/ml is 1.0000 - 0.9772 = 0.0228.

22. (a) An area of 0.2500 lies to the left of z = -0.67; an area of 0.5000 lies to the left of z = 0.00; and an area of 0.7500 lies to the left of z = 0.67. To convert these z-scores to GRE scores (x), we use x = μ + zσ. Thus $Q_1$ = 465 + (-0.67)(116) = 387.28; the median ($Q_2$) = 465 + (0.00)(116) = 465; and $Q_3$ = 465 + (0.67)(116) = 542.72. 25% of the GRE scores will be less than 387.28; 50% of the scores will be less than 465; and 75% of the scores will be less than 542.72.

    (b) An area of 0.9900 lies to the left of z = 2.33. To convert this z-scores to a GRE score (x), we use x = μ + zσ. Thus the 99th percentile is 465 + (2.33)(116) = 735.28. 99% of the GRE scores will be less than 735.28.

23. (a) 68.26% of students who took the verbal portion of the 2000 GRE scored between 349 and 581.

    (b) 95.44% of students who took the verbal portion of the 2000 GRE scored between 233 and 697.

    (c) 99.74% of students who took the verbal portion of the 2000 GRE scored between 117 and 813.

24. (a) Order the prices from smallest to largest, replacing the second 2.17 with 2.171. Obtain the normal scores for 12 observations from Table III. The result is the table below. The normal probability plot is to the right of the table.

| Price | Normal score |
|-------|--------------|
| 1.96  | -1.64 |
| 1.97  | -1.11 |
| 2.03  | -0.79 |
| 2.04  | -0.53 |
| 2.06  | -0.31 |
| 2.07  | -0.1 |
| 2.09  | 0.1 |
| 2.13  | 0.31 |
| 2.14  | 0.53 |
| 2.17  | 0.79 |
| 2.171 | 1.11 |
| 2.29  | 1.64 |

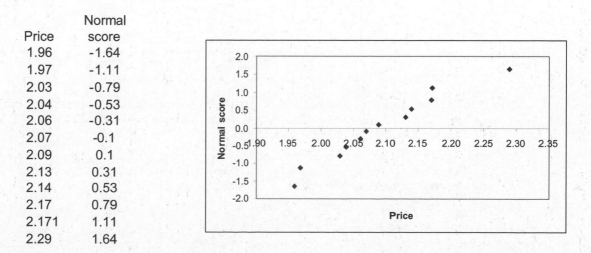

    (b) There do not appear to be any outliers.

(c) The plot is roughly linear, so it seems reasonable to assume that the data come from a normal distribution.

(d) Using Minitab, enter the data in a column named PRICE. Select **Graph ▶ Probability Plot**, select **Single** and click **OK**. Enter PRICE in the **Graph variables**, click on the **Distributions** button and the **Data Display** tab. Check the **Symbols only** option and click OK. Then click on the **Scale** button and the **Y-Scale Type** tab and select the **Score** option. Click OK twice. The resulting graph is

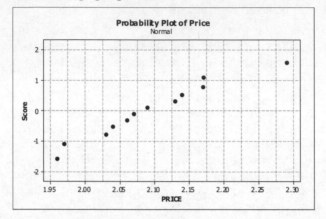

25. (a) Using Minitab, enter the data in a column named EMPLOYEES. Select **Graph ▶ Probability Plot**, select **Single** and click **OK**. Enter EMPLOYEES in the **Graph variables**, click on the **Distributions** button and the **Data Display** tab. Check the **Symbols only** option and click **OK**. Then click on the **Scale** button and the **Y-Scale Type** tab and select the **Score** option. Click OK twice. The resulting graph is

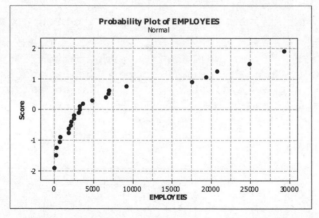

(b) There are five observations that are considerably greater than the other 20. If the distribution were otherwise normally distributed, these might be possible outliers.

(c) The graph is not close to being roughly linear. The data are very right skewed, so the five possible outliers may be just part of the pattern of right skewness.

26. For parts (a), (b), and (c), steps 1-3 are as follows:

Step 1: n = 1500; p = 0.80

Step 2: np = 1200; n(1 - p) = 300. Since both np and n(1 - p) are at least 5, the normal approximation can be used.

Step 3:

$$\mu_x = np = 1500(0.8) = 1200$$

$$\sigma_x = \sqrt{np(1-p)} = \sqrt{1500(0.8)\ (0.2)} = 15.49$$

(a)   P(x = 1225):

Step 4:   For x = 1224.5 and x = 1225.5, the z-scores are

$$z = \frac{1224.5 - 1200}{15.49} = 1.58 \text{ and } z = \frac{1225.5 - 1200}{15.49} = 1.65$$

The area to the left of z = 1.58 is 0.9429 and the area to the left of z = 1.65 is 0.9505.  The probability that the vaccine will be effective in exactly 1225 cases is approximately = 0.9505 - 0.9429 = 0.0076.

(b)   P(x ≥ 1175):

Step 4:     For x = 1174.5, the z-score is

$$z = \frac{1174.5 - 1200}{15.49} = -1.65.$$

The area to the left of z = -1.65 is 0.0495.  The probability that the vaccine will be effective in at least 1175 cases is approximately 1.0000 - 0.0495 = 0.9505.

(c)   P(1150 ≤ x ≤ 1250):

Step 4:     For x = 1149.5 and x = 1250.5, the z-scores are

$$z = \frac{1149.5 - 1200}{15.49} = -3.26 \text{ and } z = \frac{1250.5 - 1200}{15.49} = 3.26.$$

The area to the left of z = -3.26 is 0.0006 and the area to the left of z = 3.26 is 0.9994.  The probability that the vaccine will be effective in 11509 to 1250 cases, inclusive, is approximately = 0.9994 - 0.0006 = 0.9988.

# CHAPTER 7 ANSWERS

**Exercises 7.1**

**7.1** Sampling is often preferable to conducting a census because it is quicker, less costly, and sometimes it is the only practical way to get information.

**7.3** (a) $\mu$ = ∍x/N = 400/5 = 80.0 inches.

(b)

| Sample | Heights | $\bar{x}$ |
|---|---|---|
| B,D | 79, 84 | 81.5 |
| B,R | 79, 85 | 82.0 |
| B,G | 79, 78 | 78.5 |
| B,P | 79, 74 | 76.5 |
| D,R | 84, 85 | 84.5 |
| D,G | 84, 78 | 81.0 |
| D,P | 84, 74 | 79.0 |
| R,G | 85, 78 | 81.5 |
| R,P | 85, 74 | 79.5 |
| G,P | 78, 74 | 76.0 |

(c)

```
                                            B
         B  B          B  B  B          B  B   B           B
+-----+-----+-----+-----+-----+-----+-----+-----+-----+-----+-----+
74    75    76    77    78    79    80    81    82    83    84    85
```

(d) $P(\bar{x} = \mu) = P(\bar{x} = 80.0) = 0.0$

(e) $P(80.0 - 1 \leq \bar{x} \leq 80.0 + 1) = P(79.0 \leq \bar{x} \leq 81.0)$

$= P(79 \text{ or } 79.5 \text{ or } 80.5)$

$= 3/10 = 0.3$

If we take a random sample of two heights, there is a 30% chance that the mean of the sample selected will be within one inch of the population mean.

**7.5**   (b)

| Sample | Heights | $\bar{x}$ |
|--------|---------|-----------|
| B,D,R | 79,84,85 | 82.67 |
| B,D,G | 79,84,78 | 80.33 |
| B,D,P | 79,84,74 | 79.00 |
| B,R,G | 79,85,78 | 80.67 |
| B,R,P | 79,84,74 | 79.00 |
| B,G,P | 79,78,74 | 77.00 |
| D,R,G | 84,85,78 | 82.33 |
| D,R,P | 84,85,74 | 81.00 |
| D,G,P | 84,78,74 | 78.67 |
| R,G,P | 85,78,74 | 79.00 |

(c)

```
                            B
                            B
            B        B B        B B B      B B
    +-----+-----+-----+-----+-----+-----+-----+-----+-----+-----+-----+
    74    75    76    77    78    79    80    81   82    83    84    85
```

(d)  $P(\bar{x} = \mu) = P(\bar{x} = 80.0) = 0.0$

(e)  $P(80.0 - 1 \le \bar{x} \le 80.0 + 1) = P(79.0 \le \bar{x} \le 81.0)$

$= P(79.00) + P(80.33) + P(80.67) + P(81.00)$

$= 0.3 + 0.1 + 0.1 + 0.1 = 0.6$

If we take a random sample of three heights, there is a 60% chance that the mean of the sample selected will be within one inch of the population mean.

**7.7**   (b)

| Sample | Heights | |
|--------|---------|---|
| B,D,N,G,P | 79,84,85,78,74 | 80.00 |

(c)

```
                                            B
    +-----+-----+-----+-----+-----+-----+-----+-----+-----+-----+-----+
    74    75    76    77    78    79    80    81   82    83    84    85
```

(d)  $P(\bar{x} = \mu) = P(\bar{x} = 80.0) = 1.0$

(e)  $P(80.0 - 1 \le \bar{x} \le 80.0 + 1) = P(79.0 \le \bar{x} \le 81.0) = P(80.0) = 1.0$

If we take a random sample of five heights, there is a 100% chance that the mean of the sample selected will be within one inch of the population mean.

**7.9**  (a)  $\mu = \Sigma x/N = 153/6 = 25.5$ billion

(b)

| Sample | Wealth | $\bar{x}$ |
|--------|--------|-----------|
| G,B | 41, 31 | 36.0 |
| G,A | 41, 26 | 33.5 |
| G,P | 41, 20 | 30.5 |
| G,T | 41, 18 | 29.5 |
| G,E | 41, 17 | 29.0 |
| B,A | 31, 26 | 28.5 |
| B,P | 31, 20 | 25.5 |
| B,T | 31, 18 | 24.5 |
| B,E | 31, 17 | 24.0 |
| A,P | 26, 20 | 23.0 |
| A,T | 26, 18 | 22.0 |
| A,E | 26, 17 | 21.5 |
| P,T | 20, 18 | 19.0 |
| P,E | 20, 17 | 18.5 |
| T,E | 18, 17 | 17.5 |

(c)

```
  B  B B        B B  B   B B  B         B B B  B     B         B                 B
  +-----+-----+-----+-----+-----+-----+-----+-----+-----+-----+-----+-----+-----+
 17    19    21    23    25    27    29    31    33    35    37    39    41
```

(d)  $P(\bar{x} = \mu) = P(\bar{x} = 25.5) = 1/15 = 0.067$

(e)  $P(25.5 - 2 \leq \bar{x} \leq 25.5 + 2) = P(23.5 \leq \bar{x} \leq 27.5)$

$$= 3/15$$

$$= 0.20$$

If we take a random sample of two of these six rich people, there is a 20% chance that their mean wealth will be within two billion of the population mean wealth.

**7.11** (b)

| Sample | Wealth | $\bar{x}$ |
|--------|-----------|------|
| G,B,A | 41, 31, 26 | 32.7 |
| G,B,P | 41, 31, 20 | 30.7 |
| G,B,T | 41, 31, 18 | 30.0 |
| G,B,E | 41, 31, 17 | 29.7 |
| G,A,P | 41, 26, 20 | 29.0 |
| G,A,T | 41, 26, 18 | 28.3 |
| G,A,E | 41, 26, 17 | 28.0 |
| G,P,T | 41, 20, 18 | 26.3 |
| G,P,E | 41, 20, 17 | 26.0 |
| G,T,E | 41, 18, 17 | 25.3 |
| B,A,P | 31, 26, 20 | 25.7 |
| B,A,T | 31, 26, 18 | 25.0 |
| B,A,E | 31, 26, 17 | 24.7 |
| B,P,T | 31, 20, 18 | 23.0 |
| B,P,E | 31, 20, 17 | 22.7 |
| B,T,E | 31, 18, 17 | 22.0 |
| A,P,T | 26, 20, 18 | 21.3 |
| A,P,E | 26, 20, 17 | 21.0 |
| A,T,E | 26, 18, 17 | 20.3 |
| P,T,E | 20, 18, 17 | 18.3 |

(c)

```
     B   B BB  B B B   B B BBBB     BB B BB B      B
   +----+----+----+----+----+----+----+----+----+----+----+----+----+
   17   19   21   23   25   27   29   31   33   35   37   39   41
```

(d) $P(\bar{x} = \mu) = P(\bar{x} = 25.5) = 0/20 = 0.00$

(e) $P(25.5 - 2 \le \bar{x} \le 25.5 + 2) = P(23.5 \le \bar{x} \le 27.5)$

$= P(24.7) + P(25.0) + P(25.3) + P(25.7) + P(26.0) + P(26.3)$

$= 1/20 + 1/20 + 1/20 + 1/20 + 1/20 + 1/20 = 6/20$

$= 0.30$

If we take a random sample of three wealthy people, there is a 30% chance that their mean wealth will be within two billion of the population mean wealth.

**7.13** (b)

| Sample | Wealth | $\overline{x}$ |
|---|---|---|
| G,B,A,P,T | 41,31,26,20,18 | 27.2 |
| G,B,A,P,E | 41,31,26,20,17 | 27.0 |
| G,B,A,T,E | 41,31,26,18,17 | 26.6 |
| G,B,P,T,E | 41,31,20,18,17 | 25.4 |
| G,A,P,T,E | 41,26,20,18,17 | 24.4 |
| B,A,P,T,E | 31,26,20,18,17 | 22.4 |

(c)

```
            B   B  B BBB
+-----+-----+-----+-----+-----+-----+-----+-----+-----+-----+-----+-----+
17    19    21    23    25    27    29    31    33    35    37    39    41
```

(d)  $P(\overline{x} = \mu) = P(\overline{x} = 25.5) = 0/6 = 0.000$

(e)  $P(25.5 - 2 \le \overline{x} \le 25.5 + 2) = P(23.5 \le \overline{x} \le 27.5)$

$= P(24.4) + P(25.4) + P(26.6) + P(27.0) + P(27.2) = 5(1/6) = 5/6 = 0.833$

If we take a random sample of five wealthy people, there is an 83.3% chance that their mean wealth will be within two billion of the population mean wealth.

**7.15** Increasing the sample size tends to reduce the sampling error.

**7.17** (a)  If a sample of size n = 1 is taken from a population of size N, there are N possible samples.

(b)  Since each sample mean is based upon a single observation, the possible values of the sample mean and the population values are the same.

(c)  There is no difference between taking a random sample of size n = 1 from a population and selecting a member at random from the population.

**Exercises 7.2**

**7.19** Obtaining the mean and standard deviation of $\overline{x}$ is a first step in approximating the sampling distribution of the mean by a normal distribution because the normal distribution is completely determined by its mean and standard deviation.

**7.21** Yes.  The spread of the distribution of sample means becomes smaller as the sample size increases.  Since that spread is measured by the standard deviation of all possible sample means, the standard deviation also gets smaller.

**7.23** Standard error of the sample mean.  The standard deviation of $\overline{x}$ determines the amount of sampling error to be expected when a population mean is estimated by a sample mean.

**7.25** (a)  $\mu_{\overline{x}} = \dfrac{\sum x}{N} = \dfrac{800}{10} = 80.0$

(b)

$$\mu_{\bar{x}} = \frac{\sum \bar{x}}{N} = \frac{76.0 + 76.5 + 78.5 + 79.0 + 79.5 + 81.0 + 81.5 + 81.5 + 82.0 + 84.5}{10}$$

$$= \frac{800}{10} = 80.0$$

(c) $\mu_{\bar{x}} = \mu = 80.00$

**7.27** (b)

$$\mu_{\bar{x}} = \frac{\sum \bar{x}}{N} = \frac{77.00 + 78.67 + 79.00 + 79.00 + 79.33 + 80.33 + 80.67 + 81.00 + 82.33 + 82.67}{10}$$

$$= \frac{800}{10} = 80.0$$

(c) $\mu_{\bar{x}} = \mu = 80.00$

**7.29** (b)

$$\mu_{\bar{x}} = \frac{\sum \bar{x}}{N} = \frac{78.75 + 79.00 + 80.25 + 80.50 + 81.50}{5}$$

$$= \frac{400}{5} = 80.0$$

(c) $\mu_{\bar{x}} = \mu = 80.0$

**7.31** (a) The population consists of all babies born in 1991. The variable is the birth weight of the baby.

(b) $\mu_{\bar{x}} = \mu = 3369$ grams; $\sigma_{\bar{x}} = \sigma / \sqrt{n} = 581 / \sqrt{200} = 41.08$ grams

(c) $\mu_{\bar{x}} = \mu = 3369$ grams; $\sigma_{\bar{x}} = \sigma / \sqrt{n} = 10 / \sqrt{400} = 29.05$ grams

**7.33** (a) $\mu_{\bar{x}} = \mu = \$61,300$; $\sigma_{\bar{x}} = \sigma / \sqrt{n} = 7200 / \sqrt{50} = \$1018.23$

Thus, for samples of size 50, the mean and standard deviation of all possible sample means are respectively, $61,300 and $1018.23.

(b)  $\mu_{\bar{x}} = \mu = \$61,300; \ \sigma_{\bar{x}} = \sigma / \sqrt{n} = 7200 / \sqrt{100} = \$720.00$

Thus, for samples of size 100, the mean and standard deviation of all possible sample means are respectively, $61,300 and $720.

**7.35**  (a)  On the average, we would expect the sample mean of the four times to be the same as the population mean of 437 days.

(b)  The standard deviation of the sample mean is

$\sigma_{\bar{x}} = \sigma / \sqrt{n} = 399 / \sqrt{4} = 199.5 \text{ days}$ , so 99.74% of the time we would expect the sample mean to fall within 3 standard deviations of 437 days, i.e., between 437 - 3(199.5) = -161.5 and 437 + 3(199.5) = 1035.5. Since the lower limit is clearly below zero and the time between earthquakes cannot be negative, we would expect the sample mean to fall between 0 and 1035.5 days 99.74% of the time.

**7.37**  (a)  Enter the data into cells A1 to A50 of Excel. In cell B1, enter the expression =STDEVP(A1:A50) to obtain the population standard deviation of the data in B1. The result is 1362.452 Note: Using DDXL procedures will yield a SAMPLE standard deviation of 1376.284, which you could then convert to a POPULATION standard deviation by multiplying by the square root of 49/50 to obtain the same result.

(b)  Equation 7.1 is the correct formula for obtaining the standard deviation of the sample mean since we are sampling without replacement from a finite population of 50 data values.

(c)  From Equation 7.1, $\sigma_{\bar{x}} = \sqrt{\dfrac{N-n}{N-1}} \dfrac{\sigma}{\sqrt{n}} = \sqrt{\dfrac{50-30}{50-1}} \dfrac{1362.452}{\sqrt{30}} = 158.920$ .

From Equation 7.2, $\sigma_{\bar{x}} = \dfrac{\sigma}{\sqrt{n}} = \dfrac{1362.452}{\sqrt{30}} = 248.749$ .

We expect the two results to be close when the sample size is small compared to the population size, that is, when it is 5% or less of the population size. In this instance, the sample size is 30 and the population size is 50, so the sample size is 60% of the population size. This causes the first square root in Equation 7.1 to be considerably less than 1, resulting in a large discrepancy between the two results.

(d)  From Equation 7.1, $\sigma_{\bar{x}} = \sqrt{\dfrac{N-n}{N-1}} \dfrac{\sigma}{\sqrt{n}} = \sqrt{\dfrac{50-2}{50-1}} \dfrac{1362.452}{\sqrt{2}} = 953.518$ .

From Equation 7.2, $\sigma_{\bar{x}} = \dfrac{\sigma}{\sqrt{n}} = \dfrac{1362.452}{\sqrt{2}} = 963.399$ .

We expect the two results to be close when the sample size is small compared to the population size, that is, when it is 5% or less of the population size. In this instance, the sample size is 2 and the population size is 50, so the sample size is 4% of the population size. This causes the first square root in Equation 7.1 to be quite close to 1, resulting in only a small discrepancy between the two results.

(e)  In Excel, in cells C1 to C50, enter the numbers 1 to 50. In cell D1, enter the expression =SQRT((50-C1)/(50-1))*1362.452/SQRT(C1). In Cell

E1, enter the expression =1362.452/SQRT(C1). Copy cells D1 and E1 to the clipboard and then paste them into cells D2 through D50. This will produce your table in three columns. To save space, we show the table below using six columns.

| n | Eq. 7.1 | Eq. 7.2 | n | Eq. 7.1 | Eq. 7.2 |
|---|---------|---------|---|---------|---------|
| 1 | 1362.45 | 1362.45 | 26 | 187.00 | 267.20 |
| 2 | 953.52 | 963.40 | 27 | 179.64 | 262.20 |
| 3 | 770.39 | 786.61 | 28 | 172.53 | 257.48 |
| 4 | 660.04 | 681.23 | 29 | 165.63 | 253.00 |
| 5 | 583.91 | 609.31 | 30 | 158.92 | 248.75 |
| 6 | 527.08 | 556.22 | 31 | 152.38 | 244.70 |
| 7 | 482.40 | 514.96 | 32 | 145.98 | 240.85 |
| 8 | 445.97 | 481.70 | 33 | 139.70 | 237.17 |
| 9 | 415.43 | 454.15 | 34 | 133.52 | 233.66 |
| 10 | 389.27 | 430.85 | 35 | 127.42 | 230.30 |
| 11 | 366.49 | 410.79 | 36 | 121.38 | 227.08 |
| 12 | 346.36 | 393.31 | 37 | 115.37 | 223.99 |
| 13 | 328.36 | 377.88 | 38 | 109.38 | 221.02 |
| 14 | 312.11 | 364.13 | 39 | 103.37 | 218.17 |
| 15 | 297.31 | 351.78 | 40 | 97.32 | 215.42 |
| 16 | 283.73 | 340.61 | 41 | 91.19 | 212.78 |
| 17 | 271.18 | 330.44 | 42 | 84.95 | 210.23 |
| 18 | 259.51 | 321.13 | 43 | 78.53 | 207.77 |
| 19 | 248.61 | 312.57 | 44 | 71.87 | 205.40 |
| 20 | 238.38 | 304.65 | 45 | 64.88 | 203.10 |
| 21 | 228.72 | 297.31 | 46 | 57.40 | 200.88 |
| 22 | 219.58 | 290.48 | 47 | 49.17 | 198.73 |
| 23 | 210.88 | 284.09 | 48 | 39.73 | 196.65 |
| 24 | 202.58 | 278.11 | 49 | 27.81 | 194.64 |
| 25 | 194.64 | 272.49 | 50 | 0.00 | 192.68 |

When n = 2, n is less than 5% of N and the error is about 1%. When n is 10 (20% of N), the error is over 10%. When n is 25 (50% of N), the error is about 40%. We see that the larger n is relative to N, the greater the difference between the correct Equation 7.1 and the approximating Equation 7.2.

**7.39** (a)   Yes. The mean of the sampling distribution of $\bar{x}$ is always $\mu$. A demonstration of this is given in Example 7.4 and in Exercises 7.25 through 7.29.

(b)   No. For example, in Example 7.4, the population consists of five observations (76, 78, 79, 81, and 86). The population median is 79. For samples of size 2, the median and mean are identical and therefore have the same sampling distribution. The mean of the sampling distribution of the median equals the mean of the sampling distribution of the mean, which is shown to be 80 in the example. Since this is not equal to the population median, it is clear that, in general, the sample median is not an unbiased estimator of the population median.

**7.41**   (a)   $\sigma_{\bar{x}} = \sqrt{\dfrac{N-n}{N-1}}\dfrac{\sigma}{\sqrt{n}} = \sqrt{\dfrac{5-1}{5-1}}\dfrac{3.41}{\sqrt{1}} = 3.41$      $\sigma_{\bar{x}} = \dfrac{\sigma}{\sqrt{n}} = \dfrac{3.41}{\sqrt{1}} = 3.41$      20%

$\sigma_{\bar{x}} = \sqrt{\dfrac{N-n}{N-1}}\dfrac{\sigma}{\sqrt{n}} = \sqrt{\dfrac{5-2}{5-1}}\dfrac{3.41}{\sqrt{2}} = 2.09$      $\sigma_{\bar{x}} = \dfrac{\sigma}{\sqrt{n}} = \dfrac{3.41}{\sqrt{2}} = 2.41$      40%

$\sigma_{\bar{x}} = \sqrt{\dfrac{N-n}{N-1}}\dfrac{\sigma}{\sqrt{n}} = \sqrt{\dfrac{5-3}{5-1}}\dfrac{3.41}{\sqrt{3}} = 1.39$      $\sigma_{\bar{x}} = \dfrac{\sigma}{\sqrt{n}} = \dfrac{3.41}{\sqrt{3}} = 1.97$      60%

$\sigma_{\bar{x}} = \sqrt{\dfrac{N-n}{N-1}}\dfrac{\sigma}{\sqrt{n}} = \sqrt{\dfrac{5-4}{5-1}}\dfrac{3.41}{\sqrt{4}} = 0.85$      $\sigma_{\bar{x}} = \dfrac{\sigma}{\sqrt{n}} = \dfrac{3.41}{\sqrt{4}} = 1.70$      80%

$\sigma_{\bar{x}} = \sqrt{\dfrac{N-n}{N-1}}\dfrac{\sigma}{\sqrt{n}} = \sqrt{\dfrac{5-5}{5-1}}\dfrac{3.41}{\sqrt{5}} = 0.00$      $\sigma_{\bar{x}} = \dfrac{\sigma}{\sqrt{n}} = \dfrac{3.41}{\sqrt{5}} = 1.52$      100%

(b)   Equation 7.2 yields poor results when n = 2, 3, 4, or 5 because n is much larger than 5% of N.  Equation 7.2 always yields the same result as Equation 7.1 when n = 1 [See Exercise 7.40 (b)], even though n is 20% of N in this example.

**7.43**   (a)   Your answer will very likely be similar, but different from the one below, which we present for illustration.  A histogram of the sampling

distribution of $\bar{x}$ for our samples of size 4 is shown.

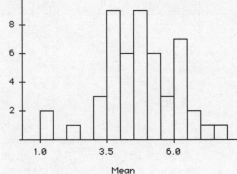

(b)   The mean of the population is

μ = 3x / N = (0 + 1 + 2 + 3 + 4 + 5 + 6 + 7 + 8 + 9)/10 = 4.5

The standard deviation of the population is

$$\sigma = \sqrt{\dfrac{\Sigma(x_i - \mu)^2}{N}} = \sqrt{\dfrac{82.5}{10}} = 2.87 .$$

Thus, for samples of size 4 with replacement, the mean of $\bar{x}$ is 4.5 and

the standard deviation of $\bar{x}$ is $2.87/\sqrt{4} = 1.435$.  We would expect the mean and standard deviation of the 50 sample means obtained in part (a) to be roughly 4.5 and 1.43 respectively.

(b)   For the 50 samples we obtained, the mean was 4.62 and the standard deviation was 1.355.

(c)   The answers in part (d) differ from the theoretical values given in part (c) as a result of random error or sampling variability.

**7.45**   (a)   Theoretically, the mean of all possible sample means is the mean of the population, 8.7, and the standard deviation is $8.7/\sqrt{10} = 2.75$ .

(b) We have Minitab take 1000 random samples of size n = 10 from an exponentially distributed population with mean 8.7 by choosing **Calc <**

**Random Data < Exponential...**, entering <u>1000</u> in the **Generate rows of data** text box, entering <u>C1-C10</u> in the **Store in column(s)** text box, entering <u>8.7</u> in the **Mean** text box, and clicking **OK**. These commands tell Minitab to place a total of 10000 observations from the population into columns C1-C10, with 1000 observations in each column.  Then, our first random sample of size n = 10 is the first row of columns C1-C10, our second random sample of size n = 10 is the second row of columns C1-C10, and so on.

(b) We compute the sample mean of each of the 1000 samples by choosing **Calc**

**< Row statistics...**, clicking on the **Mean** button, typing <u>C1-C10</u> in the **Input variables** text box and <u>XBAR</u> in the **Store result in** text box, and clicking **OK**.  This command instructs Minitab to compute the means of the 1000 rows of C1-C10 and to place those means in a column named XBAR.  Thus, the 2000 -values are now in XBAR.  (We will not print these values.)

(c) We would expect the mean of the 1000 sample means to be roughly 8.7 and the standard deviation of the sample means to be about 2.75 since we are taking a sample of 1000 means from a theoretical sampling distribution with mean and standard deviation given in part (a).

(e) The mean is obtained by choosing **Calc < Column statistics...**, clicking on the **Mean** button, selecting XBAR in the **Input variable** text box, and clicking **OK**.  The result is

    Mean of XBAR = 8.7252

Similarly, the standard deviation is obtained by choosing **Calc < Column statistics...**, clicking on the **Standard deviation** button, selecting XBAR in the **Input variable** text box, and clicking **OK**.  The result is

    Standard deviation of XBAR = 2.7595.

Thus, the mean of our 1000 means was 8.7252 and the standard deviation was 2.7595.

(f) The answers in part (e) differ from the theoretical values given in part (d) as a result of random error or sampling variability.

## Exercises 7.3

**7.47** (a) The sampling distribution of the mean is approximately normally distributed with mean

μ = 100 and standard deviation $\sigma_{\bar{x}} = 28/\sqrt{49} = 4$ .

(b) No assumptions were made about the distribution of the population.

(c) Part (a) cannot be answered if the sample size is n = 16.  Since the distribution of the population is not specified, we need a sample size of at least 30 to apply Key Fact 7.6.

**7.49** (a) The probability distribution of $\bar{x}$ is normal with mean μ and standard deviation $\sigma_{\bar{x}} = \sigma/\sqrt{n}$ .

    (b)  The answer to part (a) does not depend on how large the sample size is because the population being sampled is normally distributed.

    (c)  The mean of $\bar{x}$ is ~; its standard deviation is $\sigma_{\bar{x}} = \sigma/\sqrt{n}$ .

    (d)  No.  The mean and standard deviation of the sampling distribution of the sample mean are always as given in part (c) regardless of the distribution of the variable under consideration.

**7.51**  (a)  All four graphs are centered at the same place because the population mean of $\bar{x}$ is ~ and because normal curves are centered at their μ-parameter.

    (b)  Since $\sigma_{\bar{x}} = \sigma/\sqrt{n}$ ,we see that $\sigma_{\bar{x}}$ decreases as $n$ increases.  This results in a diminishing of the spread because the spread of a distribution is determined by its Φ-parameter.  As a consequence, we see that the larger the sample size, the greater the likelihood for small sampling error.

    (c)  The graphs in Figure 7.6(a) are bell-shaped because, for normally distributed populations, the random variable $\bar{x}$ is always normally distributed regardless of the sample size.

    (d)  The graphs in Figures 7.6(b) and 7.6(c) become bell-shaped as the sample size increases because of the central limit theorem; the probability distribution of $\bar{x}$ tends to a normal distribution as the sample size increases.

**7.53**  (a)  Because the weights themselves are normally distributed, the sampling distribution for means of samples of size 3 will also be normal and will have mean $\mu_{\bar{x}} = \mu = 1.40\,kg$ and standard deviation $\sigma_{\bar{x}} = 0.11/\sqrt{3} = 0.0635$ .  Thus, the distribution of the possible sample means for samples of three brain weights will be normal with mean 1.40 kg and standard deviation 0.0635 kg.

    (b)  Because the weights themselves are normally distributed, the sampling distribution for means of samples of size 12 will also be normal and will have mean $\mu_{\bar{x}} = \mu = 1.40\,kg$ and standard deviation $\sigma_{\bar{x}} = 0.11/\sqrt{12} = 0.0318$ .  Thus, the distribution of the possible sample means for samples of twelve brain weights will be normal with mean 1.40 kg and standard deviation 0.0318 kg.

    (c)  To facilitate the comparison of the three graphs, we have overlaid them on one set of axes.

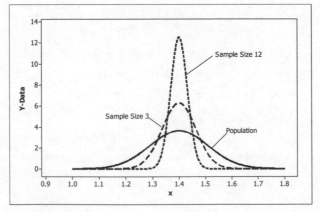

(d)  $\sigma_{\bar{x}} = 0.11/\sqrt{3} = 0.064$ ; we want P$(1.30 \leq \bar{x} \leq 1.50)$.

z-score computations:                          Area less than z:

$$\bar{x} = 1.30 \rightarrow z = \frac{1.30 - 1.40}{0.064} = -1.56 \qquad\qquad 0.0594$$

$$\bar{x} = 1.50 \rightarrow z = \frac{1.50 - 1.40}{0.064} = 1.56 \qquad\qquad 0.9406$$

Total area = 0.9406 - 0.0594 = 0.8812

Thus, 88.12% of samples of three Swedish men will have mean brain weights within 0.1 kg of the population mean brain weight of 1.40 kg.

(e)  $\sigma_{\bar{x}} = 0.11/\sqrt{12} = 0.0318$ ; we want P$(1.30 \leq \bar{x} \leq 1.50)$.

z-score computations:                          Area less than z:

$$\bar{x} = 1.30 \rightarrow z = \frac{1.30 - 1.40}{0.032} = -3.13 \qquad\qquad 0.0009$$

$$\bar{x} = 1.50 \rightarrow z = \frac{1.50 - 1.40}{0.032} = 3.13 \qquad\qquad 0.9991$$

Total area = 0.9991 - 0.0009 = 0.9982
Thus, 99.82% of samples of twelve Swedish men will have mean brain weights within 0.1 kg of the population mean brain weight of 1.40 kg.

**7.55**  (a)  The sampling distribution of the sample mean for samples of size 64 will be approximately normal with a mean of $45,871 and a standard deviation of $\sigma_{\bar{x}} = 8000/\sqrt{64} = 1000$ .  Thus, if all possible sample means for samples of size 64 were found, their distribution would be approximately normal with mean $45,871 and standard deviation $1,000.

(b)  The sampling distribution of the sample mean for samples of size 256 will be approximately normal with a mean of $45,871 and a standard deviation of $\sigma_{\bar{x}} = 8000/\sqrt{256} = \$500$ .  Thus, if all possible sample means for samples of size 256 were found, their distribution would be approximately normal with mean $45,871 and standard deviation $500.

(c)  No.  These results follow from the central limit theorem, which implies that for large samples (n $\geq$ 30), the distribution of the sample mean will be approximately normal regardless of the distribution of the original variable.

(d)  $\sigma_{\bar{x}} = 8000/\sqrt{64} = 1000$ ; we want P$(44,871 \leq \bar{x} \leq 46,871)$.

z-score computations:                          Area less than z:

$$\bar{x} = 44871 \rightarrow z = \frac{44871 - 45871}{1000} = -1.00 \qquad\qquad 0.1587$$

$$\bar{x} = 46871 \rightarrow z = \frac{46871 - 45871}{1000} = 1.00 \qquad\qquad 0.8413$$

Total area = 0.8413 - 0.1587 = 0.6826

Thus, 68.26% of samples of size 64 will have a mean that is within $1000 of the population mean, i.e., 68.26% of the samples will have a sampling error less than $1000 for samples of size 64.

(e) $\sigma_{\bar{x}} = 8000/\sqrt{256} = 500$; we want $P(44{,}871 \leq \bar{x} \leq 46{,}871)$.

z-score computations:　　　　　　　　　　Area less than z:

$$\bar{x} = 44871 \to z = \frac{44871 - 45871}{500} = -2.00 \qquad 0.0228$$

$$\bar{x} = 46871 \to z = \frac{46871 - 45871}{500} = 2.00 \qquad 0.9772$$

Total area = 0.9772 - 0.0228 = 0.9544

Thus, 95.44% of samples of size 256 will have a mean that is within $1000 of the population mean, i.e., 95.44% of the samples will have a sampling error less than $1000 for samples of size 256.

**7.57** (a) The sampling distribution of the sample mean for n = 80 will be approximately normal with some mean μ and standard deviation $\sigma_{\bar{x}} = 8.3/\sqrt{80} = 0.928$.

(b) No. The sample size of 80 is large enough so that the Central Limit Theorem applies and the sampling distribution will be approximately normal.

(c) We want $P(\mu - 2 < \bar{x} < \mu + 2)$
z-score computations:　　　　　　　　　　Area less than z:

$$\bar{x} = \mu - 2 \to z = \frac{(\mu - 2) - \mu}{0.928} = -2.16 \qquad 0.0154$$

$$\bar{x} = \mu + 2 \to z = \frac{(\mu + 2) - \mu}{0.928} = 2.16 \qquad 0.9846$$

Total area = 0.9846 - 0.0154 = 0.9692

Thus, the probability is approximately 0.9692 that the sampling error made in estimating the population mean length of stay on the intervention ward by the mean length of stay of a sample of 80 patients will be at most 2 days.

**7.59**　　$\sigma_{\bar{x}} = 40/\sqrt{250} = 2.53$; $P(|\bar{x} - \mu| \leq 5)$

z-score computations:　　　　　　　　　　Area less than z:

$$\bar{x} = \mu - 5 \to z = \frac{(\mu - 5) - \mu}{2.53} = -1.98 \qquad 0.0239$$

$$\bar{x} = \mu + 5 \to z = \frac{(\mu + 5) - \mu}{2.53} = 1.98 \qquad 0.9761$$

Required Area = 0.9761 - 0.0239 = 0.9522

Thus, there is a 0.9522 probability that the contractor's estimate will be within 5 months of the true mean.

**7.61** (a) 68.26　　(b) 95.44　　(c) 99.74　　(d) 100(1 - ∀)

**7.63** (a) No.  The original population is quite skewed.  A sample size of 4 is quite small, so we would expect that a histogram of the means of such a sample would still exhibit some degree of skewness.

(b) We'll use Excel to carry out the simulation.  First enter the values 1 through 7 in cells A1 to A7, then enter the frequencies 19.4 through 1.6 from the table in Example 7.9 into cells C1 through C7.  In C8, enter the expression =Sum(C1:C7).  Then in cell B1, enter the expression =C1/$C$8, copy this cell to the clipboard and then paste it into cells B2 through B7.  Columns A and B now contain the probability distribution of x, the household size.  Now from the Tools menu, select

**Data Analysis** < **Random Number Generation**.  Enter <u>4</u> in the **Number of Variables** text box, enter <u>1000</u> in the **Number of Random Numbers** text box, choose **Discrete** in the **Distribution** box, enter the range <u>A1:B7</u> in the **Value and Probability Input Range** text box, and enter <u>A11</u> in the **Output Range** box.  Click on **OK**.  You will get 4 columns of 100 numbers between 1 and 7 in columns A, B, C, and D, starting in row 11 and ending in row 1010.  In cell E11, enter the expression =AVERAGE(A11:D11).  Copy this cell to the clipboard and then paste it into the cells E12 through E1010.  Type the label **Mean** in cell E10.  Use the mouse to highlight cells E10 though E1010.  Now from the **DDXL** menu, select **Charts and Plots**.  In the **Function Type** box, select **Histogram**, Click on **Mean** in the **Names and Columns** box and enter it in the **Quantitative Variables** box.  Then click **OK**.  Our result follows.  Yours is very likely to be different.  Our histogram is slightly skewed to the right.

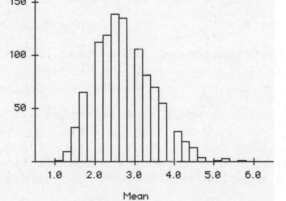

(c) We would expect this histogram to be closer to bell-shaped than the one in part (a), but 10 is still a relatively small sample size, so there may still be a trace of skewness.  Using the same procedure as in part (b), but with 10 variables, our result below is more symmetrical, but we definitely did not expect the one tall spike.  Note that the scale is different from the first histogram.  These means are more closely concentrated.

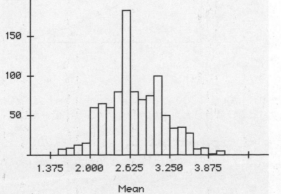

(d)  Since 100 is definitely a large sample size, we would expect the
     histogram of the sample means to be bell-shaped.  Using the same
     procedure as in part (b), but with 100 variables, our result below is
     close to bell-shaped and again is more concentrated than the first two.

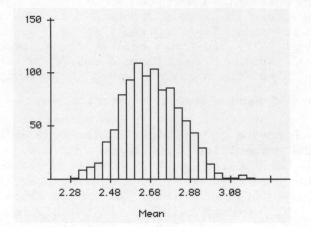

**7.65**  (a)  We will describe a procedure using Minitab to sketch the exponential
     curve for this population, although the sketch could be done using any
     spreadsheet program as well.  We begin by selecting typical values of x
     that will eventually be substituted into the exponential function
     itself to trace out values for y.  Values of x selected are the
     integers 0 through 15.  Name two columns in the Data window X and Y and

     enter the numbers 0, 1, 2, ...,15 in the X column.  Then choose **Calc** <

     **Probability Distributions** < **Exponential...,** click on the **Probability
     density** button, click in the **Mean** text box and type 8.7.  Then enter **X**
     in the **Input column** text box and **Y** in the **Optional storage** text box.

     Click **OK**.  Now choose **Graph** < **Scatterplot...,** select the **With connect
     line** version and click **OK**.  Select Y for row 1 of the **Y** column and X
     for row 1 of the **X** column.  To plot a continuous curve, click on the
     **Data view** button and check only the **Connect line** box.  Click **OK** twice.
     The result is shown below.  The exponential distribution is reverse J-
     shaped.

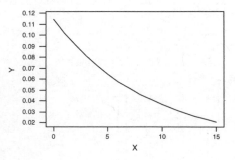

(b)  To have Minitab take 1000 random samples of size n = 4 from an
     exponentially distributed population with mean μ = 8.7, we choose **Calc**

     < **Random Data** < **Exponential...,** enter 1000 in the **Generate rows of
     data** text box, enter C1-C4 in the **Store in column(s)** text box, enter

8.7 in the **Mean** text box, and click **OK**. These commands tell Minitab to place a total of 4000 observations from the population into columns C1-C4, with 1000 observations in each column.  Then, our first random sample of size n = 4 is the first row of columns C1-C4, our second random sample of size n = 4 is the second row of columns C1-C4, and so on.

(c)  We compute the sample mean of each of the 1000 samples by choosing **Calc**

$<$ **Row statistics...**, clicking on the **Mean** button, entering C1-C4 in the **Input variables** text box and XBAR in the **Store result in** text box, and clicking **OK**.  This command instructs Minitab to compute the means of the 1000 rows of C1-C4 and to place those means in XBAR.  Thus, the 1000 -values are now in XBAR.  (We will not print these values.) Your results will differ from ours.

(d)  The mean of the 1000 sample means is obtained by choosing **Calc** $<$ **Column statistics...**, clicking on the **Mean** button, selecting XBAR in the **Input variable** text box, and clicking **OK**.  The result is

   Mean of XBAR = 8.4743

Similarly, the standard deviation of the 1000 sample means is obtained by choosing **Calc** $<$ **Column statistics...**, clicking on the **Standard deviation** button, selecting XBAR in the **Input variable** text box, and clicking **OK**.  The result is

   Standard deviation of XBAR = 4.2498

(e)  Theoretically, for all possible sample means for samples of size 4, the mean is the same as the population mean, 8.7, and the standard deviation is $\sigma / \sqrt{n} = 8.7 / \sqrt{4} = 4.35$ .  Both of the simulated values obtained in part (d) are close to these values.

(f)  To get a histogram of the 1000 sample means stored in XBAR, choose

   **Graph** $<$ **Histogram...**, selecting the **Simple** version and click **OK**.  Enter XBAR in the **Graph variables** text box, and click **OK**.  The result is shown below at the right.  The histogram is not bell-shaped; given the right-skewness of the population, we would expect for samples of only size four, that the distribution of means would be somewhat skewed to the right.

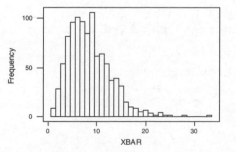

(g)  Repeating the process above for samples of size 40 (use columns C1-C40 for the samples), we find that

   Mean of XBAR = 8.7367 and Standard deviation of XBAR = 1.4204

In theory, the mean of all possible samples of size 40 is 8.7 and the standard deviation is $\sigma / \sqrt{n} = 8.7 / \sqrt{40} = 1.38$. The simulated values are very close to these.

A histogram of the means of samples of size 40 is shown following, obtained as above. The simulated distribution of sample means is close to bell-shaped as we would expect for samples over size 30.

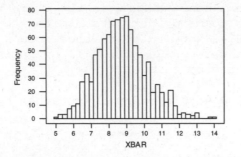

### Review Problems for Chapter 7

1. Sampling errors are errors resulting from using a sample to estimate a population characteristic.

2. The sampling distribution of a statistic is the set of all possible observations of the statistic for a sample of a given size. It is important to know the sampling distribution in order to answer questions about the accuracy of estimating a population parameter using the sample statistic.

3. Two other terms are 'sampling distribution of the sample mean' and 'the distribution of the variable $\bar{x}$.'

4. The set of possible sample means exhibits less and less variability as the sample size increases, that is, the set becomes more and more clustered about the population mean. This means that as the sample size increases, there is a greater chance that the value of the sample mean from any sample is close to the value of the population mean.

5. (a) The sampling error results from using the sample mean income tax, $\bar{x}$, of the 175,485 tax returns sampled as an estimate of the mean income tax, $\mu$, of all 2003 tax returns.

   (b) The sampling error is $8412 - $8500 = -$88.

   (c) No, not necessarily. However, increasing the sample size from 175,485 to 250,000 would increase the likelihood for a smaller sampling error.

   (d) Increase the sample size.

6. (a) $\mu = $108/6 = $18$ (thousands)

(b)

| Sample | Salaries | |
|---|---|---|
| A,B,C,D | 8, 12, 16, 20 | 14 |
| A,B,C,E | 8, 12, 16, 24 | 15 |
| A,B,C,F | 8, 12, 16, 28 | 16 |
| A,B,D,E | 8, 12, 20, 24 | 16 |
| A,B,D,F | 8, 12, 20, 28 | 17 |
| A,B,E,F | 8, 12, 24, 28 | 18 |
| A,C,D,E | 8, 16, 20, 24 | 17 |
| A,C,D,F | 8, 16, 20, 28 | 18 |
| A,C,E,F | 8, 16, 24, 28 | 19 |
| A,D,E,F | 8, 20, 24, 28 | 20 |
| B,C,D,E | 12, 16, 20, 24 | 18 |
| B,C,D,F | 12, 16, 20, 28 | 19 |
| B,C,E,F | 12, 16, 24, 28 | 20 |
| B,D,E,F | 12, 20, 24, 28 | 21 |
| C,D,E,F | 16, 20, 24, 28 | 22 |

(c)

```
                              •
              •     •     •     •     •
        •     •     •     •     •     •     •     •     •
      +----+----+----+----+----+----+----+----+
      14   15   16   17   18   19   20   21   22
```

(d)
$$P(|\bar{x} - \mu| \le 1) = P(\bar{x} = 17) + P(\bar{x} = 18) + P(\bar{x} = 19)$$
$$= 2/15 + 3/15 + 2/15 = 7/15 = 0.4667$$

(e) $\mu_{\bar{x}} = \dfrac{\sum \bar{x}}{N} = \dfrac{270}{15} = 18.0$. The mean of the means of all of the possible samples of size 4 is $18.0 thousand.

(f) Yes. The mean of the sampling distribution of $\bar{x}$ is always the same as the mean of the population, which is $18 thousand in this case.

7. (a) The population consists of all new car sales in the U.S. in 2005. The variable under consideration is the amount spent for a new car.

(b) $\mu_{\bar{x}} = \$27,958; \sigma_{\bar{x}} = \$10200/\sqrt{50} = \$1442.50$

(c) $\mu_{\bar{x}} = \$27,958; \sigma_{\bar{x}} = \$10200/\sqrt{100} = \$1020.00$

(d) The value of $\sigma_{\bar{x}}$ will be smaller than $1020 because $\sigma_{\bar{x}} = \sigma/\sqrt{n}$. Thus, the larger the sample size, the smaller the value of $\sigma_{\bar{x}}$.

8. (a) False. By the central limit theorem, the random variable $\bar{x}$ is approximately normally distributed. Furthermore, $\mu_{\bar{x}} = \mu = 45$, and

$\sigma_{\bar{x}} = \sigma/\sqrt{n} = 7/\sqrt{196} = 0.5$. Thus, P(31 $\le \bar{x} \le$ 59) equals the area under the normal curve with parameters $\mu_{\bar{x}} = 45$ and $\sigma_{\bar{x}} = 0.5$ that lies between 31 and 59.   Applying the usual techniques, we find that area to be 1.0000 to four decimal places.   Hence, there is almost a 100% chance that the mean of the sample will be between 31 and 59.

(b)  This is not possible to tell, since we do not know the distribution of the population.

(c)  True.   Referring to part (a), we see that $P(44 \leq \bar{x} \leq 46)$ equals the area under the normal curve with parameters $\mu_{\bar{x}} = 45$ and $\sigma_{\bar{x}} = 0.5$ that lies between 44 and 46.   Applying the usual techniques, we find that area to be 0.9544.   Hence, there is approximately a 95.44% chance that the mean of the sample will be between 44 and 46.

**9.**   (a)  False.   Since the population is normally distributed, so is the random variable $\bar{x}$.   Furthermore, $\mu_{\bar{x}} = \mu = 45$ and $\sigma_{\bar{x}} = \sigma / \sqrt{n} = 7 / \sqrt{196} = 0.5$.   Hence, as in Problem 8(a), we find that there is almost a 100% chance that the mean of the sample will be between 31 and 59.

(b)  True.   Since the population is normally distributed, percentages for the population are equal to areas under the normal curve with parameters $\mu = 45$ and $\Phi = 7$.   Applying the usual techniques, we find that the area under that normal curve between 31 and 59 is 0.9544.

(c)  True.   From part (a), we see that the random variable $\bar{x}$ is normally distributed with $\mu_{\bar{x}} = 45$ and $\sigma_{\bar{x}} = 0.5$.   Applying the usual techniques, we find that area to be 0.9544. Hence, we find that there is a 95.44% chance that the mean of the sample will be between 44 and 46.

**10.**   (a)  $\mu = 40 \; and \; \sigma = 12$

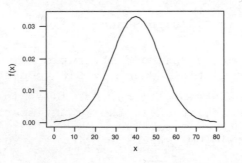

(b)  normal distribution with $\mu_{\bar{x}} = 40 \; and \; \sigma_{\bar{x}} = 12 / \sqrt{4} = 6$

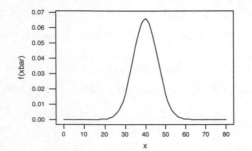

(c) normal distribution with $\mu_{\bar{x}} = 40$ *and* $\sigma_{\bar{x}} = 12/\sqrt{9} = 4$

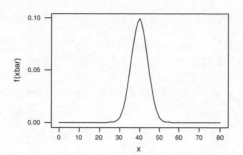

**11.**

$n = 4 : \mu_{\bar{x}} = 40$ *and* $\sigma_{\bar{x}} = 12/\sqrt{4} = 6$

(a) P(31 ≤ $\bar{x}$ ≤ 49):

z-score computations:                    Area less than z:

$$\bar{x} = 31 \rightarrow z = \frac{31-40}{6} = -1.50$$                    0.0668

$$\bar{x} = 49 \rightarrow z = \frac{49-40}{6} = 1.50$$                    0.9332

Total area = 0.9332 - 0.0668 = 0.8664, or 86.64%.

(b) The probability is 0.8664 that a sample of size four will have a mean within 9 mm of the population mean of 40 mm.

(c) There is an 86.64% chance that the sampling error when using the mean

krill length $\bar{x}$ of the four krill will be less than 9 mm.

(d) $n = 9 : \mu_{\bar{x}} = 40$ *and* $\sigma_{\bar{x}} = 12/\sqrt{9} = 4$

z-score computations:                    Area less than z:

$$\bar{x} = 31 \rightarrow z = \frac{31-40}{4} = -2.25$$                    0.0122

$$\bar{x} = 49 \rightarrow z = \frac{49-40}{4} = 2.25$$                    0.9878

Total area = 0.9878 - 0.0122 = 0.9756 or 97.56%

The probability is 0.9756 that a sample of size 9 will have a mean within 9 mm of the population mean of 40 mm.

There is an 97.56% chance that the sampling error when using the mean

krill length $\bar{x}$ of the nine krill will be less than 9 mm.

**12.**    (a) For a normally distributed population, the random variable $\bar{x}$ is normally distributed, regardless of the sample size. Also, we know that $\mu_{\bar{x}} = \mu$. Consequently, since the normal curve for a normally distributed population or random variable is centered at its $\mu$-parameter, all three curves are centered at the same place.

(b) Curve B corresponds to the larger sample size. Since $\sigma_{\bar{x}} = \sigma/\sqrt{n}$, the larger the sample size, the smaller the value of $\Phi$ and, hence, the smaller the spread of the normal curve for $\bar{x}$. Thus, Curve B, which has the smaller spread, corresponds to the larger sample size.

(c) The spread of each curve is different because $\sigma_{\bar{x}} = \sigma/\sqrt{n}$ and the spread of a normal curve is determined by $\sigma_{\bar{x}}$. Thus, different sample sizes result in normal curves with different spreads.

(d) Curve B corresponds to the sample size that will tend to produce less sampling error. The smaller the value of $\sigma_{\bar{x}}$, the smaller the sampling error tends to be.

(e) When x is normally distributed, $\bar{x}$ always has a normal distribution as well.

13. (a) Since 60 is a fairly large sample size, the distribution of the sample mean will be approximately normal with a mean of 4.60 mmol/l and a standard deviation of $\sigma_{\bar{x}} = \sigma/\sqrt{n} = 0.16/\sqrt{60} = 0.0207$.

(b) Since 120 is a large sample size, the distribution of the sample mean will be approximately normal with a mean of 4.60 mmol/l and a standard deviation of $\sigma_{\bar{x}} = \sigma/\sqrt{n} = 0.16/\sqrt{120} = 0.0146$.

(c) No. The Central Limit Theorem ensures that when n is large, the distribution of the sample mean will be approximately normal regardless of whether the blood glucose levels themselves are normally distributed.

14. (a) $n = 500: \mu_{\bar{x}} = \mu \text{ and } \sigma_{\bar{x}} = 50,900/\sqrt{500} = 2276.32$

$P(\mu - 2,000 \leq \bar{x} \leq \mu + 2,000)$:

| z-score computations: | Area less than z: |
|---|---|
| $\bar{x} = \mu - 2000 \rightarrow z = \dfrac{(\mu - 2000) - \mu}{2276.32} = -0.88$ | 0.1894 |
| $\bar{x} = \mu + 2000 \rightarrow z = \dfrac{(\mu + 2000) - \mu}{2276.32} = 0.88$ | 0.8106 |

Total area = 0.8106 - 0.1894 = 0.6212

(b) To answer part (a), it is not necessary to assume that the population is normally distributed because the sample size is large and, therefore, $\bar{x}$ is approximately normally distributed, regardless of the distribution of the population of life-insurance amounts.

If the sample size were 20 instead of 500, it would be necessary to assume normality because the sample size is small.

(c)  $n = 5000 : \mu_{\bar{x}} = \mu$ *and* $\sigma_{\bar{x}} = 50,900 / \sqrt{5000} = 719.83$

z-score computations:                                              Area less than z:

$$\bar{x} = \mu - 2000 \rightarrow z = \frac{(\mu - 2000) - \mu}{709.83} = -2.78$$          0.0027

$$\bar{x} = \mu + 2000 \rightarrow z = \frac{(\mu + 2000) - \mu}{709.83} = 2.78$$          0.9973

Total area = 0.9973 - 0.0027 = 0.9946

**15.**    (a)  P($\bar{x}$ $\leq$ 4.5):

z-score computation:                        Area less than z:

$$\bar{x} = 4.5 \rightarrow z = \frac{4.5 - 5}{0.5} = -1.00$$                0.1587

Total area = 0.1587

If the paint lasts 4.5 years, one would not consider this to be
substantial evidence against the manufacturer's claim that the paint
will last an average of five years.
Assuming the manufacturer's claim is correct, the probability is 0.1587
that the paint will last 4.5 years or less on a (randomly selected)
house painted with the paint.  In other words, there is a (fairly high)
15.87% chance that the paint would last 4.5 years or less, even if the
manufacturer's claim is correct.

(b)  P($\bar{x}$ $\leq$ 4.5):

$n = 10 : \mu_{\bar{x}} = 5$ *and* $\sigma_{\bar{x}} = 0.5 / \sqrt{10} = 0.158$

z-score computation:                        Area less than z:

$$\bar{x} = 4.5 \rightarrow z = \frac{4.5 - 5}{0.158} = -3.16$$                0.0008

Total area = 0.0008

For 10 houses, if the paint lasts an average of 4.5 years, I would
consider this to be substantial evidence against the manufacturer's
claim that the paint will last an average of five years.
Assuming the manufacturer's claim is correct, the probability is 0.0008
that the paint will last an average of 4.5 years or less for 10
(randomly selected) houses painted with the paint.  In other words,
there is less than a 0.1% chance that that would occur, if the
manufacturer's claim is correct.

(c)  P($\bar{x}$ $\leq$ 4.9):

z-score computation:                        Area less than z:

$$\bar{x} = 4.9 \rightarrow z = \frac{4.9 - 5}{0.158} = -0.63$$                0.2643

Total area = 0.2643

For 10 houses, if the paint lasts an average of 4.9 years, I would not consider this to be substantial evidence against the manufacturer's claim that the paint will last an average of five years.
Assuming the manufacturer's claim is correct, the probability is 0.2643 that the paint will last an average of 4.9 years or less for 10 (randomly selected) houses painted with the paint.  In other words, there is a (fairly high) 26.43% chance that that would occur, even if the manufacturer's claim is correct.

**16.**   (a)   We have Minitab take 1000 random samples of size n = 4 from a normally distributed population with mean 578 and standard deviation 147 by choosing **Calc < Random Data  < Normal...**, entering <u>1000</u> in the **Generate rows of data** text box, entering <u>C1-C4</u> in the **Store in column(s)** text box, entering <u>578</u> in the **Mean** text box and <u>147</u> in the **Standard deviation** text box, and clicking **OK**.

(b)   We compute the sample mean of each of the 1000 samples by choosing **Calc < Row statistics...**, clicking on the **Mean** button, typing <u>C1-C4</u> in the **Input variables** text box and <u>XBAR</u> in the **Store result in** text box, and clicking **OK**.  (We will not print these values.)

(c)   The mean is obtained by choosing **Calc < Column statistics...**, clicking on the **Mean** button, selecting XBAR in the **Input variable** text box, and clicking **OK**.  Our result is
        Mean of XBAR = 587.559

Similarly, the standard deviation is obtained by choosing **Calc < Column statistics...**, clicking on the **Standard deviation** button, selecting XBAR in the **Input variable** text box, and clicking **OK**.  Our result is
        Standard deviation of XBAR = 71.7314
To get a histogram of the 1000 sample means stored in XBAR, choose **Graph < Histogram...**, select the **Simple** version and click **OK**.  Enter XBAR in the **Graph variables** text box, and click **OK**.  The result follows.

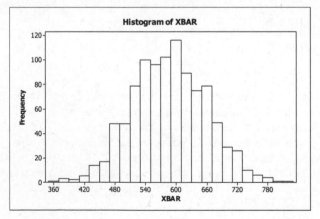

(d)   Theoretically, the distribution of all possible sample means for samples of size four from this normal population should have mean 578, standard deviation $\sigma_{\bar{x}} = \sigma / \sqrt{n} = 147 / \sqrt{4} = 73.5$, and a normal distribution.

(e)   The histogram is close to bell-shaped, is centered near 578, and most of the data lies within three standard deviations (3 x 73.5 = 220.5) of

578, i.e. between 357.5 and 798.5.  The mean of the 1000 sample means is 587.559, very close to 578, and the standard deviation of the sample means is 71.7314, very close to 73.5.

**17.**  (a)  A uniform distribution between 0 and 1:

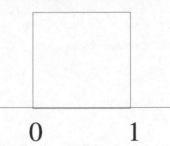

0      1

(b)  We have Minitab take 2000 random samples of size n = 2 from a uniformly distributed population between 0 and 1 by choosing **Calc < Random Data < Uniform...**, entering <u>2000</u> in the **Generate rows of data** text box, entering <u>C1-C2</u> in the **Store in column(s)** text box, entering <u>0.0</u> in the **Lower endpoint** text box and <u>1.0</u> in the **Upper endpoint** text box, and clicking **OK**.

(c)  We compute the sample mean of each of the 1000 samples by choosing **Calc < Row statistics...**, clicking on the **Mean** button, typing <u>C1-C2</u> in the **Input variables** text box and <u>XBAR</u> in the **Store result in** text box, and clicking **OK**.  (We will not print these values.)

(d)  The mean of the sample means is obtained by choosing **Calc < Column statistics...**, clicking on the **Mean** button, selecting XBAR in the **Input variable** text box, and clicking **OK**.  Our result is

  Mean of XBAR = 0.50454

Similarly, the standard deviation is obtained by choosing **Calc < Column statistics...**, clicking on the **Standard deviation** button, selecting XBAR in the **Input variable** text box, and clicking **OK**.  Our result is

  Standard deviation of XBAR = 0.20545

(e)  Theoretically, the distribution of all possible sample means for samples of size two from this uniform normal population should have mean

$(0 + 1)/2 = .5$ and standard deviation $\sigma / \sqrt{n} = ((1-0)/\sqrt{12})/\sqrt{2} = 0.2041$.  Minitab simulation results are very close to these theoretical values.

(f)  To get a histogram of the 1000 sample means stored in XBAR, choose **Graph < Histogram...**, select the **Simple** version and click **OK**.  Enter XBAR in the **Graph variables** text box, and click **OK**.  The result is shown following.

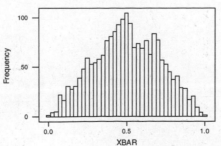

The histogram is more triangle-shaped than bell-shaped.  Since the population is far from normal and the sample size is only two, we would not expect the sampling distribution to be bell-shaped.  In fact, it can be shown using advanced methods that the sum of two uniformly distributed variables has a triangular distribution.

(g)  Repeating the process above, but using C1-C35 for the data in each sample, we obtained

Mean of XBAR = 0.49932

Standard deviation of XBAR = 0.049240

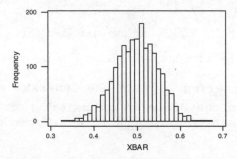

The theoretical distribution of the means has mean 0.5 and standard deviation $\sigma / \sqrt{n} = ((1-0)/\sqrt{12})/\sqrt{35} = 0.0488$.

The Minitab simulation values are very close to these.  The histogram for the means of the samples of size 35 is shown.  It is much more bell-shaped than for samples of size two, as we would expect since n > 30.

**CHAPTER 8 ANSWERS**

<u>Exercises 8.1</u>

**8.1**   The value of a statistic that is used to estimate a parameter is called a <u>point estimate</u> of the parameter.

**8.3**   (a)   $\bar{x}$ = \$526538/20 = \$26326.90

   (b)   Since some sampling error is expected, it is unlikely that the sample mean will exactly equal the population mean $\mu$.

**8.5**   (a)   The confidence interval will be

$$\bar{x}-2\sigma/\sqrt{n} \ \ to \ \ \bar{x}+2\sigma/\sqrt{n}$$

$$\$26329.90 - 2(\$8100)/\sqrt{20} \ to \ \$26329.90 + 2(\$8100)/\sqrt{20}$$

$$\$22707.47 \ to \ \$29952.33$$

   (b)   Since we know that 95.44% of all samples of 20 wedding costs have the property that the interval from $\bar{x}$ - \$3622.43 to $\bar{x}$ + \$3622.43 contains $\mu$, we can be 95.44% confident that the interval from \$22707.47 to \$29952.33 contains $\mu$.

   (c)   We can't be certain that the population mean lies in the interval, but we are 95.44% confident that it does.

**8.7**   (a)   n = 35; a point estimate for the mean fuel tank capacity is

   $\bar{x}$ = 664.9/35 = 19.00 gallons.  This number is called a point estimate because it consists of a single value.

   (b)   The confidence interval will be

$$\bar{x}-2\sigma/\sqrt{n} \ \ to \ \ \bar{x}+2\sigma/\sqrt{n}$$

$$19.00 - 2(3.50)/\sqrt{35} \ to \ 19.00 + 2(3.50)/\sqrt{35}$$

$$17.82 \ to \ 20.18$$

   (c)   We could see if a histogram looked bell-shaped or if a normal probability plot produced a relatively straight line.

   (d)   It is not necessary that the fuel tank capacities be exactly normally distributed since the sample size 35 is large enough to ensure that the interval obtained is approximately correct.

**8.9**   (a)   Using Minitab to retrieve the data from the WeissStats CD, we obtained

   the following normal probability plot by choosing **Stat ▶ Basic**

   **Statistics ▶ Normality Test**, entering LENGTHS in the **Variable** text box, and clicking **OK**.

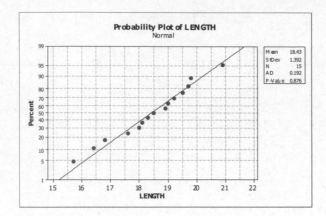

(b) Yes. There do not appear to be any outliers, and the data points fall very close to a straight line.

(c) The sample mean is 18.43 mm. Since $P(\bar{x} - 2\sigma/\sqrt{n} < \mu < \bar{x} + 2\sigma/\sqrt{n}) = 0.9544$, we can be 95.44% confident that the mean μ is somewhere between

$\bar{x} - 2\sigma/\sqrt{n}$ and $\bar{x} + 2\sigma/\sqrt{n}$. Since n = 15, $\bar{x}$ = 18.43, and $\sigma$ = 1.76, the 95.44% confidence interval is

$$\bar{x} - 2\sigma/\sqrt{n} \text{ to } \bar{x} + 2\sigma/\sqrt{n}$$
$$18.43 - 2(1.76)/\sqrt{15} \text{ to } 18.43 + 2(1.76)/\sqrt{15}$$
$$17.52 \text{ to } 19.34$$

We can be 95.44% confident that the interval 17.52 mm to 19.34 mm contains the value of the mean $\mu$.

(d) Yes. No. 18.14 lies between 17.52 and 19.34, so the interval we obtained in part (c) does contain μ. It is possible that our confidence interval will not contain μ. This should happen only about 4.56% of the time.

**8.11** Since $P(\bar{x} - 3\sigma/\sqrt{n} < \mu < \bar{x} + 3\sigma/\sqrt{n}) = 0.9974$, we can be 99.74% confident that the mean μ is somewhere between $\bar{x} - 3\sigma/\sqrt{n}$ and $\bar{x} + 3\sigma/\sqrt{n}$. Since n = 36, $\bar{x}$ = 49.28, and $\sigma$ = 7.2, the 99.74% confidence interval is (in thousands)

$$\bar{x} - 3\sigma/\sqrt{n} \text{ to } \bar{x} + 3\sigma/\sqrt{n}$$
$$59.28 - 3(7.2)/\sqrt{36} \text{ to } 59.28 + 3(7.2)/\sqrt{36}$$
$$55.68 \text{ to } 62.88$$

We can be 99.74% confident that the interval $55,680 to $62,880 contains the value of the mean $\mu$.

**Exercises 8.2**

**8.13** (a) Confidence level = 0.90; $\alpha$ = 0.10

(b) Confidence level = 0.99; $\alpha$ = 0.01

**8.15** (a) By saying that a 1-$\alpha$ confidence interval is exact, we mean that the true confidence level is equal to 1-$\alpha$.

(b) By saying that a 1-$\alpha$ confidence interval is approximately correct, we mean that the true confidence level is only approximately equal to $1-\alpha$.

**8.17**  When we use the abbreviation "normal population," we mean that the variable under consideration is normally distributed on the population of interest.

**8.19**  A statistical procedure is robust if it is insensitive to departures from the assumptions on which it is based.

**8.21**  (a)  The z-interval procedure is reasonable since the population is roughly normal and the sample size is over 15.

   (b)  The z-interval procedure is not reasonable since the sample size is too small for a population far from normal.

   (c)  The z-interval procedure is reasonable since the sampling distribution of $\bar{x}$ will be very close to normal for a sample size of 250 even though the population itself is far from normal.

**8.23**  A 95% confidence interval will give a more precise (shorter interval) estimate of $\mu$ than will a 99% confidence interval.

**8.25**  The 95% confidence interval for μ is

$$\bar{x} - z_{\alpha/2}\sigma/\sqrt{n} \quad to \quad \bar{x} + z_{\alpha/2}\sigma/\sqrt{n}$$
$$20 - 1.96(3)/\sqrt{36} \; to \; 20 + 1.96(3)/\sqrt{36}$$
$$19.02 \; to \; 20.98$$

**8.27**  The 90% confidence interval for μ is

$$\bar{x} - z_{\alpha/2}\sigma/\sqrt{n} \quad to \quad \bar{x} + z_{\alpha/2}\sigma/\sqrt{n}$$
$$30 - 1.645(4)/\sqrt{25} \; to \; 30 + 1.645(4)/\sqrt{25}$$
$$28.68 \; to \; 31.32$$

**8.29**  The 99% confidence interval for μ is

$$\bar{x} - z_{\alpha/2}\sigma/\sqrt{n} \quad to \quad \bar{x} + z_{\alpha/2}\sigma/\sqrt{n}$$
$$50 - 2.575(5)/\sqrt{16} \; to \; 50 + 2.575(5)/\sqrt{16}$$
$$46.78 \; to \; 53.22$$

**8.31**  The sample mean is $113.97 million / 18 = $6.33 million.  The 95% confidence interval for μ is

$$\bar{x} - z_{\alpha/2}\sigma/\sqrt{n} \quad to \quad \bar{x} + z_{\alpha/2}\sigma/\sqrt{n}$$
$$6.33 - 1.96(2.04)/\sqrt{18} \; to \; 6.33 + 1.96(2.04)/\sqrt{18}$$
$$\$5.39 \; to \; \$7.27 \text{(million)}$$

We can be 95% confident that the interval from $5.39 million to $7.27 million contains the population mean venture capital investment in the fiber optics business sector.

**8.33**  The sample mean is 6.31 / 12 = 0.526 ppm.  The 95% confidence interval for μ is

$$\bar{x} - z_{\alpha/2}\sigma/\sqrt{n} \quad \text{to} \quad \bar{x} + z_{\alpha/2}\sigma/\sqrt{n}$$

$$0.526 - 2.575(0.37)/\sqrt{12} \quad \text{to} \quad 0.526 + 2.575(0.37)/\sqrt{12}$$

$$0.251 \text{ to } 0.801 \text{ ppm}$$

We can be 99% confident that the interval from 0.251 to 0.801 ppm contains the population mean cadmium level in *Boletus pinicola* mushrooms.

**8.35**  $n = 32$; $\bar{x} = 33.4$; $\sigma = 42$

Step 1:  $\alpha = 0.05$; $z_{\alpha/2} = z_{0.025} = 1.96$

Step 2:

$$\bar{x} - z_{\alpha/2}\sigma/\sqrt{n} \text{ to } \bar{x} + z_{\alpha/2}\sigma/\sqrt{n}$$

$$33.4 - 1.96(42)/\sqrt{32} \quad \text{to} \quad 33.4 + 1.96(42)/\sqrt{32}$$

$$18.8 \text{ to } 48.0$$

We can be 95% confident that the mean duration of imprisonment, $\mu$, of all East German political prisoners with chronic PTSD is somewhere between 18.8 and 48.0 months.

**8.37**  $n = 18$; $\bar{x} = 6.33$; $\sigma = 2.04$

(a)  Step 1:  $\alpha = 0.01$; $z_{\alpha/2} = z_{0.005} = 2.575$

Step 2:

$$\bar{x} - z_{\alpha/2}\sigma/\sqrt{n} \text{ to } \bar{x} + z_{\alpha/2}\sigma/\sqrt{n}$$

$$6.33 - 2.575(2.04)/\sqrt{18} \quad \text{to} \quad 6.33 + 2.575(2.04)/\sqrt{18}$$

$$5.09 \text{ to } 7.57$$

(b)  The confidence interval in part (a) is longer than the one in Exercise 8.31 because we have changed the confidence level from 95% in Exercise 8.31 to 99% in this exercise.  Notice that increasing the confidence level from 95% to 99% increases the $z_{\alpha/2}$-value from 1.96 to 2.575.  The larger z-value, in turn, results in a longer interval.  In order to achieve a higher level of confidence that the interval contains the population mean, we need a longer interval.

(c)

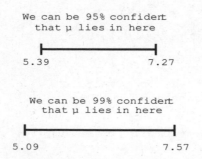

(d)  The 95% confidence interval is shorter and therefore provides a more precise estimate of μ.

**8.39**  (a)  Using Minitab, retrieve the data from the WeissStats CD.  Then choose **Stat ▶ Basic statistics ▶ 1-Sample z.**  Enter <u>TIME</u> in the **Samples in**

**Columns** text box and enter <u>30</u> in the **Standard deviation** text box.
Click on the **Graphs** button and check the boxes for **Histogram of data**
and **Boxplot of data,** then click **OK** twice.  The confidence interval
appears in the Sessions Window as

```
Variable   N     Mean   StDev  SE Mean       95% CI
TIME      20   289.950  30.741   6.708  (276.802, 303.098)
```
The histogram and boxplot are shown following.

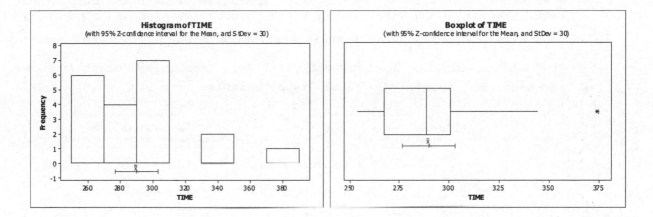

To get the probability plot and stem-and leaf diagram, choose **Stat ▶**

**Basic statistics ▶ Normality test,** enter <u>TIME</u> in the **Variable** text box

and click **OK.**  Then choose **Graph ▶ Stem-and-Leaf,** and enter <u>TIME</u> in the

**Graph Variables** text box and click **OK.**  The results are

```
Stem-and-leaf of TIME   N  = 20
Leaf Unit = 1.0


   2    25   45
   6    26   0779
   9    27   235
  10    28   7
  10    29   01257
   5    30   22
   3    31
   3    32
   3    33   3
   2    34   4
```

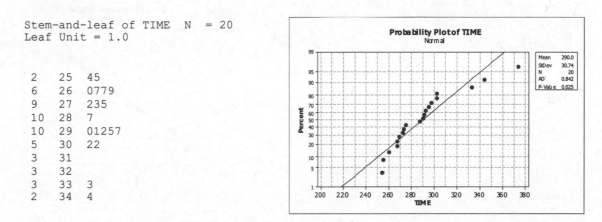

HI 374

(c)  The value 374 is a potential outlier.  Remove it from the data and
     repeat the process in part (a).  The result is

```
Variable   N     Mean   StDev  SE Mean       95% CI
TIME      19   285.526  24.174   6.882  (272.037, 299.016)
```

(d)  The graphs (not shown here) that were produced with the confidence
     interval now show that 344 is an outlier in the sample of 19
     observations.  The data are also skewed to the right both before and

after deleting the first outlier.  The relatively small sample size makes it unwise to use the z-interval procedure with these data.

8.41  (a)  We will get the confidence interval for part (c) along with the histogram and boxplot.  Using Minitab, retrieve the data from the

WeissStats CD.  Then choose **Stat ▶ Basic statistics ▶ 1-Sample z.**
Enter TEMP in the **Samples in Columns** text box and enter 0.63 in the **Standard deviation** text box.  Click on the Options button and enter 99.0 in the **Confidence level** text box and click **OK**.  Click on the **Graphs** button and check the boxes for **Histogram of data** and **Boxplot of data**, then click **OK** twice.  To get the probability plot and stem-and

leaf diagram, choose **Stat ▶ Basic statistics ▶ Normality test,** enter

TEMP in the **Variable** text box and click **OK**.  Then choose **Graph ▶ Stem-and-Leaf**, and enter TEMP in the **Graph Variables** text box and click **OK**. The results are

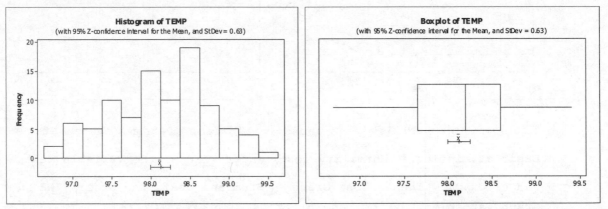

```
Stem-and-leaf of TEMP   N  = 93
Leaf Unit = 0.10

   1    96  7
   3    96  89
   8    97  00001
  13    97  22233
  19    97  444444
  26    97  6666777
  31    97  88889
  45    98  00000000000111
 (10)   98  2222222233
  38    98  4444445555
  28    98  66666666677
  17    98  8888888
  10    99  00001
   5    99  2233
   1    99  4
```

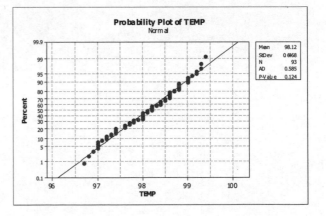

(b)  Yes.  There are no outliers, the sample size is large (93), and the distribution of the data appears to be approximately normal.

(c)  The output from part (a) in the Sessions Window is

```
Variable   N    Mean   StDev  SE Mean      99% CI
TEMP       93  98.1237  0.6468  0.0653  (97.9554, 98.2919)
```

We can be 99% confident that the interval (97.9554, 98.2919) contains the mean body temperature μ of healthy humans.  The output is surprising since the confidence interval does not inclue the generally accepted normal temperature of 98.6° F.

**8.43**  (a)  We will get the confidence interval for part (c) along with the histogram and boxplot. Using Minitab, retrieve the data from the WeissStats CD. Then choose **Stat ▶ Basic statistics ▶ 1-Sample z**. Enter SPEED in the **Samples in Columns** text box and enter 3.2 in the **Standard deviation** text box. Click on the **Options** button and enter 95.0 in the **Confidence level** text box and click **OK**. Click on the **Graphs** button and check the boxes for **Histogram of data** and **Boxplot of data**, then click **OK** twice. To get the probability plot and stem-and-leaf diagram, choose **Stat ▶ Basic statistics ▶ Normality test**, enter SPEED in the **Variable** text box and click **OK**. Then choose **Graph ▶ Stem-and-Leaf**, and enter SPEED in the **Graph Variables** text box and click **OK**. The results are

| Variable | N | Mean | StDev | SE Mean | 95% CI |
|---|---|---|---|---|---|
| SPEED | 35 | 59.5257 | 4.2739 | 0.5409 | (58.4656, 60.5859) |

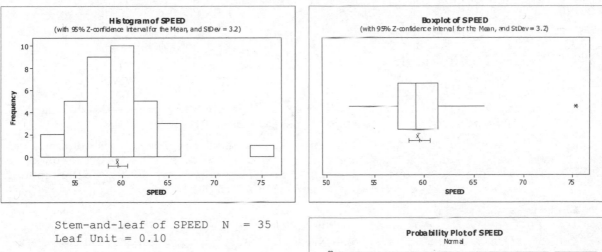

```
Stem-and-leaf of SPEED   N  = 35
Leaf Unit = 0.10

    2    52    46
    2    53
    4    54    78
    7    55    459
    8    56    5
   13    57    35688
   16    58    137
   (5)   59    02678
   14    60    12679
    9    61    36
    7    62    36
    5    63    4
    4    64
    4    65    02
    2    66    0


   HI 753
```

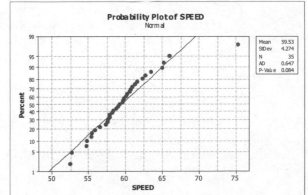

(c)  After removing 75.3 from the data, the 95% confidence interval is

| Variable | N | Mean | StDev | SE Mean | 95% CI |
|---|---|---|---|---|---|
| SPEED | 34 | 59.0618 | 3.3253 | 0.5488 | (57.9861, 60.1374) |

(d)   The outlier had an effect on the mean of about 0.46 mph and, therefore, about the same effect on the endpoints of the confidence interval.  The movement of the interval is about one-fourth the length of the interval and is significant in that sense.  On the other hand, the change in the mean represents less than a 1% error and is probably not very important, although this is a matter of judgment.  The sample size of 35 appears to be large enough that one outlier of the size observed does not have a great effect.

8.45   (a)   To increase the precision (shorten the confidence interval) without reducing the confidence level, we should increase the sample size n.

(b)   To increase our level of confidence without reducing the precision, we should increase the sample size n.

8.47   We can be $100(1 - \alpha)$% confident that the mean $\mu$ is greater than the lower confidence bound and $100(1 - \alpha)$% confident that the mean $\mu$ is less than the upper confidence bound.

8.49   (a)   The sample mean is 6.31 / 12 = 0.526 ppm.  The 99% lower confidence bound for $\mu$ is

$$\overline{x} - z_\alpha \sigma / \sqrt{n}$$
$$0.526 - 2.33(0.37)/ \sqrt{12}$$
$$0.277 \ ppm$$

We can be 99% confident that the population mean cadmium level in *Boletus pinicola* mushrooms is greater than 0.277 ppm.

(b)   This lower confidence bound is greater than the lower confidence limit of 0.251 found in Exercise 8.33.  This is because the z-value used for a one-sided 99% confidence bound is 2.33, whereas, the z-value used for a two-sided 99% confidence interval is 2.575.

## Exercises 8.3

8.51   The length of a confidence interval, and thus the precision with which $\overline{x}$ estimates $\mu$, is determined by the margin of error.

8.53   (a)   The length of the confidence interval is twice the margin of error; i.e., 2 x 3.4 = 6.8.

(b)   The confidence interval is 52.8 $\pm$ 3.4 = (49.4, 56.2).

8.55   (a)   The margin of error is 1/2 the length of the confidence interval; i.e., (1/2) x 20 = 10.

(b)   The confidence interval is 60 $\pm$ 10 = (50, 70).

8.57   (a)   True.  The length of the confidence interval is twice the margin of error.

(b)   True.  The margin of error is one-half the length of the confidence interval.

(c)   False.  One must also know the sample mean.

(d)   True.  The confidence interval is mean $\pm$ margin of error.

8.59   (a)   We want a whole number because the number of observations to be taken cannot be fractional.  It must be an integer.

(b)   The original number computed is the smallest value of n that will provide the required margin of error.  If we were to round down, the actual sample size would be slightly too small.

**8.61**  n = 10; $\bar{x}$ = 34.2 cm; $\sigma$ = 2.1 cm

(a)  $\alpha$ = 0.10; $z_{\alpha/2}$ = $z_{0.05}$ = 1.645, so the 90% confidence interval for $\mu$ is

$$\bar{x} - 1.645(2.1)/\sqrt{10} \quad to \quad \bar{x} + 1.645(2.1)/\sqrt{10}$$
$$34.2 - 1.645(2.1)/\sqrt{10} \; to \; 34.2 + 1.645(2.1)/\sqrt{10}$$
$$33.1 \, to \, 35.3 \text{cm}$$

(b)  The margin of error is $E = 1.645\sigma/\sqrt{n} = 1.645(2.1)/\sqrt{10} = 1.1\text{cm}$ .

(c)  We are 90% confident that our error in estimating $\mu$ by $\bar{x}$ is at most 1.1 cm.

(d)  $\alpha$ = 0.05; $z_{\alpha/2}$ = $z_{0.025}$ = 1.96, so $n = \left(\dfrac{z_{\alpha/2} \cdot \sigma}{E}\right)^2 = \left(\dfrac{1.96 \cdot 2.1}{0.5}\right)^2 = 67.76 \; or \; 68.$

**8.63**  (a)  E = 0.5($7.27 million- $5.39 million) = 0.94 million

(b)  $E = 1.96\sigma/\sqrt{n} = 1.96(2.04)/\sqrt{18} = \$0.94\text{million}$ .

**8.65**  (a)  E = (48.0 - 18.8)/2 = 14.6 months

(b)  We are 95% confident that the maximum error made in using $\bar{x}$ to estimate $\mu$ is 14.6 months.

(c)  The margin of error of the estimate is specified to be E = 12.0 months. Also, for a 99% confidence interval, $z_{\alpha/2}$ = $z_{0.005}$ = 2.575.

$$n = \left[\frac{z_{\alpha/2}\sigma}{E}\right]^2 = \left[\frac{2.575(42)}{12}\right]^2 = 81.2 \rightarrow 82$$

(d)

$$\bar{x} - z_{\alpha/2}\sigma/\sqrt{n} \quad to \quad \bar{x} + z_{\alpha/2}\sigma/\sqrt{n}$$
$$36.2 - 2.575(12)/\sqrt{82} \quad to \quad 36.2 + 2.575(12)/\sqrt{82}$$
$$24.3 \; to \; 48.1$$

**8.67**  (a)  The margin of error of the estimate is specified to be E = 2 years.

$$n = \left[\frac{z_{\alpha/2}\sigma}{E}\right]^2 = \left[\frac{1.96(13.36)}{2.0}\right]^2 = 171.4 \rightarrow 172$$

(b)  We used s in place of $\sigma$ because $\sigma$ was unknown.  We can do this because the sample of size 36 is large enough to provide an estimate of $\sigma$ and the variation is not likely to change much from one year to the next.

**8.69**  (a)  $\alpha$ = 0.05; $z_{\alpha/2}$ = $z_{0.025}$ = 1.96; $\sigma$ = 10

E = 1.96(10) / $\sqrt{4}$ = 9.80

(b)  $\alpha$ = 0.05; $z_{\alpha/2}$ = $z_{0.025}$ = 1.96; $\sigma$ = 10

E = 1.96(10) / $\sqrt{16}$ = 4.90

  (c)  It appears that quadrupling the sample size will halve the margin of
       error.  Therefore, increasing n from 16 to 64 will decrease the margin
       of error from 4.90 to 2.45.

**Exercises 8.4**

**8.71**  The formula for the standardized version of $\bar{x}$ uses $\sigma$ in the denominator,
          whereas the studentized version uses s.

**8.73**  (a)  The standardized version of $\bar{x}$ is
$$z = (\bar{x} - \mu)/(\sigma/\sqrt{n}) = (108 - 100)/(16/\sqrt{4}) = 1.00 \, .$$

  (b)  The studentized version of $\bar{x}$ is
$$t = (\bar{x} - \mu)/(s/\sqrt{n}) = (108 - 100)/(12/\sqrt{4}) = 1.333 \, .$$

**8.75**  (a)  Standard normal

  (b)  t-distribution with 11 degrees of freedom

**8.77**  The variation in the possible values of the standardized version of $\bar{x}$ is due

  only to the variation in $\bar{x}$ while the variation in the studentized version

  results not only from the variation in $\bar{x}$, but also from the variation in the
  sample standard deviation.

**8.79**  For df = 6:

  (a)  $t_{0.10} = 1.440$          (b)    $t_{0.025} = 2.447$          (c)    $t_{0.01} = 3.143$

**8.81**  (a)  $t_{0.10} = 1.323$                    (b)    $t_{0.01} = 2.518$

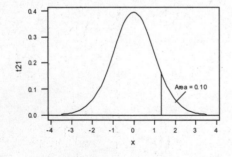

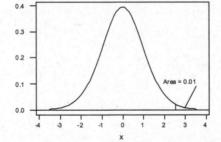

  (c)  $-t_{0.025} = -2.080$          (d)    $\pm t_{0.05} = \pm 1.721$

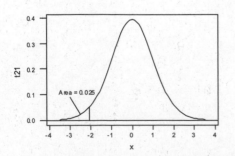

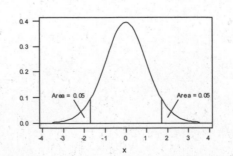

**8.83**  It is reasonable to use the t-interval procedure since the sample size is large and for large degrees of freedom (99), the t-distribution is very similar to the standard normal distribution.  Another way of expressing this is that the sampling distribution of $\bar{x}$ is approximately normal when n is large, so the standardized and studentized versions of $\bar{x}$ are essentially the same.

**8.85**  The 95% confidence interval for μ is

$$\bar{x}-t_{\alpha/2}s/\sqrt{n} \quad to \quad \bar{x}+t_{\alpha/2}s/\sqrt{n}$$

$$20-2.030(3)/\sqrt{36}\ to\ 20+2.030(3)/\sqrt{36}$$

$$18.98\ to\ 21.02$$

**8.87**  The 90% confidence interval for μ is

$$\bar{x}-t_{\alpha/2}s/\sqrt{n} \quad to \quad \bar{x}+t_{\alpha/2}s/\sqrt{n}$$

$$30-1.711(4)/\sqrt{25}\ to\ 30+1.711(4)/\sqrt{25}$$

$$28.63\ to\ 31.37$$

**8.89**  The 99% confidence interval for μ is

$$\bar{x}-t_{\alpha/2}s/\sqrt{n} \quad to \quad \bar{x}+t_{\alpha/2}s/\sqrt{n}$$

$$50-2.947(5)/\sqrt{16}\ to\ 50+2.947(5)/\sqrt{16}$$

$$46.32\ to\ 53.68$$

**8.91**  n = 30; df = 29; $t_{\alpha/2}$ = $t_{0.05}$ = 1.699; $\bar{x}$ = 27.97 minutes;  s = 10.04 minutes

(a)

$$\bar{x}-t_{\alpha/2}s/\sqrt{n}\ to\ \bar{x}+t_{\alpha/2}s/\sqrt{n}$$

$$27.97-1.699(10.04)/\sqrt{30}\ \ to\ \ 27.97+1.699(10.04)/\sqrt{30}$$

$$24.86\ to\ 31.08$$

(b)  We are 90% confident that the interval 24.86 to 31.08 minutes contains the mean, $\mu$, of all commute times for working adults in the Washington, D.C. area.

**8.93**  n = 10; df = 9; $t_{\alpha/2}$ = $t_{0.025}$ = 2.262; $\bar{x}$ = 2.33 hours;  s = 2.002 hours

(a)

$$\bar{x}-t_{\alpha/2}s/\sqrt{n}\ to\ \bar{x}+t_{\alpha/2}s/\sqrt{n}$$

$$2.33-2.262(2.002)/\sqrt{10}\ \ to\ \ 2.33+2.262(2.002)/\sqrt{10}$$

$$0.90\ to\ 3.76$$

(b)  We are 95% confident that the interval 0.90 to 3.76 hours contains the mean, $\mu$, of addition sleep obtained by a patients using laevohysocyamine hydrobromide.

**8.95**   $n = 77$; $df = 76$; $t_{\alpha/2} = t_{0.025} = 1.992$; $\bar{x} = 0.199$ m/sec; $s = 0.210$ m/sec

(a)

$$\bar{x} - t_{\alpha/2} s / \sqrt{n} \text{ to } \bar{x} + t_{\alpha/2} s / \sqrt{n}$$

$$0.199 - 1.992(0.210) / \sqrt{77} \text{ to } 0.199 + 1.992(0.210) / \sqrt{77}$$

$$0.151 \text{ to } 0.247$$

(b) We are 95% confident that the interval 0.151 m/sec to 0.247 m/sec contains the mean increase, $\mu$, of aortic-jet velocity of patiens with calcific aortic stenosis who received 80 mg of atorvasatin daily. Since this interval lies entirely above zero, we conclude that there is an increase in aortic-jet velocity for such patients.

**8.97** We used Minitab to obtain a boxplot of the data, which are on the WeissStats CD. The plot shows that the values 6.7 and 7.6 are potential outliers and the distribution is skewed right. Since the sample size is only 22, it is not reasonable to use the t-interval procedure to obtain a confidence interval for the population mean.

**8.99** We used Minitab to obtain a boxplot of the data, which are on the WeissStats CD. The plot shows no outliers and the distribution is quite symmetrical. Since the sample size is 20, it is reasonable to use the t-interval procedure to obtain a confidence interval for the population mean.

**8.101** (a) We will get the confidence interval for part (c) along with the histogram and boxplot. Using Minitab, retrieve the data from the WeissStats CD. Then choose **Stat ▶ Basic statistics ▶ 1-Sample t**.

Enter <u>DEPTH</u> in the **Samples in Columns** text box. Click on the **Options** button and enter <u>90.0</u> in the **Confidence level** text box and click **OK**. Click on the **Graphs** button and check the boxes for **Histogram of data** and **Boxplot of data**, then click **OK** twice. To get the probability plot and stem-and leaf diagram, choose **Stat ▶ Basic statistics ▶ Normality test**, enter <u>DEPTH</u> in the **Variable** text box and click **OK**. Then choose **Graph ▶ Stem-and-Leaf**, and enter <u>DEPTH</u> in the **Graph Variables** text box and click **OK**. The results are

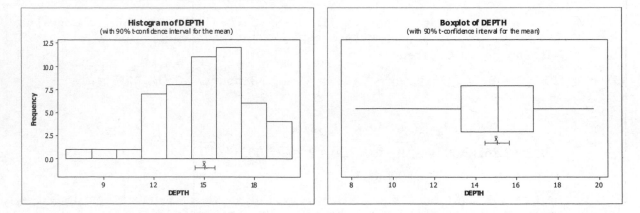

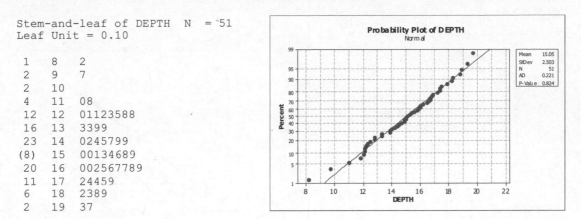

```
Stem-and-leaf of DEPTH   N  = 51
Leaf Unit = 0.10

    1     8    2
    2     9    7
    2    10
    4    11    08
   12    12    01123588
   16    13    3399
   23    14    0245799
   (8)   15    00134689
   20    16    002567789
   11    17    24459
    6    18    2389
    2    19    37
```

(b) Yes.  The plots indicate no outliers, the sample size is quite large at 51, and the distribution of the data is only a little left skewed.

(c) The output for the confidence interval obtained in part (a) is

```
Variable   N     Mean     StDev   SE Mean        90% CI
DEPTH      51   15.0510   2.5028   0.3505   (14.4636, 15.6383)
```

We can be 90% confident that the mean, $\mu$, for all burrow depths is somewhere between 14.46 cm and 15.64 cm.

**8.103** (a) We will get the confidence interval for part (c) along with the histogram and boxplot.  Using Minitab, retrieve the data from the WeissStats CD.  Then choose **Stat ▶ Basic statistics ▶ 1-Sample t.**  Enter WITHOUT in the **Samples in Columns** text box.  Click on the **Options** button and enter 95.0 in the **Confidence level** text box and click **OK**.  Click on the **Graphs** button and check the boxes for **Histogram of data** and **Boxplot of data**, then click **OK** twice.  To get the probability plot and stem-and leaf diagram, choose **Stat ▶ Basic statistics ▶ Normality test**, enter WITHOUT in the **Variable** text box and click **OK**.  Then choose **Graph ▶ Stem-and-Leaf**, and enter WITHOUT in the **Graph Variables** text box and click **OK**.  The results are

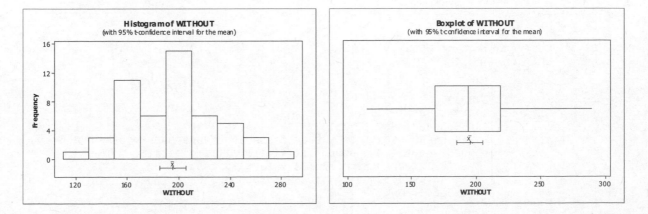

```
Stem-and-leaf of WITHOUT   N  = 51
Leaf Unit = 10

  1    1  1
  2    1  3
  9    1  4455555
 19    1  6666667777
 (8)   1  88999999
 24    2  000000000111
 12    2  2223333
  5    2  45
  3    2  66
  1    2  8
```

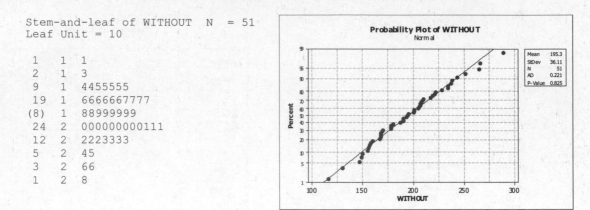

(b) Yes. The plots indicate no outliers, the sample size is quite large at 51, and the distribution of the data is quite symmetric.

(c) The output for the confidence interval obtained in part (a) is

```
Variable   N     Mean    StDev   SE Mean       95% CI
WITHOUT    51   195.275  36.110   5.056   (185.118, 205.431)
```

We can be 95% confident that the mean, $\mu$, for all plasma cholesterol concentrations of patients without evidence of heart disease is somewhere between 185.118 mg/dl and 203.431 mg/dl.

(d) Repeating the process in part (a) using the WITH data, we obtain

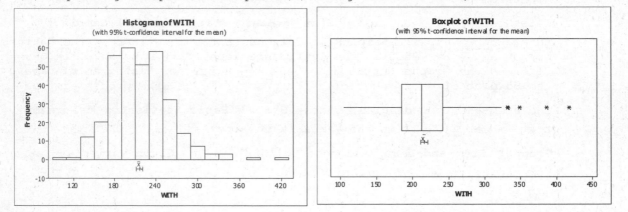

```
Stem-and-leaf of WITH   N  = 320
Leaf Unit = 10

   2    1  01
   8    1  333333
  19    1  44444455555
  63    1  66666666666666667777777777777777777777777777777
 126    1  888888888888888888888888888899999999999999999999999999999999999
 (50)   2  00000000000000000000000001111111111111111111111111111
 144    2  22222222222222222222222222233333333333333333333333333333333333
  87    2  44444444444444444444444444445555555555555555555
  44    2  6666666666666666677777777
  21    2  8888889999
  11    3  00011
   6    3  2


HI 33, 33, 34, 38, 41
```

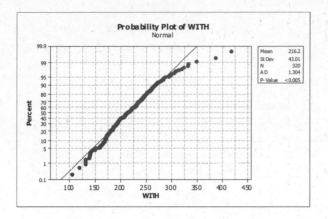

The data are quite symmetric except for five outliers on the high side. Nevertheless, the large sample size of 320 makes it reasonable to use the t-interval procedure with these data.  The results of that procedure are

```
Variable    N     Mean    StDev  SE Mean       95% CI
WITH      320   216.191   43.015   2.405   (211.460, 220.921)
```

We can be 95% confident that the mean, $\mu$, for all plasma cholesterol concentrations of patients with evidence of heart disease is somewhere between 211.460 mg/dl and 220.921 mg/dl.

**8.105** (a)  We could not provide entries for every possible degrees of freedom because the number of possibilities is infinite.

(b)  As the degrees of freedom increase, the difference in consecutive entries becomes very small, making it unnecessary to list every possibility.

(c)  Anytime the actual degrees of freedom lies between two consecutive entries in the table, we should use the table value associated with the lower number of degrees of freedom.  Thus for $t_{0.05}$ with 87 df, we should use the value associated with 80 df, 1.664; for 125 df, use 1.660; for 650 df, use 1.660; and for 3000 df, use 1.645.  This is a conservative approach, resulting in margins of error that are never smaller than we are entitled to have.

**8.107** The observed values of the studentized and standardized versions of $\bar{x}$ are the same for any sample size n whenever the sample standard deviation s is identical to the population standard deviation $\sigma$.

**8.109** (a)  Your results will vary.  To obtain the 2000 samples using Minitab,

Choose **Calc ▶ Random Data ▶ Normal...**, enter 2000 in the **Generate rows of data** text box, enter C1-C5 in the **Store in Column(s):** text box, enter .270 in the **Mean** text box, and enter .031 in the **Standard deviation:** text box.  Click **OK**.

(b)  To find the mean and sample standard deviation in each row (sample),

choose **Calc ▶ Row statistics...** and click on **Mean**. Enter C1-C5 in the **Input variable(s):** text box and enter C6 in the **Store result in:** text box.  Repeat this last process, selecting **Standard Deviation** instead of **Mean** and put the results in C7.

(c)  To obtain the Standardized version of each mean in the sample, choose

**Calc ▶ Calculator...**, enter <u>STANDARD</u> in the **Store results in variable:** text box, enter <u>(C6-.270)/(.031/SQRT(5))</u> in the **Expression:** text box and click **OK**.

(d) To facilitate a comparison in part (k),., do part (h) now.  Choose

**Graph ▶ Histogram...**, select the **Simple** version, and click **OK**.  Enter STANDARD STUDENT in the **Graph variables** text box. Click the **Multiple graphs** button, click the **On separate graphs** button and check the box for **Same X, including same bins**.  Click **OK** and click **OK**.  Our first graph is shown below, but yours will differ, yet look similar.

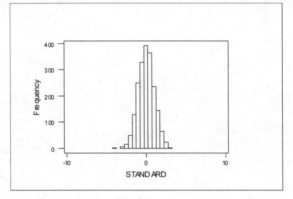

(e) In theory, the distribution of the standardized version of $\bar{x}$ is standard normal.

(f) The histogram in (d) appears to be very close to standard normal. Recall that the distribution is centered at zero and 99.74% of the data should be within 3 standard deviations of the mean.  It does appear that this is so for this simulated data.

(g) To obtain the Studentized version of each $\bar{x}$ in the sample, choose **Calc**

**▶ Calculator...**, enter <u>STUDENT</u> in the **Store results in variable:** text box, enter <u>(C6-.270)/(C7/SQRT(5)</u> in the **Expression:** text box and click **OK**.

(h) This graph was produced in produced (d).  Your graph should be similar to ours.

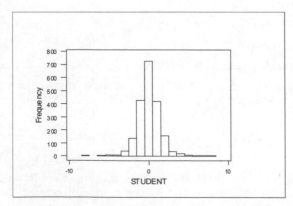

(i) In theory, the distribution of the studentized version of $\bar{x}$ is a t-distribution with 4 degrees of freedom.

(j) The distribution shown is symmetric about zero as is a t-distribution.

(k) The histogram of the studentized version in (h) is more spread out than that of the standardized version in (d). The reason is that there is more variability in the t-distribution due to the extra uncertainty arising from the use of s instead of $\sigma$.

**8.111** (a) n=30; $\bar{x}$ = 27.97; s = 10.04; df = 29; $t_\alpha = t_{0.10}$ = 1.311

$$\bar{x}+t_\alpha s/\sqrt{n}$$
$$27.97+1.311(10.04)/\sqrt{30}$$
$$30.37 \text{ minutes}$$

We can be 90% confident that the population mean commute time of all commuters in Washington, D.C. is less than 30.37 minutes.

(b) This upper confidence bound is less than the upper confidence limit of 31.08 found in Exercise 8.91. This is because the t-value used for a one-sided 90% confidence bound is 1.311, whereas, the t-value used for a two-sided 90% confidence interval is 1.699.

### Review Problems for Chapter 8

1. A point estimate of a parameter consists of a single value with no indication of the accuracy of the estimate. A confidence interval consists of an interval of numbers obtained from a point estimate of the parameter together with a percentage that specifies how confident we are that the parameter lies in the interval.

2. False. We are 95% confident that the mean lies in the interval from 33.8 to 39.0, but about 5% of the time, the procedure will produce an interval that does not contain the population mean. Therefore, we cannot say that the mean must lie in the interval.

3. No. The z-interval procedure can be used almost anytime with large samples

   because the sampling distribution of $\bar{x}$ is approximately normal for large n. The same is true for the t-interval procedure because when n is large, the t distribution is very similar to the normal distribution. However, when n is small, especially when n is 15 or less, the z-interval and t-interval procedures will not provide reliable estimates if the distribution of the underlying variable is not normal. For sample sizes in the range of 15 to 30, both procedures can be used if the data is roughly normal and has no outliers.

4. Approximately 950 of 1000 95% confidence intervals for a population mean would actually contain the true value of the mean.

5. Before applying a particular statistical inference procedure, we should look at graphical displays of the sample data to see if there appear to be any violations of the conditions required for the use of the procedure.

6. (a) Reducing the sample size from 100 to 50 will reduce the precision of the estimate (result in a longer confidence interval).

   (b) Reducing the confidence level from .95 to .90 while maintaining the sample size will increase the precision of the estimate (result in a shorter confidence interval).

7. (a) The length of the confidence interval is twice the margin of error or 2 x 10.7 = 21.4.

   (b) The confidence interval will be 75.2 $\pm$ 10.7 = (64.5, 85.9)

8. (a) $E=z_{\alpha/2}(\sigma/\sqrt{n})=1.645(12/\sqrt{9})=6.58$

(b)  To obtain the confidence interval, you also need to know $\bar{x}$ .

9.   (a)  The standardized value of $\bar{x}$ is $z = \dfrac{\bar{x} - \mu}{\sigma / \sqrt{n}} = \dfrac{262.1 - 266}{16 / \sqrt{10}} = -0.77$

(b)  The studentized value of $\bar{x}$ is $t = \dfrac{\bar{x} - \mu}{s / \sqrt{n}} = \dfrac{262.1 - 266}{20.4 / \sqrt{10}} = -0.605$

10.   (a)  standard normal distribution

(b)  t distribution with 14 degrees of freedom

11.   The curve that looks more like the standard normal curve has the larger degrees of freedom because, as the number of degrees of freedom gets larger, *t*-curves look increasingly like the standard normal curve.

12.   (a)  The t-interval procedure should be used.

(b)  The z-interval procedure should be used.

(c)  The z-interval procedure should be used.

(d)  Neither procedure should be used.

(e)  The z-interval procedure should be used.

(f)  Neither procedure should be used.

13.   $n = 36$, $\bar{x} = 58.53$, $\sigma = 13.0$, $z_{\alpha/2} = z_{0.025} = 1.96$

$$\bar{x} - z_{\alpha/2}\left(\sigma / \sqrt{n}\right) \text{ to } \bar{x} + z_{\alpha/2}\left(\sigma / \sqrt{n}\right)$$
$$58.53 - 1.96 \cdot (13.0 / \sqrt{36}) \text{ to } 58.53 + 1.96 \cdot (13.0 / \sqrt{36})$$
$$54.3 \text{ to } 62.8 \text{ years}$$

14.   A confidence-interval estimate specifies how confident we are that a (unknown) parameter lies in the interval.  This interpretation is presented correctly by (c).  A *specific* confidence interval either will or will not contain the true value of the population mean μ; the *specific* interval is either sure to contain μ or sure not to contain μ.  This interpretation is *not* presented correctly by (a), (b), and (d).

15.   $n = 461$, $= 11.9$ mm, $\sigma = 2.5$ mm, $z_{\alpha/2} = z_{0.05} = 1.645$

(a)

$$\bar{x} - z_{\alpha/2}\sigma / \sqrt{n} \text{ to } \bar{x} + z_{\alpha/2}\sigma / \sqrt{n}$$
$$11.9 - 1.645(2.5) / \sqrt{461} \text{ to } 11.9 + 1.645(2.5) / \sqrt{461}$$
$$11.71 \text{ to } 12.09$$

(b)  We can be 90% confident that the mean length, $\mu$ , of N. trivittata is somewhere between 11.71 and 12.09 mm.

(c)  Since the sample size is very large, the distribution of sample means will be approximately normal regardless of the shape of the original distribution.  It would be nice if the normal probability plot were roughly linear and did not indicate the presence of any extreme outliers, but some non-linearity and a few moderate outliers will not likely invalidate the use of the z-interval procedure.

16.   (a)  $E = z_{\alpha/2}\sigma / \sqrt{n} = 1.645(2.5) / \sqrt{461} = 0.19$

(b)  We can be 90% confident that the maximum error made in using $\bar{x}$ to estimate μ is 0.19 mm.

(c) The margin of error of the estimate is specified to be E = 0.1 mm.

$$n = \left[\frac{z_{\alpha/2}\ \sigma}{E}\right]^2 = \left[\frac{1.645\ (2.5)}{0.1}\right]^2 = 1691.3 \rightarrow 1692$$

(d)

$$\bar{x} - z_{\alpha/2}\sigma/\sqrt{n} \quad \text{to} \quad \bar{x} + z_{\alpha/2}\sigma/\sqrt{n}$$
$$12.0 - 1.645(2.5)/\sqrt{1692} \quad \text{to} \quad 12.0 + 1.645(2.5)/\sqrt{1692}$$
$$11.90 \quad \text{to} \quad 12.10$$

**17.**     (a)   $t_{0.025} = 2.101$             (b)   $t_{0.05} = 1.734$

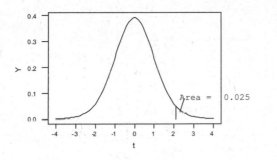
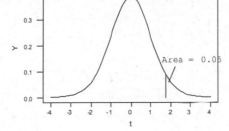

(c)   $-t_{0.10} = -1.330$             (d)   $\pm t_{0.005} = \pm\ 2.878$

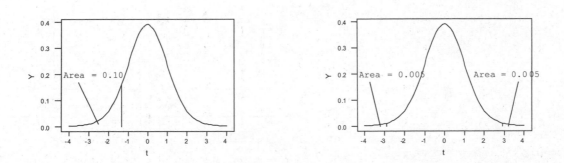

**18.**     n = 16; df = 15; $t_{\alpha/2} = t_{0.025} = 2.131$; $\bar{x} = 85.99$ mm Hg; s = 8.08 mm Hg

(a)

$$\bar{x} - t_{\alpha/2}s/\sqrt{n} \quad \text{to} \quad \bar{x} + t_{\alpha/2}s/\sqrt{n}$$
$$85.99 - 2.131(8.08)/\sqrt{16} \quad \text{to} \quad 85.99 + 2.131(8.08)/\sqrt{16}$$
$$81.69 \quad \text{to} \quad 90.29$$

We can be 95% confident that the mean arterial blood pressure $\mu$ of all children of diabetic mothers is somewhere between 81.69 and 90.29 mm

Hg.

(b)  Using Minitab, choose **Stat ▶ Basic statistics ▶ 1-Sample t...**, enter
      PRESSURE in the **Samples in columns** text box.  Click the **options…**
      button, type 95 in the **Confidence interval** text box, click the **Graphs**
      and check the boxes for **Histogram of data** and **Boxplot of data,** and

      click **OK** twice.  Then choose  **Stat ▶ Basic statistics ▶ Normality**
      **test**, enter PRESSURE in the **Variable** text box and click **OK**. Finally,
      choose **Graph ▶ Stem-and-Leaf,** enter PRESSURE in the **Graph Variables**
      text box and click **OK**. The results are

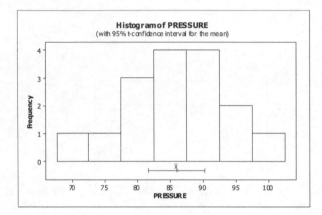

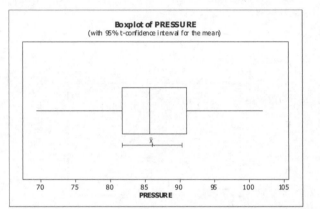

```
Stem-and-leaf of PRESSURE   N  = 16
Leaf Unit = 1.0

   1    6    9
   1    7
   3    7    58
   8    8    12244
   8    8    678
   5    9    014
   2    9    6
   1   10    1
```

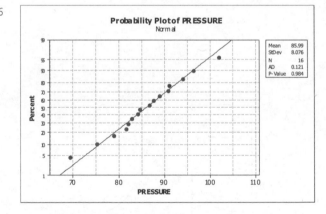

(c)  Yes.  There are no outliers and the distribution of the data is
      approximately normal.

19.   (a)  Using Minitab, choose **Stat ▶ Basic statistics ▶ 1-Sample t...**, enter
            PRICE in the **Samples in columns** text box.  Click the **options…** button,
            enter 90 in the **Confidence interval** text box and click **OK**, click the
            **Graphs** button and check the boxes for **Histogram of data** and **Boxplot of**

            **data,** and click **OK** twice.  Then choose  **Stat ▶ Basic statistics ▶**
            **Normality test**, enter PRICE in the **Variable** text box and click **OK**.

            Finally, choose **Graph ▶ Stem-and-Leaf,** enter PRICE in the **Graph**
            **Variables** text box and click **OK**. The confidence interval is

```
Variable   N    Mean    StDev  SE Mean        90% CI
PRICE     18  1964.72  206.45   48.66   (1880.07, 2049.37)
```
We can be 90% confident that the interval ($1880.07, $2049.37) contains

the mean diamond price for one-half carat diamonds.

(b)  The graphs obtained by the procedures in part (a) are

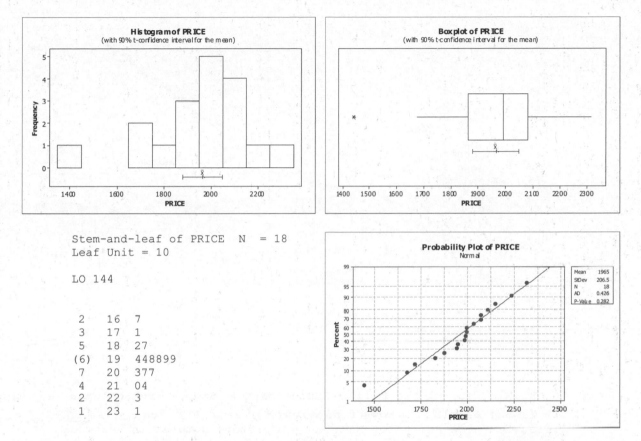

```
Stem-and-leaf of PRICE   N = 18
Leaf Unit = 10

LO 144

  2    16   7
  3    17   1
  5    18   27
 (6)   19   448899
  7    20   377
  4    21   04
  2    22   3
  1    23   1
```

(c)  No.  Since the sample size is small (18), and there is an outlier (1442), it is not reasonable to use the t-interval procedure.

20.   (a)  We will import the data into Minitab.

(b)  Choose **Stat ▶ Basic statistics ▶ 1-Sample t..,** enter <u>DURATION</u> in the **Samples in columns** text box.  Click the **options…** button, enter <u>99</u> in the **Confidence interval** text box and click **OK**, click the **Graphs** button and check the boxes for **Histogram of data** and **Boxplot of data**, and click **OK** twice.  Then choose **Stat ▶ Basic statistics ▶ Normality test**, enter <u>DURATION</u> in the **Variable** text box and click **OK**. Finally, choose **Graph ▶ Stem-and-Leaf**, enter <u>DURATION</u> in the **Graph Variables** text box and click **OK**.  The graphs are

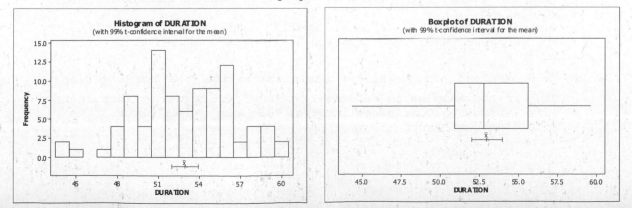

```
Stem-and-leaf of DURATION   N  = 90
Leaf Unit = 0.10

    2   44  33
    3   45  0
    3   46
    4   47  3
    9   48  00036
   16   49  0000033
   22   50  003366
   34   51  000000003333
   45   52  00000003666
   45   53  000
   42   54  000000033666
   30   55  0000336666
   20   56  000003336
   11   57  0
   10   58  00036
    5   59  00356
```

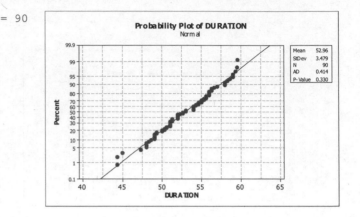

(c) Yes.  The sample size is large (90) and there are no outliers.

(d) The procedure in part (a) resulted in the following confidence interval.

```
Variable   N    Mean   StDev  SE Mean        99% CI
DURATION   90  52.9633  3.4794  0.3668  (51.9979, 53.9287)
```
We can be 99% confident that the interval (51.9979, 53.9287) contains the population mean μ of the larval duration of convict surgeonfish in days.

21.  (a) Using Minitab, choose **Calc ▶ Random data ▶ Sample from columns**, enter 35 in the **Sample ___ rows from column(s)** text box, enter <u>MILEAGE</u> in the first text box and enter <u>SAMPLE</u> in the **Store samples in** text box. Click **OK**.

   (b) Choose **Stat ▶ Basic statistics ▶ 1-Sample t..**,enter <u>SAMPLEN</u> in the **Samples in columns** text box.  Click the **options…** button, enter 95 in the **Confidence interval** text box and click **OK**.  The result is

```
Variable   N    Mean   StDev  SE Mean       95% CI
SAMPLE    35  24.4857  4.0174  0.6791  (23.1057, 25.8657)
```

   (c) Choose **Calc ▶ Column statistics** and check the box for the mean.  Enter <u>MILEAGE</u> in the **Input variable** text box and click **OK**.  The population mean is 24.8234.  This value lies in the 95% confidence interval found in part (b).  It would not necessarily have to lie in the confidence interval.  Since the sample was randomly selected, it is possible to select a sample for which the confidence interval does not contain the population mean.

22.  (a) The population consists of all eruptions of the Old Faithful Geyser. The variable under consideration is the time between eruptions (in minutes).

   (b) Using Minitab, choose **Stat ▶ Basic statistics ▶ 1-Sample t..**,enter <u>TIME</u> in the **Samples in columns** text box.  Click the **options…** button, enter 99 in the **Confidence interval** text box and click **OK**.  The resulting confidence interval is
```
Variable   N     Mean   StDev  SE Mean       99% CI
TIME      500  91.5780  9.2022  0.4115  (90.5139, 92.6421)
```

We can be 99% confident that the population mean time between eruptions of the Old Faithful Geyser lies in the interval from 90.5139 to 92.6421 minutes.

(c) Strictly speaking, the confidence interval applies to the population from which the sample was drawn.  There were no observations in the sample from the future, so the interval is not relevant to the future.  In geologic terms, however, five years is insignificant, so the confidence interval is probably still somewhat meaningful.  One should consider that unforeseen events, such as earthquakes or volcanic eruptions in the region could have a substantial effect on the time between eruptions.  In that case, the current sample would not be representative of the population in five years.

23.    (a) Your results will vary.  To obtain the 3000 samples using Minitab,

   Choose **Calc ▶ Random Data ▶ Normal...**, enter <u>3000</u> in the **Generate rows of data** text box, type <u>C1–C4</u> in the **Store in Column(s):** text box, enter <u>4.66</u> in the **Mean** text box, and enter <u>0.75</u> in the **Standard deviation:** text box.  Click **OK**.

   (b) To find the mean and sample standard deviation in each row (sample),

   choose **Calc ▶ Row statistics...** and click on **Mean**. Enter <u>C1–C4</u> in the **Input variable(s):** text box and enter <u>XBAR</u> in the **Store result in:** text box.  Repeat this last process, selecting **Standard Deviation** instead of **Mean** and put the results in <u>SD</u>.

   (c) To obtain the Standardized version of each $\bar{x}$ in the sample, choose **Calc**

   **▶ Calculator...**, enter <u>STANDARD</u> in the **Store results in variable:** text box, enter <u>('XBAR'-4.66)/(0.75/SQRT(4))</u> in the **Expression:** text box and click **OK**.

   (d) To facilitate a comparison in part (h), **do part (g) now**.  Then choose

   **Graph ▶ Histogram...**, select the **Simple** version, and click **OK**.  Enter STANDARD and STUDENT in the **Graph variables** text box. Click on the **Multiple graphs** button, then click the **On separate graphs** button and check the boxes for **Same X, including same bins** and **Same Y**.  Click **OK** twice.  The first graph is shown below.  Yours will differ, yet look similar.

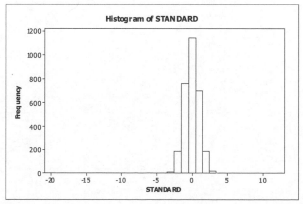

   (e) In theory, the distribution of the standardized version of $\bar{x}$ is standard normal.

   (f) The histogram in (e) appears to be very close to standard normal.  Recall that the distribution is centered at zero and 99.74% of the data should be within 3 standard deviations of the mean. It does appear that this is so for this simulated data. (g) To obtain the Studentized

version of each $\overline{x}$ in the sample, choose **Calc ▶ Calculator...**, enter <u>STUDENT</u> in the **Store results in variable** text box, enter <u>('SBAR'-4.66)/('SD'/SQRT(4)</u> in the **Expression:** text box and click **OK**.

(h) This graph was produced by the procedure in part (d). Our second graph follows, but yours will differ, yet look similar.

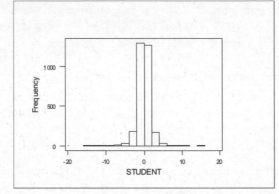

(i) In theory, the distribution of the studentized version of $\overline{x}$ is a t-distribution with 3 degrees of freedom.

(j) The distribution shown is symmetric about zero as is a t-distribution.

(k) The histogram of the studentized version in (h) is much more spread out than that of the standardized version in (d). The reason is that there is more variability in the t-distribution due to the extra uncertainty arising from the use of s instead of σ.

**Exercises 9.1**

**9.1**  A hypothesis is a statement that something is true.

**9.3**  (a)  The population mean $\mu$ is equal to some fixed amount; i.e., $\mu = \mu_0$.

(b)  The population mean $\mu$ is greater than $\mu_0$; i.e., $H_a$: $\mu > \mu_0$.

The population mean $\mu$ is less than $\mu_0$; i.e., $H_a$: $\mu < \mu_0$.

The population mean $\mu$ is unequal to $\mu_0$; i.e., $H_a$: $\mu \neq \mu_0$.

**9.5**  Let $\mu$ denote the mean cadmium level in *Boletus pinicola* mushrooms.

(a) $H_0$: $\mu = 0.5$ ppm    (b) $H_a$: $\mu > 0.5$ ppm    (c)  right-tailed test

**9.7**  Let $\mu$ denote the mean daily intake of iron by adult females under age 51.

(a) $H_0$: $\mu = 18$ mg/day  (b) $H_a$: $\mu < 18$ mg/day  (c)  left-tailed

**9.9**  Let $\mu$ denote the mean length of imprisonment for motor-vehicle theft offenders in Australia.

(a) $H_0$: $\mu = 16.7$ months      (b) $H_a$: $\mu \neq 26.7$ months

(c)  two-tailed

**9.11**  Let $\mu$ denote the mean body temperature of healthy humans.

(a) $H_0$: $\mu = 98.6°F$      (b) $H_a$: $\mu \neq 98.6°F$    (c)  two-tailed

**9.13**  Let $\mu$ denote the mean local monthly bill for cell phone users in the U.S.

(a) $H_0$: $\mu = \$47.37$        (b) $H_a$: $\mu > \$47.37$    (c)  right-tailed

**9.15**  (a)  $H_0$: $\mu = 5.6$ radios per U.S. household

$H_a$: $\mu \neq 5.6$ radios per U.S. household

(b)  If the sample mean number of radios per U.S. household $\overline{x}$ differs by too much from 5.6 radios, then we should be inclined to reject $H_0$ and

conclude that $H_a$ is true.  From the data, we compute $\overline{x}$ = 5.89 radios. The question is whether the difference of 0.29 radios between the sample mean of 5.89 and the hypothesized population mean of 5.6 can be attributed to sampling error or whether the difference is large enough to indicate that the population mean is not 5.6 radios.

(c)  The sampling distribution of $\overline{x}$ will be approximately a normal distribution.

(d)  It is quite unlikely that the sample mean $\overline{x}$ will be more than two standard deviations away from the population mean $\mu$.  If $\overline{x}$ is more than two standard deviations away from $\mu$, then reject $H_0$ and conclude that $H_a$ is true.  Otherwise, do not reject $H_0$.

(e)  We have $\sigma = 1.9$, n = 45, $\overline{x} = 5.89$, and $\mu = 5.6$ under $H_0$ true.  Thus,

$$z = (5.89 - 5.6)/(1.9/\sqrt{45}) = 1.02 .$$  Since $\overline{x}$ is less than two standard deviations away from 5.6 radios, we do not reject $H_0$ and conclude that $H_0$ is reasonable.

**9.17**  (a)  If the mean weight $\overline{x}$ of the 25 bags of pretzels sampled is more than three standard deviations away from 454 grams, then reject the null hypothesis that $\mu$ = 454 grams and conclude that the alternative hypothesis, which is $\mu \neq$ 454 grams, is true.  Otherwise, do not reject the null hypothesis.

Graphically, the decision criterion looks like:

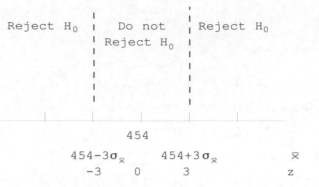

(b)

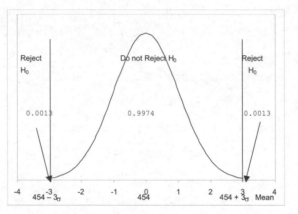

The lower figure shows that, using our decision criterion, the probability is 0.0026 (= 1 - 0.9974 = 0.0013 + 0.0013) of rejecting the null hypothesis if it is in fact true.

(c) We have $\sigma$ = 7.9, n = 25, $\bar{x}$ = 450, and $\mu$ = 454 if $H_0$ is true.  Thus,

$$z = (450 - 454)/(7.8/\sqrt{25}) = -2.56 .$$  The sample mean $\bar{x}$ is 2.56 standard deviations below the null hypothesis mean of 454 grams.  Since the mean

weight $\bar{x}$ of 25 bags of pretzels sampled is less than three standard deviations away from 454 grams, we do not reject the null hypothesis that $\mu$ = 454 grams and conclude that the null hypothesis, which is $\mu$ = 454 grams, is reasonable.  In other words, the data do not provide sufficient evidence to conclude that the packaging machine is not working properly.

### Exercises 9.2

9.19 (a) This statement is true:  If it is important not to reject a true null hypothesis, i.e., not to make a Type I error, then the hypothesis test should be performed at a small significance level.  This can be appreciated by considering the meaning of the significance level.  The significance level is equal to the probability of making a Type I error.  The smaller the significance level, the smaller the probability of rejecting a true null hypothesis.

(b)    This statement is true:  Decreasing the significance level results in an increase in the probability of making a Type II error.  This can be appreciated by considering the relation between Type I and Type II error probabilities.  For a fixed sample size, the smaller the Type I error probability (which is equal to the significance level), the larger the Type II error probability.

**9.21** A Type I error is made when a true null hypothesis is rejected.  The probability of making this error is denoted by $\alpha$.  A Type II error is made when a false null hypothesis is not rejected.  We denote the probability of a Type II error by $\beta$.

**9.23** (a)  Rejection region:  $z \geq 1.645$

(b)  Nonrejection region:  $z < 1.645$

(c)  Critical value:  $z = 1.645$

(d)  Significance level:  $\alpha = 0.05$

(e)

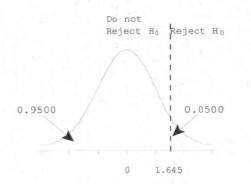

(f)   Right-tailed test

**9.25** (a)  Rejection region:  $z \leq -2.33$

(b)  Nonrejection region:  $z > -2.33$

(c)  Critical value:  $z = -2.33$

(d)  Significance level:  $\alpha = 0.01$

(e)

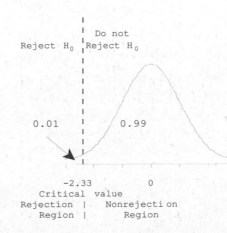

      (f)  Left-tailed test

**9.27** (a) Rejection region: $z \leq -1.645$ or $z \geq -1.645$

      (b) Nonrejection region: $-1.645 < z < -1.645$

      (c) Critical values: $z = -1.645$ and $z = 1.645$

      (d) Significance level: $\alpha = 0.10$

      (e)

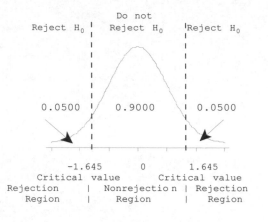

      (f)  Two-tailed test

**9.29** (a) A Type I error would occur if, in fact, $\mu = 0.5$ ppm, but the results of the sampling lead to the conclusion that $\mu > 0.5$ ppm.

      (b) A Type II error would occur if, in fact, $\mu > 0.5$ ppm, but the results of the sampling fail to lead to that conclusion.

      (c) A correct decision would occur if, in fact, $\mu = 0.5$ ppm and the results of the sampling do not lead to the rejection of that fact; or if, in fact, $\mu > 0.5$ ppm and the results of the sampling lead to that conclusion.

      (d) If, in fact, the mean cadmium level in *Boletus pinicola* mushrooms is equal to 0.5 ppm, and we do not reject the null hypothesis that $\mu = 0.5$ ppm, we made a correct decision.

      (e) If, in fact, the mean cadmium level in *Boletus pinicola* mushrooms is greater than to 0.5 ppm, and we do not reject the null hypothesis that $\mu = 0.5$ ppm, we made a Type II error.

**9.31** (a) A Type I error would occur if, in fact, $\mu = 18$ mg, but the results of the sampling lead to the conclusion that $\mu < 18$ mg.

      (b) A Type II error would occur if, in fact, $\mu < 18$ mg, but the results of the sampling fail to lead to that conclusion.

      (c) A correct decision would occur if, in fact, $\mu = 18$ mg and the results of the sampling do not lead to the rejection of that fact; or if, in fact, $\mu < 18$ mg and the results of the sampling lead to that conclusion.

      (d) If the mean iron intake equals the RDA of 18 mg, and we reject the null hypothesis that $\mu = 18$ mg, we made a Type I error.

      (e) If, in fact, the mean iron intake is less than the RDA of 18 mg, and we reject the null hypothesis that $\mu = 18$ mg, we made a correct decision.

**9.33**  (a)  A Type I error would occur if, in fact, $\mu$ = 16.7 months, but the results of the sampling lead to the conclusion that μ ≠ 16.7 months.

       (b)  A Type II error would occur if, in fact, $\mu$ ≠ 16.7 months, but the results of the sampling fail to lead to that conclusion.

       (c)  A correct decision would occur if, in fact, $\mu$ = 16.7 months and the results of the sampling do not lead to the rejection of that fact; or if, in fact, $\mu$ ≠ 16.7 months and the results of the sampling lead to that conclusion.

       (d)  If, in fact, the mean length of imprisonment equals 16.7 months, and we do not reject the null hypothesis that $\mu$ = 16.7 months, we made a correct decision.

       (e)  If, in fact, the mean length of imprisonment does not equal 16.7 months, and we do not reject the null hypothesis that $\mu$ = 16.7 months, we made a Type II error.

**9.35**  (a)  A Type I error would occur if, in fact, $\mu$ = 98.6° F, but the results of the sampling lead to the conclusion that $\mu$ ≠ 98.6° F.

       (b)  A Type II error would occur if, in fact, $\mu$ ≠ 98.6° F, but the results of the sampling fail to lead to that conclusion.

       (c)  A correct decision would occur if, in fact, $\mu$ − 98.6° F and the results of the sampling do not lead to the rejection of that fact; or if, in fact, $\mu$ ≠ 98.6° F and the results of the sampling lead to that conclusion.

       (d)  If the mean temperature of all healthy humans equals 98.6° F, and we reject the null hypothesis that $\mu$ = 98.6° F, we made a Type I error.

       (e)  If, in fact, the temperature of all healthy humans is not equal to 98.6° F, and we reject the null hypothesis that $\mu$ = 98.6° F, we made a correct decision.

**9.37**  (a)  A Type I error would occur if, in fact, $\mu$ = $47.37, but the results of the sampling lead to the conclusion that $\mu$ > $47.37.

       (b)  A Type II error would occur if, in fact, $\mu$ > $47.37, but the results of the sampling fail to lead to that conclusion.

       (c)  A correct decision would occur if, in fact, $\mu$ = $47.37 and the results of the sampling do not lead to the rejection of that fact; or if, in fact, $\mu$ > $47.37 and the results of the sampling lead to that conclusion.

       (d)  If the mean phone bill equals the 2001 mean of $47.37, and we do not reject the null hypothesis that $\mu$ = $47.37, we made a correct decision.

       (e)  If, in fact, the mean cell phone bill is greater than the 2001 mean of $47.37, and we do not reject the null hypothesis that $\mu$ = $47.37, we made a Type II error.

**9.39**  (a)  Exercise 9.31 is a situation in which it may be important to have a small $\alpha$ probability. Concluding that females under the age of 51 are, on the average, getting less than the RDA of 18 mg of iron could lead to remedial action by the nation's health providers which would be expensive and unnecessary if females under 51 are, in fact, getting the RDA of 18 mg of iron.

(b) Exercise 9.29 is a situation in which it may be important to have a small $\beta$ probability. If the mean cadmium level in the mushrooms is, in fact, higher than the government recommended limit, eating the mushrooms could have serious health consequences for those who eat them. If a hypothesis test does not lead to the conclusion that the cadmium level is too high, the population would be led to believe that the mushrooms are safe to eat when, in fact, they are not. The probability of this happening is $\beta$ and should be kept small.

(c) Exercise 9.40 provides a situation in which it is important to have both small $\alpha$ and $\beta$ probability. The null hypothesis is that the nuclear power plant is safe and the alternative is that it is not safe. A discussion concerning the desirability of $\beta$ being small is found in that exercise. Given the growing need for power and the power crises in California in 2001 and in the eastern U.S. in 2003, it would also be important not to reject a power plant (of any kind) that was actually safe. Thus we would also want $\alpha$ to be small in this case.

Another situation in which small $\alpha$ and $\beta$ are desirable involves defective products. Suppose that a manufacturing company samples items from boxes to determine if the percentage of defective items is too high. The null hypothesis in each case is that the percentage of defects is, say, 3%, while the alternative hypothesis is that the percentage is more than 3%. If the sample results in the conclusion that the percentage of defects is too high when, in fact, it is not, a Type I error has been committed and the manufacturer will unnecessarily incur the expense of shutting down an assembly line to find a nonexistent problem. If the sample results in the conclusion that the percentage of defects is acceptable when, in fact, it is too high, a Type II error has been committed and the manufacturer is likely to incur a loss of business and reputation at the hands of unhappy customers.

9.41 (a) A Type I error would occur if, in fact, the defendant is innocent, but the jury concludes that the defendant is guilty.

(b) A Type II error would occur if, in fact, the defendant is guilty, but the jury fails to conclude that the defendant is guilty.

(c) If I were a defendant, I would want $\alpha$ to be small. Given that I am innocent, I certainly want there to be a small probability of the jury rejecting my innocence (i.e., finding me guilty).

(d) If I were a prosecutor, I would want $\beta$ to be small. Given that the defendant is guilty, I want there to be a small probability that the jury would declare the defendant not guilty.

(e) If $\alpha = 0$, then an innocent person would never be declared guilty. If $\beta = 0$, then a guilty person would always be found guilty.

9.43 (a) Answers will vary. This exercise can easily be done with Minitab, but we will describe a procedure using Excel. With a blank spreadsheet, from the Menu bar, select Tools, Data Analysis, Random Number Generation. Enter 100 for the number of variables, and 25 for the Number of Random Numbers. Select Normal for the Distribution and enter 454 for the Mean and 7.8 for the Standard Deviation. Click on Output Range and enter A1. Then click OK. This will generate 100 columns of 25 random normal numbers each.

(b) At the bottom of column A in A27, enter =AVERAGE(A1:A25) and copy this formula into B27 through DV27.

(c) Then in A28, enter the formula =(A27-454)/(7.8/sqrt(25)), and copy this formula into B28 through DV28.  The null hypothesis will be rejected whenever the number in row 28 is less than or equal -2 or greater than or equal to +2.

(d) Since the significance level is 0.0456, we would expect about 4 or 5 of the 100 samples to result in the rejection of the null hypothesis.

(e) Answers will vary.  Our simulation led to rejection of the null hypothesis 7 times.

(f) If your answer to part (d) is not 4 or 5, it is most likely the result of sampling variation.  On the average, we would expect 4.56% of all samples to lead to rejection of the null hypothesis, but in any single set of 100 samples, the percentage may differ from that amount.

## Exercises 9.3

**9.45** Critical value: $z_{0.05}$ = 1.645        **9.47** Critical value: $-z_{0.05}$ = -1.645

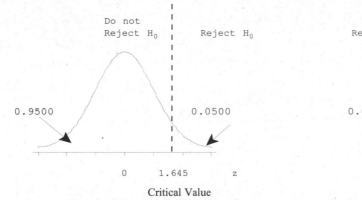

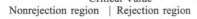

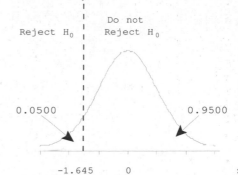

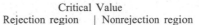

**9.49** Critical values: $\pm z_{0.025}$ = $\pm 1.96$

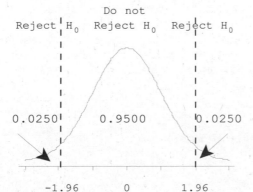

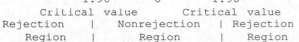

**9.51** (a) The z-test in not an appropriate method for highly skewed data when the sample size is less than 30.

(b) The z-test is appropriate for large samples with no outliers even if the data are mildly skewed.

**9.53** Reject $H_0$ if $z < -1.645$; $z = (20 - 22)/(4/\sqrt{32}) = -2.83$; therefore, reject $H_0$ and conclude that $\mu < 22$.

**9.55** Reject $H_0$ if $z > 1.645$; $z = (24 - 22)/(4/\sqrt{15}) = 1.94$; therefore, reject $H_0$ and conclude that $\mu > 22$.

**9.57** Reject $H_0$ if $z < -1.96$ or $z > 1.96$; $z = (23 - 22)/(4/\sqrt{24}) = 1.22$; therefore, do not reject $H_0$. The data do not provide sufficient evidence to support $H_a$: $\mu \neq 22$.

**9.59** $n = 12$, $\sigma = 0.37$ ppm, $\bar{x} = 6.31/12 = 0.526$ ppm

Step 1: $H_0$: $\mu = 0.5$ ppm, $H_a$: $\mu > 0.5$ ppm

Step 2: $\alpha = 0.05$

Step 3: $z = (0.526 - 0.5)/(0.37/\sqrt{12}) = 0.24$

Step 4: Critical value = 1.645

Step 5: Since $0.24 < 1.645$, do not reject $H_0$.

Step 6: At the 5% significance level, the data do not provide sufficient evidence to conclude that the mean cadmium level $\mu$ of *Boletus pinicola* mushrooms is greater than the safety limit of 0.5 ppm.

**9.61** $n = 45$, $\bar{x} = 14.68$, $\sigma = 4.2$

Step 1: $H_0$: $\mu = 18$ mg, $H_a$: $\mu < 18$ mg

Step 2: $\alpha = 0.01$

Step 3: $z = (14.68 - 18)/(4.2/\sqrt{45}) = -5.30$

Step 4: Critical value = -2.33

Step 5: Since $-5.30 < -2.33$, reject $H_0$.

Step 6: At the 1% significance level, the data provide sufficient evidence to conclude that adult females under the age of 51 are, on the average, getting less than the RDA of 18 mg of iron. Considering that iron deficiency causes anemia and that iron is required for transporting oxygen in the blood, this result could have practical significance as well.

**9.63** $n = 100$, $\bar{x} = 17.8$ months, $\sigma = 6.0$ months

Step 1: $H_0$: $\mu = 16.7$ months, $H_a$: $\mu \neq 16.7$ months

Step 2: $\alpha = 0.05$

Step 3: $z = (17.8 - 16.7)/(6.0/\sqrt{100}) = 1.83$

Step 4: Critical values = ±1.96

Step 5: Since $-1.96 < 1.83 < 1.96$, do not reject $H_0$.

Step 6: At the 5% significance level, the data do not provide sufficient evidence to conclude that the mean length of imprisonment $\mu$ of motor-vehicle theft offenders in Sydney differs from the national mean in Australia.

**9.65**  (a)  Using Minitab, with the data in a column named GAIN, we choose **Stat ▶**

**Basic Statistics ▶ 1-Sample z...,** click in the **Samples in columns** text box and specify GAIN, click in the **Standard deviation** text box and type 0.42, and click in the **Test mean** text box and type 0.2. Click the **Options...** button, enter 95 in the **Confidence level** text box, click the arrow button at the right of the **Alternative** drop-down list box and select **greater than** and click **OK**. Click on the **Graphs** button and check the boxes for **Histogram of Data** and **Boxplot of data**. Then click **OK** twice. The result of the test is

Test of mu = 0.2 vs > 0.2
The assumed standard deviation = 0.42

| Variable | N | Mean | StDev | SE Mean | 95% Lower Bound | Z | P |
|---|---|---|---|---|---|---|---|
| GAIN | 20 | 0.295000 | 0.499974 | 0.093915 | 0.140524 | 1.01 | 0.156 |

(b)  The histogram and boxplot were produced by the procedure in part (a).

Now choose **Stat ▶ Basic Statistics ▶ Normality test** and enter GAIN in

the **Variable** text box. Click **OK**. Then choose **Graph ▶ Stem-and-Leaf** , enter GAIN in the **Graph variables** text box and click **OK**. The results are

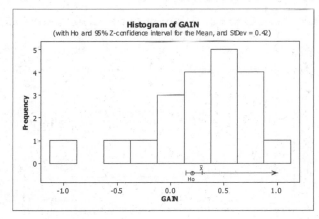

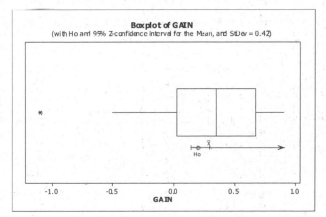

Stem-and-leaf of GAIN   N  = 20
Leaf Unit = 0.10

LO -11

```
 2    -0  5
 3    -0  2
 4    -0  1
 6     0  01
10     0  2233
10     0  45
 8     0  6667
 4     0  8889
```

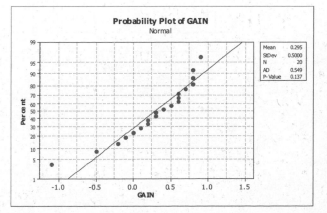

(c) Repeating the procedure of part (a), we obtain
   Test of mu = 0.2 vs > 0.2
   The assumed standard deviation = 0.42

| Variable | N | Mean | StDev | SE Mean | 95% Lower Bound | Z | P |
|---|---|---|---|---|---|---|---|
| GAIN | 19 | 0.368421 | 0.387374 | 0.096355 | 0.209932 | 1.75 | 0.040 |

(d) The original sample size is only 20.  The plots in part (b) indicate that the value −1.1 is a potential outlier.  The z-test should not be used with the original data.  This is further confirmed by the fact that using all of the data leads to z = 1.01, whereas, deleting the outlier leads to z = 1.75.  This is enough of a change to alter our conclusion from not rejecting the null hypothesis to rejecting it.  If there is no good reason for deleting the outlier, then the z-test is inappropriate for these data.

9.67 (a) Using Minitab, with the data in a column named TEMP, we choose **Stat ▶**

**Basic Statistics ▶ 1-Sample z...**, click in the **Samples in columns** text box and specify TEMP, click in the **Standard deviation** text box and enter 0.63, and click in the **Test mean** text box and enter 98.6.  Click the **Options…** button, enter 95 in the **Confidence level** text box, click the arrow button at the right of the **Alternative** drop-down list box and select **not equal** and click **OK**.  Click on the **Graphs** button and check the boxes for **Histogram of Data** and **Boxplot of data**.  Then click **OK**

twice.  Now choose **Stat ▶ Basic Statistics ▶ Normality test** and enter

CHARGE in the **Variable** text box.  Click **OK**.  Then choose **Graph ▶ Stem-and-Leaf**, enter CHARGE in the **Graph variables** text box and click **OK**.  The graphs are

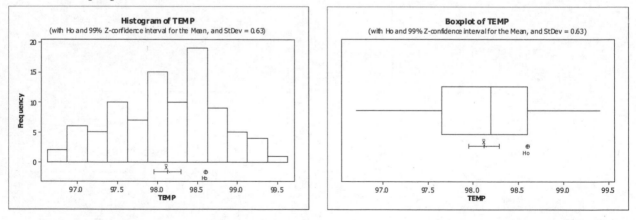

```
Stem-and-leaf of TEMP   N  = 93
Leaf Unit = 0.10

  1     96   7
  3     96   89
  8     97   00001
 13     97   22233
 19     97   444444
 26     97   6666777
 31     97   88889
 45     98   00000000000111
(10)    98   2222222233
 38     98   4444445555
 28     98   66666666677
 17     98   8888888
 10     99   00001
  5     99   2233
  1     99   4
```

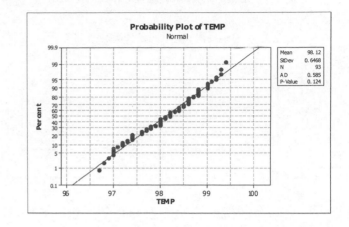

(b) Yes.  The sample size 93 is large and the distribution of the data is quite symmetric.

(c) Yes.  The procedure in part (a) also produced the results of the test which are

Test of mu = 98.6 vs not = 98.6
The assumed standard deviation = 0.63

| Variable | N | Mean | StDev | SE Mean | 99% CI | Z | P |
|---|---|---|---|---|---|---|---|
| TEMP | 93 | 98.1237 | 0.6468 | 0.0653 | (97.9554, 98.2919) | -7.29 | 0.000 |

The critical values are -2.575 and 2.575.  Since z = -7.29, we reject the null hypothesis and conclude that the mean body temperature of healthy humans is different from the generally accepted value of 98.6°F.

**9.69** (a) Using Minitab, with the data in a column named BILL, we choose **Stat ▶**

**Basic Statistics ▶ 1-Sample z...**, click in the **Samples in columns** text box and specify BILL, click in the **Standard deviation** text box and enter 25, and click in the **Test mean** text box and enter 47.37.  Click the **Options...** button, enter 95 in the **Confidence level** text box, click the arrow button at the right of the **Alternative** drop-down list box and select **greater than** and click **OK**.  Click on the **Graphs** button and check the boxes for **Histogram of Data** and **Boxplot of data**.  Then click **OK**

twice.  Now choose **Stat ▶ Basic Statistics ▶ Normality test** and enter

BILL in the **Variable** text box.  Click **OK**.  Then choose **Graph ▶ Stem-and-Leaf**, enter BILL in the **Graph variables** text box and click **OK**.  The graphs are

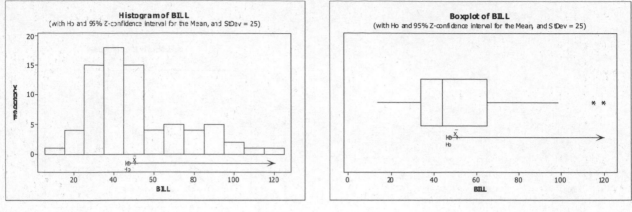

```
Stem-and-leaf of BILL   N = 75
Leaf Unit = 1.0
```

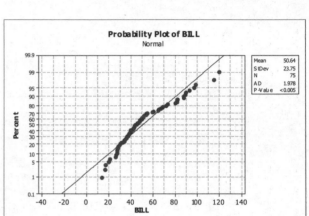

```
   1    1  3
   4    1  669
   5    2  0
  14    2  567777889
  20    3  012334
  31    3  55677889999
  (7)   4  1122333
  37    4  55678899
  29    5  0011334
  22    5  9
  21    6  044
  18    6  688
  15    7  13
  13    7
  13    8  0112
   9    8  7799
   5    9  2
   4    9  78


HI 114, 119
```

(b)  The results of the test carried out by the procedure in part (a) are

```
Test of mu = 47.37 vs > 47.37
The assumed standard deviation = 25
```

|          |    |         |         |         | 95%<br>Lower |      |       |
|----------|----|---------|---------|---------|--------------|------|-------|
| Variable | N  | Mean    | StDev   | SE Mean | Bound        | Z    | P     |
| BILL     | 75 | 50.6405 | 23.7497 | 2.8868  | 45.8922      | 1.13 | 0.129 |

The critical value for the test is z = 1.645.  Since z = 1.13, we do not reject the null hypothesis.  The data do not provide sufficient evidence that the mean local monthly cell phone bill has increased from the 2001 mean of $47.37.

After deleting the two outliers 119.61 and 114.98, the graphs and test results are

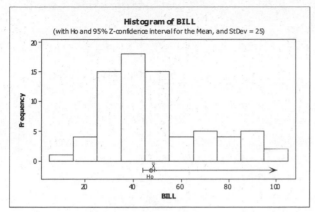

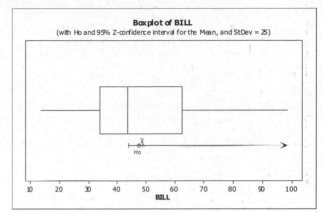

```
Stem-and-leaf of BILL   N = 73
Leaf Unit = 1.0
```

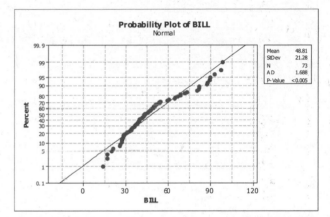

```
   1    1   3
   4    1   669
   5    2   0
  14    2   567777889
  20    3   012334
  31    3   55677889999
  (7)   4   1122333
  35    4   55678899
  27    5   0011334
  20    5   9
  19    6   044
  16    6   688
  13    7   13
  11    7
  11    8   0112
   7    8   7799
   3    9   2
   2    9   78
```

```
Test of mu = 47.37 vs > 47.37
The assumed standard deviation = 25
```

|          |    |         |         |         | 95%<br>Lower |      |       |
|----------|----|---------|---------|---------|--------------|------|-------|
| Variable | N  | Mean    | StDev   | SE Mean | Bound        | Z    | P     |
| BILL     | 73 | 48.8144 | 21.2785 | 2.9260  | 44.0015      | 0.49 | 0.311 |

Although the value of z has been changed from 1.13 to 0.49 by deleting the two outliers, the conclusion remains the same. We do not reject the null hypothesis. Intuitively, we should expect this result, since deleting two large outliers can only reduce the sample mean, making z smaller.

**9.71**  (a)  n = 45, $\overline{x}$ = 14.68, $\sigma$ = 4.2

The 99% upper level confidence bound is $14.68 + 2.33(4.2)/\sqrt{45} = 16.14$.

The hypothesized mean (18 mg) lies above the upper confidence bound, so we should reject the null hypothesis. Since the test statistic is

$z = (14.68 - 18)/(4.2 / \sqrt{45}) = -5.30$, which is less than the critical value of
-2.33, the hypothesis test also leads to the conclusion that we should
reject the null hypothesis.

(b)  n = 21, $\bar{x}$ = 52.5 years, $\sigma$ = 6.8 years

The 95% upper level confidence bound is $52.5 + 2.33(6.8) / \sqrt{21} = 55.96$ .

The hypothesized mean (55) lies below the upper confidence bound, so we
should not reject the null hypothesis.  Since the test statistic is
$z = (52.5 - 55)/(6.8/ \sqrt{21}) = -1.68$, which is greater than the critical value
of -2.33, the hypothesis test also leads to the conclusion that we
should not reject the null hypothesis.

### Exercises 9.4

**9.73** Hypothesis tests have built-in margins of error.  Errors will occur due to
the uncontrollable randomness in the data observed.

**9.75** (a)  α = significance level = P(Type I error) = P(rejecting a true null
hypothesis)

(b)  β = P(Type II error) = P(not rejecting a false null hypothesis)

(c)  1 – β = Power of the test = P(rejecting a false null hypothesis)

**9.77** Since μ is unknown, the power curve enables one to evaluate the
effectiveness of a hypothesis test for a variety of values of $\mu$ .

**9.79** If the significance level is decreased without changing the sample size, the
rejection region is made smaller (in probability terms).  This makes the
nonrejection region larger, i.e., $\beta$ gets larger.  This, in turn, makes the
power 1 - $\beta$ smaller.

**9.81**  (a)   Note:  $z = \dfrac{\bar{x} - \mu_0}{\sigma / \sqrt{n}} \Rightarrow \bar{x} = \mu_0 + z \cdot \sigma / \sqrt{n}$

Since this is a right-tailed test, we would reject $H_0$ if z ≥ 1.645; or
equivalently if $\bar{x} \geq 0.5 + 1.645(0.37) / \sqrt{12} = 0.6757$

So we reject $H_0$ if $\bar{x}$ ≥ 0.676; otherwise do not reject $H_0$.

(b)  P(Type I error) = $\alpha$ = 0.05

(c)

| True mean $\mu$ | z-score computation | P(Type II error) $\beta$ | Power $1-\beta$ |
|---|---|---|---|
| 0.55 | $z = \dfrac{0.676 - 0.55}{0.37/\sqrt{12}} = 1.18$ | 0.8810 | 0.119 0 |
| 0.60 | $z = \dfrac{0.676 - 0.60}{0.37/\sqrt{12}} = 0.71$ | 0.7611 | 0.2389 |
| 0.65 | $z = \dfrac{0.676 - 0.65}{0.37/\sqrt{12}} = 0.24$ | 0.5948 | 0.4052 |
| 0.70 | $z = \dfrac{0.676 - 0.70}{0.37/\sqrt{12}} = -0.22$ | 0.4129 | 0.5871 |
| 0.75 | $z = \dfrac{0.676 - 0.75}{0.37/\sqrt{12}} = -0.69$ | 0.2451 | 0.7549 |
| 0.80 | $z = \dfrac{0.676 - 0.80}{0.37/\sqrt{12}} = -1.16$ | 0.1230 | 0.8770 |
| 0.85 | $z = \dfrac{0.676 - 0.85}{0.37/\sqrt{12}} = -1.63$ | 0.0516 | 0.9484 |

(d)

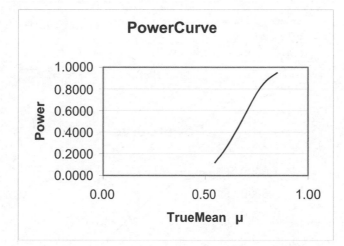

**9.83**  (a)  Note: $z = \dfrac{\bar{x} - \mu_0}{\sigma/\sqrt{n}} \Rightarrow \bar{x} = \mu_0 + z \cdot \sigma/\sqrt{n}$

Since this is a left-tailed test, we would reject $H_0$ if $z \le -2.33$; or equivalently if $\bar{x} \le 18 - 2.33(4.2)/\sqrt{45} = 16.54$

So reject $H_0$ if $\bar{x} \le 16.54$; otherwise do not reject $H_0$.

(b)  P(Type I error) = $\alpha$ = 0.01

(c)  Answers may differ slightly from those in the text due to intermediate rounding.

| True mean $\mu$ | z-score computation | P(Type II error) $\beta$ | Power $1 - \beta$ |
|---|---|---|---|
| 15.50 | $z = \dfrac{16.54 - 15.50}{4.2 / \sqrt{45}} = 1.66$ | $1.000 - 0.9515 = 0.0485$ | 0.9515 |
| 15.75 | $z = \dfrac{16.54 - 15.75}{4.2 / \sqrt{45}} = 1.26$ | $1.000 - 0.8962 = 0.1038$ | 0.8962 |
| 16.00 | $z = \dfrac{16.54 - 16.00}{4.2 / \sqrt{45}} = 0.86$ | $1.000 - 0.8051 = 0.1949$ | 0.8051 |
| 16.25 | $z = \dfrac{16.54 - 16.25}{4.2 / \sqrt{45}} = 0.46$ | $1.000 - 0.6772 = 0.3228$ | 0.6772 |
| 16.50 | $z = \dfrac{16.54 - 16.50}{4.2 / \sqrt{45}} = 0.06$ | $1.000 - 0.5239 = 0.4761$ | 0.5239 |
| 16.75 | $z = \dfrac{16.54 - 16.75}{4.2 / \sqrt{45}} = -0.34$ | $1.000 - 0.3669 = 0.6331$ | 0.3669 |
| 17.00 | $z = \dfrac{16.54 - 17.00}{4.2 / \sqrt{45}} = -0.73$ | $1.000 - 0.2327 = 0.7673$ | 0.2327 |
| 17.25 | $z = \dfrac{16.54 - 17.25}{4.2 / \sqrt{45}} = -1.13$ | $1.000 - 0.1292 = 0.8708$ | 0.1292 |
| 17.50 | $z = \dfrac{16.54 - 17.50}{4.2 / \sqrt{45}} = -1.53$ | $1.000 - 0.0630 = 0.9370$ | 0.0630 |
| 17.75 | $z = \dfrac{16.54 - 17.75}{4.2 / \sqrt{45}} = -1.93$ | $1.000 - 0.0268 = 0.9732$ | 0.0268 |

(d)

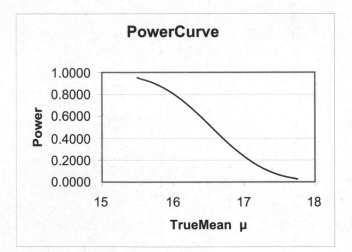

**PowerCurve**

9.85 (a) Note: $z = \dfrac{\bar{x} - \mu_0}{\sigma / \sqrt{n}} \Rightarrow \bar{x} = \mu_0 + z \cdot \sigma / \sqrt{n}$

Since this is a two-tailed test, we would reject $H_0$ if $|z| \geq 1.96$; or equivalently if $\bar{x} \geq 16.7 + 1.96(6.0) / \sqrt{100} = 17.876$ or

$$\bar{x} \le 16.7 - 1.96(6.0)/\sqrt{100} = 15.524 \, .$$

So reject $H_0$ if $\bar{x} \le 15.524$ or $\bar{x} \ge 17.876$; otherwise do not reject $H_0$.

(b) P(Type I error) = $\alpha$ = 0.05

(c)

| True mean $\mu$ | z-score computation | P(Type II error) $\beta$ | Power $1 - \beta$ |
|---|---|---|---|
| 14.0 | $z = \dfrac{15.524 - 14.0}{6.0/\sqrt{100}} = 2.54$ <br><br> $z = \dfrac{17.876 - 14.0}{6.0/\sqrt{100}} = 6.46$ | 1.0000 - 0.9945 = 0.0055 | 0.9945 |
| 14.5 | $z = \dfrac{15.524 - 14.5}{6.0/\sqrt{100}} = 1.71$ <br><br> $z = \dfrac{17.876 - 14.5}{6.0/\sqrt{100}} = 5.63$ | 1.0000 - 0.9564 = 0.0436 | 0.9564 |
| 15.0 | $z = \dfrac{15.524 - 15.0}{6.0/\sqrt{100}} = 0.87$ <br><br> $z = \dfrac{17.876 - 15.0}{6.0/\sqrt{100}} = 4.79$ | 1.0000 - 0.8078 = 0.1922 | 0.8078 |
| 15.5 | $z = \dfrac{15.524 - 15.5}{6.0/\sqrt{100}} = 0.04$ <br><br> $z = \dfrac{17.876 - 15.5}{6.0/\sqrt{100}} = 3.96$ | 1.0000 - 0.5160 = 0.4540 | 0.5160 |
| 16.0 | $z = \dfrac{15.524 - 16.0}{6.0/\sqrt{100}} = -0.79$ <br><br> $z = \dfrac{17.876 - 16.0}{6.0/\sqrt{100}} = 3.13$ | 0.9991 - 0.2148 = 0.7843 | 0.2157 |
| 16.5 | $z = \dfrac{15.524 - 16.5}{6.0/\sqrt{100}} = -1.63$ <br><br> $z = \dfrac{17.876 - 16.5}{6.0/\sqrt{100}} = 2.29$ | 0.9890 - 0.0516 = 0.9374 | 0.0626 |
| 17.0 | $z = \dfrac{15.524 - 17.0}{6.0/\sqrt{100}} = -2.46$ <br><br> $z = \dfrac{17.876 - 17.0}{6.0/\sqrt{100}} = 1.46$ | 0.9279 - 0.0069 = 0.9210 | 0.0790 |

| True mean $\mu$ | z-score computation | P(Type II error) $\beta$ | Power $1 - \beta$ |
|---|---|---|---|
| 17.5 | $z = \dfrac{15.524 - 17.5}{6.0/\sqrt{100}} = -3.29$ <br> $z = \dfrac{17.876 - 17.5}{6.0/\sqrt{100}} = 0.63$ | 0.7357 - 0.0005 = 0.7352 | 0.2648 |
| 18.0 | $z = \dfrac{15.524 - 18.0}{6.0/\sqrt{100}} = -4.13$ <br> $z = \dfrac{17.876 - 18.0}{6.0/\sqrt{100}} = -0.21$ | 0.4168 - 0.0000 = 0.4168 | 0.5832 |
| 18.5 | $z = \dfrac{15.524 - 18.5}{6.0/\sqrt{100}} = -4.96$ <br> $z = \dfrac{17.876 - 18.5}{6.0/\sqrt{100}} = -1.04$ | 0.1492 - 0.0000 = 0.1492 | 0.8508 |
| 19.0 | $z = \dfrac{15.524 - 19.0}{6.0/\sqrt{100}} = -5.79$ <br> $z = \dfrac{17.876 - 19.0}{6.0/\sqrt{100}} = -1.87$ | 0.0307 - 0.0000 = 0.0307 | 0.9693 |

(d)

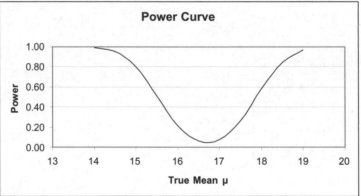

9.87 (a) Note: $z = \dfrac{\bar{x} - \mu_0}{\sigma/\sqrt{n}} \Rightarrow \bar{x} = \mu_0 + z \cdot \sigma / \sqrt{n}$

Since this is a right-tailed test, we would reject $H_0$ if $z \geq 1.645$; or equivalently if $\bar{x} \geq 0.5 + 1.645(0.37)/\sqrt{20} = 0.636$

So we reject $H_0$ if $\bar{x} \geq 0.636$; otherwise do not reject $H_0$.

(b) P(Type I error) = $\alpha$ = 0.05

(c) Answers may differ slightly from those in the text due to intermediate rounding.

| True mean $\mu$ | z-score computation | P(Type II error) $\beta$ | Power $1 - \beta$ |
|---|---|---|---|
| 0.55 | $z = \dfrac{0.636 - 0.55}{0.37/\sqrt{20}} = 1.04$ | 0.8508 | 0.1492 |
| 0.60 | $z = \dfrac{0.636 - 0.60}{0.37/\sqrt{20}} = 0.44$ | 0.6700 | 0.3300 |
| 0.65 | $z = \dfrac{0.636 - 0.65}{0.37/\sqrt{20}} = -0.17$ | 0.4325 | 0.5675 |
| 0.70 | $z = \dfrac{0.636 - 0.70}{0.37/\sqrt{20}} = -0.77$ | 0.2206 | 0.7794 |
| 0.75 | $z = \dfrac{0.636 - 0.75}{0.37/\sqrt{20}} = -1.38$ | 0.0838 | 0.9162 |
| 0.80 | $z = \dfrac{0.636 - 0.80}{0.37/\sqrt{20}} = -1.98$ | 0.0239 | 0.9761 |
| 0.85 | $z = \dfrac{0.636 - 0.85}{0.37/\sqrt{20}} = -2.59$ | 0.0048 | 0.9952 |

(d)

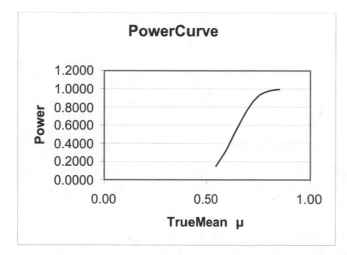

The power curve with n = 20 rises more quickly as the true mean μ increases, resulting in a higher power at any given value of $\mu$ than for n = 12. This illustrates the principle that a larger sample size has a higher probability of rejecting the null hypothesis when the null hypothesis is false and the significance level remains the same.

9.89   (a)   Note:   $z = \dfrac{\bar{x} - \mu_0}{\sigma/\sqrt{n}} \Rightarrow \bar{x} = \mu_0 + z \cdot \sigma / \sqrt{n}$

Since this is a two-tailed test, we would reject $H_0$ if $|z| \geq 1.96$; or equivalently if $\bar{x} \geq 16.7 + 1.96(6.0)/\sqrt{40} = 18.559$ or

$\bar{x} \leq 16.7 - 1.96(6.0)/\sqrt{40} = 14.841$.

So reject $H_0$ if $\bar{x} \leq 14.841$ or $\bar{x} \geq 18.559$; otherwise do not reject $H_0$.

(b)  P(Type I error) = $\alpha$ = 0.05

(c)  The details of the z-score computation are the same as in Exercise 9.85, with 14.841 replacing 15.524, 18.559 replacing 17.876, and 40 replacing 100.  The results of the computations are

| $\mu$ | z-left | z-right | P(z<z-left) | P(z<z-rt) | $\beta$ | Power $1-\beta$ |
|-------|--------|---------|-------------|-----------|---------|-----------------|
| 14.0 | 0.89 | 4.81 | 0.8133 | 1.0000 | 0.1867 | 0.8133 |
| 14.5 | 0.36 | 4.28 | 0.6406 | 1.0000 | 0.3594 | 0.6406 |
| 15.0 | -0.17 | 3.75 | 0.4325 | 0.9999 | 0.5674 | 0.4326 |
| 15.5 | -0.69 | 3.22 | 0.2451 | 0.9994 | 0.7543 | 0.2457 |
| 16.0 | -1.22 | 2.70 | 0.1112 | 0.9965 | 0.8853 | 0.1147 |
| 16.5 | -1.75 | 2.17 | 0.0401 | 0.9850 | 0.9449 | 0.0551 |
| 17.0 | -2.28 | 1.64 | 0.0113 | 0.9495 | 0.9382 | 0.0618 |
| 17.5 | -2.80 | 1.12 | 0.0026 | 0.8686 | 0.8661 | 0.1339 |
| 18.0 | -3.33 | 0.59 | 0.0004 | 0.7224 | 0.7220 | 0.2780 |
| 18.5 | -3.86 | 0.06 | 0.0001 | 0.5239 | 0.5239 | 0.4761 |
| 19.0 | -4.38 | -0.46 | 0.0000 | 0.3228 | 0.3228 | 0.6772 |

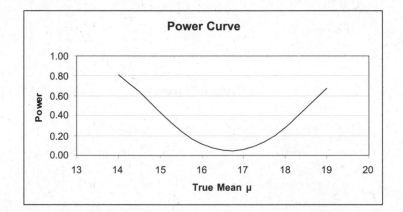

The power curve with n = 40 rises less quickly as the true mean $\mu$ increases or decreases from 16.7, resulting in a lower power at any given value of $\mu$ than for n = 100.  This illustrates the principle that a larger sample size has a higher probability of rejecting the null hypothesis when the null hypothesis is false and the significance level remains the same.

**9.91**  (a)

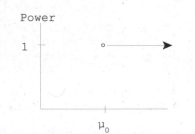

(b)  The curve in part (a) portrays that, ideally, one desires the value for the power for any given $\mu_a$ to be as close to 1 as possible.

**9.93** (a)

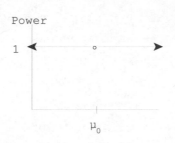

(b)  The curve in (a) portrays that, ideally, one desires the value for the power for any given $\mu_a$ to be as close to 1 as possible.

### Exercises 9.5

**9.95**  (1) It allows the reader to assess significance at any desired level, and (2) it permits the reader to evaluate the strength of the evidence against the null hypothesis.

**9.97**  In the **critical value** approach, we determine critical values based on the significance level.  The critical values determine where the rejection and nonrejection regions lie for the test statistic.  If the value of the test statistic falls in the rejection region, the null hypothesis is rejected.  In the **P-value** approach, the test statistic is computed and then the probability of observing a value as extreme or more extreme than the value obtained is determined.  If the P-value is smaller than the significance level, the null hypothesis is rejected.  Reporting a P-value allows a reader to draw his/her own conclusion based on the strength of the evidence.

**9.99**  True

**9.101** (a)  Do not reject the null hypothesis.

(b)  Reject the null hypothesis.

(c)  Do not reject the null hypothesis.

**9.103** (a)  Strength of the evidence against the null hypothesis is moderate.

(b)  There is weak or no evidence against the null hypothesis.

(c)  Strength of the evidence against the null hypothesis is strong.

(d)  Strength of the evidence against the null hypothesis is very strong.

**9.105** (a)  $z = (\bar{x} - \mu)/(\sigma/\sqrt{n}) = (20 - 22)/(4/\sqrt{32}) = -2.83$; P-value = P(z < -2.83) = 0.0023

(b)  The evidence against the null hypothesis is very strong.

**9.107** (a)  $z = (\bar{x} - \mu)/(\sigma/\sqrt{n}) = (24 - 22)/(4/\sqrt{15}) = 1.94$; P-value = P(z > 1.94) = 0.0262

(b)  The evidence against the null hypothesis is strong.

**9.109** (a)  $z = (\bar{x} - \mu)/(\sigma/\sqrt{n}) = (23 - 22)/(4/\sqrt{24}) = 1.22$; P-value = 2P(z > |1.22|) = 0.2224

(b)  The evidence against the null hypothesis is weak or none.

**9.111** (a)  z = 2.03,     P-value = 1.0000 - 0.9788 = 0.0212

(b)  z = -0.31,    P-value = 1.0000 - 0.3783 = 0.6217

**9.113** (a)  z = -0.74,    P-value = 0.8770

(b)  z = 1.16,     P-value = 0.0329

**9.115** (a)  z = -1.66,    Left-tail probability = 0.0485

P-value = 0.0485 x 2 = 0.0970

(b) $z = 0.52$,      Right-tail probability $= 1.0000 - 0.6985 = 0.3015$

P-value $= 0.3015 \times 2 = 0.6030$

**9.117** (See Exercise 9.59 for classical approach results.)

Step 1: $H_0$: $\mu = 0.5$ ppm, $H_a$: $\mu > 0.5$ ppm

Step 2: $\alpha = 0.05$

Step 3: $z = 0.24$

Step 4: P-value $= P(z \geq 0.24) = 1.0000 - 0.5948 = 0.4052$

Step 5: Since $0.4052 > 0.05$, do not reject $H_0$.

Step 6: At the 5% significance level, the data do not provide sufficient evidence to conclude that the mean cadmium level $\mu$ of *Boletus pinicola* mushrooms is greater than the safety limit of 0.5 ppm.

Using Table 9.12, we classify the strength of evidence against the null hypothesis as weak or none because P > 0.10.

**9.119** (See Exercise 9.61 for classical approach results.)

Step 1: $H_0$: $\mu = 18$ mg, $H_a$: $\mu < 18$ mg

Step 2: $\alpha = 0.01$

Step 3: $z = -5.30$

Step 4: P-value $= P(z < -5.30) = 0.0000$ (to four decimal places)

Step 5: Since $0.0000 < 0.01$, reject $H_0$.

Step 6: At the 1% significance level, the data provide sufficient evidence to conclude that adult females under the age of 51 are, on the average, getting less than the RDA of 18 mg of iron.

Using Table 9.12, we classify the strength of evidence against the null hypothesis as very strong because P < 0.01.

**9.121** (See Exercise 9.63 for classical approach results.)

(a) Step 1: $H_0$: $\mu = 16.7$, $H_a$: $\mu \neq 16.7$

Step 2: $\alpha = 0.05$

Step 3: $z = 1.83$

Step 4: P-value $= P(z \leq -1.83 \text{ or } z \geq 1.83) = 2(0.0336) = 0.0672$

Step 5: Since $0.0672 > 0.05$, do not reject $H_0$.

Step 6: At the 5% significance level, the data do not provide sufficient evidence to conclude that the mean length of imprisonment for motor-vehicle theft offenders in Sydney, Australia differs from the Australian national mean of 16.7 months.

Using Table 9.12, we classify the strength of evidence against the null hypothesis as moderate since $0.05 < P < 0.10$.

**9.123** (a) Using Minitab, with the data in a column named GAIN, we choose **Stat** ▶

**Basic Statistics** ▶ **1-Sample z...**, click in the **Samples in columns** text box and specify GAIN, click in the **Standard deviation** text box and type 0.42, and click in the **Test mean** text box and type 0.2. Click the **Options…** button, enter 95 in the **Confidence level** text box, click the arrow button at the right of the **Alternative** drop-down list box and select **greater than** and click **OK**. Click on the **Graphs** button and check the boxes for **Histogram of Data** and **Boxplot of data**. Then click **OK** twice. The result of the test is

Test of mu = 0.2 vs > 0.2
The assumed standard deviation = 0.42

```
                                        95%
                                       Lower
Variable   N      Mean      StDev   SE Mean    Bound    Z      P
GAIN      20   0.295000   0.499974  0.093915  0.140524  1.01  0.156
```

The P-value is reported in the last line of the output as 0.156.  Since 0.156 > 0.05, we do not reject the null hypothesis.  The evidence against the null hypothesis is weak or nonexistent.

(b)  The histogram and boxplot were produced by the procedure in part (a).

Now choose **Stat** ▶ **Basic Statistics** ▶ **Normality test** and enter GAIN in the **Variable** text box.  Click **OK.**  Then choose **Graph** ▶ **Stem-and-Leaf** , enter GAIN in the **Graph variables** text box and click **OK.**  The results are

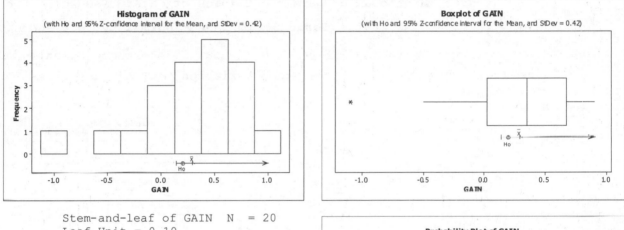

```
Stem-and-leaf of GAIN   N  = 20
Leaf Unit = 0.10

LO -11

    2    -0   5
    3    -0   2
    4    -0   1
    6     0   01
   10     0   2233
   10     0   45
    8     0   6667
    4     0   8889
```

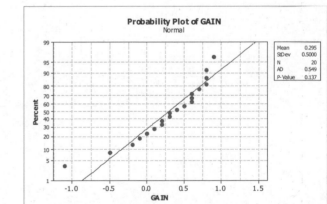

(c)  Repeating the procedure of part (a) after removing the outlier, we obtain

Test of mu = 0.2 vs > 0.2
The assumed standard deviation = 0.42

```
                                        95%
                                       Lower
Variable   N      Mean      StDev   SE Mean    Bound    Z      P
GAIN      19   0.368421   0.387374  0.096355  0.209932  1.75  0.040
```

Now the P-value is 0.04, which is less than the significance level, leading to rejection of the null hypothesis.  Now the data do provide sufficient evidence to conclude that, on average, the net percentage gain exceeds 0.2.

(d) The original sample size is only 20.  The plots in part (b) indicate that the value -1.1 is a potential outlier.  The z-test should not be used with the original data.  This is further confirmed by the fact that using all of the data leads to z = 1.01, whereas, deleting the outlier leads to z = 1.75.  This is enough of a change to alter our conclusion from not rejecting the null hypothesis to rejecting it.  If there is no good reason for deleting the outlier, then the z-test is inappropriate for these data.

9.125 (a) Using Minitab, with the data in a column named TEMP, we choose **Stat** ▶

**Basic Statistics** ▶ **1-Sample z...**, click in the **Samples in columns** text box and specify TEMP, click in the **Standard deviation** text box and enter 0.63, and click in the **Test mean** text box and enter 98.6.  Click the **Options**... button, enter 95 in the **Confidence level** text box, click the arrow button at the right of the **Alternative** drop-down list box and select **not equal** and click **OK**.  Click on the **Graphs** button and check the boxes for **Histogram of Data** and **Boxplot of data**.  Then click **OK**

twice.  Now choose **Stat** ▶ **Basic Statistics** ▶ **Normality test** and enter

CHARGE in the **Variable** text box.  Click **OK**.  Then choose **Graph** ▶ **Stem-and-Leaf**, enter CHARGE in the **Graph variables** text box and click **OK**.
The graphs are

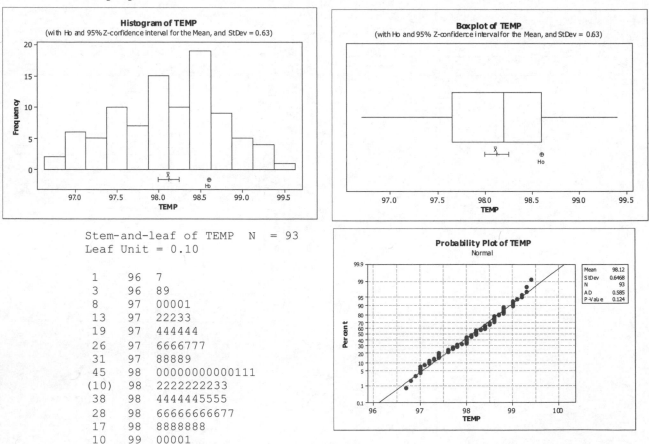

```
Stem-and-leaf of TEMP   N  = 93
Leaf Unit = 0.10

    1    96   7
    3    96   89
    8    97   00001
   13    97   22233
   19    97   444444
   26    97   6666777
   31    97   88889
   45    98   00000000000111
  (10)   98   2222222233
   38    98   4444445555
   28    98   66666666677
   17    98   8888888
   10    99   00001
    5    99   2233
    1    99   4
```

(b) Yes.  The sample size 93 is large and the distribution of the data is quite symmetric.

(c) Yes.  The procedure in part (a) also produced the results of the test which are

```
Test of mu = 98.6 vs not = 98.6
The assumed standard deviation = 0.63

Variable   N     Mean   StDev  SE Mean       99% CI          Z      P
TEMP      93  98.1237  0.6468  0.0653  (97.9554, 98.2919)  -7.29  0.000
```
The P-value is shown in the last line of the output as 0.000.  This is less than the significance level of 0.01, leading to rejection of the null hypothesis.  The evidence against the null hypothesis is categorized as very strong since P < 0.01.

**9.127** (a)  Using Minitab, **w**ith the data in a column named BILL, we choose **Stat ▶**

**Basic Statistics ▶ 1-Sample z...,** click in the **Samples in columns** text box and specify BILL, click in the **Standard deviation** text box and enter 25, and click in the **Test mean** text box and enter 47.37.  Click the **Options...** button, enter 95 in the **Confidence level** text box, click the arrow button at the right of the **Alternative** drop-down list box and select **greater than** and click **OK**.  Click on the **Graphs** button and check the boxes for **Histogram of Data** and **Boxplot of data**.  Then click **OK** twice.  Now choose **Stat ▶ Basic Statistics ▶ Normality test** and enter

BILL in the **Variable** text box.  Click **OK**.  Then choose **Graph ▶ Stem-and-Leaf**, enter BILL in the **Graph variables** text box and click **OK**.

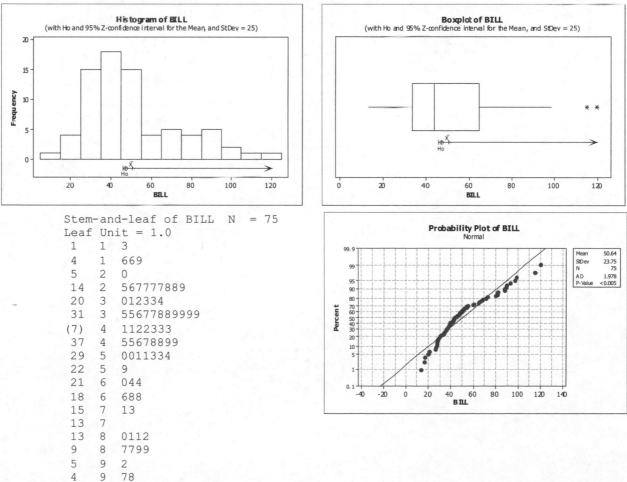

```
Stem-and-leaf of BILL   N  = 75
Leaf Unit = 1.0
  1    1  3
  4    1  669
  5    2  0
 14    2  567777889
 20    3  012334
 31    3  55677889999
 (7)   4  1122333
 37    4  55678899
 29    5  0011334
 22    5  9
 21    6  044
 18    6  688
 15    7  13
 13    7
 13    8  0112
  9    8  7799
  5    9  2
  4    9  78
 HI 114, 119
```

(b)     The results of the test carried out by the procedure in part (a) are
        Test of mu = 47.37 vs > 47.37
        The assumed standard deviation = 25

|          |    |         |         |         | 95%<br>Lower |      |       |
|----------|----|---------|---------|---------|--------------|------|-------|
| Variable | N  | Mean    | StDev   | SE Mean | Bound        | Z    | P     |
| BILL     | 75 | 50.6405 | 23.7497 | 2.8868  | 45.8922      | 1.13 | 0.129 |

The P-value is shown in the last line as 0.129, which is greater than the significance level of 0.05, leading to nonrejection of the null hypothesis. The data do not provide sufficient evidence to conclude that the mean local monthly cell phone bill has increased from the 2001 mean of $47.37. The evidence against the null hypothesis is weak or nonexistent.

After deleting the two outliers 119.61 and 114.98, the graphs and test results are

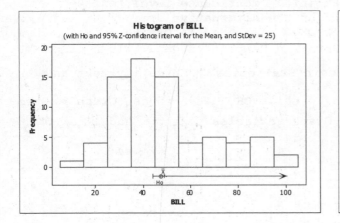

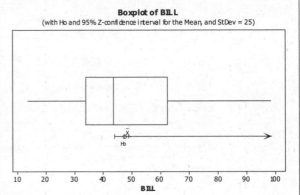

Stem-and-leaf of BILL   N  = 73
Leaf Unit = 1.0

```
 1    1   3
 4    1   669
 5    2   0
14    2   567777889
20    3   012334
31    3   55677889999
(7)   4   1122333
35    4   55678899
27    5   0011334
20    5   9
19    6   044
16    6   688
13    7   13
11    7
11    8   0112
 7    8   7799
 3    9   2
 2    9   78
```

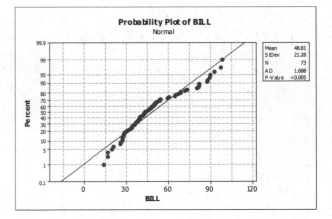

Test of mu = 47.37 vs > 47.37
The assumed standard deviation = 25

|          |    |         |         |         | 95%<br>Lower |      |       |
|----------|----|---------|---------|---------|--------------|------|-------|
| Variable | N  | Mean    | StDev   | SE Mean | Bound        | Z    | P     |
| BILL     | 73 | 48.8144 | 21.2785 | 2.9260  | 44.0015      | 0.49 | 0.311 |

Although the P-value has been changed from 0.129 to 0.311 by deleting the two outliers, the conclusion remains the same.  We do not reject the null hypothesis. Intuitively, we should expect this result, since deleting two large outliers can only reduce the sample mean, making the P-value larger.

**9.129** (a) Left-tailed: P-value $= P(z \le z_0) = \Phi(z_0)$

(b) Right-tailed: P-value $= P(z \ge z_0) = 1 - P(z < z_0) = 1 - \Phi(z_0)$

(c) Two-tailed: P-value $= P(|z| \ge |z_0|)$

$= P(z \ge |z_0|) + P(z \le - |z_0|)$

By symmetry $= P(z \ge |z_0|) + P(z \ge |z_0|)$

$= 2 \cdot P(z \ge |z_0|) = 2 \cdot [1 - P(z < |z_0|)]$

$= 2 \cdot [1 - \Phi(|z_0|)]$

**9.131** The P-value approach provides the actual significance of the hypothesis test, that is, the smallest significance level at which the results of the test are significant. It also allows the reader to judge the strength of the evidence against the null hypothesis for himself/herself. The only disadvantage of the P-value approach is that one must compute the P-value (one extra step) after computing the z-value.

### Exercises 9.6

**9.133** (a)  $0.01 < P < 0.025$

(b) Reject $H_0$ for $\alpha \ge 0.025$; Do not reject $H_0$ for $\alpha \le 0.01$; Undecided for $0.01 < \alpha < 0.025$.

**9.135** (a)  $P < 0.005$

(b) Reject $H_0$ for $\alpha \ge 0.005$; Undecided for $\alpha < 0.005$.

**9.137** (a)  $0.01 < P < 0.02$

(b) Reject $H_0$ for $\alpha \ge 0.02$; Do not reject $H_0$ for $\alpha \le 0.01$; Undecided for $0.01 < \alpha < 0.02$.

**9.139** (a)  df $= 31$; t $= -2.828$

(b)  $P < 0.005$; Reject $H_0$; Evidence against $H_0$ is very strong.

**9.141** (a)  df $= 14$; t $= 1.936$

(b)  $0.05 < P < 0.10$; Do not reject $H_0$; Evidence against $H_0$ is moderate.

**9.143** (a)  df $= 23$; t $= 1.225$

(b)  $P > 0.20$; Do not reject $H_0$; Evidence against $H_0$ is weak or none.

**9.145** $H_0$: $\mu = 4.66$, $H_a$: $\mu \ne 4.66$, $\alpha = 0.10$; Critical values: $\pm 1.729$

$$t = (\bar{x} - \mu)/(\sigma / \sqrt{n}) = (4.835 - 4.66)/(2.291 / \sqrt{20}) = 5.123 .$$

Since $5.123 > 1.729$, we reject $H_0$. The data provide sufficient evidence to conclude that the amount of television watched per day last year by the average person differed from that in 2002.

**9.147** n $= 10$, df $= 9$, $\bar{x} = 2.5$, s $= 0.149$

Step 1:  $H_0$: $\mu = 2.3$, $H_a$: $\mu > 2.3$

Step 2:  $\alpha = 0.01$

Step 3: $t = \dfrac{2.5 - 2.3}{0.149/\sqrt{10}} = 4.251$

Step 4: Critical value = 2.821

Step 5: Since 4.251 > 2.821, reject $H_0$. Note: For the P-value approach, P-value < 0.01. So, since the P-value < $\alpha$, reject $H_0$.

Step 6: At the 1% significance level, the data do provide sufficient evidence to conclude that the mean available limestone in soil treated with 100% MMBL effluent is greater than 2.30%. The practical significance of this result probably depends on what crop is to be grown in the soil.

**9.149** n = 187, df = 186, $\bar{x}$ = 0.64, s = 0.15

Step 1: $H_0$: $\mu$ = 0.9, $H_a$: $\mu$ < 0.9

Step 2: $\alpha$ = 0.05

Step 3: $t = \dfrac{0.64 - 0.90}{0.15/\sqrt{187}} = -23.703$

Step 4: Critical value = -1.653

Step 5: Since -23.703 < -1.653, reject $H_0$. Note: For the p-value approach, P < 0.005. So, since the p-value < $\alpha$, reject $H_0$.

Step 6: At the 5% significance level, the data provide sufficient evidence to conclude that the mean ABI $\mu$ for women with peripheral arterial disease is less than the healthy ABI of 0.9. Thus, we conclude that such women do have an unhealthy ABI. The practical significance of this result is that the ABI may be a good tool for determining the possibility of peripheral arterial disease in women. There could also be other causes of a low ABI, so this test by itself may not be able to determine the precise ailment.

**9.151** We used Minitab to produce the following histogram.

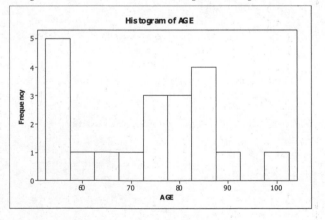

The sample size is only 20. Although there are no outliers, the distribution is not very close to being normally distributed. It does not appear to be reasonable to use a t-test with these data.

**9.153** We used Minitab to produce the following normal probability plot.

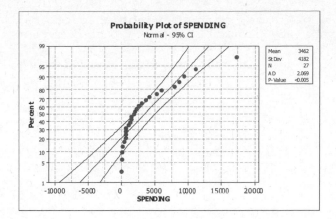

9.155 (a) Using Minitab, with the data in a column named PRESSURE, we choose **Stat**
▶ **Basic Statistics** ▶ **1-Sample t...**, click in the **Samples in columns**
text box and specify <u>PRESSURE</u>, click in the **Test mean** text box and
enter <u>80</u>. Click the **Options...** button, enter <u>90</u> in the **Confidence level**
text box, click the arrow button at the right of the **Alternative** drop-
down list box and select **greater than** and click **OK**. Click on the
**Graphs** button and check the boxes for **Histogram of Data** and **Boxplot of
data**. Then click **OK** twice. Now choose **Stat** ▶ **Basic Statistics** ▶
**Normality test** and enter <u>PRESSURE</u> in the **Variable** text box. Click **OK**.

Then choose **Graph** ▶ **Stem-and-Leaf** , enter <u>PRESSURE</u> in the **Graph
variables** text box and click **OK**. The results are

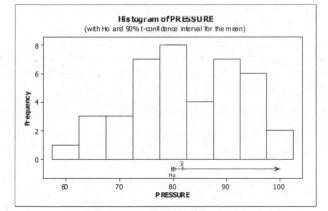

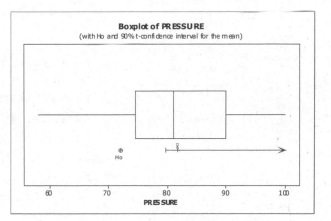

```
Stem-and-leaf of PRESSURE   N  = 41
Leaf Unit = 1.0

     1   5   8
     2   6   3
     5   6   569
    10   7   00334
    17   7   5677999
    (8)  8   00111334
    16   8   5899
    12   9   0001334
     5   9   5559
     1  10   0
```

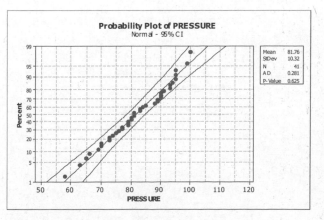

(b) The large sample size, lack of potential outliers, and the nearly linear probability plot all indicate that a t-test is reasonable for these data.

(c) The first procedure in part (a) also yielded the following test results:

Test of mu = 80 vs > 80

| | | | | 90% Lower | | |
|---|---|---|---|---|---|---|
| Variable | N | Mean | StDev | SE Mean | Bound | T | P |
| PRESSURE | 41 | 81.7561 | 10.3242 | 1.6124 | 79.6551 | 1.09 | 0.141 |

We see that t = 1.09 and the P-value is 0.141. Since the P-value is greater than the significance level 0.10, we do not reject the null hypothesis and conclude that the data do not provide evidence that the mean diastolic blood pressure of bus drivers in Stockholm exceeds the normal pressure of 88 mm Hg.

**9.157** (a) Using Minitab, with the data in a column named DISTANCE, we choose **Stat ▶ Basic Statistics ▶ 1-Sample t...**, click in the **Samples in columns** text box and specify RENT, click in the **Test mean** text box and enter 692. Click the **Options...** button, enter 95 in the **Confidence level** text box, click the arrow button at the right of the **Alternative** drop-down list box and select **greater than** and click **OK**.

Test of mu = 692 vs > 692

| | | | | 95% Lower | | |
|---|---|---|---|---|---|---|
| Variable | N | Mean | StDev | SE Mean | Bound | T | P |
| RENT | 100 | 703.960 | 89.846 | 8.985 | 689.042 | 1.33 | 0.093 |

The P-value for the test is 0.093, larger than the significance lever 0.05, so we do not reject the null hypothesis. The data do not provide sufficient evidence to conclude that the mean rent for a two-bedroom unit in Maine is greater than the FMR of $692.

(b) After removing the outlier 405 and following the procedure in part (a), the results are

Test of mu = 692 vs > 692

| | | | | 95% Lower | | |
|---|---|---|---|---|---|---|
| Variable | N | Mean | StDev | SE Mean | Bound | T | P |
| RENT | 99 | 706.980 | 85.049 | 8.548 | 692.786 | 1.75 | 0.041 |

(c) Now the P-value is 0.041, leading to rejection of the null hypothesis. The sample mean increased by $3.02 and the standard deviation decreased by about 4.8.

(d) The sample size is large in both cases, yet the effect of the outlier is considerable. Caution should be used and perhaps the data should be analyzed using a method that is not influenced by outliers.

**9.159** (a) n = 20, df = 19, $t_{\alpha/2}$ = 1.729, s = 2.291, $\bar{x}$ = 4.835

The 90% confidence interval is $4.835 \pm 1.729(2.291)/\sqrt{20} = (3.949, 5.721)$.

The hypothesized mean (4.66) lies within the confidence interval, so we should not reject the null hypothesis. Since the test statistic is

$t = (4.835 - 4.66)/(2.291/\sqrt{20}) = 0.342$, which is less than the critical value of 1.729, the hypothesis test also leads to the conclusion that we should not reject the null hypothesis.

(b)   n = 25, df = 24, $t_{\alpha/2}$ = 2.064, s = \$350.90, $\bar{x}$ = \$1935.76
      The 95% confidence interval is

$$1935.76 \pm 2.064(350.90)/\sqrt{25} = (1790.91, 2080.61).$$

The hypothesized mean (\$1749) lies outside the confidence interval, so we should reject the null hypothesis.  Since the test statistic is

$t = (1935.76 - 1749)/(350.90/\sqrt{25}) = 2.661$, which is greater than the critical value of 2.064, the hypothesis test also leads to the conclusion that we should reject the null hypothesis.

**9.161** (a)   n = 6, df = 5, $t_\alpha$ = 2.015, s = 2.7, $\bar{x}$ = 182.7

The 95% lower confidence bound is $182.7 - 2.015(2.7)/\sqrt{6} = 180.479$.

The hypothesized mean (180) lies below the lower confidence bound, so we should reject the null hypothesis.  Since the test statistic is

$t = (182.7 - 180)/(2.7/\sqrt{6}) = 2.449$, which is greater than the critical value of 2.015, the hypothesis test also leads to the conclusion that we should reject the null hypothesis.

The 99% lower confidence bound is $182.7 - 3.365(2.7)/\sqrt{6} = 178.991$.

The hypothesized mean (180) lies above the lower confidence bound, so we should not reject the null hypothesis.  Since the test statistic is

$t = (182.7 - 180)/(2.7/\sqrt{6}) = 2.449$, which is less than the critical value of 3.365, the hypothesis test also leads to the conclusion that we should not reject the null hypothesis.

(b)   n = 10, df = 9, $t_\alpha$ = 2.821, s = 0.149, $\bar{x}$ = 2.5

The 99% lower confidence bound is $2.5 - 2.821(0.149)/\sqrt{10} = 2.367$.

The hypothesized mean (2.30) lies below the lower confidence bound, so we should reject the null hypothesis.  Since the test statistic is

$t = (2.5 - 2.3)/(0.149/\sqrt{10}) = 4.245$, which is greater than the critical value of 2.821, the hypothesis test also leads to the conclusion that we should reject the null hypothesis.

### Exercises 9.7

**9.163** The advantages of nonparametric methods are that they do not require normality, they make use of fewer and simpler calculations than do parametric methods, and they are resistant to outliers.  The disadvantage of nonparametric methods is that they tend to give less accurate results than parametric methods when the assumptions underlying the parametric methods are actually met.

**9.165** Because the D-value for such a data value equals 0, we cannot attach a sign to the rank of |D|.

**9.167** (a)  Wilcoxon signed-rank test (b)    Wilcoxon signed-rank test
       (c)  Neither

**9.169** (a)  $W_{0.05}$ = 30        (b)    $W_{0.95}$ = 8(8 + 1)/2 - 30 = 6

(c) $W_{0.025} = 32$        $W_{0.975} = 8(8 + 1)/2 - 32 = 4$

**9.171** (a) $W_{0.10} = 128$    (b) $W_{0.90} = 19(19 + 1)/2 - 128 = 62$
(c) $W_{0.05} = 136$        $W_{0.995} = 19(19 + 1)/2 - 136 = 54$

**9.173** $H_0: \mu = 124.9$ days, $H_a: \mu < 124.9$ days; $n = 8$; $\alpha = 0.05$,
$W_{0.95} = 8(8 + 1)/2 - 30 = 6$

| Days | D | \|D\| | Rank \|D\| | R |
|------|------|------|------|------|
| 103 | -21.9 | 21.9 | 4 | -4 |
| 80 | -44.9 | 44.9 | 5 | -5 |
| 79 | -45.9 | 45.9 | 7 | -7 |
| 135 | 10.1 | 10.1 | 2 | 2 |
| 134 | 9.1 | 9.1 | 1 | 1 |
| 77 | -47.9 | 47.9 | 8 | -8 |
| 80 | -44.9 | 44.9 | 5 | -5 |
| 111 | -13.9 | 13.9 | 3 | -3 |

The sum of the positive ranks is $W = 3$. This is less than the critical value 6, so we reject the null hypothesis and conclude that the data provide sufficient evidence that the mean number of days of ice cover on Lake Wingra is less than it was in the late 1800s.

**9.175** $\eta_0 = 35.7$, $n = 10$, $\alpha = 0.01$

Step 1:  $H_0: \eta = 35.7$, $H_a: \eta > 35.7$

Step 2:  $\alpha = 0.01$

Step 3:

| x | $x - \eta_0 = D$ | \|D\| | Rank of \|D\| | Signed Rank R |
|------|------|------|------|------|
| 42 | 6.3 | 6.3 | 3 | 3 |
| 45 | 9.3 | 9.3 | 4 | 4 |
| 62 | 26.3 | 26.3 | 10 | 10 |
| 49 | 13.3 | 13.3 | 6 | 6 |
| 14 | -21.7 | 21.7 | 8 | -8 |
| 39 | 3.3 | 3.3 | 2 | 2 |
| 57 | 21.3 | 21.3 | 7 | 7 |
| 11 | -24.7 | 24.7 | 9 | -9 |
| 36 | 0.3 | 0.3 | 1 | 1 |
| 26 | -9.7 | 9.7 | 5 | -5 |

Step 4:  W = sum of the + ranks = 33

Step 5:  Critical value = 50

Step 6:  Since W < 50, do not reject $H_0$.

Step 7:  At the 1% significance level, the data do not provide sufficient evidence to conclude that the median age has increased over the 2002 median age of 35.7 years. The P-value is 0.305.

**9.177** $\mu_0 = \$12,850$, $n = 10$, $\alpha = 0.10$

Step 1:  $H_0: \mu = \$12,850$, $H_a: \mu < \$12,850$

Step 2:  $\alpha = 0.10$

Step 3:

| Price | D | |D| | Rank |D| | R |
|-------|------|------|----------|-----|
| 13480 | 630 | 630 | 8 | 8 |
| 12992 | 142 | 142 | 3 | 3 |
| 12988 | 138 | 138 | 2 | 2 |
| 12800 | -50 | 50 | 1 | -1 |
| 12599 | -251 | 251 | 5 | -5 |
| 12499 | -351 | 351 | 7 | -7 |
| 11500 | -1350 | 1350 | 9 | -9 |
| 10400 | -2450 | 2450 | 10 | -10 |
| 12500 | -350 | 350 | 6 | -6 |
| 12600 | -250 | 250 | 4 | -4 |

Step 4:  W = sum of the + ranks = 13

Step 5:  Critical value = 10(11)/2 - 41 = 14

Step 6:  Since W < 14, reject $H_0$.

Step 7:  At the 10% significance level, the data do provide sufficient evidence to conclude that the mean asking price for a 2003 Ford Mustang is less than the *2006 Kelly Blue Book* value.  The P-value is 0.077.

**9.179** $\mu_0$ = 2.30, n = 10, $\alpha$ = 0.01

(a)  Step 1:    $H_0$: $\mu$ = 2.30, $H_a$: $\mu$ > 2.30

Step 2:    $\alpha$ = 0.01

Step 3:

| x | $x - \mu_0 = D$ | |D| | Rank of |D| | Signed Rank R |
|------|------|------|------|------|
| 2.41 | 0.11 | 0.11 | 3 | -3 |
| 2.60 | 0.30 | 0.30 | 8 | 8 |
| 2.31 | 0.01 | 0.01 | 1 | 1 |
| 2.51 | 0.21 | 0.21 | 5.5 | 5.5 |
| 2.54 | 0.24 | 0.24 | 7 | 7 |
| 2.51 | 0.21 | 0.21 | 5.5 | 5.5 |
| 2.28 | -0.02 | 0.02 | 2 | -2 |
| 2.42 | 0.12 | 0.12 | 4 | 4 |
| 2.72 | 0.42 | 0.42 | 10 | 10 |
| 2.70 | 0.40 | 0.40 | 9 | 9 |

Step 4:    W = sum of the + ranks = 53.0

Step 5:    Critical value = 50

Step 6:    Since W $\geq$ 50, reject $H_0$.

Step 7:    At the 1% significance level, the data provide sufficient evidence to conclude that the mean available limestone in soil treated with 100% MMBL effluent exceeds 2.30%.  The P-value is 0.005.

(b)  A Wilcoxon signed-rank test is permissible because a normally distributed population is symmetric.

(a) Step 1:  $H_0$:  $\mu = 310$, $H_a$:  $\mu < 310$

Step 2:  $\alpha = 0.05$

Step 3:  $t = \dfrac{306 - 310}{8.671/\sqrt{16}} = -1.845$

Step 4: Critical value = -1.753

Step 5: Since -1.845 < -1.753, reject $H_0$. Note: For the P-value approach, 0.025 < p-value < 0.05. So, since P-value < $\alpha$, reject $H_0$.

Step 6: At the 5% significance level, the data do provide sufficient evidence to conclude that the mean content, μ, is less than the advertised content of 310 ml.

(b) Step 1:  $H_0$:  $\mu = 310$, $H_a$:  $\mu < 310$

Step 2:  $\alpha = 0.05$

Step 3:

| X | $x - \mu_0 = D$ | \|D\| | Rank of \|D\| | Signed Rank R |
|-----|------|-----|------|-------|
| 297 | -13 | 13 | 14 | -14 |
| 311 | 1 | 1 | 2 | 2 |
| 322 | 12 | 12 | 12.5 | 12.5 |
| 315 | 5 | 5 | 7 | 7 |
| 318 | 8 | 8 | 9 | 9 |
| 303 | -7 | 7 | 8 | -8 |
| 307 | -3 | 3 | 5 | -5 |
| 296 | -14 | 14 | 15 | -15 |
| 306 | -4 | 4 | 6 | -6 |
| 291 | -19 | 19 | 16 | -16 |
| 312 | 2 | 2 | 4 | 4 |
| 309 | -1 | 1 | 2 | -2 |
| 300 | -10 | 10 | 10.5 | -10.5 |
| 298 | -12 | 12 | 12.5 | -12.5 |
| 300 | -10 | 10 | 10.5 | -10.5 |
| 311 | 1 | 1 | 2 | 2 |

Step 4: W = sum of the + ranks = 36.5

Step 5: Critical value = 36

Step 6: Since W > 36, do not reject $H_0$.

Step 7: At the 5% significance level, the data do not provide sufficient evidence to conclude that the mean content, μ, is less than the advertised content of 310 ml. The P-value is 0.054.

(c) Since the population is normally distributed, the t-test is more powerful than the Wilcoxon signed-rank test; that is, the t-test is more likely to detect a false null hypothesis.

9.183 (a) Using Minitab, with the data in a column named SCORES, we choose **Stat** ▶

**Nonparametrics** ▶ **1-Sample Wilcoxon...**, select SCORES in the **Variables** text box, click in the **Test median** box and type 5, click the arrow

button at the right of the **Alternative** drop down list box, and select
**Less than,** and click **OK.**   The result is

Test of median = 5.000 versus median < 5.000

|  | N | N for Test | Wilcoxon Statistic | P | Estimated Median |
|---|---|---|---|---|---|
| SCORE | 156 | 84 | 777.0 | 0.000 | 4.500 |

Since the P-value of 0.000 is less than the significance level of 0.01,
we reject the null hypothesis and conclude that there is very strong
evidence that professional golfers score better than par on the *Hole
O'Cross Out.*

(b)   To perform the t-test with the original data, we choose **Stat ▶ Basic
statistics ▶ 1-Sample t...,** select the **Samples in columns** box, enter
SCORES in the **Samples in Columns** text box, enter 5 in the **Test Mean**
text box, click on the **Options** button, enter 99.0 in the **Confidence
level** text box, select **less than** in the **Alternative** box, and click **OK**
twice.   The results in the Session Window are

Test of mu = 5 vs < 5

| Variable | N | Mean | StDev | SE Mean | 99% Upper Bound | T | P |
|---|---|---|---|---|---|---|---|
| SCORES | 156 | 4.66667 | 0.76482 | 0.06123 | 4.81061 | -5.44 | 0.000 |

Again, since the P-value of 0.000 is less than the significance level
of 0.01, we reject the null hypothesis and conclude that there is very
strong evidence that professional golfers score better than par on the
*Hole O'Cross Out.*

(c)   The results are the same for the two tests.  The Wilcoxon test is the
more appropriate test, however, since the data are discrete and cannot
have a normal distribution.

**9.185** Using Minitab, with the data in a column named DISTANCE, we choose **Stat ▶
Nonparametrics  ▶ 1-Sample Wilcoxon...,** select DISTANCE in the **Variables**
text box, click in the **Test median** box and type 11.9, click the arrow button
at the right of the **Alternative** drop down list box, and select **not equal,**
and click **OK.**   The result is

Test of median = 11.90 versus median not = 11.90

|  | N | N for Test | Wilcoxon Statistic | P | Estimated Median |
|---|---|---|---|---|---|
| DISTANCE | 500 | 497 | 49193.0 | 0.000 | 10.75 |

Since the P-value of 0.000 is less than the significance level 0.05, we
reject the null hypothesis and conclude that the mean miles driven by cars
last year differs from the mean distance driven in 2000.

**9.187** The Wilcoxon signed-rank test is likely to give better results. The
distribution of marriage durations is unlikely to be normal, possibly not
even symmetric.  Given that duration cannot be less than 0 years and there
are likely to be some fairly long marriages which might look like outliers,
it would be better to use the Wilcoxon signed-rank test which is insensitive
to outliers than to use the t-test which assumes normality and is sensitive
to outliers.

**9.189** (a) If John is not unlucky, he should expect to wait 15 minutes for the train, on the average.

(b) If John is not unlucky, the distribution of the times he waits for the trains should be a uniform distribution over the interval from 0 to 30 minutes.

(c) Step 1: $H_0$: $\eta = 15$, $H_a$: $\eta > 15$

Step 2: $\alpha = 0.10$

Step 3:

| x | $x - \eta_0 = D$ | \|D\| | Rank of \|D\| | Signed Rank R |
|---|---|---|---|---|
| 24 | 9 | 9 | 6.5 | 6.5 |
| 26 | 11 | 11 | 9.5 | 8.5 |
| 20 | 5 | 5 | 3.5 | 3.5 |
| 4 | −11 | 11 | 9.5 | −9.5 |
| 20 | 5 | 5 | 3.5 | 3.5 |
| 3 | −12 | 12 | 11 | −11 |
| 19 | 4 | 4 | 2 | 4 |
| 5 | −10 | 10 | 8 | −8 |
| 28 | 13 | 13 | 12 | 12 |
| 16 | 1 | 1 | 1 | 1 |
| 22 | 7 | 7 | 5 | 5 |
| 24 | 9 | 9 | 6.5 | 6.5 |

Step 4: W = sum of the + ranks = 50.5

Step 5: Critical value = 56

Step 6: Since W = 50.5 < 56, do not reject $H_0$.

Step 7: At the 10% significance level, the data do not provide sufficient evidence to conclude that John waits more than 15 minutes for the train, on the average.

(d) Since the population is uniform (which is symmetric), the Wilcoxon test is appropriate.

(e) Since the population is symmetric and nonnormal, the Wilcoxon signed-rank test is more powerful than the t-test and more appropriate than the t-test, which assumes normality.

**9.191** (a) Step 1: $H_0$: $\eta = 7.4$ lb, $H_a$: $\eta \neq 7.4$ lb

Step 2: $\alpha = 0.05$

Step 3: See Step 3 in the solution to Exercise 9.178.

Step 4: From Step 4 in the solution to Exercise 9.178, W = 57.5. Now:

$$z = \frac{W - n(n+1)/4}{\sqrt{n(n+1)(2n+1)/24}} = \frac{57.5 - 13(13+1)/4}{\sqrt{13(13+1)(2 \cdot 13 + 1)/24}} = 0.84$$

Step 5: Critical values = ±1.96

Step 6: Since −1.96 < 0.84 < 1.96, do not reject $H_0$.

Step 7: At the 5% significance level, the data do not provide sufficient evidence to conclude that this year's median birth weight differs from that in 2000.

(b) Neither the Wilcoxon signed-rank test nor the normal approximation led to rejection of the null hypothesis.

**9.193** Summing the ranks corresponding to the "+" signs in each row results in a

value for W.  All sixteen possible values for W are presented in the last
column.

(a)                                                          (b)        $1/16 = 0.0625$

|        | Rank |   |   |    |
|--------|------|---|---|----|
| 1 | 2 | 3 | 4 | W |
| + | + | + | + | 10 |
| + | + | + | − | 6 |
| + | + | − | + | 7 |
| + | + | − | − | 3 |
| + | − | + | + | 8 |
| + | − | + | − | 4 |
| + | − | − | + | 5 |
| + | − | − | − | 1 |
| − | + | + | + | 9 |
| − | + | + | − | 5 |
| − | + | − | + | 6 |
| − | + | − | − | 2 |
| − | − | + | + | 7 |
| − | − | + | − | 3 |
| − | − | − | + | 4 |
| − | − | − | − | 0 |

(c)

| W | P(W) |
|---|------|
| 0 | 0.0625 |
| 1 | 0.0625 |
| 2 | 0.0625 |
| 3 | 0.1250 |
| 4 | 0.1250 |
| 5 | 0.1250 |
| 6 | 0.1250 |
| 7 | 0.1250 |
| 8 | 0.0625 |
| 9 | 0.0625 |
| 10 | 0.0625 |

(d)

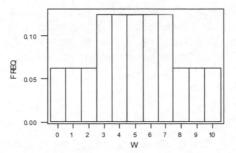

(e)  For a left-tailed test with n = 4 and $\alpha$ = 0.125, the critical value $W_\ell$
     equals 1 since W = 0 or 1 has probability 0.125.

**9.195** (a)  The Wilcoxon signed-rank test takes into account the sign and the
     absolute size of each difference from the median, whereas, the sign
     test only considers the sign.  Since the Wilcoxon test uses more
     information than the sign test, it is more likely to be able to detect
     a false null hypothesis.

(b)  The sign test can be used with any distribution since the probability
     that an observation exceeds the median is always 0.5, regardless of the
     shape of the distribution.  The Wilcoxon signed-rank test is based on

an assumption that the underlying distribution of the data is symmetric, a slightly more restrictive assumption than for the sign test. **9.197** (a) For the sign test, x = 2 of the n = 8 observations are above 124.9. The P-value is P(x ≤ 2) for a binomial distribution with n = 8 and p = 0.5. From Table XII, this probability is 0.004 + 0.031 + 0.109 = 0.144. Since this is greater than the 0.05 significance level, the data do not provide sufficient evidence that the number of days the lake was frozen over is less now than in the late 1800s.

(b) The Wilcoxon signed rank test had W = 3.0 with a P-value of 0.021. Since this is less than the 0.05 significance level, the data do provide sufficient evidence that the number of days the lake was frozen over is less now than in the late 1800s.

**9.199** (a) For the sign test, x = 7 of the n = 10 observations are above 35.7. The P-value is P(x ≥ 7)} for a binomial distribution with n = 10 and p = 0.5. From Table XII, this probability is 0.117 + 0.044 + 0.010 + 0.001 = 0.172. Since this is greater than the 0.05 significance level, the data do not provide sufficient evidence that the median age of today's U.S. residents has increased from the 2002 median of 35.7 years.

(b) The Wilcoxon signed rank test had W = 33.0 with a P-value of 0.363. Since this is greater than the 0.05 significance level, the data do not provide sufficient evidence that the median age of today's U.S. residents has increased from the 2002 median of 35.7 years. Same conclusion, larger P-value.

**9.201** (a) For the sign test, x = 3 of the n = 10 observations are above $12850. The P-value is P(x ≤ 3)} for a binomial distribution with n = 10 and p = 0.5. From Table XII, this probability is 0.001 + 0.044 + 0.010 + 0.001 = 0.172. Since this is greater than the 0.10 significance level, the data do not provide sufficient evidence that the mean asking price for 2003 Ford Mustang coupes in Phoenix is less than the *2006 Kelley Blue Book* retail value of $12850.

(b) The Wilcoxon signed rank test had W = 13.0 with a P-value of 0.077. Since this is less than the 0.10 significance level, the data do provide sufficient evidence that the mean asking price for 2003 Ford Mustang coupes in Phoenix is less than the *2006 Kelley Blue Book* retail value of $12850. Different conclusion, smaller P-value.

### Exercises 9.8

**9.203** (a) One-mean z-test, One-mean t-test, Wilcoxon signed-rank test

(b) The z-test assumes that $\sigma$ is known, and that the population is normal or the sample is large. The t-test assumes that $\sigma$ is unknown, and that the population is normal or the sample is large. The signed-rank test assumes only that the population is symmetric.

(c) $z = (\bar{x} - \mu_0)/(\sigma/\sqrt{n})$       $t = (\bar{x} - \mu_0)/(s/\sqrt{n})$

W = the sum of the positive ranks

**9.205** (a) Yes. The t-test can be used when the sample size is large. It is almost equivalent to the z-test in this situation.

(b) Yes. The Wilcoxon signed-rank test can be used when the population distribution is symmetric.

(c) The Wilcoxon signed-rank test is preferable in this situation since it is more powerful (more likely to detect a false null hypothesis) when the population is symmetric, but nonnormal.

**9.207** Since we have normality and $\sigma$ is known, use the z-test.

**9.209** Since we have a large sample with no outliers and $\sigma$ is unknown, use the t-test.

**9.211** Since we have a symmetric non-normal distribution, use the Wilcoxon signed-rank test.

**9.213** The distribution looks skewed and the sample size is not large.  Consult a statistician.

## Review Problems for Chapter 9

1.  (a) A null hypothesis always specifies a single value for the parameter of a population which is of interest.

    (b) The alternative hypothesis reflects the purpose of the hypothesis test, which can be to determine that the parameter of interest is greater than, less than, or different from the single value specified in the null hypothesis.

    (c) The test statistic is a quantity calculated from the sample, under the assumption that the null hypothesis is true, which is used as a basis for deciding whether or not to reject the null hypothesis.

    (d) The rejection region is a set of values of the test statistic that lead to rejection of the null hypothesis.

    (e) The nonrejection region is a set of values of the test statistic that lead to not rejecting the null hypothesis.

    (f) The critical values are values of the test statistic that separate the rejection region from the nonrejection region.

2.  (a) The statement is expressing the fact that there is variability in the net weights of the boxes' content and some boxes may actually contain less than the printed weight on the box. However, the net weights for each day's production will average a bit more than the printed weight.

    (b) To test the truth of this statement, we would use a null hypothesis that stated that the population mean net weight of the boxes was <u>equal</u> to the printed weight and an alternative hypothesis that stated that the population mean net weight of the boxes was <u>greater than</u> the printed weight.

    (c) Null hypothesis: Population mean net weight = 76 oz

    Alternative hypothesis: Population mean net weight > 76 oz

    or

    $H_0$:    $\mu = 76$          $H_a$:    $\mu > 76$

3.  (a) Roughly speaking, there is a range of values of the test statistic which one could reasonably expect to occur if the null hypothesis were true.  If the value of the test statistic is one that would not be expected to occur when the null hypothesis is true, then we reject the null hypothesis.

    (b) To make this procedure objective and precise, we specify the probability with which we are willing to reject the null hypothesis when it is actually true.  This is called the significance level of the test and is usually some small number like 0.05 or 0.01.  Specifying the significance level allows us to determine the range of values of the test statistic that will lead to rejection of the null hypothesis.

If the computed value of the test statistic falls in this "rejection region," then the null hypothesis is rejected. If it does not fall in the rejection region, then the null hypothesis is not rejected.

4.    We would use the alternative hypothesis $\mu \neq \mu_0$ if we wanted to determine whether the population mean were <u>different from</u> the value $\mu_0$ specified in the null hypothesis. We would use the alternative hypothesis $\mu > \mu_0$ if we wanted to determine whether the population mean were <u>greater than</u> from the value $\mu_0$ specified in the null hypothesis. We would use the alternative hypothesis $\mu < \mu_0$ if we wanted to determine whether the population mean were <u>less than</u> the value $\mu_0$ specified in the null hypothesis.

5.    (a)  A Type I error is made whenever the null hypothesis is true, but the value of the test statistic leads us to reject the null hypothesis. A Type II error is made whenever the null hypothesis is false, but the value of the test statistic leads us to not reject the null hypothesis.

      (b)  The probability of a Type I error is represented by $\alpha$ and that of a Type II error by $\beta$.

      (c)  If the null hypothesis is true, the test statistic can lead us to either reject or not reject the null hypothesis. The first is the correct decision, while the latter constitutes a Type I error. Thus a Type I error is the only type of error possible when the null hypothesis is true.

      (d)  If the null hypothesis is not rejected, a correct decision has been made if the null hypothesis is, in fact, true. But if the null hypothesis is false, we have made a Type II error. Thus a Type II error is the only type of error possible when the null hypothesis is not rejected.

6.    Assuming that the null hypothesis is true, find the value of the test statistic for which the probability of obtaining a value greater than the specified value is 0.05.

7.    (a)  If the population standard deviation is unknown, and the population is normal or the sample size is large, we can use the one-mean t-statistic, $t = (\bar{x} - \mu_0)/(s/\sqrt{n})$.

      (b)  If the population standard deviation is known, and the population is normal or the sample size is large, we can use the one-mean z-statistic, $z = (\bar{x} - \mu_0)/(\sigma/\sqrt{n})$.

      (c)  If the population is symmetric, we can use the Wilcoxon signed-rank statistic W = the sum of the positive ranks.

8.    (a)  A hypothesis test is exact if the actual significance level is the same as the one that is stated.

      (b)  A hypothesis test is approximately correct if the actual significance level only approximately equals $\alpha$.

9.    A statistically significant result occurs when the value of the test statistic falls in the rejection region. A result has practical significance when it is statistically significant <u>and</u> the result also is different enough from results expected under the null hypothesis to be important to the consumer of the results. By taking large enough sample sizes, almost any result can be made statistically significant due to the increased ability of the test to detect a false null hypothesis, but small differences from the conditions expressed by the null hypothesis may not be important, that is, they may not have practical significance.

10. The probability of a Type II error is increased when the significance level is decreased for a fixed sample size.

11. (a) The power of a hypothesis test is the probability of rejecting the null hypothesis when the null hypothesis is false.

    (b) The power of a test increases when the sample size is increased while keeping the significance level constant.

12. (a) The P-value of a hypothesis test is the probability, assuming that the null hypothesis is true, of getting a value of the test statistic that is as extreme or more extreme than the one actually obtained.

    (b) True. If the null hypothesis were true, a value of the test statistic with a P-value of 0.02 would be more extreme than one with a P-value of 0.03.

    (c) True. If the P-value is 0.74, this means that 74% of the time when the null hypothesis is true, the value of the test statistic would be more extreme than the one actually obtained.

    (d) The P-value of a hypothesis test is also called the observed significance level since it represents the smallest possible significance level at which the null hypothesis could have been rejected.

13. In the critical-value approach, the null hypothesis is rejected if the value of the test statistic falls in the rejection region that is determined by the chosen significance level. In the P-value approach, the test statistic is computed and then the probability of obtaining a value as extreme or more extreme than the one actually obtained is found. This is the P-value. The advantages of providing the P-value are that the observed significance level of the of the test is given and the reader of the results can determine for him/herself whether the results are strong enough evidence against the null hypothesis to reject it.

14. Non-parametric methods have the advantages of involving fewer and simpler calculations than parametric methods and are more resistant to outliers and other extreme values. Parametric methods are preferred when the population is normal or the sample size is large since they are more powerful than non-parametric methods and thus tend to give more accurate results than non-parametric methods under those conditions.

15. Let $\mu$ denote last year's mean cheese consumption by Americans.

    (a) $H_0$: $\mu = 30.0$ lb

    (b) $H_a$: $\mu > 30.0$ lb

    (c) This is a right-tailed test.

16. (a) Rejection region: $z \geq 1.28$

    (b) Nonrejection region: $z < 1.28$

    (c) Critical value: $z = 1.28$

    (d) Significance level: $\alpha = 0.10$

(e)

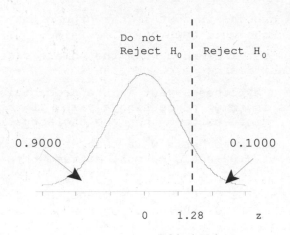

Critical Value
Nonrejection region |Rej ectionreg ion

(f) Right-tailed test

**17.** (a) A Type I error would occur if, in fact, $\mu$ = 30.0 lb, but the results of the sampling lead to the conclusion that $\mu$ > 30.0 lb.

(b) A Type II error would occur if, in fact, $\mu$ > 30.0 lb, but the results of the sampling fail to lead to that conclusion.

(c) A correct decision would occur if, in fact, $\mu$ = 30.0 lb and the results of the sampling do not lead to the rejection of that fact; or if, in fact, $\mu$ > 30.0 lb and the results of the sampling lead to that conclusion.

(d) If, in fact, last year's mean consumption of cheese for all Americans has not increased over the 2001 mean of 30.0 lb, and we do not reject the null hypothesis that $\mu$ = 30.0 lb, we made a correct decision.

(e) If, in fact, last year's mean consumption of cheese for all Americans has increased over the 2001 mean of 30.0 lb, and we fail to reject the null hypothesis that $\mu$ = 30.0 lb, we made a Type II error.

**18.** (a) P(Type I error) = significance level = $\alpha$ = 0.10

(b) The distribution of $\bar{x}$ will be approximately normal with a mean of 30.0 and a standard deviation of $6.9/\sqrt{35} = 1.17$ .

(c) Note: $z = \dfrac{\bar{x} - \mu_0}{\sigma / \sqrt{n}} \Rightarrow \bar{x} = \mu_0 + z \cdot \sigma / \sqrt{n}$

Since this is a right-tailed test, we would reject $H_0$ if $z \geq 1.28$; or equivalently if $\bar{x} \geq 30.00 + 1.28(6.9)/\sqrt{35} = 31.49$ .

So reject $H_0$ if $\bar{x} \geq 31.49$; otherwise do not reject $H_0$.
If $\mu$ = 30.5, then

P(Type II error) $= P(\bar{x} \leq 31.49)$

$$= P(z \le (31.49 - 30.50)/(6.9/\sqrt{35}\,)$$

$$= P(z \le 0.85) = 0.8023$$

(d-e) Assuming that the true mean μ is one of the values listed, the distribution of $\bar{x}$ will be approximately normal with that mean and with a standard deviation of 1.166.  The computations of $\beta$ and the power 1 - $\beta$ are shown in the table below.

| True mean $\mu$ | z-score computation | P(Type II error) $\beta$ | Power $1 - \beta$ |
|---|---|---|---|
| 30.5 | $\dfrac{31.49 - 30.50}{6.9/\sqrt{35}} = 0.85$ | 0.8023 | 0.1977 |
| 31.0 | $\dfrac{31.49 - 31.00}{6.9/\sqrt{35}} = 0.42$ | 0.6628 | 0.3372 |
| 31.5 | $\dfrac{31.49 - 31.50}{6.9/\sqrt{35}} = -0.01$ | 0.4960 | 0.5040 |
| 32.0 | $\dfrac{31.49 - 32.00}{6.9/\sqrt{35}} = -0.44$ | 0.3300 | 0.6700 |
| 32.5 | $\dfrac{31.49 - 32.50}{6.9/\sqrt{35}} = -0.87$ | 0.1922 | 0.8078 |
| 33.0 | $\dfrac{31.49 - 33.00}{6.9/\sqrt{35}} = -1.29$ | 0.0985 | 0.9015 |
| 33.5 | $\dfrac{31.49 - 33.50}{6.9/\sqrt{35}} = -1.72$ | 0.0427 | 0.9573 |
| 34.0 | $\dfrac{31.49 - 34.00}{6.9/\sqrt{35}} = -2.15$ | 0.0158 | 0.9842 |

(f)

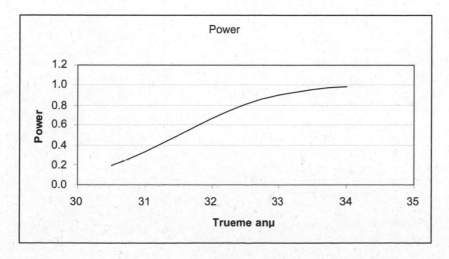

(g)  The distribution of $\bar{x}$ will be approximately normal with a mean of 30.0
     and a standard deviation of $6.9 / \sqrt{60} = 0.891$.

(h)  Note:  $z = \dfrac{\bar{x} - \mu_0}{\sigma / \sqrt{n}} \Rightarrow \bar{x} = \mu_0 + z \cdot \sigma / \sqrt{n}$

     Since this is a right-tailed test, we would reject $H_0$ if $z \geq 1.28$; or
     equivalently if $\bar{x} \geq 30.00 + 1.28(6.9)/\sqrt{60} = 31.14$.

     So reject $H_0$ if $\bar{x} \geq 31.14$; otherwise do not reject $H_0$.
     If $\mu = 30.5$, then

     $P(\text{Type II error}) \quad = P(\bar{x} \leq 31.14)$

     $\qquad\qquad = P(z \leq (31.14 - 30.5)/(6.9/\sqrt{60})$

     $\qquad\qquad = P(z \leq 0.72) = 0.7642$

(i-j) Assuming that the true mean $\mu$ is one of the values listed, the

     distribution of $\bar{x}$ will be approximately normal with that mean and with
     a standard deviation of 0.891.  The computations of $\beta$ and the power $1 -$
     $\beta$ are shown in the following table.

| True mean $\mu$ | z-score computation | P(Type II error) $\beta$ | Power $1 - \beta$ |
|---|---|---|---|
| 30.5 | $z = \dfrac{31.14 - 30.50}{6.9/\sqrt{60}} = 0.72$ | 0.7642 | 0.2358 |
| 31.0 | $z = \dfrac{31.14 - 31.00}{6.9/\sqrt{60}} = 0.16$ | 0.5636 | 0.4364 |
| 31.5 | $z = \dfrac{31.14 - 31.50}{6.9/\sqrt{60}} = -0.40$ | 0.3446 | 0.6554 |
| 32.0 | $z = \dfrac{31.14 - 32.00}{6.9/\sqrt{60}} = -0.97$ | 0.1660 | 0.8340 |
| 32.5 | $z = \dfrac{31.14 - 32.50}{6.9/\sqrt{60}} = -1.53$ | 0.0630 | 0.9370 |
| 33.0 | $z = \dfrac{31.14 - 33.00}{6.9/\sqrt{60}} = -2.09$ | 0.0183 | 0.9817 |
| 33.5 | $z = \dfrac{31.14 - 33.50}{6.9/\sqrt{60}} = -2.65$ | 0.0040 | 0.9960 |
| 34.0 | $z = \dfrac{31.14 - 34.00}{6.9/\sqrt{60}} = -3.21$ | 0.0007 | 0.9993 |

(k)

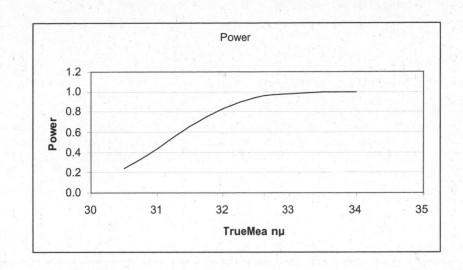

(l)    The principle being illustrated is that increasing the sample size for a hypothesis test without changing the significance level $\alpha$ increases the power.

**19.**    (a)    $n = 35$, $\bar{x} = 1078/35 = 30.8$, $\sigma = 6.9$

Step 1:    $H_0$: $\mu = 30.0$ lb, $H_a$: $\mu > 30.0$ lb

Step 2:    $\alpha = 0.10$

Step 3:    $z = (30.8 - 30.0)/(6.9/\sqrt{35}) = 0.69$

Step 4:    Critical value $= 1.28$

Step 5:    Since $0.69 < 1.28$, do not reject $H_0$.

Step 6:    At the 10% significance level, the data do not provide sufficient evidence to conclude that last year's mean cheese consumption $\mu$ for all Americans has increased over the 2001 mean of 30.0 lb.

(b)    Given the conclusion in part (a), if an error has been made, it must be a Type II error.  This is because, given that the null hypothesis was not rejected, the only error that could be made is the error of not rejecting a false null hypothesis.

**20.**    (a)    Step 1:    $H_0$: $\mu = 30.0$, $H_a$: $\mu > 30.0$

Step 2:    $\alpha = 0.10$

Step 3:    $z = 0.69$

Step 4:    $P = 1 - 0.7549 = 0.2451$

Step 5:    Since $0.2451 > 0.10$, do not reject $H_0$.

Step 6:    At the 10% significance level, the data do not provide sufficient evidence to conclude that last year's mean cheese consumption $\mu$ for all Americans has increased over the 2001 mean of 30.0 lb.

(b)    Using Table 9.12, we classify the strength of evidence against the null hypothesis as weak or none because $P > 0.10$.

**21.**    $n = 12$, $\bar{x} = \$284.10$, $s = \$86.90$

Step 1:  $H_0$: $\mu$ = \$332, $H_a$: $\mu$ < \$332

Step 2:  $\alpha$ = 0.05

Step 3:  t = (284.1 − 332)/(86.90/$\sqrt{12}$) = −1.909

Step 4:  Critical value = −1.782

Step 5:  Since −1.909 < −1.782, reject $H_0$.

Step 6:  At the 5% significance level, the data do provide sufficient evidence to conclude that the mean value lost because of purse snatching has decreased from the 2002 mean of \$332.

22.  (a)  n=12

$H_0$: $\eta$ = 332; $H_a$: $\eta$ < 332

$\alpha$ = 0.05

Critical value = 12(13)/2 − 61 = 17

| x | $x-\eta_0$=D | \|D\| | Rank of \|D\| | Signed Rank R |
|---|---|---|---|---|
| 207 | −125 | 125 | 10 | −10 |
| 237 | −95 | 95 | 7 | −7 |
| 422 | 90 | 90 | 6 | 6 |
| 226 | −106 | 106 | 9 | −9 |
| 272 | −60 | 60 | 3 | −3 |
| 205 | −127 | 127 | 11 | −11 |
| 362 | 30 | 30 | 2 | 2 |
| 348 | 16 | 16 | 1 | 1 |
| 165 | −167 | 167 | 12 | −12 |
| 266 | −66 | 66 | 5 | −5 |
| 269 | −63 | 63 | 4 | −4 |
| 430 | 98 | 98 | 8 | 8 |

W = sum of the positive signed ranks = 17

Since 17 is less than or equal to the critical value of 17, we reject the null hypothesis and conclude that there is evidence that last year's mean value lost to purse snatching has decreased from the 2002 mean.

(b)  In performing the Wilcoxon signed-rank test, we are assuming that the distribution of last year's values lost to purse snatching is symmetric.

(c)  If the distribution of values lost is, in fact, a normal distribution, it is permissible to use the Wilcoxon test since a normal distribution is also symmetric.

23.  If the values lost last year do have a normal distribution, the t-test is the preferred procedure for performing the hypothesis test since it is the more powerful test when the distribution is normal, that is, it has a greater chance of rejecting a false null hypothesis.

24.  (a)  If the odds-makers are estimating correctly, the mean point-spread error is zero.

(b)  It seems reasonable to assume that the distribution of point spread errors is approximately normal.  In any case, the sample size of 2109 is very large, so the t-test of $H_0$: $\mu = 0$ vs. $H_a$: $\mu \neq 0$ is appropriate. At the 5% significance level, the critical values are $\pm 1.960$. Since $t = (-0.2 - 0.0)/(10.9 / \sqrt{2109}) = 0.843$, we do not reject $H_0$.

(c) There is not sufficient evidence to conclude that the mean point-spread is different from zero.

25. Since the distribution is symmetric, use the Wilcoxon signed-rank test. The sample size is 50 and $\sigma$ is known, so the z-test may be appropriate. Use caution, however, since it appears that there may be outliers.

26. The distribution is far from normal, being left-skewed. However, since the sample size is 37 and $\sigma$ is unknown, it is probably reasonable to use the t-test.

27. (a) The Wilcoxon signed-rank test is appropriate in Exercise 25 since the distribution is symmetric. It is not appropriate in Exercise 26 since the distribution is highly skewed to the left.

   (b) The sample size is large (n=50) and the distribution is symmetric in Exercise 26, so either the z-test or the Wilcoxon signed-rank test could be used. Since the distribution appears to be more peaked with longer tails than a normal distribution would have, the Wilcoxon test is preferable.

28. Using Minitab, we choose **Stat ▶ Basic Statistics ▶ 1-Sample t...**, click in the **Samples in columns** text box, enter COST in the **Samples in columns** text box, click in the **Test mean** text box and enter 168 in the **Test mean** text box, click on the **Options** button, type 90.0 in the **Confidence level** text box, click on the arrow to the right of the **Alternative** box and select **greater than**, and click **OK**. Then choose **Stat ▶ Nonparametrics ▶ 1-Sample Wilcoxon...**, enter COST in the **Variables** text box, click on the **Test median** button and type 168 in its text box, select **greater than** in the **Alternative** text box, and click **OK**. Click on the **Graphs** button and check the boxes for **Histogram of data** and **Boxplot of data**. Click **OK** twice. Then choose **Stat ▶ Basic Statistics ▶ Normality test**, enter COST in the **Variable** text box and click **OK**. Finally, choose **Graph ▶ Stem-and-Leaf** and enter COST in the **Graph Variables** text box and click **OK**. The results are

   (a) Test of mu = 168 vs > 168

|          |    |         |        |         | 95% Lower |      |       |
|----------|----|---------|--------|---------|-----------|------|-------|
| Variable | N  | Mean    | StDev  | SE Mean | Bound     | T    | P     |
| COST     | 11 | 185.182 | 55.496 | 16.733  | 154.855   | 1.03 | 0.164 |

   The p-value is greater than the significance level of 0.10, so we do not reject the null hypothesis. There is not sufficient evidence to claim that the average cost of a private room in a nursing home exceeded $168 per day.

   (b) Test of median = 168.0 versus median > 168.0

|      |    | N for Test | Wilcoxon Statistic | P     | Estimated Median |
|------|----|------------|--------------------|-------|------------------|
| COST | 11 | 11         | 48.0               | 0.099 | 186.8            |

The p-value is less than the significance level of 0.10, so we reject the null hypothesis. There is sufficient evidence to claim that the average cost of a private room in a nursing home exceeded $168 per day.

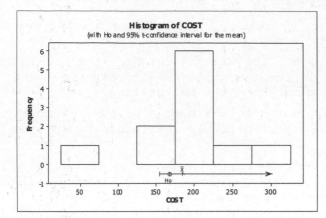

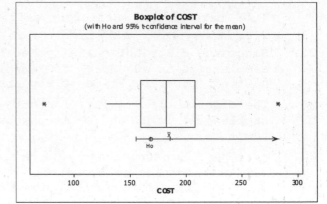

Stem-and-leaf of COST  N = 11 Leaf Unit = 10

```
LO 7

 2   1  2
 3   1  5
 3   1
(5)  1  88899
 3   2  0
 2   2
 2   2  5


HI 28
```

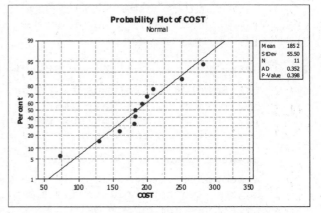

(d)    Although the distribution may be symmetric, it is not bell-shaped and it contains two potential outliers (see boxplot). Since the t-test should not be used when the sample size is small (11) and there are outliers, the Wilcoxon test is the more appropriate one to use.

29.   (a,b)  Using Minitab, choose **Stat ▶ Basic Statistics ▶ 1-Sample t**, enter CONSUMPTION in the **Samples in columns** text box and 64.5 in the **Test mean** text box. Click on the **Graphs** button and check the boxes for **Histogram of data** and **Boxplot of data**. Click **OK**. Click on the **Options** button, enter 95 in the **Confidence level** text box and select **Less than** from the **Alternative** drop down box. Click **OK** twice. Then choose **Stat ▶ Basic Statistics ▶ Normality test** and enter CONSUMPTION

in the **Variable** text box and click **OK**. Finally, choose **Graph ▶ Stem-and-Leaf** and enter CONSUMPTION in the **Graph Variables** text box and click **OK**. The results are

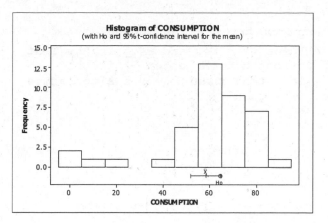

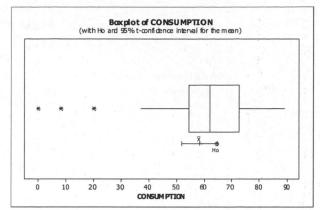

```
Stem-and-leaf of CONSUMPTION   N  = 40
Leaf Unit = 1.0

LO 0, 0, 8, 20

    5    3   7
    5    4
    7    4   79
   10    5   014
   16    5   666667
  (7)    6   0112223
   17    6   57789
   12    7   1234
    8    7   5567789
    1    8
    1    8   9
```

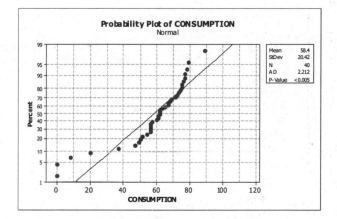

```
Test of mu = 64.5 vs not = 64.5

Variable       N      Mean     StDev  SE Mean        95% CI         T      P
CONSUMPTION   40   58.4000   20.4172   3.2282  (51.8703, 64.9297)  -1.89  0.066
```

The P-value of the test is 0.066, which is greater than the 0.05 significance level.  We do not reject $H_0$.  The data do not provide sufficient evidence to conclude that the mean beef consumption this year is less than the 2002 mean of 64.5 lbs.

(c)  After removing the four outliers, the test results are

```
Test of mu = 64.5 vs not = 64.5

Variable       N      Mean     StDev  SE Mean        95% CI         T      P
CONSUMPTION   36   64.1111   11.0163   1.8360  (60.3837, 67.8385)  -0.21  0.833
```

(d)  The outliers had a very large effect on the test results.  Although the sample size was 40, four outliers is too many for the t-test to be appropriate.  If the outliers are not recording errors, they represent legitimate observations from the population and should not be deleted.  If they were deleted, the test results would not yield valid conclusions about the population.  With so many outliers, it is just possible that the population itself is quite left skewed as is the sample.

(e)   This is more appropriately done with a non-parametric test that is insensitive to outliers.

30.    (a)   Using Minitab, choose **Stat ▶ Nonparametrics ▶ 1-Wilcoxon**, enter CONSUMPTION in the **Samples in columns** text box and 64.5 in the **Test mean** text box.   Click on the **Options** button, enter 95 in the **Confidence level** text box and select **Less than** from the **Alternative** drop down box.   Click **OK** twice.   The results are

Test of median = 64.50 versus median < 64.50

|  | N | N for Test | Wilcoxon Statistic | P | Estimated Median |
|---|---|---|---|---|---|
| CONSUMPTION | 40 | 40 | 324.5 | 0.127 | 62.00 |

After removing the outliers and repeating part (a), the results are

Test of median = 64.50 versus median < 64.50

|  | N | N for Test | Wilcoxon Statistic | P | Estimated Median |
|---|---|---|---|---|---|
| CONSUMPTION | 36 | 36 | 324.5 | 0.450 | 64.50 |

(b)   The results are similar to those in Problem 29 in that the P-value increases substantially.   The results are different in that the P-value increases from 0.066 to 0.833 for the t-test, and only from 0.127 to 0.450 for the Wilcoxon test when the four outliers are deleted.   In both cases, we should expect the P-value to increase since the four smallest observations were deleted and the alternative hypothesis was $\mu < 64.5$.   At the 5% significance level, the conclusion does not change in either case.

(c)   While the Wilcoxon test is less sensitive to outliers, it is based on an assumption that the data are symmetrically distributed.   That assumption does not appear to be reasonable for these data, so caution is advised in using the Wilcoxon test.   The sign test would be a better choice.

31.    (a)   Using Minitab, choose **Stat ▶ Basic Statistics ▶ 1-Sample z**, enter BMI in the **Samples in columns** text box and 25 in the **Test mean** text box. Click on the **Graphs** button and check the boxes for **Histogram of data** and **Boxplot of data**.  Click **OK**.  Click on the **Options** button, enter 95 in the **Confidence level** text box and select **Greater than** from the **Alternative** drop down box.  Click **OK** twice.  Then choose **Stat ▶ Basic Statistics ▶ Normality test** and enter BMI in the **Variable** text box and click **OK**.  Finally, choose **Graph ▶ Stem-and-Leaf** and enter BMI in the **Graph Variables** text box and click **OK**.  The graphic results are

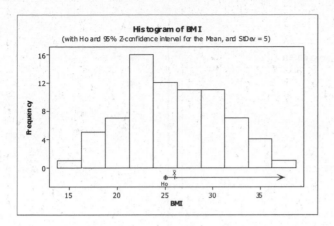

Stem-and-leaf of BMI   N = 75
Leaf Unit = 1.0

```
    1   1  4
    4   1  677
    6   1  88
   16   2  0000111111
   29   2  2222222233333
   (9)  2  444444455
   37   2  666666777
   28   2  88888899999
   17   3  0000111
   10   3  22233
    5   3  4555
    1   3  7
```

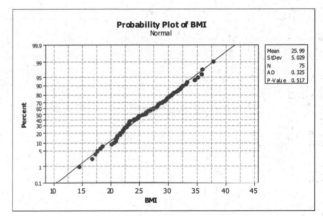

(b)  Yes.  There are no outliers, the sample size is large, and the data is reasonably normally distributed.

(c)  The process in part (a) yielded the following test results:

```
Test of mu = 25 vs > 25
The assumed standard deviation = 5
```

|          |    |         |        |         | 95% Lower |      |       |
|----------|----|---------|--------|---------|-----------|------|-------|
| Variable | N  | Mean    | StDev  | SE Mean | Bound     | Z    | P     |
| BMI      | 75 | 25.9867 | 5.0293 | 0.5774  | 25.0370   | 1.71 | 0.044 |

Since the P-value of 0.044 is less than the significance level 0.05, we reject the null hypothesis.  The data do provide sufficient evidence to conclude that the mean BMI of U.S. adults is greater than that for a healthy weight.

32.  (a)  Using Minitab, we choose **Graph ▶ Histogram**, choose the **Simple** version, enter BEER in the **Graph variables** text box, and click **OK**.  The result is

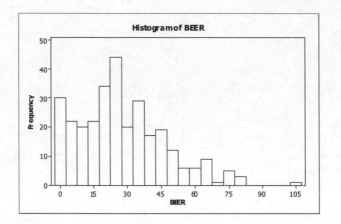

(b)  There may be an outlier at about 105.

(c)  Since the sample size is very large (300), we can use the z-test or the
     t-test.  Since the standard deviation is unknown, we use the t-test.
     Using Minitab, we obtained the following result using all of the data.
     Test of mu = 22 vs > 22

| Variable | N | Mean | StDev | SE Mean | 99% Lower Bound | T | P |
|---|---|---|---|---|---|---|---|
| BEER | 300 | 27.5200 | 19.4265 | 1.1216 | 24.8967 | 4.92 | 0.000 |

After eliminating the potential outlier (106), we repeated the process
and obtained

     Test of mu = 22 vs > 22

| Variable | N | Mean | StDev | SE Mean | 99% Lower Bound | T | P |
|---|---|---|---|---|---|---|---|
| BEER | 299 | 27.2575 | 18.9187 | 1.0941 | 24.6985 | 4.81 | 0.000 |

Clearly, elimination of the value 106 had little effect on the value of
t or on the P-value.  In either case, we reject the null hypothesis.
The data do provide sufficient evidence to claim that the mean annual
consumption of beer in Washington, D.C. exceeds the national mean.

CHAPTER 10 ANSWERS

**Exercises 10.1**

**10.1** Answers will vary.

**10.3** (a) $\mu_1$, $\sigma_1$, $\mu_2$, and $\sigma_2$ are parameters; $\bar{x}_1$, $s_1$, $\bar{x}_2$, and $s_2$ are statistics.

(b) $\mu_1$, $\sigma_1$, $\mu_2$, and $\sigma_2$ are fixed numbers; $\bar{x}_1$, $s_1$, $\bar{x}_2$, and $s_2$ are random variables.

**10.5** It is the sampling distribution of the difference of the two sample means that allows us to determine whether the difference of the sample means can reasonably be attributed to sampling error or whether it is large enough for us to conclude that the population means are different.

**10.7** $H_0$: $\mu_1 = \mu_2$, $H_a$: $\mu_1 < \mu_2$

**10.9** (a) The variable is systolic blood pressure (SBP).

(b) The two populations are the adolescent offspring of diabetic mothers (ODM) and of nondiabetic mothers (ODN).

(c) $H_0$: $\mu_1 = \mu_2$, $H_a$: $\mu_1 > \mu_2$ where $\mu_1$ is the mean SBP for ODM and $\mu_2$ is the mean SBP for ODN.

(d) Right tailed

**10.11** (a) The variable is household vehicle miles traveled (VMT).

(b) The two populations are households in the Midwest and in the South.

(c) $H_0$: $\mu_1 = \mu_2$, $H_a$: $\mu_1 \neq \mu_2$ where $\mu_1$ is the mean VMT for households in the Midwest and $\mu_2$ is the mean VMT for households in the South.

(d) two-tailed

**10.13** (a) The variable is operative times in minutes.

(b) The two populations are operations performed using a dynamic system (Z-plate) and operations performed using a static system (ALPS plate).

(c) $H_0$: $\mu_1 = \mu_2$, $H_a$: $\mu_1 < \mu_2$ where $\mu_1$ is the mean time of operations performed using the dynamic system and $\mu_2$ is the mean time of operations performed using the static system.

(d) left-tailed

**10.15** (a) The mean of $\bar{x}_1 - \bar{x}_2$ is $\mu_1 - \mu_2 = 40 - 40 = 0$

The standard deviation of $\bar{x}_1 - \bar{x}_2$ is

$$\sigma_{\bar{x}_1-\bar{x}_2} = \sqrt{\frac{\sigma_1^2}{n_1} + \frac{\sigma_2^2}{n_2}} = \sqrt{\frac{12^2}{9} + \frac{6^2}{4}} = 5$$

(b) No. The determination of the mean of $\bar{x}_1 - \bar{x}_2$ is the same for all populations. The formula for the standard deviation of $\bar{x}_1 - \bar{x}_2$ is dependent on the samples being independent, but not on the populations from which they came.

(c) No. It is not known that the two populations are normally distributed (which would lead to $\bar{x}_1 - \bar{x}_2$ being normally distributed), and the sample sizes of 9 and 4 are too small to claim that $\bar{x}_1$ and $\bar{x}_2$ are normally distributed (which would also lead to $\bar{x}_1 - \bar{x}_2$ being normally distributed).

**10.17** (a) The mean of $\bar{x}_1 - \bar{x}_2$ is $\mu_1 - \mu_2 = 40 - 40 = 0$

The standard deviation of $\bar{x}_1 - \bar{x}_2$ is $\sigma_{\bar{x}_1 - \bar{x}_2} = \sqrt{\dfrac{\sigma_1^2}{n_1} + \dfrac{\sigma_2^2}{n_2}} = \sqrt{\dfrac{12^2}{9} + \dfrac{6^2}{4}} = 5$

(b)  Yes.  Since $x_1$ and $x_2$ are each normally distributed, then so are $\bar{x}_1$ and $\bar{x}_2$.  Since the difference of two normally distributed random variables is also normally distributed, so is $\bar{x}_1 - \bar{x}_2$. See also Key Fact 10.1.

(c)  Since $\bar{x}_1 - \bar{x}_2$ is normally distributed with mean 0 and standard deviation 5, $P(-10 < \bar{x}_1 - \bar{x}_2 < 10) = P([-10 - 0]/5 < z < [10 - 0]/5)$
     $= P(-2 < z < 2) = 0.9772 - 0.0228 = 0.9544$, so 95.44% of the differences in sample means from samples of size 9 and 4 from the two populations will lie between -10 and 10.

**10.19** (a)  We have Minitab take 2000 observations from a normally distributed population having mean 100 and standard deviation 16 by choosing **Calc ▶**

**Random Data ▶ Normal...**, typing 2000 in the **Generate rows of data** text box, typing X1 in the **Store in columns** text box, typing 100 in the **Mean** text box and 16 in the **Standard deviation** text box, and clicking **OK**.

(b)  We repeat part (a) for a normally distributed population having mean 120 and standard deviation 12 by choosing **Calc ▶ Random Data ▶ Normal...**, typing 2000 in the **Generate rows of data** text box, typing X2 in the **Store in columns** text box, typing 120 in the **Mean** text box and 12 in the **Standard deviation** text box, and clicking **OK**.

(c)  To determine the difference between each pair of observations in parts (a) and (b), we choose **Calc ▶ Calculator...**, type DIFF in the **Store results in variable** text box, type X1 - X2 in the **Expression** text box, and click **OK**.  DIFF will be stored in C3.

(d)  To obtain a histogram of the 2000 differences found in part (c), we choose **Graph ▶ Histogram...**, select the **Simple** version and click **OK**. Enter DIFF in the **Graph variables** text box and click **OK**.

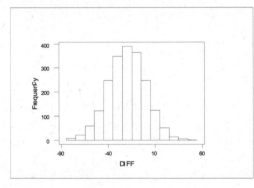

The histogram is bell-shaped because the random variable named DIFF is the difference of two random variables which themselves are normally distributed.  A random variable which is the difference of two normally distributed random variables is itself normally distributed.  Thus, its histogram is bell-shaped.

**Section 10.2**

**10.21** (a)  The four conditions required for using the pooled t-procedure are (i) both samples are simple random samples, (ii) independence of the two samples, (iii) the variable under consideration is normally distributed in each of the two populations or both samples are large, and (iv) the population standard deviations are equal.

(b)  Independence and randomness are absolutely essential. Moderate violations of the normality assumption are not serious problems even for small or moderate size samples. Moderate violations of the equal standard deviation requirement are not serious if the two sample sizes are roughly equal.

**10.23** Not Reasonable. One sample standard deviation is more than twice the other one. This is a good indication that the population standard deviations are not equal. Since the two sample sizes are not roughly equal (6 and 14), nor are both of them large, there appears to be a serious violation of the equal standard deviations requirement of the pooled t-test.

**10.25** Reasonable. The standard deviations are roughly equal and both sample sizes are large.

**10.27** (a)  $s_p = \sqrt{\dfrac{14(2.1)^2 + 14(2.3)^2}{15+15-2}} = \sqrt{4.85} = 2.2023$

Step 1:  $H_0$: $\mu_1 = \mu_2$,  $H_a$: $\mu_1 \neq \mu_2$

Step 2:  $\alpha = 0.05$

Step 3:  $t = \dfrac{10-12}{2.2023\sqrt{(1/15)+(1/15)}} = -2.487$

Step 4:  df = 28, Critical values = $\pm 2.048$

Step 5:  Since $-2.487 < -2.048$, reject $H_0$. $0.01 < $ P-value $< 0.02$.

(b)  95% CI = $(10-12) \pm 2.048(2.2023)\sqrt{(1/15+1/15)} = (-3.647, -0.353)$

**10.29** (a)  $s_p = \sqrt{\dfrac{9(4)^2 + 14(5)^2}{10+15-2}} = \sqrt{21.47826} = 4.6345$

Step 1:  $H_0$: $\mu_1 = \mu_2$,  $H_a$: $\mu_1 > \mu_2$

Step 2:  $\alpha = 0.05$

Step 3:  $t = \dfrac{20-18}{4.6345\sqrt{(1/10)+(1/15)}} = 1.057$

Step 4:  df = 23, Critical value = 1.714

Step 5:  Since $1.057 < 1.714$, do not reject $H_0$. P-value $> 0.10$.

(b)  90% CI = $(20-18) \pm 1.714(4.6345)\sqrt{(1/10+1/15)} = (-1.243, 5.243)$

**10.31** (a)  $s_p = \sqrt{\dfrac{19(4)^2 + 14(5)^2}{20+15-2}} = \sqrt{19.81818} = 4.4518$

Step 1:  $H_0$: $\mu_1 = \mu_2$,  $H_a$: $\mu_1 < \mu_2$

Step 2:  $\alpha = 0.05$

Step 3:  $t = \dfrac{20-24}{4.4518\sqrt{(1/20)+(1/15)}} = -2.631$

Step 4:  df = 33, Critical value = -1.697

Step 5:  Since $-2.631 < -1.697$, reject $H_0$. $0.005 < $ P-value $< 0.010$.

(b)  90% CI = $(20-24) \pm 1.697(4.4518)\sqrt{(1/20+1/15)} = (-6.580, -1.420)$

**10.33** Population 1: Fraud, $n_1 = 10$, $\bar{x}_1 = 10.12$, $s_1 = 4.90$
Population 2: Firearms, $n_2 = 10$, $\bar{x}_2 = 18.78$, $s_2 = 4.64$

$$s_p = \sqrt{\frac{9(4.90)^2 + 9(4.64)^2}{10 + 10 - 2}} = \sqrt{22.7698} = 4.7718$$

Step 1:     $H_0: \mu_1 = \mu_2$,    $H_a: \mu_1 < \mu_2$
Step 2:     $\alpha = 0.05$

Step 3:     $t = \dfrac{10.12 - 18.78}{4.7718\sqrt{(1/10) + (1/10)}} = -4.058$

Step 4:     df = 18, Critical value = -1.734
Step 5:     Since -4.06 < -1.734, reject $H_0$. P-value < 0.005.
Step 6:     At the 5% significance level, the data do provide sufficient evidence to conclude that, on the average, the mean time served for fraud is less than that served for firearms offenses.

**10.35** Population 1:  Fortified, $n_1 = 14$, $\bar{x}_1 = 9.0$, $s_1 = 37.4$
Population 2:  Unfortified, $n_2 = 12$, $\bar{x}_2 = 1.6$, $s_2 = 34.6$

$$s_p = \sqrt{\frac{13(37.4)^2 + 11(34.6)^2}{24}} = \sqrt{1306.36} = 36.1436$$

Step 1:     $H_0: \mu_1 = \mu_2$,    $H_a: \mu_1 > \mu_2$
Step 2:     $\alpha = 0.05$

Step 3:     $t = \dfrac{9.0 - 1.6}{36.1436\sqrt{1/14 + 1/12}} = 0.520$

Step 4:     Critical value = 1.711
Step 5:     Since 0.520 < 1.711, do not reject $H_0$.
Step 6:     At the 5% significance level, the data do not provide sufficient evidence that drinking fortified orange juice reduces PTH level more than drinking unfortified orange juice.
            For the P-value approach, P-value > 0.10.  Therefore, because the P-value is larger than the significance level, do not reject $H_0$.

**10.37** Population 1:  Cropland, $n_1 = 126$, $\bar{x}_1 = 14.06$, $s_1 = 4.83$
Population 2:  Wetland, $n_2 = 98$, $\bar{x}_2 = 15.36$, $s_2 = 4.95$

$$s_p = \sqrt{\frac{125(4.83)^2 + 97(4.95)^2}{222}} = \sqrt{23.84169} = 4.8828$$

Step 1:     $H_0: \mu_1 = \mu_2$,    $H_a: \mu_1 \neq \mu_2$
Step 2:     $\alpha = 0.05$

Step 3:     $t = \dfrac{14.06 - 15.36}{4.8828\sqrt{1/126 + 1/98}} = -1.977$

Step 4:     Critical values = ±1.982
Step 5:     Since -1.982 < -1.977 < 1.982, do not reject $H_0$.
Step 6:     At the 5% significance level, the data do not provide sufficient evidence that there is a difference in the mean number of native species in the two regions.

For the P-value approach, $0.05 < 2\{P(t < -1.977)\} > 0.10$. Therefore, because the P-value is larger than the significance level, do not reject $H_0$.

In each of Exercises 10.39, 41, 43, the value of $s_p$ has been obtained in Exercise 10.33, 35, 37, respectively.

**10.39** $t_{\alpha/2} = 1.734, 90\%\text{CI} = (10.12-18.78) \pm 1.734(4.7718)\sqrt{1/10+1/10} = (-12.36, -4.96)$

**10.41** $t_{\alpha/2} = 1.711, 90\%\text{CI} = (9.0-1.6) \pm 1.711(36.1436)\sqrt{1/14+1/12} = (-16.93, 31.73)$

**10.43** $t_{\alpha/2} = 1.982, 95\%\text{CI} = (14.06-15.36) \pm 1.982(4.8828)\sqrt{1/126+1/98} = (-2.603, 0.003)$

**10.45** (a)  Using Minitab, we choose **Graph ▶ Probability Plot**, select **Single** and click **OK**.  Enter VEGETARIAN and OMNIVORE in the **Graph Variables** text box.  Click on the **Multiple Graphs** button and select **On separate graphs**.  Click **OK** twice.  Note that the standard deviations are shown on the plots.  Then choose **Graph ▶ Boxplot**, click on the **Simple** plot in the **Multiple Y's** row and click **OK**..Enter VEGETARIAN and OMNIVORE in the **Graph Variables** text box and click **OK**.  The results are

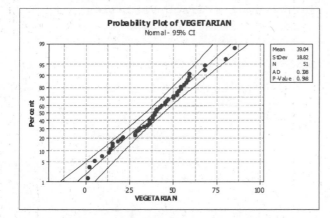

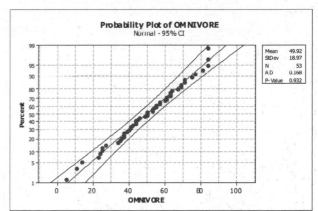

The standard deviations are 18.82 for Vegetarians and 18.97 for Omnivores.

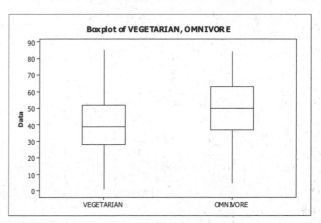

(b)  Choose **Stat ▶ Basic Statistics ▶ 2-Sample t**, click on **Samples in different columns** and enter VEGETARIANS in the **First** textbox and OMNIVORES in the **Second** text box.  Check the box for **Assume equal variances** and click on the **Options** button.  Enter 99 in the **Confidence level** textbox, 0 in the **Test difference** text box and select **not equal** in the **Alternative drop down box**.  Click **OK** twice.  The results are

```
Two-sample T for VEGETARIAN vs OMNIVORE

              N   Mean   StDev   SE Mean
VEGETARIAN   51   39.0   18.8     2.6
OMNIVORE     53   49.9   19.0     2.6

Difference = mu (VEGETARIAN) - mu (OMNIVORE)
Estimate for difference:  -10.8853
99% CI for difference:  (-20.6147, -1.1559)
T-Test of difference = 0 (vs not =): T-Value = -2.94   P-Value = 0.004
DF = 102
Both use Pooled StDev = 18.8964
```

Since the P-value of 0.004 is less than the significance level of 0.01, the data do provide sufficient evidence that there is a difference between the mean daily protein intakes of female vegetarians and female omnivores.

(c)   The 99% confidence interval is (-20.6147, -1.1559).

(d)   Yes.  The probability plots indicate that normality is reasonable, the sample sizes are large, and there are no outliers.  Furthermore, the two sample standard deviations are almost equal, making the pooled t

**10.47** (a)   Using Minitab, we choose **Graph ▶ Probability Plot**, select **Single** and click **OK**.  Enter STINGER and REGULAR in the **Graph Variables** text box.  Click on the **Multiple Graphs** button and select **On separate graphs**.  Click **OK** twice.  Note that the standard deviations are shown on the plots.  Then choose **Graph ▶ Boxplot**, click on the **Simple** plot in the **Multiple Y's** row and click **OK**...Enter STINGER and REGULAR in the **Graph Variables** text box and click **OK**.  The results are

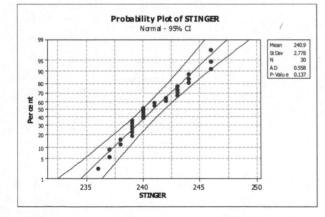
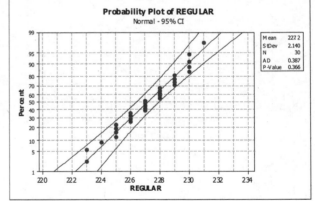

The standard deviations are
2.778 for STINGER and
2.140 for REGULAR.

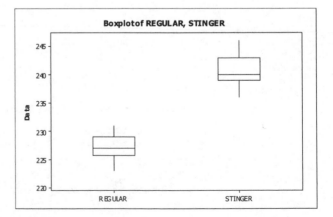

    (b)  Choose **Stat ▶ Basic Statistics ▶ 2-Sample t**, click on **Samples in different columns** and enter <u>STINGER</u> in the **First** text box and <u>REGULAR</u> in the **Second** text box. Check the box for **Assume equal variances** and click on the **Options** button. Enter <u>99</u> in the **Confidence level** textbox, <u>0</u> in the **Test difference** text box and select **greater than** in the **Alternative drop down** box. Click **OK** twice. The results are

```
Two-sample T for STINGER vs REGULAR

            N    Mean   StDev  SE Mean
STINGER    30  240.93    2.78     0.51
REGULAR    30  227.20    2.14     0.39

Difference = mu (STINGER) - mu (REGULAR)
Estimate for difference:  13.7333
99% lower bound for difference:  12.2015
T-Test of difference = 0 (vs >): T-Value = 21.45  P-Value = 0.000  DF = 58
Both use Pooled StDev = 2.4798
```

        Since the P-value of 0.000 is less than the significance level of 0.01, the data do provide sufficient evidence that the Stinger tee improves total distance traveled.

    (c)  Repeat the process in part (b), but select **not equal** in the **Alternative drop down** box. The confidence interval is

          99% CI for difference:  (12.0281, 15.4386)

    (d)  Yes. The probability plots indicate that normality is reasonable, the sample sizes are large, and there are no outliers. Furthermore, the two sample standard deviations are almost equal, making the pooled t procedure a good choice.

**10.49** If $n_1 = n_2$, say both equal n, then

$$s_p^2 = \frac{(n_1 - 1)s_1^2 + (n_2 - 1)s_2^2}{n_1 + n_2 - 2} = \frac{(n-1)s_1^2 + (n-1)s_2^2}{n + n - 2}$$

$$= \frac{(n-1)(s_1^2 + s_2^2)}{2n - 2} = \frac{(n-1)(s_1^2 + s_2^2)}{2(n-1)} = \frac{(s_1^2 + s_2^2)}{2}$$

    Thus, if $n_1 = n_2$, then $s_p^2$ is just the mean of $s_1^2$ and $s_2^2$.

**10.51** (a)  The test value of t = -0.927 fell between the critical values ±2.052, leading to nonrejection of the null hypothesis. The 95% confidence interval was (-4.69, 1.77). This includes zero, so the null hypothesis is not rejected.

    (b)  The test value of t = -1.977 fell between the critical values ±1.982, leading to nonrejection of the null hypothesis. The 95% confidence interval was (-2.603, 0.003). This includes zero, so the null hypothesis is not rejected. Note: These results are so close to rejection that if intermediate rounding is done during the calculations, it is possible to obtain results rejecting the null hypothesis and getting a confidence interval that does not include zero.

**10.53** (a)  The test value of t = 0.520 fell to the left of the critical value 1.711, leading to nonrejection of the null hypothesis. The 95% lower confidence bound is -16.928. This is not positive, so the null hypothesis is not rejected.

    (b)  The test value of t = = 2.707 fell to the right of the critical value 2.362, leading to rejection of the null hypothesis. The 99% lower confidence bound is 0.790. This is positive, so the null hypothesis is rejected.

**Exercises 10.3**

**10.55** (a) Pooled t-test.  When the standard deviations are equal, the pooled t-test is more powerful than the non-pooled t-test.

(b) Nonpooled t-test.  Use of the pooled t-test would result in a larger Type I error probability than the one specified.

(c) Pooled t-test.  The closeness of the sample standard deviations indicates that the population standard deviations are probably close to being equal.  Using the pooled t-test will not cause serious problems under these conditions when the sample sizes are equal.

(d) Nonpooled t-test.  The sample standard deviations are enough different

**10.57** (a)  Step 1:   $H_0: \mu_1 = \mu_2$,   $H_a: \mu_1 \neq \mu_2$

Step 2:   $\alpha = 0.05$

Step 3:   $t = \dfrac{10-12}{\sqrt{(2^2/15)+(5^2/15)}} = -1.438$

Step 4:   $\Delta = \dfrac{\left[\left(\frac{s_1^2}{n_1}\right)+\left(\frac{s_2^2}{n_2}\right)\right]^2}{\dfrac{\left(\frac{s_1^2}{n_1}\right)}{n_1-1}+\dfrac{\left(\frac{s_2^2}{n_2}\right)}{n_2-1}} = \dfrac{\left[\left(\frac{2^2}{15}\right)+\left(\frac{5^2}{15}\right)\right]^2}{\dfrac{\left(\frac{2^2}{15}\right)}{15-1}+\dfrac{\left(\frac{5^2}{15}\right)}{15-1}} = 18.37; \; df = 18$

Critical values = $\pm 2.101$

Step 5:   Since $-2.101 < -1.438 < 2.101$, do not reject $H_0$. $0.100 <$ P-value $< 0.200$.

(b)   95% CI = $(10-12)\pm 2.101\sqrt{(2^2/15+5^2/15)} = (-4.921, 0.921)$

**10.59** (a)  Step 1:   $H_0: \mu_1 = \mu_2$,   $H_a: \mu_1 > \mu_2$

Step 2:   $\alpha = 0.05$

Step 3:   $t = \dfrac{20-18}{\sqrt{(4^2/10)+(5^2/15)}} = 1.107$

Step 4:   $\Delta = \dfrac{\left[\left(\frac{s_1^2}{n_1}\right)+\left(\frac{s_2^2}{n_2}\right)\right]^2}{\dfrac{\left(\frac{s_1^2}{n_1}\right)}{n_1-1}+\dfrac{\left(\frac{s_2^2}{n_2}\right)}{n_2-1}} = \dfrac{\left[\left(\frac{4^2}{10}\right)+\left(\frac{5^2}{15}\right)\right]^2}{\dfrac{\left(\frac{4^2}{10}\right)}{10-1}+\dfrac{\left(\frac{5^2}{15}\right)}{15-1}} = 22.10; \; df = 22$

Critical value = 1.717

Step 5:   Since $1.107 < 1.717$, do not reject $H_0$. P-value $> 0.100$.

(b)   90% CI = $(20-18)\pm 1.717\sqrt{(4^2/10+5^2/15)} = (-1.103, 5.103)$

**10.61** (a)  Step 1:   $H_0: \mu_1 = \mu_2$,   $H_a: \mu_1 < \mu_2$

Step 2:   $\alpha = 0.05$

Step 3:   $t = \dfrac{20-24}{\sqrt{(6^2/20)+(2^2/15)}} = -2.782$

Step 4: $\Delta = \dfrac{\left[\left(\dfrac{s_1^2}{n_1}\right)+\left(\dfrac{s_2^2}{n_2}\right)\right]^2}{\dfrac{\left(\dfrac{s_1^2}{n_1}\right)}{n_1-1}+\dfrac{\left(\dfrac{s_2^2}{n_2}\right)}{n_2-1}} = \dfrac{\left[\left(\dfrac{6^2}{20}\right)+\left(\dfrac{2^2}{15}\right)\right]^2}{\dfrac{\left(\dfrac{6^2}{20}\right)}{20-1}+\dfrac{\left(\dfrac{2^2}{15}\right)}{15-1}} = 11.77;\ df = 11$

Critical value = -1.796

Step 5: Since -2.782 < -1.796, reject H$_0$.
0.005 < P-value < 0.010.

(b)   90% CI $= (20-24)\pm1.796\sqrt{(6^2/20 + 2^2/15)} = (-6.582,\ -1.418)$

**10.63** Population 1: Chronic, $n_1 = 32$, $\bar{x}_1 = 28.8$, $s_1 = 9.2$

Population 2: Remitted, $n_2 = 20$, $\bar{x}_2 = 22.1$, $s_2 = 5.7$

Step 1:   H$_0$: $\mu_1 = \mu_2$,   H$_a$: $\mu_1 \neq \mu_2$

Step 2:   $\alpha = 0.10$

Step 3:   $t = \dfrac{25.8-22.1}{\sqrt{(9.2^2/32)+(5.7^2/20)}} = 1.791$

Step 4: $\Delta = \dfrac{\left[\left(\dfrac{s_1^2}{n_1}\right)+\left(\dfrac{s_2^2}{n_2}\right)\right]^2}{\dfrac{\left(\dfrac{s_1^2}{n_1}\right)}{n_1-1}+\dfrac{\left(\dfrac{s_2^2}{n_2}\right)}{n_2-1}} = \dfrac{\left[\left(\dfrac{9.2^2}{32}\right)+\left(\dfrac{5.7^2}{20}\right)\right]^2}{\dfrac{\left(\dfrac{9.2^2}{32}\right)}{32-1}+\dfrac{\left(\dfrac{5.7^2}{20}\right)}{20-1}} = 49.99995;\ df = 49$

Critical values = ±1.677

Step 5: Since 1.791 > 1.677, reject H$_0$.
0.050 < P-value < 0.100.

At the 10% significance level, the data do provide sufficient evidence to conclude that there is a difference between the mean age at arrest of East German prisoners with chronic PTSD and remitted PTSD. For the P-value approach, $0.05 < 2\{P(t > 1.791)\} < 0.10$. Therefore, because the P-value is smaller than the significance level, reject H$_0$.

**10.65** Population 1: Dynamic, $n_1 = 14$, $\bar{x}_1 = 7.36$, $s_1 = 1.22$

Population 2: Static, $n_2 = 6$, $\bar{x}_2 = 10.50$, $s_2 = 4.59$

H$_0$: $\mu_1 = \mu_2$,   H$_a$: $\mu_1 < \mu_2$

Step 2: $\alpha = 0.05$

Step 3: $t = \dfrac{7.36-10.50}{\sqrt{(1.22^2/14)+(4.59^2/6)}} = -1.651$

Step 4: $\Delta = \dfrac{\left[\left(\dfrac{s_1^2}{n_1}\right)+\left(\dfrac{s_2^2}{n_2}\right)\right]^2}{\dfrac{\left(\dfrac{s_1^2}{n_1}\right)}{n_1-1}+\dfrac{\left(\dfrac{s_2^2}{n_2}\right)}{n_2-1}} = \dfrac{\left[\left(\dfrac{1.22^2}{14}\right)+\left(\dfrac{4.59^2}{6}\right)\right]^2}{\dfrac{\left(\dfrac{1.22^2}{14}\right)}{14-1}+\dfrac{\left(\dfrac{4.59^2}{6}\right)}{6-1}} = 5.305;\ df = 5$

Critical value = -2.015

Step 5: Since $-1.651 > -2.015$, do not reject $H_0$.
$\qquad$ $0.050 < $ P-value $< 0.100$.
At the 5% significance level, the data do not provide sufficient evidence to conclude that the mean number of acute postoperative days in the hospital is smaller with the dynamic system than with the static system.
For the P-value approach, $0.05 < P(t<-1.651) < 0.10$. Therefore, since the P-value is larger than the significance level, do not reject $H_0$.

**10.67** Population 1: Psychotic patients, $n_1 = 10$, $\bar{x}_1 = 0.02426$, $s_1 = 0.00514$

Population 2: Non-psychotic patients, $n_2 = 15$, $\bar{x}_2 = 0.01643$, $s_2 = 0.00470$

(a) $H_0$: $\mu_1 = \mu_2$, $\quad H_a$: $\mu_1 > \mu_2$

Step 2: $\alpha = 0.01$

Step 3: $t = \dfrac{0.02426 - 0.01643}{\sqrt{(0.00514^2 / 10) + (0.00470^2 / 15)}} = 3.860$

Step 4: $\Delta = \dfrac{\left[\left(\dfrac{s_1^2}{n_1}\right) + \left(\dfrac{s_2^2}{n_2}\right)\right]^2}{\dfrac{\left(\dfrac{s_1^2}{n_1}\right)}{n_1 - 1} + \dfrac{\left(\dfrac{s_2^2}{n_2}\right)}{n_2 - 1}} = \dfrac{\left[\left(\dfrac{0.00514^2}{10}\right) + \left(\dfrac{0.01643^2}{15}\right)\right]^2}{\dfrac{\left(\dfrac{0.00514^2}{10}\right)}{10 - 1} + \dfrac{\left(\dfrac{0.01643^2}{15}\right)}{15 - 1}} = 18.20;\ df = 18$

Critical value = 2.552 (approximately)

Step 5: Since $3.860 > 2.552$, reject $H_0$.
$\qquad$ P-value $< 0.005$.
At the 1% significance level, the data do provide sufficient evidence to conclude that dopamine activity is higher, on average, in psychotic patients than in non-psychotic patients.
For the P-value approach, $P(t > 3.860) < 0.005$. Therefore, since the P-value is smaller than the significance level, reject $H_0$.

**10.69** $t_{\alpha/2} = 1.677$; 90% CI= $(25.8-22.1) \pm 1.677\sqrt{9.2^2/32 + 5.7^2/20} = (0.235, 7.165)$

**10.71** $t_{\alpha/2} = 2.015$; 90% CI= $(7.36-10.50) \pm 2.015\sqrt{1.22^2/14 + 4.59^2/6} = (-6.973, 0.693)$

**10.73** $t_{\alpha/2} = 2.552$; 98% CI= $(0.02426-0.01643) \pm 2.552\sqrt{0.00514^2/10 + 0.00470^2/15}$
$\qquad = (0.00265, 0.01301)$

**10.75** (a) Choose nonpooled t-prodedures. There is clearly much more variability in the data for males than in the data for females.

(b) No. The two sample sizes are small and moderate. The data for the males is not normally distributed or has a large outlier at 39.2. Neither t-procedure is reasonable under these circumstances.

**10.77** Population 1: Dynamic, $n_1 = 14$, $\bar{x}_1 = 7.36$, $s_1 = 1.22$

Population 2: Static, $n_2 = 6$, $\bar{x}_2 = 10.50$, $s_2 = 4.59$

(a) $s_p = \sqrt{\dfrac{13(1.22)^2 + 5(4.59)^2}{18}} = \sqrt{6.9272} = 2.6320$, $df = 18$

Step 1: $\qquad$ $H_0$: $\mu_1 = \mu_2$, $\quad H_a$: $\mu_1 < \mu_2$

Step 2: $\qquad$ $\alpha = 0.05$

Step 3: $\qquad$ $t = \dfrac{7.36 - 10.50}{2.6320\sqrt{1/14 + 1/6}} = -2.445$

Step 4: Critical value = $-1.734$

Step 5:  Since −2.445 < −1.734, reject $H_0$.
Step 6:  At the 5% significance level, the data do provide sufficient evidence to conclude that the mean number of acute postoperative days in the hospital is smaller with the dynamic system than with the static system.
For the P-value approach, $0.01 < P(t<-2.445) < 0.025$.  Therefore, since the P-value is smaller than the significance level, reject $H_0$.

(b)  The pooled t-test resulted in rejecting the null hypothesis while the nonpooled t-test resulted in not rejecting the null hypothesis.

(c)  The nonpooled t-test is more appropriate.  One sample standard deviation is almost four times as large as the other, making it highly unlikely that the two population standard deviations are equal.  The fact that the two sample sizes are also quite different makes it essential that the pooled t-test not be used.

**10.79** (a)  Pooled t-test.  Both populations are normally distributed and the population standard deviations are equal.

(b)  Nonpooled t-test.  Both populations are normally distributed, but the population standard deviations are unequal.

(c)  Neither.  Both populations are skewed and it is given that both sample sizes are small.

(d)  Neither.  Only one population is normally distributed; the other is skewed.  Since the sample sizes are small, the nonnormality of one population rules out the use of either t-test.

**10.81** (a)  Using Minitab, we choose **Graph ▶ Probability Plot**, select **Single** and click **OK**.  Enter REGULAR and STINGER in the **Graph Variables** text box.  Click on the **Multiple Graphs** button and select **On separate graphs**.  Click **OK** twice.  Note that the standard deviations are shown on the plots.  Then choose **Graph ▶ Boxplot**, click on the **Simple** plot in the **Multiple Y's** row and click **OK**...Enter REGULAR and STINGER in the **Graph Variables** text box and click **OK**.  The results are

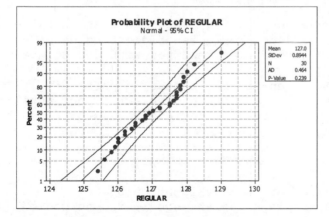

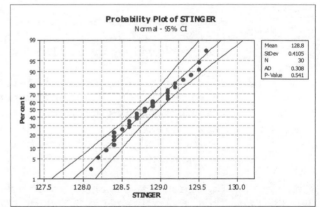

Standard deviations are 0.8944 for REGULAR and 0.4105 for STINGER.

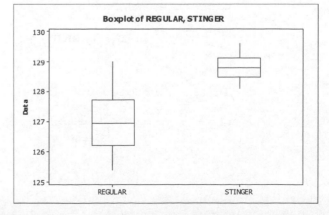

(b) Nonpooled t-procedure.  The samples are both large enough (30), normality is reasonable, there are no potential outliers, but one standard deviation is more than twice the other, making it unlikely that the population standard deviations are equal.

(c) Choose **Stat ▶ Basic Statistics ▶ 2-Sample t**, click on **Samples in different columns** and enter <u>REGULAR</u> in the **First** text box and <u>STINGER</u> in the **Second** text box.  Check the box for **Assume equal variances** and click on the **Options** button.  Enter <u>95</u> in the **Confidence level** textbox, <u>0</u> in the **Test difference** text box and select **Less than** in the **Alternative** drop down box.  Click **OK** twice.  Then repeat this procedure with the **Assume equal variances** box unchecked.   The results are

```
Two-sample T for REGULAR vs STINGER

             N     Mean  StDev  SE Mean
REGULAR     30  127.007  0.894     0.16
STINGER     30  128.833  0.410    0.075

Difference = mu (REGULAR) - mu (STINGER)
Estimate for difference:  -1.82667
95% upper bound for difference:  -1.52634
T-Test of difference = 0 (vs <): T-Value = -10.17  P-Value = 0.000  DF = 58
Both use Pooled StDev = 0.6959
Two-sample T for REGULAR vs STINGER

             N     Mean  StDev  SE Mean
REGULAR     30  127.007  0.894     0.16
STINGER     30  128.833  0.410    0.075


Difference = mu (REGULAR) - mu (STINGER)
Estimate for difference:  -1.82667
95% upper bound for difference:  -1.52413
T-Test of difference = 0 (vs <): T-Value = -10.17  P-Value = 0.000  DF = 40
```

Both procedures result in a P-value of 0.000, which is less than the significance level 0.05.  The data do provide sufficient evidence that the ball velocity is less with the regular tee than with the Stinger tee.

(d) To get the confidence intervals, repeat the process in part (b), entering <u>90</u> in the **Confidence level** text box and selecting **not equal** in the **Alternative** drop down box.  The confidence interval portions of the results are

Pooled      90% CI for difference:  (-2.12700, -1.52634)
Nonpooled   90% CI for difference:  (-2.12921, -1.52413)

The results of the two procedures are almost identical.

**10.83** (a) From Exercise 10.49, when the sample sizes are equal, in the pooled procedure, we have

$$s_p^2 = \frac{s_1^2 + s_2^2}{2}; \text{if } n_1 = n_2, \text{then } t = \frac{\bar{x}_1 - \bar{x}_2}{\sqrt{\frac{s_1^2 + s_2^2}{2}(\frac{1}{n} + \frac{1}{n})}} = \frac{\bar{x}_1 - \bar{x}_2}{\sqrt{\frac{s_1^2 + s_2^2}{2}(\frac{2}{n})}} = \frac{\bar{x}_1 - \bar{x}_2}{\sqrt{\frac{s_1^2 + s_2^2}{n}}}.$$

In the nonpooled procedure, we have

$$t = \frac{\bar{x}_1 - \bar{x}_2}{\sqrt{\frac{s_1^2}{n} + \frac{s_2^2}{n}}} = \frac{\bar{x}_1 - \bar{x}_2}{\sqrt{\frac{s_1^2 + s_2^2}{n}}}, \text{ which is identical to the pooled t-statistic.}$$

   (b)  If the standard deviations are not equal, the computed degrees of
        freedom in the nonpooled t-procedure could be quite different from the
        degrees of freedom in the pooled t-procedure.  This will lead to
        different critical values and/or P-values and possibly to different
        conclusions.

**Exercises 10.4**

**10.85** All normal distributions are symmetric about the population mean, and the
        standard deviation determines the spread (and hence the shape).  Thus two
        normal distributions with the same standard deviation have the same shape.

**10.87** (a)  Use the pooled t-test.  It is slightly more powerful than the Mann-
             Whitney when the conditions for its use (normal distributions with
             equal stand deviations) are met.
       (b)  Use the Mann-Whitney test since the distributions have the same shape,
             but are not normal.

**10.89** (a)  90      (b)  $8(8 + 9 + 1) - 90 = 54$
       (c)  Right tail = 93, Left tail = $8(8 + 9 + 1) - 93 = 51$

**10.91** (a)  95      (b)  $9(9 + 8 + 1) - 95 = 67$
       (c)  Right tail = 99, Left tail = $9(9 + 8 + 1) - 99 = 63$

**10.93** Step 1:  $H_0: \mu_1 = \mu_2$,  $H_a: \mu_1 \neq \mu_2$
       Step 2:  $\alpha = 0.05$
       Step 3:

| Friese | Overall Rank | Cockerell | Overall Rank |
|--------|--------------|-----------|--------------|
| 188    | 7            | 169       | 1            |
| 190    | 8            | 178       | 2            |
| 225    | 9            | 180       | 3.5          |
| 235    | 10           | 180       | 3.5          |
|        |              | 182       | 5            |
|        |              | 185       | 6            |
|        | 34           |           |              |

       Step 4:  M = 34
       Step 5:  The critical values are 32 and $4(4 + 6 + 1) - 32 = 12$.
       Step 6:  Since M = 34 > 32, we reject the null hypothesis.  The data do
                provide sufficient evidence that there is a difference between the
                mean wing stroke frequencies of the two species of Euglossine bees.

**10.95** Population 1:  Students with fewer than two years of high-school algebra;
       Population 2:  Students with two or more years of high-school algebra.

       Step 1:  $H_0: \mu_1 = \mu_2$,  $H_a: \mu_1 < \mu_2$, $n_1 = 6$, $n_2 = 9$
       Step 2:  $\alpha = 0.05$
       Step 3:  Construct a worktable based upon the following:  First, rank all
                the data from both samples combined.  Adjacent to each column of
                data as it is presented in the Exercise, record the overall
                rank.  Assign tied rankings the average of the ranks they would
                have had if there were no ties.  For example, the two 81s in the
                table below are tied for eleventh smallest.  Thus, each is
                assigned the rank $(11 + 12)/2 = 11.5$.

| Fewer Than Two Years of High-School Algebra | Overall Rank | Two or More Years of High-School Algebra | Overall Rank |
|---|---|---|---|
| 58 | 3 | 84 | 14 |
| 81 | 11.5 | 67 | 7 |
| 74 | 8.5 | 65 | 6 |
| 61 | 4 | 75 | 10 |
| 64 | 5 | 74 | 8.5 |
| 43 | 1 | 92 | 15 |
| | | 83 | 13 |
| | | 52 | 2 |
| | | 81 | 11.5 |
| | 33 | | |

Step 4:    The value of the test statistic is the sum of the ranks for the sample data from Population 1:
    $M = 3 + 11.5 + 8.5 + 4 + 5 + 1 = 33$.

Step 5:    We have $n_1 = 6$ and $n_2 = 9$.  Since the hypothesis test is left-tailed with $\alpha = 0.05$, we use Table VI to obtain the critical value, which is $M_1 = 6(6 + 9 + 1) - 63 = 33$.  Thus, we reject $H_0$ if $M \leq 33$.

Step 6:    Since $M = 33$ equals the critical value, reject $H_0$.

Step 7:    At the 5% significance level, the data provide sufficient evidence to conclude that students with fewer than two years of high-school algebra have a lower mean semester average in this teacher's chemistry courses than do students with two or more years of high-school algebra.

10.97 Population 1:  Weekly earnings of males;
Population 2:  Weekly earnings of females.

Step 1:    $H_0$:  $\mu_1 = \mu_2$ ,  $H_a$:  $\mu_1 > \mu_2$

Step 2:    $\alpha = 0.05$

Step 3:    To construct a worktable, first rank all the data from both samples combined.  Adjacent to each column of data, record the overall rank.  Assign tied rankings the average of the ranks they would have had if there were no ties.

| Male | Overall Rank | Female | Overall Rank |
|------|------|------|------|
| 382 | 4 | 353 | 1 |
| 401 | 5 | 369 | 2 |
| 411 | 7 | 370 | 3 |
| 571 | 9 | 407 | 6 |
| 812 | 11 | 505 | 8 |
| 920 | 12 | 589 | 10 |
| 1884 | 15 | 1176 | 13 |
| 2617 | 18 | 1440 | 14 |
|  |  | 2073 | 16 |
|  |  | 2188 | 17 |
|  | 81 |  |  |

Step 4:  The value of the test statistic is the sum of the ranks for the sample data from Population 1:  $M = 81$

Step 5:  We have $n_1 = 8$ and $n_2 = 10$.  Since the hypothesis test is right-tailed with $\alpha = 0.05$, we use Table VI to obtain the critical value, which is $M_r = 95$.  Thus, we reject $H_0$ if $M \geq 95$.

Step 6:  Since $M < 95$, do not reject $H_0$.

Step 7:  At the 5% significance level, the data do not provide sufficient evidence to conclude that the mean weekly earnings of male full-time wage and salary workers is greater than that for females.

**10.99** (a)  Population 1: Release times for prisoners with fraud offenses;
Population 2: Release times for prisoners with firearms offenses.

Step 1:  $H_0$: $\mu_1 = \mu_2$, $H_a$: $\mu_1 < \mu_2$

Step 2:  $\alpha = 0.05$

Step 3:  To construct a worktable, first rank all the data from both samples combined.  Adjacent to each column of data, record the overall rank.  Assign tied rankings the average of the ranks they would have had if there were no ties.  For example, the two 17.9s in the following table are tied for thirteenth smallest.  Thus, each is assigned the rank $(13 + 14)/2 = 13.5$.

| Fraud | Overall Rank | Firearms | Overall Rank |
|------|------|------|------|
| 3.6 | 1 | 10.4 | 6 |
| 5.3 | 2 | 13.3 | 9 |
| 5.9 | 3 | 16.1 | 11 |
| 7.0 | 4 | 17.9 | 13.5 |
| 8.5 | 5 | 18.4 | 15 |
| 10.7 | 7 | 19.6 | 16 |
| 11.8 | 8 | 20.9 | 17 |
| 13.9 | 10 | 21.9 | 18 |
| 16.6 | 12 | 23.8 | 19 |
| 17.9 | 13.5 | 25.5 | 20 |
|  | 65.5 |  |  |

Step 4:  The value of the test statistic is the sum of the ranks for the sample data from Population 1:  $M = 65.5$

Step 5:  We have $n_1 = 10$ and $n_2 = 10$.  Since the hypothesis test is left-tailed with $\alpha = 0.05$, we use Table VI to obtain the critical value, which is $M_1 = 10(10 + 10 + 1) - 127 = 83$. Thus, we reject $H_0$ if $M \leq 83$.

Step 6:  Since $M < 83$, reject $H_0$.

Step 7:  At the 5% significance level, the data provide sufficient evidence to conclude that the mean time served for fraud offenses is less than that for firearms offenses.

(b)  Normal distributions with the same standard deviations have the same shape, thus meeting the requirement for using the Mann-Whitney test. If the distributions are, in fact, normal with equal standard deviations, it is better to use the pooled t-test since it is slightly more powerful than the Mann-Whitney test in this situation.

**10.101** (a)  Since the populations are normally distributed and have the same shape, use the pooled t-test.

(b)  No assumptions are met for any of the tests; therefore, none of these tests can be performed.

(c)  Since both samples are large, use the non-pooled t-test.

**10.103** (a) Using Minitab, we choose **Graph ▶ Probability Plot**, select **Single** and click **OK**.  Enter MEN and WOMEN in the **Graph Variables** text box.  Click on the **Multiple Graphs** button and select **On separate graphs**.  Click **OK** twice.  Then choose **Graph ▶ Boxplot**, click on the **Simple** plot in the **Multiple Y's** row and click **OK**...Enter MEN and WOMEN in the **Graph Variables** text box and click **OK**.  The results are

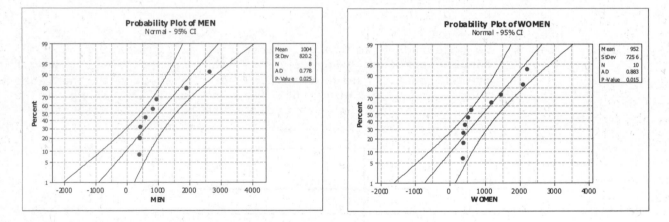

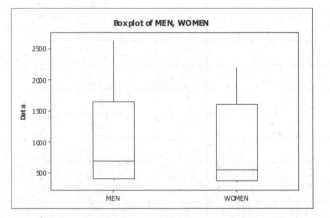

    (b)   No.   The sample sizes are small – 10 and 8, and the samples are both right skewed.

    (c)   Yes.   The sample distributions are both right skewed, about the same shape.

**10.105** (a)   Using Minitab with the data in two columns named LAB1 and LAB2, we choose **Graph ▶ Probability Plot...**, select the **Single** version and click **OK**. Enter LAB1 and LAB2 in the **Graph variables** text box and click **OK**. Then choose **Graph ▶ Boxplot**, click on the **Simple** plot in the **Multiple Y's** row and click **OK**...Enter LAB1 and LAB2 in the **Graph Variables** text box and click **OK**.   The results are

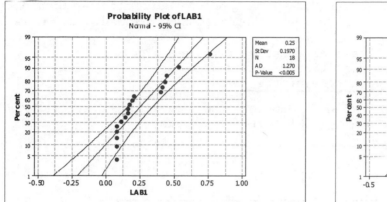
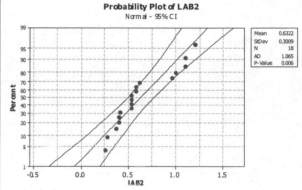

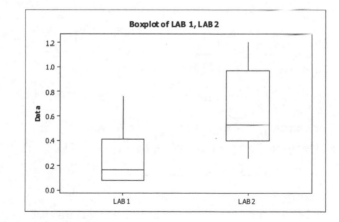

    (b)   Neither set of data appears to be normally distributed.   Since the sample sizes are only moderate, the better choice for comparing the two means is the Mann Whitney test.

    (c)   Choose **Stat ▶ Nonparametrics ▶ Mann-Whitney...**, enter LAB1 in the **First sample** text box and LAB2 in the **Second sample** text box.   Enter 95 in the **Confidence level** text box and choose **not equal** in the **Alternative** box.   The results are

### Mann-Whitney Test and CI: LAB1, LAB2

```
        N  Median
LAB1   18  0.1650
LAB2   18  0.5300
```

Point estimate for ETA1-ETA2 is -0.3500
95.2 Percent CI for ETA1-ETA2 is (-0.4799,-0.1999)

W = 213.5
Test of ETA1 = ETA2 vs ETA1 not = ETA2 is significant at 0.0002
The test is significant at 0.0002 (adjusted for ties)

Since the P-value is 0.0002, the data provide sufficient evidence that there is a difference in the median formaldehyde exposure in the two labs at the 5% significance level.

**10.107** Step 1:   State the null and alternative hypotheses.

Step 2:   Decide on the significance level $\alpha$.

Step 3:   Construct a worktable of the form.

| Sample from Population 1 | Overall Rank | Sample from Population 2 | Overall Rank |
|---|---|---|---|
| . | . | . | . |
| . | . | . | . |
| . | . | . | . |

Step 4:   Compute the value of the test statistic

$$z = \frac{M - n_1(n_1 + n_2 + 1)/2}{\sqrt{n_1 n_2 (n_1 + n_2 + 1)/12}}$$

where M is the sum of the ranks for Population 1.

Step 5:   The critical value(s):

(a)   for a two-tailed test are $\pm z_{\alpha/2}$.

(b)   for a left-tailed test is $-z_\alpha$.

(c)   for a right-tailed test is $z_\alpha$.

Use Table II to find the critical value(s)

Step 6:   If the value of the test statistic falls in the rejection region, reject $H_0$; otherwise, do not reject $H_0$.

Step 7:   State the conclusion in words.

**10.109**   (a)                                    (b)     5%

| Rank | | | | | | | | | M | P(M) |
|---|---|---|---|---|---|---|---|---|---|---|
| 1 | 2 | 3 | 4 | 5 | 6 | M | | | | |
| A | A | A | B | B | B | 6 | | | 6 | 0.05 |
| A | A | B | A | B | B | 7 | | | 7 | 0.05 |
| A | A | B | B | A | B | 8 | | | 8 | 0.10 |
| A | A | B | B | B | A | 9 | | | 9 | 0.15 |
| A | B | A | A | B | B | 8 | | | 10 | 0.15 |
| A | B | A | B | A | B | 9 | | | 11 | 0.15 |
| A | B | A | B | B | A | 10 | | | 12 | 0.15 |
| A | B | B | A | A | B | 10 | | | 13 | 0.10 |
| A | B | B | A | B | A | 11 | | | 14 | 0.05 |
| A | B | B | B | A | A | 12 | | | 15 | 0.05 |
| B | A | A | A | B | B | 9 | | | | |
| B | A | A | B | A | B | 10 | | | | |
| B | A | A | B | B | A | 11 | | | | |
| B | A | B | A | A | B | 11 | | | | |
| B | A | B | A | B | A | 12 | | | | |
| B | A | B | B | A | A | 13 | | | | |
| B | B | A | A | A | B | 12 | | | | |
| B | B | A | A | B | A | 13 | | | | |
| B | B | A | B | A | A | 14 | | | | |
| B | B | B | A | A | A | 15 | | | | |

(c)   Each row in part (a) has a $1/20 = 0.05$ chance of occurring.  This results in the probability distribution above at the right for M for the case when $n_1 = 3$ and $n_2 = 3$.

(d)   A histogram for the probability distribution of M is shown at the right for the case when $n_1 = 3$ and $n_2 = 3$.

(e)   From part (c), we see that $P(M \le 6) = 0.05$ and $P(M \ge 15) = 0.05$.  These results correspond with the entries in Table VI for $n_1 = 3$ and $n_2 = 3$.  That is, $M_l = 6$ at $\alpha = 0.05$, and $M_r = 15$ at $\alpha = 0.05$.

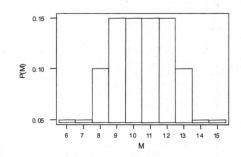

**Exercises 10.5**

**10.111** Paired sampling helps to remove extraneous sources of variation, resulting in a smaller sampling error for the estimated difference between the sample means.  This, in turn, makes it more likely that differences between the means will be detected when those differences actually exist.

**10.113** The samples must be paired (this is essential), and the population of all paired differences must be normally distributed or the sample must be large (the procedure works reasonably well even for small or moderate size samples if the paired-difference variable is not normally distributed provided that the deviation from normality is small).

**10.115** (a)  Age      (b)   Married men and married women (c)   Married couples
(d)  Paired difference variable = Age of Man – Age of Woman

(e)  $H_0$:  $\mu_M = \mu_W$ ;  $H_a$:   $\mu_M > \mu_W$      (f)   Right-tailed

**10.117** (a)  Price of a home
(b)  Homes near a sports stadium
(c)  Prices of homes before (1) and after (2) construction of a sports

        stadium

(d)   Paired difference variable = Price before construction – Price after construction

(e)   $H_0$: $\mu_1 = \mu_2$; $H_a$: $\mu_1 \neq \mu_2$   (f)   Two-tailed

**10.119** (a)   The variable is the height of the plant.

(b)   The two populations are the cross-fertilized plants and the self-fertilized plants.

(c)   The paired-difference variable is the difference in heights of the cross-fertilized plants and the self-fertilized plants grown in the same pot.

(d)   Yes.  They represent the difference in heights of two plants grown under the same conditions (same pot), one from each category of plants.

(e)   Step 1:      $H_0$: $\mu_1 = \mu_2$, $H_a$: $\mu_1 \neq \mu_2$

            Step 2:      $\alpha = 0.05$

            Step 3:      The paired differences are given.

            Step 4:

$$t = \frac{\bar{d}}{s_d / \sqrt{n}} = \frac{20.93}{37.74 / \sqrt{15}} = 2.148$$

            Step 5:      df = 14; critical values = ±2.145

            Step 6:      Since 2.148 > 2.145, reject $H_0$.  The data do provide sufficient evidence at the 5% significance level that there is a difference between the mean heights of cross-fertilized and self-fertilized plants.

(f)   For the 0.01 significance level, the critical values are ±2.977.  Since t = 2.148 is not in the rejection region, do not reject the null hypothesis at the 0.01 significance level.

**10.121** Population 1:      Weights after treatment for anorexia nervosa

      Population 2:      Weights before treatment for anorexia nervosa

            Step 1:      $H_0$: $\mu_1 = \mu_2$, $H_a$: $\mu_1 > \mu_2$

            Step 2:      $\alpha = 0.05$

            Step 3:      The paired differences 'Weight after - Weight before' are 11.0, 5.5, 9.4, 13.6, -2.9, 10.7, -0.1, 7.4, 21.5, -5.3, -3.8, 11.4, 13.4, 13.1, 9.0, 3.9, 5.7

                 For these differences, $\sum d = 123.5$ and $\sum d^2 = 1716.9$

            Step 4:

$$\bar{d} = \sum d / n = 123.5 / 17 = 7.26$$

$$s_d = \sqrt{\frac{\sum d^2 - \left(\sum d\right)^2 / n}{n-1}} = \sqrt{\frac{1716.9 - 123.5^2 / 17}{17-1}} = 7.16$$

$$t = \frac{\bar{d}}{s_d / \sqrt{n}} = \frac{7.26}{7.16 / \sqrt{17}} = 4.181$$

            Step 5:      df = 16; critical values = 1.746

            Step 6:      Since 4.181 > 1.746, reject $H_0$.  The data do provide sufficient evidence at the 5% significance level that family therapy is effective in helping anorexic young women gain weight.

For the P-value approach, P(t > 4.181) < 0.005.  Since the P-value is smaller than the significance level, reject H$_0$.

**10.123** Population 1: Corneal thickness in normal eyes

Population 2: Corneal thickness in glaucoma eyes

Step 1:     H$_0$: $\mu_1 = \mu_2$,   H$_a$: $\mu_1 > \mu_2$

Step 2:     $\alpha$ = 0.10

Step 3:     The paired differences, d = x$_1$ - x$_2$, are

-4, 0 12, 18, -4, -12, 6, 16

For these differences, $\sum d = 32$ and $\sum d^2 = 936$

Step 4:

$$\bar{d} = \sum d / n = 32/8 = 4.0$$

$$s_d = \sqrt{\frac{\sum d^2 - \left(\sum d\right)^2 / n}{n-1}} = \sqrt{\frac{936 - 32^2/8}{8-1}} = 10.7438$$

$$t = \frac{\bar{d}}{s_d / \sqrt{n}} = \frac{4.0}{10.7438 / \sqrt{8}} = 1.053$$

Step 5:     df = 7, Critical value = 1.415

Step 6:     Since 1.053 < 1415, do not reject H$_0$.  The data do not provide sufficient evidence at the 10% significance level that the mean corneal thickness is greater in normal eyes than in eyes with glaucoma.

For the P-value approach, P(t > 1.053) > 0.10.  Since the P-value is larger than the significance level, do not reject H$_0$.

**10.125** From Exercise 10.119, df = 14.

(a)

$$20.93 \pm 2.145 \cdot \frac{37.74}{\sqrt{15}} = 20.93 \pm 20.90 = 0.03 \text{to } 41.83 \text{eighth sof aninch}$$

We can be 95% confident that the difference, $\mu_1 - \mu_2$, between the mean heights of cross-fertilized and self-fertilized Zea mays is somewhere between 0.03 and 41.83 eighths of an inch.

(b)

$$20.93 \pm 2.977 \cdot \frac{37.74}{\sqrt{15}} = 20.93 \pm 29.01 = -8.08 \text{to } 49.94 \text{eighth sof aninch}$$

We can be 99% confident that the difference, $\mu_1 - \mu_2$, between the mean heights of cross-fertilized and self-fertilized Zea mays is somewhere between -8.08 and 49.94 eighths of an inch.

**10.127** From Exercise 10.121, df = 16.

$$7.26 \pm 1.746 \cdot \frac{7.16}{\sqrt{17}} = 7.26 \pm 3.03 = 4.23 \text{to } 10.29 \text{ pounds}$$

We can be 90% confident that the mean weight gain, $\mu_1 - \mu_2$, resulting from family-therapy treatment by anorexic young women is somewhere between 4.23 and 10.29 pounds.

**10.129**From Exercise 10.123, , df = 7.

(a)
$$4.0 \pm 1.415 \cdot \frac{10.7434}{\sqrt{8}} = 4.0 \pm 5.37 = (-1.37, 9.37) \text{ microns}$$

We can be 80% confident that the difference, $\mu_1 - \mu_2$, in corneal thickness of patients who had glaucoma in one eye but not the other is between -1.37 and 9.37 microns.

**10.131**Using Minitab, with the differences in a column named DIFF, choose **Graph ▶ Probability plot...**, select the **Single** version and click **OK**.  Enter DIFF in the **Graph variables text box** and click **OK**.  Then choose **Graph ▶ Boxplot**, select the **One Y Simple** version and click **OK**.  Enter DIFF in **Graph variables** text box and click **OK**.

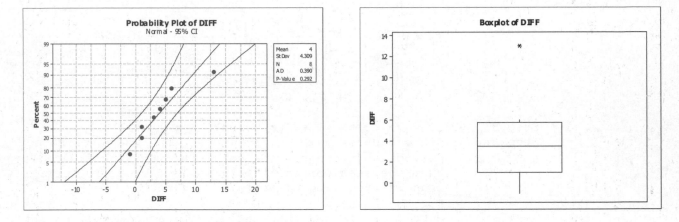

The sample size is very small (8) and there is one outlier (13) in the differences.  Since the t-test is not robust to outliers, the t-test is not appropriate for this data.

**10.133**Using Minitab, enter DIFF at the top of column C3, then choose **Calc ▶ Calculator**, enter DIFF in the **Store result in variable** box, enter 'RESOLUTION'-'ONSET' in the Expression box, and click **OK**.  Choose **Graph ▶ Probability plot...**, select the **Single** version and click **OK**.  Enter ONSET RESOLUTION DIFF in the **Graph variables** text box and click **OK**.  Then choose **Graph ▶ Boxplot**, select the **Multiple Y's Simple** version and click **OK**.  Enter ONSET and RESOLUTION in the **Graph Variables** box and click **OK**.

Finally, choose **Graph ▶ Boxplot**, select the **One Y Simple** version and click **OK**.  Enter DIFF in **Graph variables** text box and click **OK**.  The results are

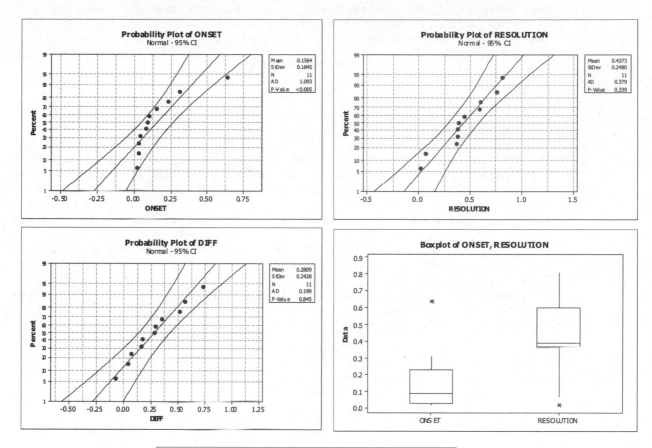

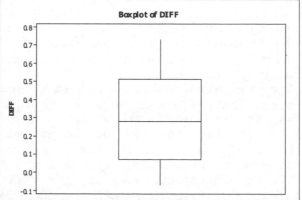

(b)  No.  The sample size is small, there is an outlier, and the normal probability plot is not linear.

(c)  No.  The sample size is small and there is an outlier.  The probability plot is closer to linear than for the ONSET data, but overall, t-procedures are not advisable.

(d)  Yes.  There are no outliers and the probability plot is reasonably linear.

(e)  They imply that the assumptions for using a paired t-procedure may be met even though the assumptions for using a t-procedure may not be met for the original variables.

10.135 (a)  Using Minitab, choose **Stat ▶ Basic Statistics   ▶ Paired t...,**enter <u>MEN</u>

in the **First sample** box and <u>WOMEN</u> in the **Second Sample** box.  Click on the **Options** button and enter <u>99</u> in the **Confidence** level box, <u>0.0</u> in the

**Test mean** box, and **greater than** in the **Alternative** drop down box. Click **OK** twice.   The results are
```
Paired T for MEN - WOMEN

                  N     Mean    StDev  SE Mean
MEN              75  35.4800   9.7695   1.1281
WOMEN            75  32.0800   7.0265   0.8114
Difference       75  3.40000  7.23542  0.83547

99% lower bound for mean difference: 1.41341
```
T-Test of mean difference = 0 (vs > 0): T-Value = 4.07   P-Value = 0.000
The P-value is 0.000, which is less than the significance level 0.01. Therefore, the data do provide sufficient evidence that the mean age of Norwegian men at the time of marriage exceeds that of Norwegian women.

(b)   The 99% confidence interval is found by repeating the procedure of part (a) using **not equal** in the **Alternative** droop down box.   The result is
```
99% CI for mean difference: (1.19108, 5.60892)
```
We can be 99% confident that the mean difference in ages of Norwegian men and women at the time of marriage lies somewhere between 1.19 and 5.61 years.

(c)   We remove the pairs (18,32) and (53,29) from the data and repeat parts (a) and (b).   The results are
```
Paired T for MEN - WOMEN

                  N     Mean    StDev  SE Mean
MEN              73  35.4795   9.4650   1.1078
WOMEN            73  32.1233   7.1140   0.8326
Difference       73  3.35616  6.61095  0.77375

99% lower bound for mean difference: 1.51520
T-Test of mean difference = 0 (vs > 0): T-Value = 4.34   P-Value = 0.000

99% CI for mean difference: (1.30893, 5.40340)
```
By removing the outliers, the t-value increases from 4.07 to 4.34, the P-value remains at 0.000, and the conclusion remains the same.   The 99% confidence interval is shortened by about 0.3 years and the mean difference changes from 3.40 to 3.36.

**10.137** Data is obtained from two populations whose members can be naturally paired. By letting d represent the difference between the values in each pair, we can reduce the data set from pairs of numbers to a single number representing each pair.   Since the mean of the paired differences is the same as the difference of the two population means, we can test the equality of the two population means by testing to see if the mean of the paired differences is zero (or some other value if appropriate).   If the paired differences are normally distributed, then the one-sample t-test can be used to carry out the test.

**10.139** (a)   Population 1: Additive used, $n_1 = 10$, $\bar{x}_1 = 18.910$, $s_1 = 7.4721$

Population 2: Additive not used, $n_2 = 10$, $\bar{x}_2 = 18.250$, $s_2 = 7.4188$
Using the formula in Step 3 of the nonpooled-t procedure, we get df = 17.

Step 1:     $H_0$: $\mu_1 = \mu_2$,   $H_a$: $\mu_1 > \mu_2$
Step 2:     $\alpha = 0.05$
Step 3:

$$t = \frac{18.910 - 18.250}{\sqrt{\dfrac{7.4721^2}{10} + \dfrac{7.4188^2}{10}}} = 0.198$$

Step 4:        Since 0.198 < 1.740, do not reject $H_0$.
Step 5:        Critical value = 1.740
Step 6:        At the 5% significance level, the data do not provide
               sufficient evidence to conclude that the mean gas mileage
               when the additive is used is greater than the mean gas
               mileage when the additive is not used.

(b)   It is inappropriate to perform the hypothesis test as presented in
      part (a) because the data are paired and so the samples are not
      independent.

(c)   In Example 10.15, where the appropriate procedure was used, we
      rejected $H_0$.  On the other hand, in part (a), where an inappropriate
      procedure was used, we did not reject $H_0$.

## Exercises 10.6

**10.141** (a)  No.  Unless the sample size is large, the paired t-test should only be
      used when the distribution is normal.

(b)   Yes.  When the sample size is large, the distribution of the sample
      mean will be approximately normal regardless of the shape of the
      distribution and the t-test will be nearly the same as a z-test.

(c)   Yes.  The paired Wilcoxon signed-rank test requires only that the
      distribution of differences be symmetric.

(d)   When the paired-difference variable is far from normally distributed,
      the paired Wilcoxon signed-rank test is preferred because it is usually
      more powerful in this situation.

**10.143** (a)  Paired t-test.  With approximately normally distributed data, the
      t-test will be the more powerful test.

(b)   Neither.  The Wilcoxon signed-rank test is based on an assumption of
      symmetry in the distribution and the paired t-test is based on an
      assumption of normality or large samples.  Neither assumption appears
      to be valid.

(c)   Wilcoxon signed-rank test.  The data are not normally distributed, but
      they are symmetric.

**10.145** (a)  Population 1: Cross-fertilized; Population 2: Self-fertilized

Step 1:   $H_0$: $\mu_1 = \mu_2$,   $H_a$: $\mu_1 \neq \mu_2$
Step 2:   $\alpha = 0.05$
Step 3:

| d | \|d\| | Rank | Signed-rank |
|---|---|---|---|
| 49 | 49 | 11 | 11 |
| -67 | 67 | 14 | -14 |
| 8 | 8 | 2 | 2 |
| 16 | 16 | 4 | 4 |
| 6 | 6 | 1 | 1 |
| 23 | 23 | 5 | 5 |
| 28 | 28 | 7 | 7 |
| 41 | 41 | 9 | 9 |
| 14 | 14 | 3 | 3 |
| 29 | 29 | 8 | 8 |
| 56 | 56 | 12 | 12 |
| 24 | 24 | 6 | 6 |
| 75 | 75 | 15 | 15 |
| 60 | 60 | 13 | 13 |
| -48 | 48 | 10 | -10 |

Step 4:        n = 15
Step 5:        The critical values are $W_1 = 25$, $W_r = 95$
Step 6:        Work table prepared in Step 3.

Step 7:        The value of the test statistic is the sum of the positive
               ranks.
               W = 11 + 2 + 4 + 1 + 5 + 7 + 9 + 3 + 8 + 12 + 6 + 15 + 13
               = 96
Step 8:        Since 96 > 95, reject $H_0$.
Step 9:        At the 5% significance level, the data provide enough
               evidence to conclude that there is a difference in the
               mean heights of cross-fertilized Zea mays and self-
               fertilized Zea mays.

(b)   All of the computations above are the same for a 1% significance level
      except for the critical values which are now $W_1$ = 16 and
      $W_r$ = 104.  Since W = 96 is between the two critical values, we do not
      reject the null hypothesis at the 1% significance level.  Thus at the
      1% level, the data do not provide enough evidence to conclude that
      there is a difference in the mean heights of cross-fertilized Zea mays
      and self-fertilized Zea mays.

**10.147** (a)  Population 1: Weight before therapy
           Population 2: Weight after therapy

Step 1:   $H_0$: $\mu_1 = \mu_2$,   $H_a$: $\mu_1 < \mu_2$
Step 2:   $\alpha$ = 0.05
Step 3:

| D | \|d\| | Rank | Signed-rank |
|---|---|---|---|
| −11.0 | 11.0 | 12 | −12 |
| −5.5 | 5.5 | 6 | −6 |
| −9.4 | 9.4 | 10 | −10 |
| −13.6 | 13.6 | 16 | −16 |
| 2.9 | 2.9 | 2 | 2 |
| −10.7 | 10.7 | 11 | −11 |
| 0.1 | 0.1 | 1 | 1 |
| −7.4 | 7.4 | 8 | −8 |
| −21.5 | 21.5 | 17 | −17 |
| 5.3 | 5.3 | 5 | 5 |
| 3.8 | 3.8 | 3 | 3 |
| −11.4 | 11.4 | 13 | −13 |
| −13.4 | 13.4 | 15 | −15 |
| −13.1 | 13.1 | 14 | −14 |
| −9.0 | 9.0 | 9 | −9 |
| 3.9 | 3.9 | 4 | −4 |
| −5.7 | 5.7 | 7 | −7 |

Step 4:   n = 17
Step 5:   The critical value: $W_1$ = (17)(18)/2 − 112 = 41
Step 6:   Work table prepared in Step 3.
Step 7:   The value of the test statistic is the sum of the positive
          ranks.
               W = 1 + 2 + 3 + 5 = 11
Step 8:   Since 11 < 41, reject $H_0$.
Step 9:   At the 5% significance level, the data provide enough evidence
          to conclude that the family therapy is effective in helping
          anorexic young women gain weight.

**10.149**   Population 1: Corneal thickness in normal eyes
             Population 2: Corneal thickness in eyes with Glaucoma

Step 1:  $H_0$: $\mu_1 = \mu_2$,  $H_a$: $\mu_1 > \mu_2$

Step 2:  $\alpha = 0.10$

Step 3:

| d | \|d\| | Rank | Signed-rank |
|---|---|---|---|
| -4 | 4 | 1.5 | -1.5 |
| 0 | 0 | --- | --- |
| 12 | 12 | 4.5 | 4.5 |
| 18 | 18 | 7 | 7 |
| -4 | 4 | 1.5 | -1.5 |
| -12 | 12 | 4.5 | -4.5 |
| 6 | 6 | 3 | 3 |
| 16 | 16 | 6 | 6 |

Step 4:  n = 7

Step 5:  The critical value: $W_r = 22$

Step 6:  Work table prepared in Step 3.

Step 7:  The value of the test statistic is the sum of the positive ranks.

$$W = 4.5 + 7 + 3 + 6 = 20.5$$

Step 8:  Since 20.5 < 22, do not reject $H_0$.

Step 9:  At the 10% significance level, the data do not provide enough evidence to conclude that the mean corneal thickness is greater in normal eyes than in eyes with glaucoma.

**10.151** The necessary plots for this exercise were produced in Exercise 10.131.  We display them again here.

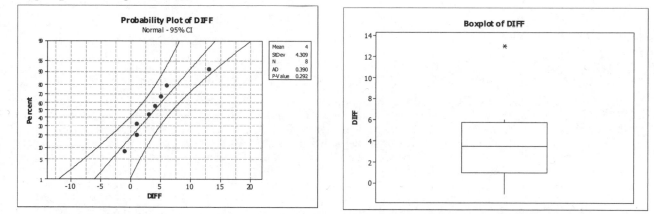

From the boxplot, we see that the distribution of differences is quite symmetric with the exception of the outlier at 13.  However, the 1-sample Wilcoxon test will be little affected by one outlier since the value of W would be the same if the difference of 13 were in fact a difference of 7 since the next largest difference is only 6.  We conclude that the paired Wilcoxon signed-rank test is appropriate for these difference data.

**10.153**(a)   Using Minitab, we prepared a normal probability plot and a boxplot of the difference (Initial - Final) data.

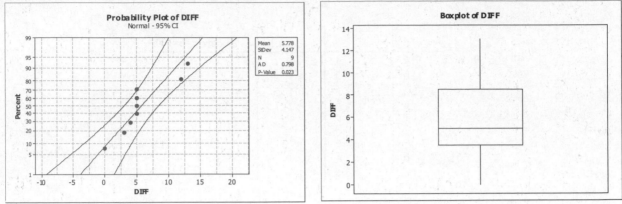

It is not reasonable to apply the paired-t test.  The sample size is small and the probability plot is far from linear.
(b)   The paired Wilcoxon signed-rank test is reasonable.  We see from the box plot that the data is nearly symmetric.

**10.155**(a)   Using Minitab, choose **Calc ▶ Calculator**, enter `'DIFF'` in the **Store result in variable** text box, and `'MEN' - 'WOMEN'` in the Expression box and click **OK**.  **Stat ▶ Nonparametrics ▶ 1-Sample Wilcoxon**, enter DIFF in the **V**ariables text box, enter 0.0 in the **Test media**n box and **greater than** in the **Alternative** drop down box, click **OK**.  The result is

```
Test of median = 0.000000 versus median > 0.000000

                    N
                  for  Wilcoxon            Estimated
         N   Test  Statistic       P       Median
DIFF    75    68     1763.0     0.000      3.000
```
The P-value = 0.000, so reject the null hypothesis.  The data do provide sufficient evidence that the mean age of Norwegian men at the time of marriage exceeds that of Norwegian women.
(c)   In Exercise 10.135, the P-value was also zero, so the conclusions are the same.
(d)   Paired Wilcoxon signed-rank test.  The difference data have two outliers, one on the left and one on the right.  These will not affect the mean difference much, but they will affect the standard deviation.  They will have no effect on the Wilcoxon test.  The sample size of 73 is quite large, so if the Wilcoxon test has any advantage over the paired t-test, that advantage is small.

**10.157**   Since the difference of the two population means, $\mu_1 - \mu_2$, is the same as the mean of the differences, $\mu_d$, the null hypothesis that $\mu_1 = \mu_2$ or $\mu_1 - \mu_2 = 0$ is the same as $\mu_d = 0$.

**10.159**(a)   The paired Wilcoxon signed-rank test makes use of the magnitude of each difference as well as the sign of the difference.  If the only criterion for using the Wilcoxon test, symmetry of the distribution, is satisfied, the paired Wilcoxon singed-rank test will be more powerful than the paired sign test, which does not require symmetry.
(b)   The paired sign test can be used for any data, regardless of the shape of the distribution.  It does not require symmetry for the distribution, as does the Wilcoxon signed-rank test.

**10.161**(a)   $H_0$: $\eta_d = 0$; $H_0$: $\eta_d \neq 0$; $\alpha = 0.05$
n = 15, Number of '+' signs = 14
Using Table XII, P-value = $2P(X \geq 14) = 2(0.000 + 0.000) = 0.000 < \alpha$

Reject $H_0$. The data provide sufficient evidence that the mean heights of cross-fertilized and self-fertilized Zea Mays differ.

(b)  For the paired Wilcoxon signed-rank test, $W = 96$ and $0.02 < $ P-value $< 0.05$. Reject $H_0$ at $\alpha = 0.05$. The conclusion is same as in part (a).

**10.163** (a)  $H_0: \eta_d = 0$; $H_0: \eta_d < 0$; $\alpha = 0.05$
$n = 17$, Number of '+' signs $= 4$
Using Excel, in any cell enter =BINOMDIST(4,17,0.5,1) . The last '1' indicates that you want the cumulative probability from 0 through 4. The result is 0.0245. Thus, P-value $= P(X \leq 4) = 0.0245 < \alpha$.
Reject $H_0$. The data provide sufficient evidence that the mean weights of anorexic young women after receiving a family therapy treatment are greater than the mean weights before treatment.

(b)  For the paired Wilcoxon signed-rank test, $W = 11$ and P-value $< 0.005$. Reject $H_0$ at $\alpha = 0.05$. The conclusion is same as in part (a).

**10.165** (a)  $H_0: \eta_d = 0$; $H_0: \eta_d > 0$; $\alpha = 0.10$
$n = 7$, Number of '+' signs $= 4$ (There was one 0 which was dropped.)
Using Table XII, P-value $= P(X \geq 4) = (0.273 + 0.164 + 0.055 + 0.008) = 0.500 > \alpha$.
Do not reject $H_0$. The data do not provide sufficient evidence that the mean corneal thickness of normal eyes is greater than in eyes with glaucoma.

(b)  For the paired Wilcoxon signed-rank test, $W = 20.5$ and P-value $> 0.100$. Do not reject $H_0$ at $\alpha = 0.10$. The conclusion is same as in part (a).

### Exercises 10.7

**10.167** (a)  Pooled t-test, nonpooled t-test, Mann-Whitney test

(b)  **Pooled t-test:** Independent simple random samples, normal populations or large samples, and equal population standard deviations
**Nonpooled t-test:** Independent simple random samples, and normal populations or large samples
**Mann-Whitney signed-rank test:** Independent simple random samples, same shape populations, and $n_1 \leq n_2$

(c)  **Pooled t-test**

$$t = \frac{\bar{x}_1 - \bar{x}_2}{s_p\sqrt{\dfrac{1}{n_1} + \dfrac{1}{n_2}}} \quad \text{where} \quad s_p = \sqrt{\frac{(n_1 - 1)s_1^2 + (n_2 - 1)s_2^2}{n_1 + n_2 - 2}}$$

**Nonpooled t-test**

$$t = \frac{\bar{x}_1 - \bar{x}_2}{\sqrt{\dfrac{s_1^2}{n_1} + \dfrac{s_2^2}{n_2}}} \quad \text{where the degrees of freedom are given by}$$

$$\Delta = \frac{[s_1^2/n_1 + s_2^2/n_2]^2}{\dfrac{(s_1^2/n_1)^2}{n_1 - 1} + \dfrac{(s_2^2/n_2)^2}{n_2 - 1}} \,.$$ This number is truncated to an integer if

it contains a fractional part.
**Mann-Whitney test**
$M = $ the sum of the ranks for sample data from Population 1

**10.169** (a)  One could use the pooled t-test, nonpooled t-test, or Mann-Whitney test.

(b)  The pooled t-test is the most appropriate since its conditions are satisfied and it will be the most powerful under these circumstances.

**10.171** (a)  Because the sample sizes are large, one could use the pooled t-test, nonpooled t-test, or Mann-Whitney test.

(b)  Since the populations have the same shape but are not normally distributed, the Mann-Whitney test is the most appropriate since its conditions are satisfied and it will be the most powerful under these circumstances.

**10.173** (a)  One could use the paired t-test or the paired Wilcoxon signed-rank test (paired W-test).

(b)  The paired W-test is the more appropriate test when the distribution is symmetric, but nonnormal.

**10.175**   To determine which procedure should be used to perform the hypothesis test, ask the following questions in sequence according to the flowchart in Figure 10.10:

|  | Answer to question | |
|---|---|---|
| Question to ask | Yes | No |
| Are the samples paired? | | x |
| Are the populations normal? <br> Are the populations <br> the same shape? <br> Is the sample large? | x | <br><br> x <br> x |

Since the two independent populations are normal, use the nonpooled t-test.

**10.177**   To determine which procedure should be used to find the required confidence interval, ask the following questions in sequence according to the flowchart in Figure 10.10:

|  | Answer to question | |
|---|---|---|
| Question to ask | Yes | No |
| Are the samples paired? | | x |
| Are the populations normal? | | x |
| Are the samples large? | x | |

It appears that one of the populations is neither normal nor symmetric, but the sample size is large.  Thus, use the nonpooled t-test.

**10.179**   To determine which procedure should be used to perform the hypothesis test, ask the following questions in sequence according to the flowchart in Figure 10.10:

|  | Answer to question | |
|---|---|---|
| Question to ask | Yes | No |
| Are the samples paired? | x | |
| Are the differences normal? <br> Are the differences <br> symmetric? <br> Is the sample large? | | x <br><br> x <br> x |

It appears the paired differences are neither normal or symmetric. Thus, a statistician must be consulted.

**Review Problems for Chapter 10**

1.  Randomly sample independently from both populations, compute the means of both samples and reject the null hypothesis if the sample means differ by too much. Otherwise, do not reject the null hypothesis.

2.  Sample independent pairs of observations from the two populations, compute the difference of the two observations in each pair, compute the mean of the differences, and reject the null hypothesis if the sample mean of the differences differs from zero by too much. Otherwise, do not reject the null hypothesis.

3.  (a) The pooled t-test requires that the population standard deviations be equal whereas the nonpooled t-test does not.
    (b) It is absolutely essential that the assumption of independence be satisfied.
    (c) The normality assumption is especially important for both t-tests for small samples. With large samples, the Central Limit Theorem applies and the normality assumption is less important.
    (d) Unless we are quite sure that the population standard deviations are equal, the nonpooled t-procedures should be used instead of the pooled t-procedures.

4.  (a) No. If the two distributions are normal and have the same shape, then they have the same population standard deviations, and the pooled t-test should be used. If the two distributions are not normal, but have the same shape, the Mann-Whitney test is preferred.
    (b) The pooled t-test is preferred to the Mann-Whitney test if both populations are normally distributed with equal standard deviations.

5.  A paired sample may reduce the estimate of the standard error of the mean of the differences, making it more likely that a difference of a given size will be judged significant.

6.  The paired t-test is preferred to the paired Wilcoxon signed-rank test if the distribution of differences is normal, or if the sample size is large and the distribution of differences is not symmetric. If the distribution of differences is symmetric but not normal, then the paired Wilcoxon signed-rank test is preferred.

7.  (a) Population 1: Male; $n_1 = 13$, $\bar{x}_1 = 2127$, $s_1 = 513$

    Population 2: Female; $n_2 = 14$, $\bar{x}_2 = 1843$, $s_2 = 446$

    Step 1: $H_0$: $\mu_1 = \mu_2$, $H_a$: $\mu_1 > \mu_2$
    Step 2: $\alpha = 0.05$,

    Step 3:

    $$s_p = \sqrt{\frac{12(513)^2 + 13(446)^2}{25}} = 479.33$$

    $$t = \frac{2127 - 1843}{479.33\sqrt{\frac{1}{13} + \frac{1}{14}}} = 1.538$$

    Step 4: df = 25, Critical value = 1.708
    Step 5: Since 1.538 < 1.708, do not reject $H_0$.
    Step 6: At the 5% significance level, the data do not provide enough evidence to conclude that the mean right-leg strength of males exceeds that of females.
    (b) For the P-value approach, $0.05 < P(t > 1.538) < 0.10$. Therefore, because the P-value is larger than the significance level, do not reject $H_0$.

8.

$$2127 - 1843 \pm 1.708(479.33)\sqrt{\frac{1}{13} + \frac{1}{14}}$$

$$284 \pm 315.33$$

$$-31.33 \text{ to } 599.33$$

We can be 90% confident that the difference, $\mu_1 - \mu_2$, between the mean right-leg strengths of males and females is between -31.3 and 599.3 newtons.

9.    Population 1: Florida, $n_1 = 24$, $\bar{x}_1 = 5.46$, $s_1 = 1.59$

Population 2: Virginia, $n_2 = 44$, $\bar{x}_2 = 7.59$, $s_2 = 2.68$

$$\Delta = \frac{[s_1^2 / n_1 + s_2^2 / n_2]^2}{\frac{(s_1^2 / n_1)^2}{n_1 - 1} + \frac{(s_2^2 / n_2)^2}{n_2 - 1}} = \frac{[5.46^2 / 24 + 7.59^2 / 44]^2}{\frac{(5.46^2 / 24)^2}{24 - 1} + \frac{(7.59^2 / 44)^2}{44 - 1}} = 65.45 = 65 = df$$

Step 1:       $H_0: \mu_1 = \mu_2$,   $H_a: \mu_1 < \mu_2$
Step 2:       $\alpha = 0.01$
Step 3:

$$t = \frac{5.4583 - 7.5909}{\sqrt{\frac{1.5874^2}{24} + \frac{2.6791^2}{44}}} = -4.119$$

Step 4:   Critical value = -2.385
Step 5:   Since -4.119 < -2.385, reject $H_0$.
Step 6:   At the 1% significance level, the data provide sufficient evidence to conclude that the average litter size of cottonmouths in Florida is less than that in Virginia.

For the P-value approach, P(t < -4.110) < 0.005.  Therefore, because the P-value is smaller than the significance level, reject $H_0$.

10.

$$(5.46 - 7.49) \pm 2.385\sqrt{\frac{1.59^2}{24} + \frac{2.68^2}{44}}$$

$$-2.13 \pm 1.24$$

$$-3.37 \text{ to } -0.89$$

We can be 98% confident that the average difference, $\mu_1 - \mu_2$, between cottonmouth litter sizes in Florida and Virginia is somewhere between -3.37 and -0.89.

11.    Population 1: Home prices in New York City;
Population 2: Home prices in Los Angeles.

Step 1:       $H_0: \mu_1 = \mu_2$,   $H_a: \mu_1 \neq \mu_2$
Step 2:       $\alpha = 0.05$
Step 3:       Construct a worktable based upon the following:  First, rank all the data from both samples combined.  Adjacent to each column of data as it is presented in the Exercise, record the overall rank.

| NYC | Overall Rank | LA | Overall Rank |
|---|---|---|---|
| 365.3 | 6 | 411.1 | 12 |
| 754.4 | 20 | 651.6 | 18 |
| 367.4 | 7 | 534.1 | 17 |
| 387.7 | 9 | 395.6 | 10 |
| 528.8 | 16 | 722.4 | 19 |
| 344.3 | 2 | 399.4 | 11 |
| 423.3 | 14 | 354.8 | 3 |
| 324.1 | 1 | 416.3 | 13 |
| 381.5 | 8 | 359.4 | 4 |
| 433.2 | 15 | 363.3 | 5 |

Step 4:     The value of the test statistic is the sum of the ranks for the sample data from Population 1:
M = 1 + 2 + 6 + 7 + 8 + 9 + 14 + 15 + 16 + 20 = 98.

Step 5:     We have $n_1 = 10$ and $n_2 = 10$. Since the hypothesis test is two-tailed with $\alpha = 0.05$, we use Table VI to obtain the critical values, which are $M_l = 10(21) - 131 = 79$ and $M_r = 131$. Thus, we reject $H_0$ if $M \geq 131$ or $M \leq 79$.

Step 6:     Since $79 < M < 131$, do not reject $H_0$.

Step 7:     At the 5% significance level, the data do not provide sufficient evidence to conclude that the mean costs for existing single-family homes differ in New York City and Los Angeles.

12.    Step 1:     $H_0$: $\mu_1 = \mu_2$,    $H_a$: $\mu_1 \neq \mu_2$

Step 2:     $\alpha = 0.10$

Step 3:     Paired differences, $d = x_1 - x_2$:

| 12 | 7 | 1 | 0 |
|---|---|---|---|
| -11 | 0 | 6 | -4 |

(a)  We used Minitab to produce the following normal probability plot and boxplot of the differences.

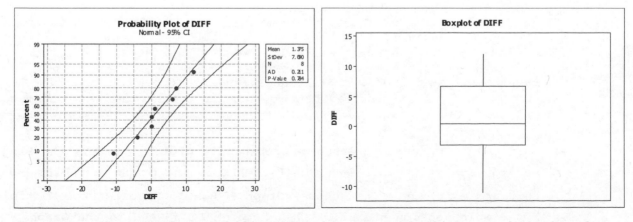

(b)  Yes. The normal probability plot is reasonably linear and the boxplot shows no outliers.

Step 4:   From the probability plot, we see that the sample mean of the differences is 1.375 and the standard deviation is 7.090. Therefore,

$$t = \frac{1.375}{7.090/\sqrt{8}} = 0.549$$

Step 5:      Critical values = ±1.860
Step 6:      Since −1.860 < 0.549 < 1.860, do not reject H₀.
Step 7:      At the 10% significance level, the data do not provide sufficient evidence to conclude that there is a difference in the mean length of time that ice stays on the two lakes.
For the P-value approach,  2{P(t > 0.549)} > 0.20.  Therefore, because the P-value is larger than the significance level, do not reject H₀.

13.

$$1.375 \pm 1.860 \cdot 7.090/\sqrt{8} = 1.375 \pm 4.749 = -3.374 \text{to } 6.124 \text{days}$$

We can be 90% confident that the difference, $\mu_1 - \mu_2$, between the mean numbers of days ice stays on Lakes Mendota and Mononar is somewhere between −3.374 and 6.124 days.

14.   Population 1:  Eyepiece;  Population 2:  TV;
Step 1:      H₀: $\mu_1 = \mu_2$,  Hₐ: $\mu_1 > \mu_2$
Step 2:      $\alpha$ = 0.05

Step 3:

| Eyepiece | TV | Difference | Rank | Signed Rank |
|----------|-------|-----------|------|-------------|
| 182.2 | 177.8 | 4.4 | 4 | 4 |
| 118.5 | 116.6 | 1.9 | 2 | 2 |
| 100.0 | 924.0 | 7.6 | 5 | 5 |
| 161.3 | 145.0 | 16.3 | 7 | 7 |
| 42.7 | 38.9 | 3.8 | 3 | 3 |
| 299.1 | 226.3 | 72.8 | 10 | 10 |
| 547.8 | 514.6 | 33.2 | 9 | 9 |
| 437.3 | 458.1 | −20.8 | 8 | −8 |
| 174.4 | 159.2 | 15.2 | 6 | 6 |
| 85.4 | 86.6 | −1.2 | 1 | −1 |

Step 4:      n = 10
Step 5:      The critical value, $W_r$ = 44
Step 6:      Worktable shown in Step 3.
Step 7:      The value of the test statistic is the sum of the positive ranks.
                   W = 4 + 2 + 5 + 7 + 3 + 10 + 9 + 6 = 46
Step 8:      Since 46 > 44, reject H₀.
Step 9:      At the 5% significance level, the data do provide sufficient evidence to conclude that the eyepiece method gives a greater fiber-density reading than the TV-screen method.

15.   (a) Using Minitab, choose  **Graph ▶ Probability plot...**, select the **Single** version and click **OK**.  Enter <u>MALE FEMALE</u> in the **Graph Variables** text box, and click **OK**.  Then choose **Graph ▶ Boxplot**, select the **Multiple Y's Simple** version and click **OK**.  Enter <u>MALE FEMALE</u> **in the Graph variables** text box, and click **OK**.  The standard deviations will be

produced on the probability plot legend. The results are

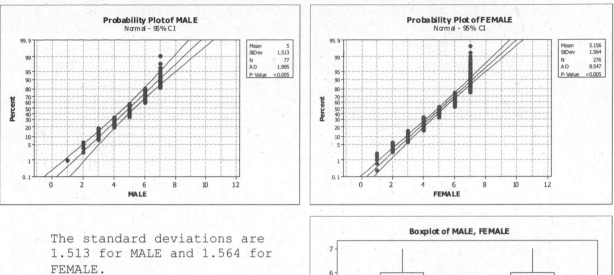

The standard deviations are 1.513 for MALE and 1.564 for FEMALE.

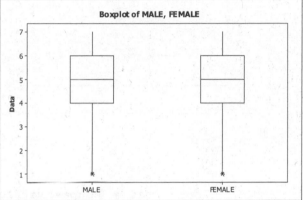

(b)   Pooled t-procedures. The sample sizes are both large, there are no potential outliers in either sample, and the sample standard deviations are nearly equal.

(c)   To carry out the pooled t-test, choose **Stat ▶ Basic Statistics ▶ 2-Sample t**, click on **Samples in different columns**, enter <u>MALE</u> in the **First** text box and <u>FEMALE</u> in the **Second** text box, select **not equal** in the **Alternative** box, enter <u>90</u> in the **Confidence level** text box, click on **Assume equal variances** to make certain that there is an **X** in the box, and click **OK**. The results are

Two-sample T for MALE vs FEMALE

```
              N   Mean   StDev   SE Mean
MALE         77   5.00   1.51    0.17
FEMALE      276   5.16   1.56    0.094
```

Difference = mu (MALE) - mu (FEMALE)
Estimate for difference: -0.155797
95% CI for difference: (-0.549387, 0.237793)
T-Test of difference = 0 (vs not =): T-Value = -0.78   P-Value = 0.437   DF = 351
Both use Pooled StDev = 1.5528

The P-value is -0.437, which is greater than the significance level 0.05. Do not reject the null hypothesis. The data do not provide sufficient evidence to conclude that there is a difference in mean dating satisfaction of male and female college students.

(d)   A 95% confidence interval for the difference between mean dating satisfaction of male and female college students was produced as part of the process in part (b).  The interval is (-0.55, 0.24).

(e) Yes.  Although the data for both genders are left-skewed and discrete (possible values include only the whole numbers from 1 to 7), the large sample sizes justify the use of the pooled t-procedures.

16.   (a) Using Minitab, choose  **Graph ▶ Histogram...**, select the **Simple** version and click **OK**.  Enter <u>MALE FEMALE</u> in the **Graph Variables** text box, clcik on the scale button and the Y-Scle type tab.  Select Percent and click **OK** twice.  The results are

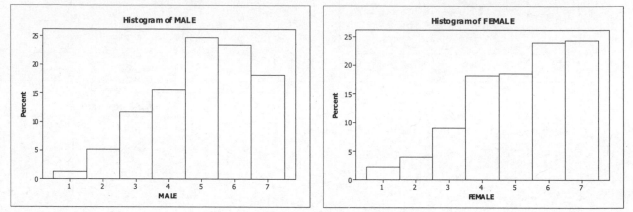

(b)   Yes.  All that is required for the Mann-Whitney test is that both populations be the same shape.  We can see from the histograms that this is the case.

(c)   Choose **Stat ▶ Nonparametrics ▶ Mann-Whitney**, enter <u>MALE</u> in the **First Sample** text box and <u>FEMALE</u> in the **Second Sample** text box.  Enter <u>95</u> in the **Confidence level** text box and select **not equal** in the **Alternative** drop down box.  The results are

### Mann-Whitney Test and CI: MALE, FEMALE
```
            N   Median
MALE       77   5.0000
FEMALE    276   5.0000

Point estimate for ETA1-ETA2 is -0.0000
95.0 Percent CI for ETA1-ETA2 is (-1.0000,0.0001)
W = 12917.0
Test of ETA1 = ETA2 vs ETA1 not = ETA2 is significant at 0.3689
The test is significant at 0.3591 (adjusted for ties)
```

The P-value is 0.3591, which is greater than the significance level 0.05.  Do not reject the null hypothesis.  The data do not provide sufficient evidence that there is a difference in mean dating satisfaction for mal and female college students.  This is the same conclusion we reached in Problem 15.

17.   (a) Using Minitab, choose  **Graph ▶ Probability plot...**, select the **Single** version and click **OK**.  Enter <u>MALE FEMALE</u> in the **Graph Variables** text box, and click **OK**.  Then choose **Graph ▶ Boxplot**, select the **Multiple Y's Simple** version and click **OK**.  Enter <u>MALE FEMALE</u> **in the Graph variables** text box, and click **OK**.  The standard deviations will be produced on the probability plot legend.  The results are

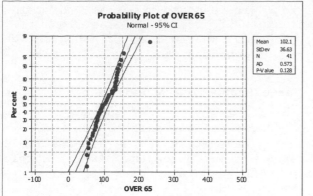

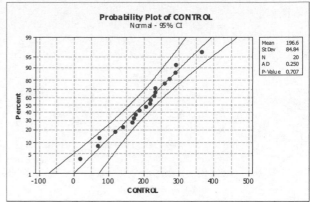

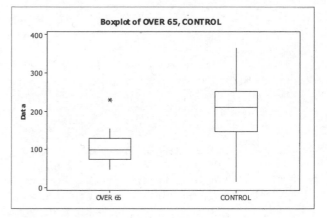

The standard deviations are 36.63 for OVER 65 and 84.84 for CONTROL,

(b) The nonpooled t-procedure is preferable since the ratio of the standard deviations is greater than 2.

(c) To carry out the nonpooled t-test, choose **Stat ▶ Basic Statistics ▶ 2-Sample t**, click on **Samples in different columns**, enter 'OVER 65' in the **First** text box and CONTROL in the **Second** text box, and leave the **Assume equal variances** box unchecked. Click on the **Options** button, enter 99 in the **Confidence level** text box, select **less than** in the **Alternative** box, and click **OK** twice. The results are

```
Two-sample T for OVER 65 vs CONTROL

            N    Mean   StDev   SE Mean
OVER 65    41   102.1    36.6       5.7
CONTROL    20   196.6    84.8        19


Difference = mu (OVER 65) - mu (CONTROL)
Estimate for difference:  -94.4768
99% upper bound for difference:  -44.7729
T-Test of difference = 0 (vs <): T-Value = -4.77   P-Value = 0.000   DF = 22
```

Since the P-value = 0.000, which is less than the significance level 0.01, reject the null hypothesis. The data do provide sufficient evidence to conclude that, on average, men over 65 have a lower IGF-1 level than younger men.

(d) To obtain the 99% confidence interval, follow the procedure in part (c), but select not equal in the Alternative drop down box. The confidence interval is given by
    99% CI for difference:  (-150.3322, -38.6215)
We can be 99% confident that the difference in IGF-1 levels, on average, between men over 65 and younger men is somewhere between -150.3 and -38.6.

(e)   Yes.   The CONTROL sample of size 20 is reasonably normally distributed
      with no outliers.  The OVER 65 sample of size 41 is also close to
      normal with one outlier.  The effect of the outlier is not great enough
      to alter the conclusion (Deleting it changes the value of t from −4.77
      to −4.99).

18.   (a)  Using Minitab, choose **Stat ▶ Basic Statistics ▶ Paired t**, click on
           **Samples in columns**, enter MEN in the **First** text box and WOMEN in the
           **Second** text box.  Click on the **Options** button, enter 95 in the
           **Confidence level** text box, select **greater than** in the **Alternative** box,
           and click **OK**.  Click on the **Graphs** button, check the box for **Boxplot of
           differences**, and click **OK** twice.  The results are

           Paired T for MEN - WOMEN

           |            | N  | Mean    | StDev  | SE Mean |
           |------------|----|---------|--------|---------|
           | MEN        | 50 | 1003.96 | 820.04 | 115.97  |
           | WOMEN      | 50 | 860.88  | 818.12 | 115.70  |
           | Difference | 50 | 143.080 | 93.758 | 13.259  |

           95% lower bound for mean difference: 120.850
           T-Test of mean difference = 0 (vs > 0): T-Value = 10.79   P-Value = 0.000

      Since the P-value = 0.000, which is less than the significance level
      0.05, reject the null hypothesis.  The data do provide sufficient
      evidence to conclude that, on average, the weekly earnings of male
      full-time and salary workers exceed that of women.

      (b)  Follow the procedure in part (a), but enter 90 in the **Confidence level**
           text box and select **not equal** from the **Alternative** drop down box.  The
           resulting interval is
             90% CI for mean difference: (120.850, 165.310)
           We can be 90% confident that the mean difference in weekly earnings of
           male and female full-time wage and salary workers is somewhere between
           $120.85 and $165.31.

      (c)  To create a column of differences, click in the **Sessions Window** and
           then click on the **Editor** tab and on **Enable Commands**.  An MTB> prompt
           will appear at the bottom of the Sessions Window.  After the prompt,
           enter Let c3=c1-c2 and press ENTER.  Click at the top of Column 3 and
           name the column DIFF.  Choose **Graph ▶ Probability plot...**, select the
           **Single** version and click **OK**.  Enter DIFF in the **Graph Variables** text
           box, and click **OK**.  The standard deviations will be produced on the
           probability plot legend.  A boxplot was produced by the procedure in
           part (a).  Now choose **Graph ▶ Stem-and-Leaf...**, enter DIFF in the **Graph
           Variables** text box, and click **OK**.  The results are

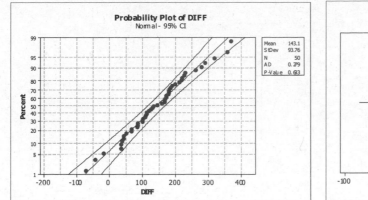

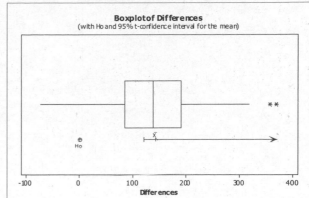

```
Stem-and-leaf of DIFF  N  = 50
Leaf Unit = 10

    1    -0  7
    3    -0  41
    8     0  33344
   15     0  5668889
  (11)    1  00011112334
   24     1  5666777788889
   11     2  11122
    6     2  678
    3     3  1
    2     3  56
```

(d)   Yes.  The sample size is large (50), there are only two mild outliers
      on the right, and the normal probability plot is close to linear.  The
      outliers may cause some concern since they are in direction that helps
      to reject the null hypothesis.  If so, a nonparametric method may be
      used.

**19.**  (a)   Using Minitab, choose **Stat ▶ Nonparametrics ▶ 1-Sample Wilcoxon**,

enter <u>DIFF</u> in the **Variables** text box, Click on the **Test median**, enter
<u>0</u> in the **Test median** text box, select **greater than** in the **Alternative**
box, and click **OK**.   The results are
        Test of median = 0.000000 versus median > 0.000000

```
                    N
                   for   Wilcoxon             Estimated
              N  Test   Statistic      P      Median
      DIFF   50    50      1256.0   0.000       141.0
```
Since the P-value = 0.000, which is less than the significance level
0.05, reject the null hypothesis.  The data do provide sufficient
evidence to conclude that, on average, the weekly earnings of male
full-time and salary workers exceed that of women.

(b)   The P-value and conclusion are the same as those in Problem 18(a).

(c)   See Problem 18(c) for the boxplot and stem-and-leaf diagram.  For the

histogram, choose **Graph ▶ Histogram...**, select the **Simple** version and
click **OK**.  Enter <u>DIFF</u> in the **Graph Variables** text box, and click **OK**.
The result is

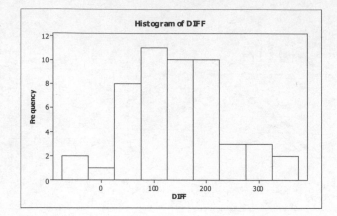

(d)     Yes.  The only assumption for the Wilcoxon signed-rank test is that the population be symmetric.  All of the plots indicate that this is a reasonable assumption.

## Chapter 11 Answers

### Exercises 11.1

**11.1** A variable has a chi-square distribution if its distribution has the shape of a right-skewed curve called a chi-square curve.

**11.3** The curve with 20 degrees of freedom more closely resembles a normal distribution. By Property 4 of Key Fact 11.1, as the degrees of freedom increase, the distributions look increasing like normal distribution curves.

**11.5** (a) $\chi^2_{0.025} = 32.852$   (b) $\chi^2_{0.95} = 10.117$

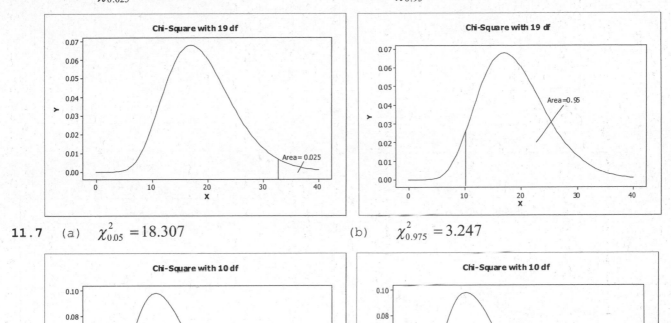

**11.7** (a) $\chi^2_{0.05} = 18.307$   (b) $\chi^2_{0.975} = 3.247$

**11.9**

(a)   A left area of 0.01 is equivalent to a right area of 0.99.

$$\chi^2_{0.99} = 1.646$$

(b)   A left area of 0.95 is equivalent to a right area of 0.05.

$$\chi^2_{0.95} = 15.507$$

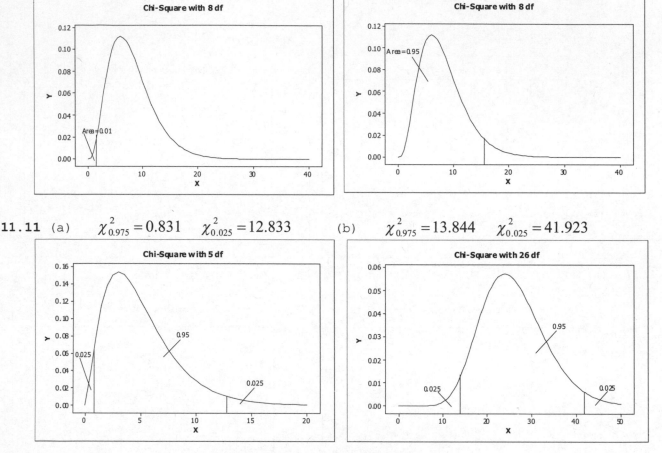

**11.11** (a)   $\chi^2_{0.975} = 0.831 \quad \chi^2_{0.025} = 12.833$   (b)   $\chi^2_{0.975} = 13.844 \quad \chi^2_{0.025} = 41.923$

**11.13** The chi-square test for one population standard deviation is not robust to moderate violations of the normality assumption.

**11.15** (a)   Critical value = 3.325;   $\chi^2 = (n-1)s^2 / \sigma_0^2 = (9)3^2 / 4^2 = 5.0625$;   Left tail test

Since this is greater than 3.325, do not reject $H_0$.

(b)   90% Confidence Interval =

$$(s\sqrt{\frac{(n-1)}{\chi^2_{0.05}}}, s\sqrt{\frac{(n-1)}{\chi^2_{0.95}}}) = (3\sqrt{\frac{(10-1)}{16.919}}, 3\sqrt{\frac{(10-1)}{3.325}}) = (2.188, 4.936)$$

**11.17** (a)   Critical value = 44.314;   $\chi^2 = (n-1)s^2 / \sigma_0^2 = (25)7^2 / 5^2 = 49.00$;   Right tail test

Since this is greater than 44.314, reject $H_0$.

(b)   98% Confidence Interval =

$$(s\sqrt{\frac{(n-1)}{\chi^2_{0.01}}}, s\sqrt{\frac{(n-1)}{\chi^2_{0.99}}}) = (7\sqrt{\frac{(26-1)}{44.314}}, 7\sqrt{\frac{(26-1)}{11.524}}) = (5.257, 10.310)$$

**11.19** (a)   Critical values = 8.907 and 32.852;

$$\chi^2 = (n-1)s^2 / \sigma_0^2 = (19)5^2 / 6^2 = 13.194;$$   Two tail test

Since 8.907 < 13.194 < 32.852, do not reject $H_0$.

(b)  95% Confidence Interval =

$$(s\sqrt{\frac{(n-1)}{\chi^2_{0.025}}}, s\sqrt{\frac{(n-1)}{\chi^2_{0.975}}}) = (5\sqrt{\frac{(20-1)}{32.852}}, 5\sqrt{\frac{(20-1)}{8.907}}) = (3.802, 7.303)$$

**11.21** $H_0$: $\sigma$ = 8.45    $H_a$: $\sigma \neq$ 8.45; df = 28 - 1 = 27
Critical values = 16.151 and 40.113;

$$\chi^2 = (n-1)s^2/\sigma_0^2 = (27)9.229^2/8.45^2 = 32.208;$$ Two tail test

Since 16.151 < 32.208 < 40.113, do not reject $H_0$.

**11.23** Step 1: $H_0$: $\sigma$ = 0.27    $H_a$: $\sigma\sigma$ > 0.27
Step 2: $\alpha$ = 0.01

Step 3:  $$\chi^2 = \frac{n-1}{\sigma_0^2}s^2 = \frac{9}{.27^2}0.75638^2 = 70.631$$

Step 4: The critical value with n-1 = 9 df is 21.666.
Step 5: Since 70.560 > 21.666, reject $H_0$.
Step 6: There is sufficient evidence at the 0.01 level to claim that
the process variation for this piece of equipment exceeds the
analytical capability of 0.27.  Using Excel, in any cell, enter
=chidist(70.631,9) to obtain P($\chi^2$ > 70.631) = 1.15 x 10$^{-11}$.  The
P-value is 1.15 x 10$^{-11}$.

**11.25** Step 1: $H_0$: $\sigma$ = 0.2    $H_a$: $\sigma$ < 0.2
Step 2: $\alpha$ = 0.05

Step 3:  $$\chi^2 = \frac{n-1}{\sigma_0^2}s^2 = \frac{14}{0.2^2}0.15416^2 = 8.318$$

Step 4: The critical value with n-1 = 14 df is 6.571.
Step 5: Since 8.318 > 6.571, do not reject $H_0$.
Step 6: There is insufficient evidence at the 0.05 level to claim that
the standard deviation of the amounts dispensed is less than 0.2
fluid ounces.   Using Excel, in any cell, enter =1-
chidist(8.318,14) to obtain P($\chi^2$ < 8.318) = 0.1279.  The P-
value is 0.1279.

**11.27** The 90% confidence interval for $\sigma$ of prices of this year's agriculture
books is

$$\left(\sqrt{\frac{n-1}{\chi^2_{0.025}}}\cdot s, \sqrt{\frac{n-1}{\chi^2_{0.975}}}\cdot s\right) = \left(\sqrt{\frac{27}{40.113}}9.229, \sqrt{\frac{27}{16.151}}9.229\right) = (7.57, 11.93)\text{ dollars}$$

**11.29** The 98% confidence interval for the process variation of the piece of
equipment under consideration is

$$\left(\sqrt{\frac{n-1}{\chi^2_{0.025}}}\cdot s, \sqrt{\frac{n-1}{\chi^2_{0.975}}}\cdot s\right) = \left(\sqrt{\frac{9}{21.666}}\cdot 0.756, \sqrt{\frac{9}{2.0886}}\cdot 0.756\right) = (0.487, 1.570)$$

**11.31** The 90% confidence interval for the standard deviation of coffee
dispensed is

$$\left(\sqrt{\frac{n-1}{\chi^2_{0.05}}}\cdot s, \sqrt{\frac{n-1}{\chi^2_{0.95}}}\cdot s\right) = \left(\sqrt{\frac{14}{23.685}}\cdot 0.154, \sqrt{\frac{14}{6.571}}\cdot 0.154\right) = (0.1184, 0.2248)\text{ fluid oz.}$$

**11.33**

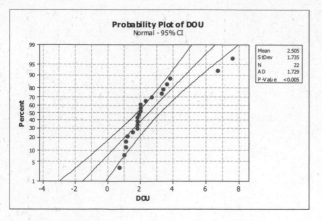

Applying one-standard deviation $x^2$-procedures is not reasonable for these data in Exercise 11.33.  The probability plot is not linear and there are at least two outliers.

**11.35**

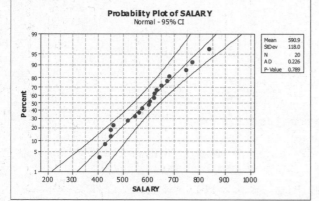

Applying one-standard deviation $x^2$-procedures is reasonable for these data in Exercise 11.35.  The probability plot is close to linear and there are no outliers.

**11.37** (a) Using Minitab, choose **Graph ▶ Probability Plot...,** select the **Single** version and click **OK.**  Enter TEMP in the **Graph variables** text box and click **OK.**  Then choose **Graph ▶ Boxplot...,** select the **One Y Simple** version and click **OK.**  Enter TEMP in the **Graph variables** text box and click **OK.**  Now choose **Graph ▶ Histogram...,** select the **Simple** version and click **OK.**  Enter TEMP in the **Graph variables** text box and click **OK.**  Then choose **Graph ▶ Stem-and-Leaf...** enter TEMP in the **Graph variables** text box and click **OK.**  The results are

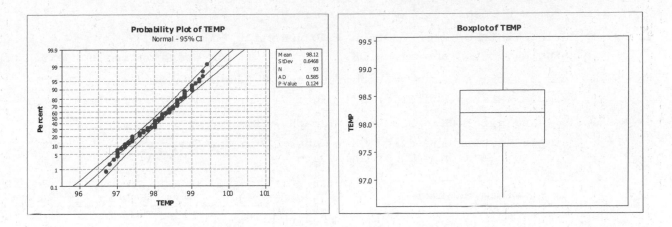

```
Stem-and-leaf of TEMP   N = 93
Leaf Unit = 0.10
```

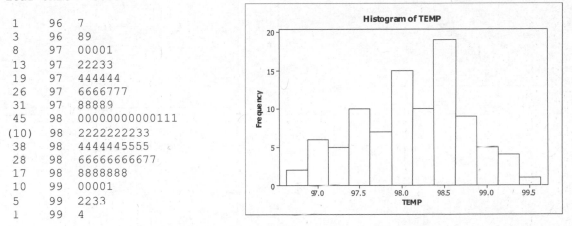

```
  1     96   7
  3     96   89
  8     97   00001
 13     97   22233
 19     97   444444
 26     97   6666777
 31     97   88889
 45     98   00000000000111
(10)    98   2222222233
 38     98   4444445555
 28     98   66666666677
 17     98   8888888
 10     99   00001
  5     99   2233
  1     99   4
```

(b)  Yes.  The probability plot is very close to linear and there are no
     outliers.

(c)  Choose **Stat ▶ Basic Statistics ▶ 1-Variance**, enter <u>TEMP</u> in the **Samples in**
     **Columns** text box, check the box for **Perform hypothesis test**, enter the variance
     0.3969 (the square of 0.63) in the **Hypothesized Variance** text box, click on the
     **Options** button, enter <u>0.95</u> in the **Confidence level** text box and select **not**
     **equal** in the **Alternative** drop down box, and click **OK** twice.  The result is

     Test of sigma squared = 0.3969 vs not = 0.3969

     Chi-Square Method (Normal Distribution)

     Variable   N   Variance      95% CI       Chi-Square     P
     TEMP       93    0.418    (0.320, 0.571)     96.97    0.683

     Since the P-value = 0.683, which is larger than the significance
     level 0.05, do not reject the null hypothesis.  The data do not
     provide sufficient evidence that the standard deviation differs
     from 0.63 degrees.

(d)  The 95% confidence interval for the variance was produced in part
     (c) as (0.320,0.571).  Taking the square root of each confidence
     limit, this is equivalent to a 95% confidence interval for the
     standard deviation of (0.57, 0.76).  We can be 95% confident that

the population standard deviation lies somewhere in the interval from 0.57 degrees to 0.76 degrees (F).

**11.39** (a) Using Minitab, choose **Graph ▶ Probability Plot...**, select the **Single** version and click **OK**. Enter <u>IQ</u> in the **Graph variables** text box and click **OK**. Then choose **Graph ▶ Boxplot...**, select the **One Y Simple** version and click **OK**. Enter <u>IQ</u> in the **Graph variables** text box and click **OK**. Now choose **Graph ▶ Histogram...**, select the **Simple** version and click **OK**. Enter <u>IQ</u> in the **Graph variables** text box and click **OK**. The results are

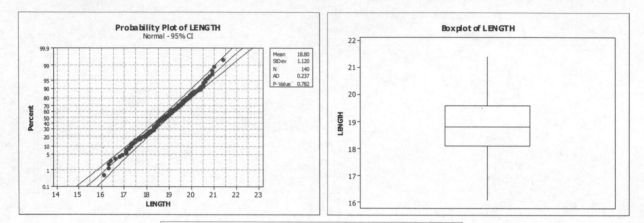

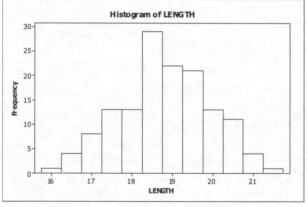

(b) Yes. None of the plots exhibit departures from normality and there are no potential outliers.

(c) Choose **Stat ▶ Basic Statistics ▶ 1-Variance**, enter <u>IQ</u> in the **Samples in Columns** text box, uncheck the box for **Perform hypothesis test**, click on the **Options** button, enter <u>0.95</u> in the **Confidence level** text box and select <u>0</u>**less than** in the **Alternative** drop down box, and click **OK** twice. The result is

```
Chi-Square Method (Normal Distribution)

Variable   N   Variance      99% CI
LENGTH    140    1.26     (1.01, 1.61)
```

The 95% confidence interval for the variance is (1.01, 1.61). Taking the square root of each confidence limit, this is equivalent to a 95% confidence interval for the standard deviation of (1.005, 1.269). We can be 95% confident that the population standard deviation $\sigma$ lies somewhere in the interval from 1.005 to 1.269 inches.

**11.41** If the standard deviation is too large, some cups will be filled with too little coffee (making for dissatisfied customers), and some cups may overflow or be in danger of being spilled by the customer. Customers don't appreciate hot coffee spilled on them; in addition, the company loses money (or makes less) if too much coffee is dispensed.

**11.43** (a)    $P(\text{satisfactory}) = P(9.7 \le X \le 10.3) = 1 - P(\text{defective})$
Since it is desired to have $P(\text{defective}) < 0.001$,
$P(\text{satisfactory}) > 1 - 0.001 = 0.999$, or
$P(9.7 \le X \le 10.3) > 0.001$

(b)

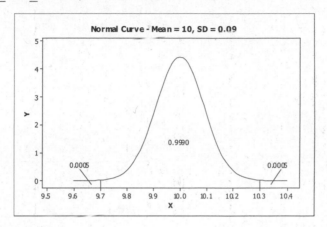

(c)    $0.9990 < P(9.7 < X < 10.3) = P((9.7-10)/\sigma < Z < (10.3-10)/\sigma)$
$= P(-0.3/\sigma < Z < 0.3/\sigma)$

On the other hand, since $z_{0.0005} = 3.30$, we have

$0.9990 = P(-3.30 < Z < 3.30)$.    Setting $0.3/\sigma = z_{0.0005}$, we have $3.30 = 0.3/\sigma$ which implies that $\sigma = 0.3/3.30 = 0.09$.

(d)    If $\sigma$ were smaller than 0.09, the probability that the bolt diameter lies between 9.7 and 10.3 would be larger yet.  Thus setting the tolerance specifications at $\pm 0.3$mm with a probability of 0.9990 is equivalent to saying that the standard deviation must be less than 0.09.

### Exercises 11.2

**11.45** We identify an F-distribution and its corresponding F-curve by stating its two numbers of degrees of freedom.

**11.47** $F_{0.05}$, $F_{0.025}$, $F_\alpha$

**11.49** (a)    12
(b)    7

**11.51** (a)    $F_{0.05} = 1.89$    (b)    $F_{0.01} = 2.47$    (c)    $F_{0.025} = 2.14$

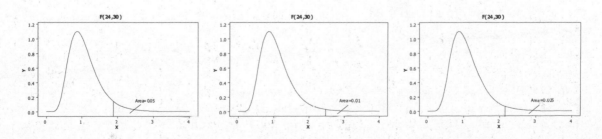

**11.53** (a)     $F_{0.01} = 2.88$          (b)     $F_{0.05} = 2.10$          (c)     $F_{0.10} = 1.78$

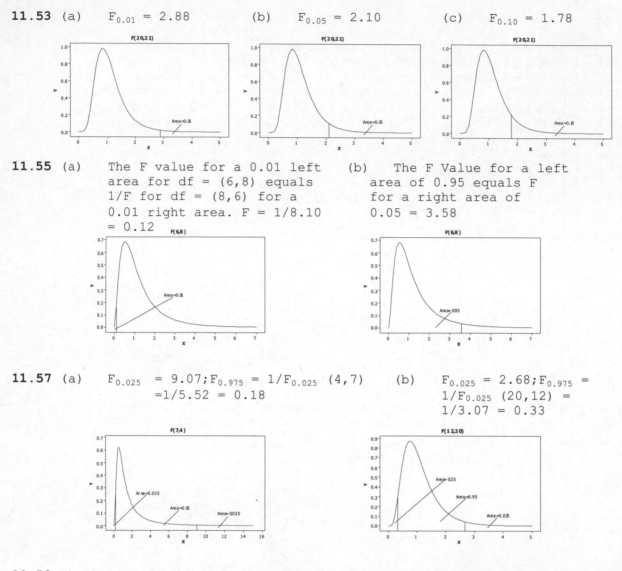

**11.55** (a)  The F value for a 0.01 left       (b)  The F Value for a left
              area for df = (6,8) equals             area of 0.95 equals F
              1/F for df = (8,6) for a               for a right area of
              0.01 right area. F = 1/8.10            0.05 = 3.58
              = 0.12

**11.57** (a)     $F_{0.025} = 9.07; F_{0.975} = 1/F_{0.025}$ (4,7)       (b)     $F_{0.025} = 2.68; F_{0.975} =$
                 $= 1/5.52 = 0.18$                                               $1/F_{0.025}$ (20,12) =
                                                                                 $1/3.07 = 0.33$

**11.59** The F procedures are extremely nonrobust to even moderate violations of the
normality assumption for both populations.

**11.61** Critical Value = 2.84, $F = s_1^2 / s_2^2 = 19.4^2 / 10.5^2 = 3.414$.   This is greater than
2.84, so reject $H_0$.

**11.63** Critical Value = 0.37 {$1/F_{12,7,0.01} = 1/2.67$),   $F = s_1^2 / s_2^2 = 28.82^2 / 38.97^2 = 0.547$.
This is greater than 0.37, so do not reject $H_0$.

**11.65** Critical Values = 0.26 {$1/F_{8,10,0.025} = 1/3.85$) and 4.30,
$F = s_1^2 / s_2^2 = 14.5^2 / 30.4^2 = 0.228$.   This is less than 0.26, so reject $H_0$.

**11.67** 1=Control, 2=Experimental

Step 1: $H_0$: $\sigma_1 = \sigma_2$     $H_a$: $\sigma_1 > \sigma_2$
Step 2: $\alpha = 0.05$
Step 3: F = $7.813^2 / 5.286^2 = 2.185$
Step 4: The critical value is $F_{0.05}$ with df = (40,19) or 2.03
Step 5: Since 2.185 > 2.03, reject the null hypothesis.
Step 6:   There is sufficient evidence at the 0.05 significance level to
          claim that the variation in the control group is greater than that
          in the experimental group.  In Excel, enter =FDIST(2.1853,40,19) in
          any cell to obtain P(F >2.185) = 0.0348.   The P-value is 0.0348.

**11.69** 1=Relaxation tapes, 2=Neutral tapes

Step 1: $H_0$: $\sigma_1 = \sigma_2$    $H_a$: $\sigma_1 \neq \sigma_2$

Step 2: $\alpha$ = 0.10

Step 3: $F = 10.154^2/9.197^2 = 1.2189$

Step 4: The critical values are $F_{0.05}$ with df = (30,24) or 1.94, and $F_{0.95}$
= $1/F_{0.05}$ with df =(24,30) = 1/1.89 = 0.53

Step 5: Since 0.53 < 1.2189 < 1.94, do not reject the null hypothesis.

Step 6: There is not sufficient evidence at the 0.10 significance level to
claim that the variation in anxiety test scores for patients seeing
videotapes showing progressive relaxation exercises is different
from that in patients seeing neutral videotapes.  In Excel, enter
=FDIST(1.2189,30,24) in any cell to obtain P(F >1.22) = 0.31217.
The P-value is 2(0.31217) = 0.62434.

**11.71** 1=Stinger, 2=Regular

Step 1: $H_0$: $\sigma_1 = \sigma_2$    $H_a$: $\sigma_1 < \sigma_2$

Step 2: $\alpha$ = 0.01

Step 3: $F = 0.410^2/0.894^2 = 0.210$

Step 4: The critical value is $F_{0.99}$ with df = (29,29) or 0.41 (1/2.42).

Step 5: Since 0.210 < 0.41, reject the null hypothesis.

Step 6: There is sufficient evidence at the 0.01 significance level to claim
that the standard deviation of ball velocity is less with the
Stinger tee than with the regular tee.  In Excel, enter
=1-FDIST(0.210,29,29) in any cell to obtain P(F < 0.210) = 0.000035.
The P-value is 0.000035.

**11.73** (a) **Using Minitab**, choose Stat ▶ Basic Statistics ▶ 2-Variances, click on
**Samples in different columns**, enter ITALIANS in the **First** box and
ETRUSCANS in the **Second** box, and click **OK**.  The results are

### Test for Equal Variances: ITALIANS, ETRUSCANS

95% Bonferroni confidence intervals for standard deviations

|          | N  | Lower   | StDev   | Upper   |
|----------|----|---------|---------|---------|
| ITALIANS | 70 | 4.82562 | 5.74995 | 7.08966 |
| ETRUSCANS| 84 | 5.08293 | 5.97051 | 7.21503 |

F-Test (Normal Distribution)
Test statistic = 0.93, p-value = 0.750

Levene's Test (Any Continuous Distribution)
Test statistic = 0.05, p-value = 0.826

Since the P-value for the F test = 0.750, which is larger than the
significance level 0.05, do not reject the null hypothesis.  The data
do not provide sufficient evidence to conclude that there is a
difference in the variation of skull measurements of the two
populations.

(b) Choose **Graph ▶ Probability Plot...**,select the **Single** version and click
**OK**.  Enter ITALIANS and ETRUSCANS in the **Graph** variables text box and
click on the Multiple graphs button.  Select **In separate panels of the
same graph** and click **OK**.  The result is

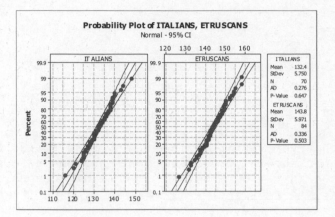

(c)   Yes.  Both plots are very close to linear with only one minor outlier
       at the low end.  The sample sizes are also large (70 and 84).

**11.75**   (a)   Although Minitab does not have a procedure for the comparison of
       two standard deviations, there is a macro on the WeissStats CD that
       will do it.  With that CD in Drive D, click in the **Sessions Win**dow
       and after the MTB> prompt (If there isn't one, select **Editor** form
       the Tool bar and check **Enable Commands.**), enter
       %d:\MINITAB_MACROS\2STDEV.MAC 'CONTROL' 'MONITOR' [There must be a space
       after **mac** and between the variable names.].  You will be asked a
       series of questions.  Type in your answer and hit the ENTER key
       each time.  We have underlined your responses.  This will produce
       the following dialogue and results.

```
MTB > %d:\MINITAB_MACROS\2STDEV.MAC 'CONTROL' 'MONITOR'
Executing from file: d:\MINITAB_MACROS\2STDEV.MAC

This macro performs a hypothesis test and/or obtains
a confidence interval for two population standard deviations.

Do you want to perform a hypothesis test (Y/N)?

Y

Enter 0, 1, or -1, respectively, for a two-tailed, right-tailed,
or left-tailed test.

DATA> -1

F-Test of sigma1 = sigma2 (vs<)

Row  Variable  n   StDev   F     P
  1  CONTROL   47  25.026  0.73  0.141
  2  MONITOR   46  29.374

Do you want a confidence interval (Y/N)?

N
```

The P-value = 0.141, which is larger than the significance level
0.05, so do not reject the null hypothesis.  The data do not provide
sufficient evidence that the variation in hemoglobin level is less
without the in-line blood gas and chemistry monitor.

(b)   Choose **Graph ▶ Probability Plot...**, select the **Single** version and click
       **OK.**  Enter <u>CONTROL</u> and <u>MONITOR</u> in the **Graph** variables text box and

click on the **Multiple graphs** button.  Select **In separate panels of the same graph** and click **OK**.  The result is

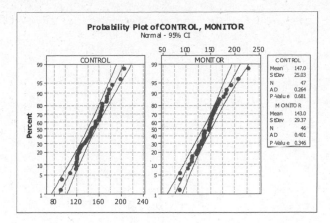

(c)  Yes.  Both probability plots are linear and there are no outliers.

**11.77**  Step 1:  df = (40,19).  $F_{0.05}$ = 2.03.  $F_{0.95}$ = $1/F_{0.05}$ for df = (19,40) = 1/1.85 = 0.54.

Step 2:  The 90% confidence interval for $\sigma_1/\sigma_2$ is

$$\frac{1}{\sqrt{F_{0.05}}}\cdot\frac{s_1}{s_2}\,to\,\frac{1}{\sqrt{F_{0.95}}}\cdot\frac{s_1}{s_2} = \frac{1}{\sqrt{2.03}}\cdot\frac{7.813}{5.286}\,to\,\frac{1}{\sqrt{0.54}}\cdot\frac{7.813}{5.286} = (1.037, 2.011)$$

**11.79**  Step 1:  df = (30,24).  $F_{0.05}$ is 1.94; $F_{0.95}$ = $1/F_{0.05}$ for df = (24,30) = 1/1.89 = 0.529.

Step 2:  The 90% confidence interval for $\sigma_1/\sigma_2$ is

$$\frac{1}{\sqrt{F_{0.05}}}\cdot\frac{s_1}{s_2}\,to\,\frac{1}{\sqrt{F_{0.95}}}\cdot\frac{s_1}{s_2} = \frac{1}{\sqrt{1.94}}\cdot\frac{10.154}{9.197}\,to\,\frac{1}{\sqrt{0.529}}\cdot\frac{10.154}{9.197} = (0.79, 1.52)$$

**11.81**  Step 1:  df = (29,29).  $F_{0.01}$ for df = (29,29) is 2.42;
For $F_{0.99}$ with df = (29,29), we use $1/F_{0.01}$ with df = (29,29).  Thus $F_{0.99}$ = 1/2.42 = 0.413

Step 2:  The 98% confidence interval for $\sigma_1/\sigma_2$ is

$$\frac{1}{\sqrt{F_{0.01}}}\cdot\frac{s_1}{s_2}\,to\,\frac{1}{\sqrt{F_{0.99}}}\cdot\frac{s_1}{s_2} = \frac{1}{\sqrt{2.42}}\cdot\frac{0.410}{0.894}\,to\,\frac{1}{\sqrt{0.413}}\cdot\frac{0.410}{0.894} = (0.295, 0.714)$$

### Review problems for Chapter 11

1.  Chi-square distribution ($\chi^2$)

2.  (a)  A $\chi^2$-curve is <u>right</u> skewed.

   (b)  A $\chi^2$-curve looks increasingly like a <u>normal</u> curve as the number of degrees of freedom becomes larger.

3.  The variable must be normally distributed.  That assumption is very important because the $\chi^2$ procedures are not robust to even moderate violations of the normality assumption.

4.  (a)  6.408        (b)    33.409        (c)    27.587        (d)    8.672
   (e)  7.564 and 30.191

5.  The F distribution is used when making inferences comparing two population standard deviations.

6.  (a)  An F-curve is <u>right</u>-skewed.

(b)  reciprocal, (5,14)

(c)  0

7.     Both variables must be normally distributed.  This is very important since the F procedures are not robust to violations of the normality assumption.

8.     (a)  7.01

(b)  $F_{0.99} = 1/F_{0.01}$ where $F_{0.01}$ has df = (8,4).
     Thus $F_{0.99} = 1/14.80 = 0.068$

(c)  3.84

(d)  The F value with 0.05 to its left = $F_{0.95} = 1/F_{0.05}$ where $F_{0.05}$ has df = (8,4).
     Thus $F_{0.95} = 1/6.04 = 0.166$

(e)  The F value with 0.025 to its left = $F_{0.975} = 1/F_{0.025}$ where $F_{0.025}$ has df = (8,4).
     Thus $F_{0.975} = 1/8.98 = 0.111$; $F_{0.025} = 5.05$

9.     (a)  Step 1: $H_0$: $\sigma = 16$    $H_a$: $\sigma \neq 16$
     Step 2: $\alpha = 0.10$

     Step 3: $\chi^2 = \dfrac{n-1}{\sigma_0^2} \cdot s^2 = \dfrac{24}{16^2} \cdot 15.006^2 = 21.111$

     Step 4: The critical values with n-1 = 24 degrees of freedom are 13.848 and 36.415.
     Step 5: Since 13.848 < 21.111 < 36.415, we do not reject $H_0$.
     Step 6: There is insufficient evidence at the 0.10 level to claim that σ for IQs measured on the Stanford revision of the Binet-Simon Intelligence Scale is different from 16.

(b)  Normality is crucial for the hypothesis test in (a) since the procedure is not robust to moderate deviations from the normality assumption.

10.    The 90% confidence interval for σ of IQs measured on the Stanford Revision of the Binet-Simon Intelligence Scale is

$$\left( \sqrt{\frac{n-1}{\chi_{0.05}^2}} \cdot s, \sqrt{\frac{n-1}{\chi_{0.95}^2}} \cdot s \right) = \left( \sqrt{\frac{24}{36.415}} \cdot 15.006, \sqrt{\frac{24}{13.848}} \cdot 15.006 \right) = (12.182, 19.755))$$

11.    (a)  F distribution with df = (14,19)

(b)  1=Runners, 2=Others

     Step 1: $H_0$: $\sigma_1 = \sigma_2$    $H_a$: $\sigma_1 < \sigma_2$
     Step 2: $\alpha = 0.01$
     Step 3: F = $1.798^2/6.606^2 = 0.074$
     Step 4: For df = (14,19), the critical value is $F_{0.99} = 1/F_{0.01} = 1/3.53 = 0.28$
     Step 5: Since 0.074 < 0.28, reject the null hypothesis.
     Step 6: There is sufficient evidence at the 0.01 significance level to claim that the variation in skinfold thickness among runners is less than that among others.

(c)  We are assuming that skinfold thickness is a normally distributed variable. This assumption can be checked by looking at a normal probability plot for each set of data.  If both plots are linear, the normality assumption is reasonable.

(d)  We also assume that the samples are independent.

12.    (a)  Using Minitab, choose **Graph ▶ Probability Plot**, select the **Single** version and click **OK**.  Enter <u>BMI</u> in the **Graph variables** text box, and click **OK**.  Choose **Graph ▶ Boxplot**, select the **One-y Simple** version and click **OK**.  Enter <u>BMI</u> in the **Graph variables** text box, and click **OK**.

Choose **Graph ▶ Histogram**, select the **Simple** version and click **OK**.
Enter <u>BMI</u> in the **Graph variables** text box, and click **OK**. The results
are

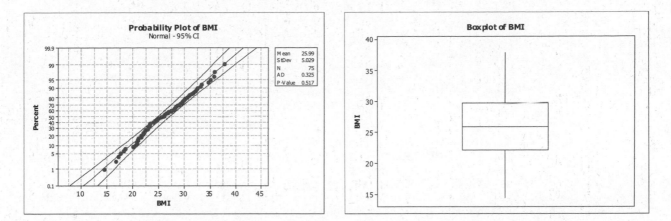

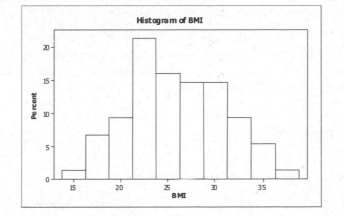

(b) Yes. The probability plot is nearly linear and there are no outliers.
(c) In Minitab, click on the Sessions Window. If there is no MTB> prompt
showing, click on the Editor pull-down menu and on Enable commands.
After the Mtb> prompt( and with the Weiss CD in Drive D), enter
%D:\Minitab_Macros\1stdev.mac 'BMI' and press ENTER. We have
underlined your responses to the questions that appear in the following
dialog.

```
MTB > %d:\minitab_macros\1stdev.mac 'bmi'
Executing from file: d:\minitab_macros\1stdev.mac

This macro performs a hypothesis test and/or obtains
a confidence interval for one population standard deviation.

Do you want to perform a hypothesis test (Y/N)?

Y

Enter the null hypothesis population standard deviation.

DATA> 5

Enter 0, 1, or -1, respectively, for a two-tailed, right-tailed,
or left-tailed test.

DATA> 0
```

```
Test of sigma =      5.00000 vs sigma not =      5.00000


Row  Variable   n   StDev  Chi-Sq  P
  1  BMI        75  5.029  74.871  0.900

Do you want a confidence interval (Y/N)?

Y

Enter the confidence level, as a percentage.

DATA> 95

Row  Variable   n   StDev  Level  CI for sigma
  1  BMI        75  5.029  95.0%  (4.333,5.994)
```

The P-value = 0.900, which is greater than the significance level 0.05, so do
not reject the null hypothesis.  The data do not provide sufficient evidence
that the standard deviation is not 5.0.

13.  The 95% confidence interval was obtained in Problem 12 as (4.333, 5.994).

14.  We used Minitab to obtain a histogram of each set of data.  With the data in

columns named GSOD and PSOD, choose **Graph ▶ Histogram...**, select the **Simple**
version, enter GSOD and PSOD in the **Graph variables** text box, click on the
**Multiple graphs** button and select **On separate graphs, Same Y,** and **Same X,
including bins,** and click **OK** twice.  The results are

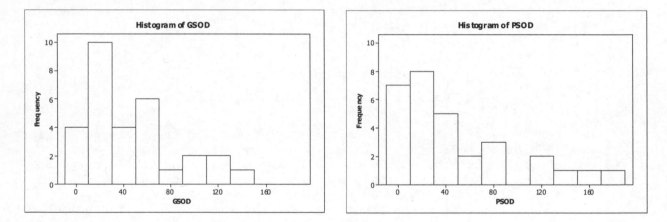

It is clear from both histograms that the data are very right-skewed.  Since
the F-test is very sensitive to non-normality, the F-test should not be used
with these data.

15.  First, check to ensure that the data distributions are reasonably normal.
With the data in columns in Minitab named BRAND A and BRAND B, choose **Graph**

**▶ Probability Plot,** select the **Single** version and click **OK.**  Enter BRAND A
and BRAND B in the **Graph variables** text box, and click **OK.**  The results are

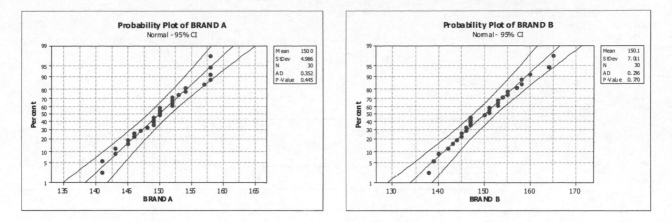

Both plots indicate that normality is a reasonable assumption. Therefore we can proceed with an F test to determine whether BRAND A has a smaller standard deviation than BRAND B. After an MTB> prompt, enter %D:\Minitab_Macros\2stdev.mac 'BRAND A' 'BRAND B' and press ENTER. The dialogue follows with your responses underlined.

```
MTB > %D:\Minitab_Macros\2stdev.mac 'BRAND A'   'BRAND B'
Executing from file: D:\Minitab_Macros\2stdev.mac

This macro performs a hypothesis test and/or obtains
a confidence interval for two population standard deviations.

Do you want to perform a hypothesis test (Y/N)?

Y

Enter 0, 1, or -1, respectively, for a two-tailed, right-tailed,
or left-tailed test.

DATA> -1

F-Test of sigma1 = sigma2 (vs <)

Row   Variable   n    StDev   F      P
  1   BRAND A    30   4.986   0.51   0.036
  2   BRAND B    30   7.011

Do you want a confidence interval (Y/N)?

N
```

The P-value for the test is 0.036, which is less than the significance level of 0.05. Therefore, we reject the null hypothesis of equal standard deviations. There is sufficient evidence at the 0.05 level that BRAND A has a smaller standard deviation than BRAND B and therefore has a more consistent popping time.

**Exercises 12.1**

**12.1** Answers will vary.

**12.3** A population proportion *p* is a parameter since it is a descriptive measure for a population. A sample proportion $\hat{p}$ is a statistic since it is a descriptive measure for a sample.

**12.5** (a)  p = 2/5 = 0.4

(b)

| Sample | Number of females X | Sample proportion $\hat{p}$ |
|--------|---------------------|-----------------------------|
| J, G | 1 | 0.5 |
| J, P | 0 | 0.0 |
| J, C | 0 | 0.0 |
| J, F | 1 | 0.5 |
| G, P | 1 | 0.5 |
| G, C | 1 | 0.5 |
| G, F | 2 | 1.0 |
| P, C | 0 | 0.0 |
| P, F | 1 | 0.5 |
| C, F | 1 | 0.5 |

(c)  The population proportion is marked by the vertical line below.

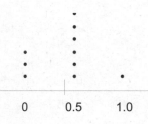

(d)  $\mu_{\hat{p}} = \sum \hat{p}/10 = 4.0/10 = 0.4$

(e)  The answers to (a) and (d) are the same. $\hat{p}$ is a sample mean. The mean of the sampling distribution of $\hat{p}$ is the same as the mean of the population, which is p.

**12.7** (b)

| Sample | Number of females x | Sample proportion $\hat{p}$ |
|---|---|---|
| J.P.C | 0 | 0 |
| J.P.G | 1 | 1/3 |
| J.P.F | 1 | 1/3 |
| J.C.G | 1 | 1/3 |
| J.C.F | 1 | 1/3 |
| J.G.F | 2 | 2/3 |
| P.C.G | 1 | 1/3 |
| P.C.F | 1 | 1/3 |
| P.G.F | 2 | 2/3 |
| C.G.F | 2 | 2/3 |

(c)  The population proportion is marked by the vertical line below.

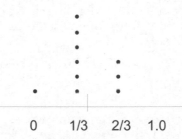

$$\begin{array}{cccc} 0 & 1/3 & 2/3 & 1.0 \end{array}$$

(d)   $\mu_{\hat{p}} = \sum \hat{p}/10 = (12/3)/10 = 0.4$

(e)  The answers to (a) and (d) are the same. $\hat{p}$ is a sample mean.  The mean
of the sampling distribution of $\hat{p}$ is the same as the mean of the
population, which is p.

**12.9** (b)

| Sample | Number of females X | Sample proportion $\hat{p}$ |
|---|---|---|
| J,P,C,G,F | 2 | 0.4 |

(c)  The population proportion is marked by the vertical line below.

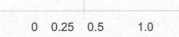

$$\begin{array}{cccc} 0 & 0.25 & 0.5 & 1.0 \end{array}$$

(d)  $\mu_{\hat{p}} = \sum \hat{p}/1 = 0.4/1 = 0.4$

(e)  The answers to (a) and (d) are the same. $\hat{p}$ is a sample mean.  The mean of the sampling distribution of $\hat{p}$ is the same as the mean of the population, which is p.

**12.11** (a)  The population consists of all #1 NBA draft picks since 1966.

(b)  The specified attribute is being a center.

(c)  The 45% is a population proportion since all of the #1 picks are known. There is no need to sample this group.

**12.13** (a)  $\hat{p}$ = 0.79; $z_{0.005}$ = 2.575; the margin of error is

$$E = z_{0.005} \cdot \sqrt{\hat{p}(1-\hat{p})/n} = 2.575\sqrt{0.79(0.21)/21355} = 0.00718$$

(b)  The margin of error will be smaller for a 90% confidence interval. Specifically, 2.575 will be replaced by 1.645 in the formula for E and everything else stays the same.  More generally speaking, in order to have a higher level of confidence in an interval, one needs to have a wider interval.

**12.15** (a)  0.4  (b)    0.2   (c)    0.5

(d)  For (a), $0.4 < \hat{p} < 0.6$; For (b), $0.2 < \hat{p} < 0.8$; For (c), none

**12.17** (a)  $\hat{p} = 8/40 = 0.2$

(b)  x = 8 and n - x = 32.  Both are at least 5, so the one-proportion z-interval procedure is appropriate.

(c)  The 95% confidence interval is

$$\hat{p} \pm z_{\alpha/2}\sqrt{\hat{p}(1-\hat{p})/n} = 0.2 \pm 1.96\sqrt{0.2(1-0.2))/40} = (0.176, 0.324)$$

**12.19** (a)  $\hat{p} = 35/50 = 0.70$

(b)  x = 35 and n - x = 15.  Both are at least 5, so the one-proportion z-interval procedure is appropriate.

(c)  The 99% confidence interval is

$$\hat{p} \pm z_{\alpha/2}\sqrt{\hat{p}(1-\hat{p})/n} = 0.70 \pm 2.575\sqrt{0.70(1-0.70))/50} = (0.533, 0.867)$$

**12.21** (a)  $\hat{p} = 16/20 = 0.8$

(b)  x = 16 and n - x = 4.  n - x is less than  5, so the one-proportion z-interval procedure is not appropriate.

**12.23** $\hat{p} = 410/603 = 0.68$; The 95% confidence interval is

$$\hat{p} \pm z_{\alpha/2}\sqrt{\hat{p}(1-\hat{p})/n} = 0.68 \pm 1.96\sqrt{0.68(1-0.68))/603} = (0.643, 0.717)$$

**12.25** n = 500, x = 38, and n - x = 462.  Both x and n - x are at least 5.

$\hat{p}$ = x/n = 38/500 = 0.076; $z_{\alpha/2}$ = $z_{0.025}$ = 1.96

(a)
$$0.076 - 1.96\sqrt{0.076(1-0.076)/500} \ to \ 0.076 + 1.96\sqrt{0.076(1-0.076)/500}$$
$$0.0528 \ to \ 0.0992$$

(b)  We can be 95% confident that the proportion of U.S. asthmatics who are allergic to sulfites is somewhere between 0.0528 and 0.0992.

**12.27** n = 1000, x = 800, and n - x = 200.  Both x and n - x are at least 5.

$\hat{p}$ = x/n = 800/1000 = 0.80; $z_{\alpha/2}$ = $z_{0.005}$ = 2.575

(a)
$$0.80 - 2.575\sqrt{0.80(1-0.80)/1000} \ to \ 0.80 + 2.575\sqrt{0.80(1-0.80)/1000}$$
$$0.767 \ to \ 0.833$$

(b)  We can be 99% confident that the percentage of all registered voters who favor the creation of standards on CAFO pollution is somewhere between 76.7% and 83.3%.

**12.29** Here, n = 100, x = 96, and n - x = 4.   To use Procedure 12.1, both x and

n - x must be at least 5.  Since n - x < 5, Procedure 12.1 should not have been used.

**12.31**  $\hat{p}$ = 647/1043 = 0.620, E = 0.029

$\hat{p}$ - E     to     $\hat{p}$ + E

0.620 - 0.029 to 0.620 + 0.029      In percentage terms, this is
  59.1%      to      64.9%

**12.33** From Exercise 12.25, $\hat{p}$ = 0.076; $z_{\alpha/2}$ = 1.96; n = 500

(a)  $E = z_{\alpha/2}\sqrt{\hat{p}(1-\hat{p})/n} = 1.96\sqrt{0.076(1-0.076)/500} = 0.0232$

(b)   E = 0.01; $z_{\alpha/2}$ = $z_{0.025}$ = 1.96;

$$n = \hat{p}(1-\hat{p})\frac{z_{\alpha/2}^2}{E^2} = 0.5(1-0.5)\frac{1.96^2}{0.01^2} = 9604$$

(c)   n = 9604; $\hat{p}$ = 0.071; $z_{\alpha/2}$ = $z_{0.025}$ = 1.96

$$0.071 - 1.96\sqrt{0.071(1-0.071)/9604} \text{ to } 0.3192 + 1.96\sqrt{0.071(1-0.071)/9604}$$

$$0.0659 \text{ to } 0.0761$$

(d)  The margin of error for the estimate is 0.0051.  As expected, this is less than the required margin of error of 0.01 specified in part (b).

(e)  $\hat{p}_g$ = 0.10; $z_{\alpha/2}$ = $z_{0.025}$ = 1.96

part (b)      E = 0.01, $z_{0.025}$ = 1.96; sample size is:

$$n = \hat{p}(1-\hat{p})\frac{z_{\alpha/2}^2}{E^2} = 0.10(1-0.10)\frac{1.96^2}{0.01^2} = 3457.44 \rightarrow 3458$$

Thus the required sample size is n = 3458.
part (c)

$$0.071 - 1.96\sqrt{0.071(1-0.071)/3458} \text{ to } 0.3192 + 1.96\sqrt{0.071(1-0.071)/3458}$$

$$0.0624 \text{ to } 0.0796$$

part (d) The margin of error is 0.0086.  As expected, this is less than what was specified in part (b).

(f)  By employing the guess for $\hat{p}$ in part (d) we can reduce the required sample size by more than 6000 (from 9604 to 3458), saving considerable time and money.  Moreover, the margin of error only rises from 0.0051 to 0.0086.  The risk of using the guess 0.10 for $\hat{p}$ is that if the actual value of $\hat{p}$ turns out to be between 0.10 and 0.90, then the achieved margin of error will exceed the specified 0.01.

**12.35** (a)  The confidence interval from Exercise 12.27 was (76.7%, 83.3%).  To find the error, take the width of the confidence interval and divide by 2.   E = (83.3% - 76.7%)/2 = 3.3%.

(b)  E = 0.015; $z_{\alpha/2}$ = $z_{0.005}$ = 2.575;

$$n = \hat{p}(1-\hat{p})\frac{z_{\alpha/2}^2}{E^2} = 0.5(1-0.5)\frac{2.575^2}{0.015^2} = 7367.36 \rightarrow 7368$$

(c)  n = 7368;   = 0.822; $z_{\alpha/2}$ = $z_{0.005}$ = 2.575

$$0.822 - 2.575\sqrt{0.822(1-0.822)/7368} \;\; to \;\; 0.822 + 2.575\sqrt{0.822(1-0.822)/7368}$$
$$0.811 \; to \; 0.833$$

(d)  The margin of error for the estimate is 0.011, which is less than the 0.015 required in part (b).

(e)  $\hat{p}_g$ is between 0.75 and .85;   $z_{\alpha/2}$ = $z_{0.005}$ = 2.575

part (b) E = 0.015, $z_{0.005}$ = 2.575; sample size is:

$$n = \hat{p}(1-\hat{p})\frac{z_{\alpha/2}^2}{E^2} = 0.75(1-0.75)\frac{2.575^2}{0.015^2} = 5525.52 \rightarrow 5526$$

Thus the required sample size is n = 5526.
part (c)

$$0.822 - 2.575\sqrt{0.822(1-0.822)/5526} \;\; to \;\; 0.822 + 2.575\sqrt{0.822(1-0.822)/5526}$$
$$0.809 \; to \; 0.835$$

part (d)
 The margin of error is 0.013, which is less than the 0.015 specified in part (b).

(f)  By employing the guess for $\hat{p}$ in part (d) we can reduce the required sample size by 1842, from 7368 to 5526, saving considerable time and money.  Moreover, the margin of error only rises from 0.011 to 0.013. The risk of using the guess 0.75 for $\hat{p}$ is that if the actual value of $\hat{p}$ turns out to be between 0.25 and 0.75, then the achieved margin of error will exceed the specified 0.015.

**12.37** (a)  E = 0.01; $z_{\alpha/2}$ = $z_{0.025}$ = 1.96;

$$n = \hat{p}(1-\hat{p})\frac{z_{\alpha/2}^2}{E^2} = 0.5(1-0.5)\frac{1.96^2}{0.01^2} = 9604$$

(b)  $\hat{p}_g$ is between 0.5% and 4.9%;   $z_{\alpha/2}$ = $z_{0.025}$ = 1.96; E = 0.01

$$n = \hat{p}(1-\hat{p})\frac{z_{\alpha/2}^2}{E^2} = 0.049(1-0.049)\frac{1.96^2}{0.01^2} = 1790.15 \rightarrow 1791$$

Thus the required sample size is n = 1,791.

(c)  The sample size has decreased from 9604 to 1791.

(d)  If the actual value of $\hat{p}$ turns out to be between 0.049 and 0.951, then the achieved margin of error will exceed the specified error of 0.01.

**12.39** $\hat{p}$= 444/1010 = 0.440; $z_{0.025}$ = 1.96; the 95% confidence interval is

$$0.440 - 1.96\sqrt{0.440(1-0.440)/\;1010} \;\; to \;\; 0.440 + 1.96\sqrt{0.440(1-0.440)/\;1010}$$
$$0.409 \; to \; 0.470$$

We can be 95% confident that the proportion of U.S. adults who approved of the way that President George W. Bush was doing his job was somewhere between 0.409 and 0.470.

**12.41** $\hat{p}$= 276/1063 = 0.260; $z_{0.05}$ = 1.645; the 90% confidence interval is

$$0.260 - 1.645\sqrt{0.260(1-0.260)/\ 1063}\ to\ 0.260 + 1.645\sqrt{0.260(1-0.260)/\ 1063}$$

$$0.238\ to\ 0.282$$

We can be 90% confident that the proportion of U.S. adults who would purchase or lease a new car from a manufacturer that had declared bankruptcy is somewhere between 0.238 and 0.282.

**12.43** The Central Limit Theorem

**12.45** The sample size is directly proportional to the quantity p(1 - p), but the final margin of error is directly proportional to $\hat{p}$(1 - $\hat{p}$). The quantity

p(1 - p) takes on its maximum value when p = 0.5. If $\hat{p}$ is closer to .5 than

is p, then $\hat{p}$(1 - $\hat{p}$) will be greater than p(1 - p) and the confidence

interval using $\hat{p}$(1 - $\hat{p}$) will be larger than the desired one associated with

p(1 - p).

**Exercises 12.2**

**12.47** Procedure 12.2 is a special case of the one-mean z-test for a population mean. If y is a variable taking on the values 0 for a failure and 1 for a success, then $\bar{y}=(\sum y)/\ n=x/n=\hat{p}$. Applying Procedure 9.1 (z-test for a population mean) is the same as Procedure 12.2.

**12.49** (a) The sample proportion is $\hat{p}$ = x/n = 8/40 = 0.2.

(b) $np_o$ = 40(0.3) = 12; n(1 - $p_0$) = 40(1-0.30) = 28
Since both are at least 5, we can employ Procedure 12.2.

(c) $z = \dfrac{0.2-0.3}{\sqrt{.03(1-0.3)/40}} = -1.38$; For $\alpha$ = 0.10, the critical value is

$z_u$ = -1.28. Since -1.38 < -1.28, reject $H_0$.

**12.51** (a) The sample proportion is $\hat{p}$ = x/n = 35/50 = 0.70.

(b) $np_o$ = 50(0.6) = 30; n(1 - $p_0$) = 50(1-0.60) = 20
Since both are at least 5, we can employ Procedure 12.2.

(c) $z = \dfrac{0.70-0.60}{\sqrt{0.6(1-0.6)/50}} = 1.44$; For $\alpha$ = 0.05, the critical value is

$z_\alpha$ = -1.645. Since 1.44 < 1.645, do not reject $H_0$.

**12.53** (a) The sample proportion is $\hat{p}$ = x/n = 16/20 = 0.80.

(b) $np_o$ = 50(0.7) = 35; n(1 - $p_0$) = 50(1-0.70) = 15
Since both are at least 5, we can employ Procedure 12.2.

(c) $z = \dfrac{0.80-0.70}{\sqrt{0.7(1-0.7)/20}} = 0.98$; For $\alpha$ = 0.05, the critical values are

$z_\alpha$ = ±1.96. Since -1.96 < 0.98 < 1.96, do not reject $H_0$

**12.55** (a) The sample proportion is $\hat{p}$ = x/n = 459/850 = 0.540.

(b) $\alpha$ = 0.05, $p_0$ = 0.50
$np_o$ = 850(0.50) = 425; n(1 - $p_0$) = 850(1-0.50) = 425
Since both are at least 5, we can employ Procedure 12.2.
Step 1: $H_0$: p = 0.50, $H_a$: p > 0.50
Step 2: $\alpha$ = 0.05

Step 3:  $z = \dfrac{0.540 - 0.500}{\sqrt{0.5(1-0.5)/850}} = 2.33$

Step 4:  Since $\alpha$ = 0.05, the critical value is $z_\alpha$ = 1.645

Step 5:  Since 2.33 > 1.645, reject $H_0$.  Note, for the p-value approach, $P(z > 2.33) = 0.0099$; so p-value < $\alpha$.  Thus we reject $H_0$.

Step 6:  The test results are statistically significant at the 5% level; that is, at the 5% significance level, the data do provide sufficient evidence to conclude that a majority of Generation Y Web users use the Internet to download music.

**12.57** The sample proportion is $\hat{p}$ = x/n = 205/1283 = 0.1598.

$\alpha$ = 0.10, $p_0$ = 0.136
$np_o$ = 1283(0.136) = 174.5; $n(1 - p_0)$ = 1283(1 - 0.136) = 1108.5
Since both are at least 5, we can employ Procedure 12.2.

Step 1:  $H_0$: p = 0.136, $H_a$: p $\neq$ 0.136
Step 2:  $\alpha$ = 0.10

Step 3:  $z = \dfrac{0.1598 - 0.1360}{\sqrt{0.136(1-0.136)/1283}} = 2.49$

Step 4:  Since $\alpha$ = 0.10, the critical values are $z_\alpha$ = $\pm$1.645

Step 5:  Since 2.49 > 1.645, reject $H_0$.  Note, for the p-value approach, $2P(Z > 2.49) = 2(0.0064) = 0.0128$; so p-value < $\alpha$.  Thus we reject $H_0$.

Step 6:  The test results are statistically significant at the 10% level; that is, at the 10% significance level, the data provide sufficient evidence to conclude that the percentage of 18-25 year-olds who currently use marijuana or hashish has changed from the 2000 percentage of 13.6%.

**12.59** (a)  The sample proportion is $\hat{p}$ = 0.650.

$\alpha$ = 0.05, $p_0$ = 0.72
$np_o$ = 1003(0.72) = 722.2; $n(1 - p_0)$ = 1003(1 - 0.72) = 280.8
Since both are at least 5, we can employ Procedure 12.2.

Step 1:  $H_0$: p = 0.72, $H_a$: p < 0.72
Step 2:  $\alpha$ = 0.05

Step 3:  $z = \dfrac{0.65 - 0.72}{\sqrt{0.72(1-0.72)/1003}} = -4.94$

Step 4:  Since $\alpha$ = 0.05, the critical value is $z_\alpha$ = -1.645

Step 5:  Since -4.94 < -1.645, reject $H_0$.  Note, for the p-value approach, $P(Z < -4.94) = 0.0000$; so p-value < $\alpha$.  Thus we reject $H_0$.

Step 6:  The test results are statistically significant at the 5% level; that is, at the 5% significance level, the data do provide sufficient evidence to conclude that the percentage of Americans who approve of labor unions has decreased since 1936.

(b)  The sample proportion is $\hat{p}$ = 0.650.

$\alpha$ = 0.05, $p_0$ = 2/3 = 0.667
$np_o$ = 1003(0.67) = 672; $n(1 - p_0)$ = 1003(1 - 0.67) = 331
Since both are at least 5, we can employ Procedure 12.2.

Step 1:   $H_0$: p = 0.67, $H_a$: p < 0.67
Step 2:   $\alpha$ = 0.05

Step 3:   $z = \dfrac{0.65 - 0.67}{\sqrt{0.67(1 - 0.67)/1003}} = -1.35$

Step 4:   Since $\alpha$ = 0.05, the critical value is $z_\alpha$ = -1.645

Step 5:   Since -1.35 > -1.645, do not reject $H_0$.  Note, for the p-value approach, P(Z < -1.35) = 0.0885; so p-value > $\alpha$ .  Thus we do not reject $H_0$.

Step 6:   The test results are not statistically significant at the 5% level; that is, at the 5% significance level, the data do not provide sufficient evidence to conclude that the percentage of Americans has decreased since 1963.

**12.61** n = 609, $p_0$ = 0.5, x =341, $\hat{p}$ = 341/609 = 0.560

$np_0$ = 304.5, $n(1 - p_0)$ = 304.5
Since both are at least 5, we can employ Procedure 12.2.

Step 1:   $H_0$: p = 0.5, $H_a$: p > 0.5
Step 2:   $\alpha$ = 0.01

Step 3:   $z = \dfrac{0.560 - 0.500}{\sqrt{0.500(1 - 0.500)/609}} = 2.96$

Step 4:   Since $\alpha$ = 0.01, the critical value is $z_\alpha$ = 2.33

Step 5:   Since 2.96 > 2.33, reject $H_0$. Note for the p-value approach, P(z > 2.96) = 0.0015; so the p-value < $\alpha$ .  Thus reject $H_0$.

Step 6:   The test results are statistically significant at the 1% level; that is, at the 1% significance level, the data do provide sufficient evidence to conclude that the headline is justified.

**12.63** n = 11, $p_0$ = 0.5, x =5, $\hat{p}$ = 5/11 = 0.455

$np_0$ = 5.5, $n(1 - p_0)$ = 5.5
Since both are at least 5, we can employ Procedure 12.2.
Step 1:   $H_0$: p = 0.5, $H_a$: p < 0.5
Step 2:   $\alpha$ = 0.05

Step 3:   $z = \dfrac{0.455 - 0.5}{\sqrt{0.5(1 - 0.5)/11}} = -0.30$

Step 4:   Since $\alpha$ = 0.05, the critical value is $-z_\alpha = -1.645$

Step 5:   Since -0.30 > -1.645, do not reject $H_0$. Note for the p-value approach, P(z < -0.30) = 0.3821; so the p-value = 0.3821 > $\alpha$ .  Thus do not reject $H_0$.

Step 6:   The test results are not statistically significant at the 5% level; that is, at the 5% significance level, the data do not provide sufficient evidence to conclude that less than half of all young children drowning in Victorian dams located on farms are girls.

**Exercises 12.3**

**12.65** We need to decide whether the difference between two sample proportions can reasonably be attributed to sampling error or are the population proportions really different.

**12.67** (a)  Using sunscreen before going out in the sun
       (b)  Teen-age girls and teen-age boys

(c) The two proportions are sample proportions. The reference is specifically to those teen-age girls and boys who were surveyed

**12.69** (a) The parameters are $p_1$ and $p_2$. The rest are statistics.

(b) The fixed numbers are $p_1$ and $p_2$. The rest are variables.

**12.71** (a) $\hat{p}_1 = x_1 / n_1 = 18/40 = 0.45$; $\hat{p}_2 = x_2 / n_2 = 30/40 = 0.75$;

$\hat{p}_p = (x_1 + x_2)/(n_1 + n_2) = 48/80 = 0.6$

(b) $x_1 = 18, n_1 - x_1 = 22, x_2 = 30, n_2 - x_2 = 10$.

All are at least 5, so z-procedures are appropriate.

(c) $z = \dfrac{0.45 - 0.75}{\sqrt{0.6(1 - 0.6)[(1/40) + (1/40)]}} = -2.74$

$\alpha = 0.10$, critical value is $-1.28$; since $-2.74 < -1.28$, reject $H_0$.

(d) The 80% confidence interval is

$$\hat{p}_1 - \hat{p}_2 \pm z_{\alpha/2} \sqrt{\hat{p}_1(1 - \hat{p}_1)/n_1 + \hat{p}_2(1 - \hat{p}_2)/n_2}$$

$$(0.45 - 0.75) \pm 1.28 \sqrt{0.45(1 - 0.45)/40 + 0.75(1 - 0.75)/40} = (-0.43, \ -0.17)$$

**12.73** (a) $\hat{p}_1 = x_1 / n_1 = 15/20 = 0.75$; $\hat{p}_2 = x_2 / n_2 = 18/30 = 0.60$;

$\hat{p}_p = (x_1 + x_2)/(n_1 + n_2) = 33/50 = 0.66$

(b) $x_1 = 15, n_1 - x_1 = 5, x_2 = 18, n_2 - x_2 = 12$.

All are at least 5, so z-procedures are appropriate.

(c) $z = \dfrac{0.75 - 0.60}{\sqrt{0.66(1 - 0.66)[(1/20) + (1/30)]}} = 1.10$

$\alpha = 0.05$, critical value is $1.645$; since $1.10 < 1.645$, do not reject $H_0$.

(d) The 90% confidence interval is

$$\hat{p}_1 - \hat{p}_2 \pm z_{\alpha/2} \sqrt{\hat{p}_1(1 - \hat{p}_1)/n_1 + \hat{p}_2(1 - \hat{p}_2)/n_2}$$

$$(0.75 - 0.60) \pm 1.645 \sqrt{0.75(1 - 0.75)/20 + 0.60(1 - 0.60)/30} = (-0.067, \ 0.367)$$

**12.75** (a) $\hat{p}_1 = x_1 / n_1 = 30/80 = 0.375$; $\hat{p}_2 = x_2 / n_2 = 15/20 = 0.750$;

$\hat{p}_p = (x_1 + x_2)/(n_1 + n_2) = 45/100 = 0.45$

(b) $x_1 = 30, n_1 - x_1 = 50, x_2 = 15, n_2 - x_2 = 5$.

All are at least 5, so z-procedures are appropriate.

(c) $z = \dfrac{0.375 - 0.750}{\sqrt{0.45(1 - 0.45)[(1/80) + (1/20)]}} = -3.02$

$\alpha = 0.05$, critical values are $\pm 1.96$; since $-3.02 < -1.96$, reject $H_0$.

(d) The 95% confidence interval is

$$\hat{p}_1 - \hat{p}_2 \pm z_{\alpha/2} \sqrt{\hat{p}_1(1 - \hat{p}_1)/n_1 + \hat{p}_2(1 - \hat{p}_2)/n_2}$$

$$(0.375 - 0.750) \pm 1.96 \sqrt{0.375(1 - 0.375)/80 + 0.750(1 - 0.750)/20}$$

$$= (-0.592, \ -0.158)$$

**12.77** (a) Population 1: Women who took multivitamins containing folic acid

$$\hat{p}_1 = 35/2701 = 0.01296$$

Population 2: Women who received only trace elements

$$\hat{p}_2 = 47/2052 = 0.02290$$

$$\hat{p}_p = (35 + 47)/(2,701 + 2,052) = 0.01725$$

Step 1:  $H_0$: $p_1 = p_2$,   $H_a$: $p_1 < p_2$
Step 2:  $\alpha = 0.01$

Step 3:  $z = \dfrac{0.01296 - 0.0220}{\sqrt{0.01725(1 - 0.01725)[(1/2701) + (1/2052)]}} = -2.61$

Step 4:  Since $\alpha = 0.01$, the critical value is $-z_\alpha = -2.33$

Step 5:  Since -2.61 < -2.33, reject $H_0$. Note: For the P-value approach, $P(z < -2.61) = 0.0045$; so P-value $< \alpha$. Thus, we reject $H_0$.

Step 6:  The test results are significant at the 1% level; that is, at the 1% significance level, the data do provide sufficient evidence to conclude that the women who take folic acid are at lesser risk of having children with major birth defects than those women who do not.

(b)  This is a designed experiment. The researchers decided which women would take daily multivitamins.

(c)  Yes. By using the basic principles of design (control, randomization, and replication) the doctors can conclude that the reduction in the rates of major birth defects in the folic acid group is likely caused by the folic acid.

**12.79**  Population 1: Drivers of age 25-34, $\hat{p}_1 = 270/1000 = 0.270$

Population 2: Drivers of age 45-64, $\hat{p}_2 = 330/1100 = 0.300$

$$\hat{p}_p = (270 + 330)/(1000 + 1100) = 0.286$$

Step 1:  $H_0$: $p_1 = p_2$,   $H_a$: $p_1 \neq p_2$
Step 2:  $\alpha = 0.10$

Step 3:  $z = \dfrac{0.270 - 0.300}{\sqrt{0.286(1 - 0.286)[(1/1000) + (1/1000)]}} = -1.52$

Step 4:  Since $\alpha = 0.10$, the critical values is $\pm z_\alpha = \pm1.645$

Step 5:  Since $-1.645 < -1.52 < 1.645$, do not reject $H_0$. Note: For the P-value approach, $2P(z < -1.52) = 0.1286$; so P-value $> \alpha$. Thus, we do not reject $H_0$.

Step 6:  The test results are not significant at the 10% level; that is, at the 10% significance level, the data do not provide sufficient evidence to conclude that there is a difference in seat-belt usage between drivers 25-34 years old and those 45-64 years old.

**12.81**  Population 1: Bachelors degree, $\hat{p}_1 = 386/750 = 0.5147$

Population 2: Graduate degree, $\hat{p}_2 = 237/500 = 0.4740$

$$\hat{p}_p = (386 + 237)/(750 + 500) = 0.4984$$

(a)  The assumptions for using the two-sample z-test are simple random samples, independent samples, and $x_1$, $n_1 - x_1$, $x_2$, and $n_2 - x_2$ must all be greater than or equal to 5.

(b)  Step 1:    $H_0$: $p_1 = p_2$,   $H_a$: $p_1 > p_2$
     Step 2:    $\alpha = 0.05$

     Step 3:    $z = \dfrac{0.5147 - 0.4740}{\sqrt{0.4984(1-0.4984)[(1/750)+(1/500)]}} = 1.41$

     Step 4:    Since $\alpha = 0.05$, the critical value is $z_\alpha = 1.645$

     Step 5:    Since $1.41 < 1.645$, do not reject $H_0$. Note:  For the P-value
                approach, $P(z > 1.41) = 0.0793$; so P-value $> \alpha$.  Thus, we do
                not reject $H_0$.
     Step 6:    The test results are not significant at the 5% level; that is,
                at the 5% significance level, the data do not provide
                sufficient evidence to conclude that a higher percentage of
                adults with Bachelors degrees are overweight than of adults
                with graduate degrees.

(c)  At the 10% significance level, the P-value of 0.0793 is less than the
     significance level, so we reject the null hypothesis.  The data do
     provide sufficient evidence that a higher percentage of adults with
     Bachelors degrees are overweight than of adults with graduate degrees.

**12.83** From Exercise 12.77, the 99% confidence interval is

$$(0.012958 - 0.022904) \pm 2.33\sqrt{\frac{0.012958(1-0.012958)}{2701} + \frac{0.022904(1-0.022904)}{2052}}$$

$$-0.009946 \pm 0.009215 \ or \ -0.019161 \ to \ -0.000731$$

We can be 98% confident that the difference $p_1 - p_2$ between the rates of
major birth defects for babies born to women who have taken folic acid and
those born to women who have not taken folic acid is somewhere between -
0.019161 and -0.000731.

**12.85** From Exercise 12.79, the 90% confidence interval is

$$\hat{p}_1 - \hat{p}_2 \pm z_{\alpha/2}\sqrt{\hat{p}_1(1-\hat{p}_1)/n_1 + \hat{p}_2(1-\hat{p}_2)/n_2}$$

$$(0.270 - 0.300) \pm 1.645\sqrt{0.270(1-0.270)/1000 + 0.300(1-0.300)/100}$$

$$= (-0.0624, \ 0.0024)$$

We can be 90% confident that the difference between proportions of seat belt
users for drivers in the age groups 25-34 years and 45-64 years is somewhere
between -0.0624- and 0.0024.

**12.87** (a)   From Exercise 12.81, the 90% confidence interval is

$$\hat{p}_1 - \hat{p}_2 \pm z_{\alpha/2}\sqrt{\hat{p}_1(1-\hat{p}_1)/n_1 + \hat{p}_2(1-\hat{p}_2)/n_2}$$

$$(0.515 - 0.474) \pm 1.645\sqrt{0.515(1-0.515)/750 + 0.474(1-0.474)/500}$$

$$= (-0.007, \ 0.088)$$

(b)   The 80% confidence interval is

$$\hat{p}_1 - \hat{p}_2 \pm z_{\alpha/2}\sqrt{\hat{p}_1(1-\hat{p}_1)/n_1 + \hat{p}_2(1-\hat{p}_2)/n_2}$$

$$(0.515 - 0.474) \pm 1.28\sqrt{0.515(1-0.515)/750 + 0.474(1-0.474)/500}$$

$$= (0.004, \ 0.078)$$

**12.89** (a)  Using Minitab, choose **Stat ▶ Basic statistics ▶ 2 Proportions...**,
              select the **Summarized data** option button, click in the **Trials** text box
              for **First sample** and enter 300, click in the **Successes** text box for

**First sample** and type <u>219</u>, click in the **Trials** text box for **Second sample** and type <u>250</u>, click in the **Successes** text box for **Second sample** and type <u>174</u>, click the **Options...** button, click in the **Confidence level** text box and type <u>95</u>, click in the **Test difference** text box and type <u>0</u>, click the arrow button at the right of the **Alternative** dropdown list box and select **not equal**, select the **Use pooled estimate of p for test** check box, click **OK**, and click **OK**. The resulting output is

```
Test and CI for Two Proportions

Sample    X    N   Sample p
1        219  300  0.730000
2        174  250  0.696000


Difference = p (1) - p (2)
Estimate for difference:  0.034
95% CI for difference:  (-0.0419934, 0.109993)
Test for difference = 0 (vs not = 0):  Z = 0.88  P-Value = 0.379
```

Since the P-value = 0.379, which is greater than the significance level 0.05, do not reject the null hypothesis. The data do not provide sufficient evidence to conclude that there is a difference between the labor-force participation rates of U.S. and Canadian women.

(b) The confidence interval is part of the output for part (a). We can be 95% confident that the difference in labor-force participation rates between of U.S. and Canadian women is somewhere between -0.042 and 0.110.

**12.91** (a) $E = (-0.031 - (-0.129))/2 = 0.049$; We can be 90% confident that when we estimate $p_1 - p_2$ with $\hat{p}_1 - \hat{p}_2$ that our estimate is not off by more than 0.049.

(b) $$E = 1.645\sqrt{\frac{0.369(1-0.369)}{747} + \frac{0.449(1-0.449)}{434}} = 0.049$$

(c) $$n = 0.5\frac{1.645^2}{0.01^2} = 13530.125 \rightarrow 13531$$

Thus, the sample size is 13,531.

(d) $n = 13,531$, $\hat{p}_1 = 0.223$, $\hat{p}_2 = 0.272$

$$(0.383 - 0.437) \pm 1.645\sqrt{\frac{0.383(1-0.383)}{13531} + \frac{0.437(1-0.437)}{13531}}$$

$$-0.054 \pm 0.010 \ or \ -0.064 \ to \ -0.044$$

(e) $E = (-0.044 - (-0.064))/2 = 0.010$ (if we carried more significant digits, we would find that $E = 0.0098$); this is just slightly smaller than the specified error in part (c). This is expected since 0.383 and 0.437 are both considerably smaller than the assumed 0.5 probability in part (c).

(f) $\hat{p}_1 = 0.41$, $\hat{p}_2 = 0.49$

part (c)

$$n = \{0.41(1-0.41) + 0.49(1-0.49)\} \cdot \frac{1.645^2}{0.01^2} = 13308.23 \rightarrow 13309$$

Thus, the sample size is 13309.

part (d)

n = 13309, $\hat{p}_{1g}$ = 0.383, $\hat{p}_{2g}$ = 0.437

$$(0.383-0.437)\pm 1.645\sqrt{\frac{0.383(1-0.383)}{13309}+\frac{0.437(1-0.437)}{1339}}$$
$$-0.0540\pm 0.0099 \ or \ -0.064 \ to \ -0.044$$

part (e)

E = (-0.044 -(-0.064)/2 = 0.010.  As you can see from the calculation above, the actual margin of error is 0.0099, just slightly smaller than the required 0.01.

(g) By employing the guesses of $\hat{p}_1$ and $\hat{p}_2$, the sample size is reduced from 13,531 to 13,309.  Moreover, the margin of error only rises from 0.0098 to 0.0099.  The confidence interval is unchanged for all practical purposes.

## Chapter 12 Review Problems

1. (a) Feeling that marijuana should be legalized for medicinal use in patients with cancer and other painful and terminal diseases.
   (b) Americans
   (c) Proportion of all Americans who feel that marijuana should be legalized for medicinal use in patients with cancer and other painful and terminal diseases.
   (d) Proportion of Americans in the sample who feel that marijuana should be legalized for medicinal use in patients with cancer and other painful and terminal diseases.  Clearly the population of all Americans is much larger than 83,957.

2. It is often impossible to take a census of an entire population.  It is also expensive and time-consuming.

3. (a) "Number of successes" stands for the number of members of the sample that exhibit the specified attribute.
   (b) "Number of failures" stands for the number of members of the sample that do not exhibit the specified attribute.

4. (a) population proportion
   (b) normal
   (c) number of successes, number of failures, 5

5. The margin of error for the estimate of a population proportion tells us what the maximum difference between the sample proportion and the population proportion is <u>likely</u> to be.  It is not an absolute maximum, and how likely it is depends on the confidence level used.

6. (a) Getting the "holiday blues"
   (b) Men and women
   (c) The proportion of men in the population who get the "holiday blues" and the proportion of women in the population who get the "holiday blues"
   (d) The proportion of men in the sample who get the "holiday blues" and the proportion of women in the sample who get the "holiday blues"
   (e) They are sample proportions since the information came from a poll, not a census.  Also, it could not be a population proportion because I was not asked.

7. (a) The mean of all possible differences between the two sample proportions equals the <u>difference of the population proportions</u>.
   (b) For large samples, the possible differences between the two sample proportions have approximately a <u>normal</u> distribution.

8. The 95% confidence interval for the percentage of U.S. adults who would get a smallpox shot if it were available is 40% ± 3.0% or from 37% to 43%.

**9.** (a) $n = 0.5 \dfrac{1.96^2}{0.01^2} = 19208$

(b) $n = \{0.75(1-0.75) + 0.75(1-0.75)\} \dfrac{1.96^2}{0.01^2} = 14406$

**10.** n = 1218, $\hat{p}$ = 733/1218 = 0.60, $z_{\alpha/2}$ = $z_{0.025}$ = 1.96

(a) $0.602 - 1.96\sqrt{0.602(1-0.602)/1218}$ *to* $0.602 + 1.96\sqrt{0.602(1-0.602)/1218}$

$$0.574 \text{ to } 0.629$$

(b) We can be 95% confident that the proportion, p, of students who expect difficulty finding a job is somewhere between 0.574 and 0.629.

**11.** (a) The error is found by taking the width of the confidence interval and dividing by 2. So E = (0.629 - 0.575)/2 = 0.027.

(b) E = 0.02; $z_{\alpha/2}$ = $z_{0.025}$ = 1.96; $\hat{p}_g$ = 0.50

$$n = 0.5^2 \frac{z_{\alpha/2}^2}{E^2} = 0.25 \cdot \frac{1.96^2}{0.02^2} = 2401$$

(c) n = 2401; $\hat{p}$ = 0.587; $z_{\alpha/2}$ = $z_{0.025}$ = 1.96

$$0.587 - 1.96\sqrt{0.587(1-0.587)/2401} \text{ to } 0.587 + 1.96\sqrt{0.587(1-0.587)/2401}$$

$$0.567 \text{ to } 0.607$$

(d) The margin of error for the estimate is 0.020 (actually it's 0.197), the same as what is required in part (b).

(e) $\hat{p}_g$ = 0.56; $z_{\alpha/2}$ = $z_{0.025}$ = 1.96

part (b)   E = 0.02, $z_{0.025}$ = 1.96; sample size is:

$$n = 0.56(1-0.56) \frac{z_{\alpha/2}^2}{E^2} = 0.2464 \cdot \frac{1.96^2}{0.02^2} = 2366.42 \rightarrow 2367$$

Thus the required sample size is n = 2367.

part (c)
$$0.587 - 1.96\sqrt{0.587(1-0.587)/2367} \text{ to } 0.587 + 1.96\sqrt{0.587(1-0.587)/2367}$$

$$0.567 \text{ to } 0.607$$

part (d)   The margin of error is 0.02 (actually it's 0.0198), which is the same as what is specified in part (b).

(f) By employing the guess for $\hat{p}$ in part (d) we can reduce the required

sample size by 34, from 2401 to 2367, saving a little time and money. Moreover, the margin of error stays the same. The risk of using the guess 0.56 for $\hat{p}$ is that if the actual value of $\hat{p}$ turns out to be

between .44 and .56, then the achieved margin of error will exceed the specified 0.02.

**12.** n = 2512, x = 578, $\alpha$ = 0.05, $\hat{p}$ = 578/2512 = 0.2301

$np_o$ = 2512(0.25) = 628; $n(1 - p_0)$ = 2512(1-0.25) = 1884
Since both are at least 5, we can employ Procedure 12.2.

(a) Step 1: $H_0$: p = 0.25, $H_1$: p < 0.25
Step 2: $\alpha$ = 0.05

Step 3: $z = \dfrac{0.23 - 0.25}{\sqrt{0.25(1-0.25)/2512}} = -2.31$

Step 4: Since $\alpha = 0.05$, the critical value is $z_\alpha = -1.645$

Step 5: Since -2.31 < -1.645, reject $H_0$.
Step 6: The test results are statistically significant at the 5% level; that is, at the 5% significance level, the data do provide sufficient evidence to conclude that less than one in four Americans believe that juries "almost always" convict the guilty and free the innocent. $P(z < -2.31) = 0.0104$. Since P-value < $\alpha$, reject $H_0$.

(b) The strength of the evidence against the null hypothesis is strong.

13. (a) Observational study. The researchers had no control over any of the factors of the study.

(b) Height may not be the only factor to be considered. Although there does appear to be an association, we don't know if it is a direct association.

14. Population 1: first poll, $\hat{p}_1 = 0.48$,

Population 2: second poll, $\hat{p}_2 = 0.60$

$$\hat{p}_p = (0.48 + 0.60)/2 = 0.54$$

(a) Step 1: $H_0$: $p_1 = p_2$,  $H_a$: $p_1 < p_2$
Step 2: $\alpha = 0.01$

Step 3: $z = \dfrac{0.48 - 0.60}{\sqrt{0.54(1-0.54)}\sqrt{(1/600)+(1/600)}} = -4.17$

Step 4:   Since $\alpha = 0.01$, the critical value $-z_\alpha = -2.33$

Step 5: Since -4.17 < -2.33, reject $H_0$.
Step 6: The test is significant at the 1% level; that is the data do provide evidence to conclude that the percentage of Maricopa County residents who thought that the state's economy would improve over the next 2 years was less during the time of the first poll than during the time of the second poll.
$P(z < -4.17) = 0.0000$. So P-value < $\alpha$. Thus, we reject $H_0$.

(b) The strength of the evidence against the null hypothesis is very strong.

15. (a) From Exercise 14

$$(0.48 - 0.60) \pm 2.326\sqrt{\dfrac{0.48(1-0.48)}{600} + \dfrac{0.60(1-0.60)}{600}}$$

$-0.120 \pm 0.066$ *or* $-0.186$ *to* $0.054$

(b) We can be 95% confident that the difference $p_1 - p_2$ between the proportions of Maricopa County residents who thought that the state's economy would improve over the next 2 years during the time of the 1992 poll and during the time of the poll is somewhere between -0.186 and -0.054.

16. (a) E = (-0.054 - (-0.186))/2 = 0.066
We can be 98% confident that the error in estimating the difference between the two population proportions, $p_1 - p_2$, by the difference between the two sample proportions, -0.12, is at most 0.066.

(b)  $E = 2.326\sqrt{\dfrac{0.48(1-0.48)}{600} + \dfrac{0.60(1-0.60)}{600}} = 0.066$

(c)  E = 0.03, $\alpha$ = 0.02

$n = 0.5 \cdot \dfrac{2.326^2}{0.03^2} = 3005.715 \rightarrow 3006$

(d)  n = 3006, $\hat{p}_1$ = 0.475, $\hat{p}_2$ = 0.603

$$(0.475 - 0.603) \pm 2.326\sqrt{\dfrac{0.475(1-0.475)}{3006} + \dfrac{0.603(1-0.603)}{3006}}$$

$$-0.128 \pm 0.030 \;\; or \;\; -0.158 \;\; to \;\; -0.098$$

(e)  E = 0.030 which is the same as that required in part (c).

17.  The discrepancy between the two methods is primarily due to using a z-interval when the conditions for using such an interval are not met.  To get a good approximation to an exact interval, both x and n – x should be at least 5.  In this case, x is only 4.

18.  Using Minitab, choose **Stat ▶ Basic statistics ▶ 1 Proportion...**, select the **Summarized data** option button, click in the **Number of trials** text box and enter 2435, click in the **Number of events** text box and enter 317, click the **Options...** button, click in the **Confidence level** text box and enter 95, ensure that the **Alternative** drop down box contains **not equal**, check the box for **Use test and interval based on normal distribution**, click **OK**, and click **OK**. The resulting output is

Test and CI for One Proportion

Test of p = 0.5 vs p not = 0.5

```
Sample    X      N   Sample p        95% CI          Z-Value  P-Value
  1      317   2435  0.130185  (0.116819, 0.143551)   -36.50   0.000
```

We can be 95% confident that the **percentage** of U.S. adults who would participate in an office pool for March Madness is somewhere between 11.68% and 14.36%.

19.  Using Minitab, choose **Stat ▶ Basic statistics ▶ 1 Proportion...**, select the **Summarized data** option button, click in the **Number of trials** text box and enter 1961, click in the **Number of events** text box and enter 1137, click the **Options...** button, click in the **Confidence level** text box and enter 95, enter 0.5 in the Test proportion text box, select **greater than** from the **Alternative** drop down box, check the box for **Use test and interval based on normal distribution**, click **OK**, and click **OK**. The resulting output is

Test of p = 0.5 vs p > 0.5

```
                                95%
                               Lower
Sample    X      N   Sample p   Bound   Z-Value  P-Value
  1      1137   1961  0.579806  0.561472   7.07    0.000
```

Since the P-value = 0.000, which is less than the significance level 0.05, reject the null hypothesis.  The data provide sufficient evidence to conclude that a majority of U.S. adults do not believe that abstinence programs are effective in reducing or preventing AIDS.

20.  Using Minitab, choose **Calc ▶ Basic statistics ▶ 2 Proportions...**, select the **Summarized data** option button, click in the **Trials** text box for **First sample** and enter 56, click in the **Successes** text box for **First sample** and enter 32, click in the **Trials** text box for **Second sample** and enter 70, click

in the **Successes** text box for **Second sample** and enter <u>9</u>, click the
**Options...** button, click in the **Confidence level** text box and enter <u>95</u>,
click in the **Test difference** text box and type <u>0</u>, click the arrow button at
the right of the **Alternative** drop-down list box and select **not equal**, select
the **Use pooled estimate of p for test** check box, click **OK**, and click **OK**. The
resulting output is

      Test and CI for Two Proportions

      Sample   X   N   Sample p
      1      32  56  0.571429
      2       9  70  0.128571

      Difference = p (1) - p (2)
      Estimate for difference:  0.442857
      95% CI for difference:  (0.291371, 0.594343)
      Test for difference = 0 (vs not = 0):  Z = 5.27  P-Value = 0.000

(a) Since the P-value = 0.000, which is less than the significance level
0.05, reject the null hypothesis.  The data provide sufficient evidence
that there is a difference in the cure rates of the two types of
treatment.

(b) The 95% confidence interval for the difference in cure rates is found
in the output for part (a) as (0.291, 0.594).

**Exercises 13.1**

**13.1** A variable has a chi-square distribution if its distribution has the shape of a special type of right-skewed curve, called a chi-square curve.

**13.3** The $\chi^2$-curve with 20 degrees of freedom more closely resembles a normal curve. This follows from Property 4 of Key Fact 13.1, "As the number of degrees of freedom becomes larger, $\chi^2$-curves look increasingly like normal curves."

**13.5** (a) $\chi^2_{0.025} = 32.852$ (b) $\chi^2_{0.95} = 10.117$

**13.7** (a) $\chi^2_{0.05} = 18.307$ (b) $\chi^2_{0.975} = 3.247$

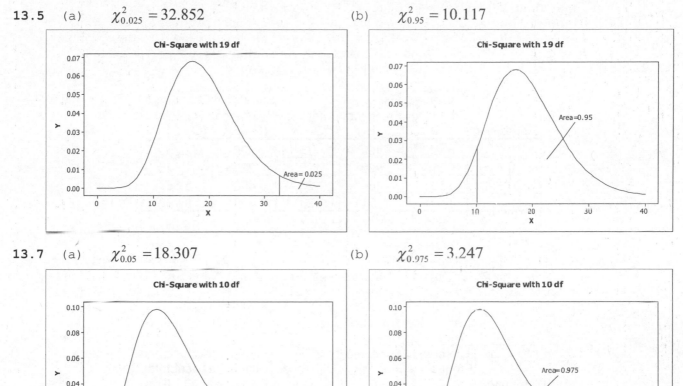

**13.9** The $\chi^2$-value having area 0.05 to its left has 0.95 to its right. Table VII gives values with specified areas to their right. If df = 26, $\chi^2_{0.95} = 15.379$

**Exercises 13.2**

**13.11** The term "goodness-of-fit" is used to describe the type of hypothesis test considered in this section because the test is carried out by determining how well the observed frequencies match or fit the expected frequencies.

**13.13** The assumptions are satisfied. The expected frequencies are np = 65, 30, and 5. All are 1 or more, and none are less than 5.

**13.15** The assumptions are satisfied. The expected frequencies are np = 10, 10, 12.5, 15, and 2.5. All are 1 or more, and exactly 20% are less than 5.

**13.17** The assumptions are not satisfied. The expected frequencies are np = 11, 11, 12.5, 15, and 0.5. One of them is not 1 or more.

**13.19** (a) The population consists of occupied housing units built after 2000. The variable under consideration is the primary heating fuel of each unit.

(b) The assumptions are not satisfied when n = 200.   The expected
frequencies are np = 103, 19.6, 61.4, 11.4, 3.8, 0.8.   One of the
frequencies is less than 1, and 33% are less than 5.
The assumptions are not satisfied when n = 250.   The expected
frequencies are np = 128.75, 24.5, 76.75, 14.25, 4.75, 1.0.   None of
the frequencies is less than 1, but two (33%) are less than 5.

The assumptions are satisfied when n = 300.   The expected frequencies
are np = 154.5, 29.4, 92.1, 17.1, 5.7, 1.2.   None of the frequencies is
less than 1 and only one (17%) is less than 5.

(c) We need the count for Wood and other fuel to be at least 5.   Thus we
must have N(0.019) > 5.   This implies that N must be greater than
5/0.019 = 263.15.   Thus the smallest possible value of N is 264.

**13.21** The procedure is summarized in the following table.

| Distribution | Observed Frequency O | Expected Frequency E | Difference O - E | Square of Difference $(O - E)^2$ | Chi-square subtotal $(O - E)^2/E$ |
|---|---|---|---|---|---|
| 0.2 | 85 | 100 | -15 | 225 | 2.250 |
| 0.4 | 215 | 200 | 15 | 225 | 1.125 |
| 0.3 | 130 | 150 | -20 | 400 | 2.667 |
| 0.1 | 70 | 50 | 20 | 400 | 8.000 |
| | 500 | | | | 14.042 |

df = 3; From Table VII, critical Value = 7.815.   $\chi^2$ = 14.0422, which is
greater than the critical value, so reject the null hypothesis.   The data
provide sufficient evidence that the variable differs from the given
distribution.

**13.23** The procedure is summarized in the following table.

| Distribution | Observed Frequency O | Expected Frequency E | Difference O - E | Square of Difference $(O - E)^2$ | Chi-square subtotal $(O - E)^2/E$ |
|---|---|---|---|---|---|
| 0.2 | 9 | 10 | -1 | 1 | 0.100 |
| 0.1 | 7 | 5 | 2 | 4 | 0.800 |
| 0.1 | 1 | 5 | -4 | 16 | 3.200 |
| 0.3 | 12 | 15 | -3 | 9 | 0.600 |
| 0.3 | 21 | 15 | 6 | 36 | 2.400 |
| | 50 | | | | 7.100 |

df = 4; From Table VII, critical Value = 7.779.   $\chi^2$ = 7.100, which is less
than the critical value, so do not reject the null hypothesis.   The data do
not provide sufficient evidence that the variable differs from the given
distribution.

**13.25** The procedure is summarized in the following table.

| Distribution | Observed Frequency O | Expected Frequency E | Difference O - E | Square of Difference $(O - E)^2$ | Chi-square subtotal $(O - E)^2/E$ |
|---|---|---|---|---|---|
| 0.5 | 147 | 175 | -28 | 784 | 4.480 |
| 0.3 | 115 | 105 | 10 | 100 | 0.952 |
| 0.2 | 88 | 70 | 18 | 324 | 4.629 |
| | 350 | | | | 10.061 |

df = 2;   From Table VII, critical Value = 9.210.   $\chi^2$ = 10.061, which is greater than the critical value, so reject the null hypothesis. The data provide sufficient evidence that the variable differs from the given distribution.

**13.27** (a)   The population consists of this year's incoming college freshmen in the U.S. The variable under consideration is their political view.

(b)

| Political View | Distribution | Observed Frequency O | Expected Frequency E | Difference O - E | Square of Difference $(O - E)^2$ | Chi-square subtotal $(O - E)^2/E$ |
|---|---|---|---|---|---|---|
| Liberal | 0.277 | 160 | 138.500 | 21.500 | 462.250 | 3.338 |
| Moderate | 0.519 | 246 | 259.500 | -13.500 | 182.250 | 0.702 |
| Conservative | 0.204 | 94 | 102.000 | -8.000 | 64.000 | 0.627 |
| | | 500 | | | | 4.667 |

Step 1:   $H_0$:   The distribution of political views for this year's incoming freshmen is the same as the 2000 distribution.

   $H_a$:   The distribution of political views for this year's incoming freshmen is different from the 2000 distribution.

Step 2:   Expected frequencies are presented in column 4 of the table.

Step 3:   Assumptions 1 and 2 are satisfied since all expected frequencies are at least 5.

Step 4:   $\alpha = 0.05$

Step 5:   $\chi^2 = 4.667$   (See column 7 of the table.)

Step 6:   df = 2; From Table VII, critical value = 5.991

Step 7:   Since 4.667 < 5.991, do not reject $H_0$.

Step 8:   The data do not provide sufficient evidence at the 5% level to conclude that the distribution of political views for this year's incoming freshmen is different from the 2000 distribution.

For the P-value approach, $0.05 < P(\chi^2 > 4.667) < 0.10$. Since the P-value is greater than the significance level, do not reject $H_0$.

(c)   Since the P-value is less than 0.10, reject $H_0$ at the 10% significance level. The data do provide sufficient evidence at the 10% level to conclude that the distribution of political views for this year's incoming freshmen is different from the 2000 distribution.

**13.29**

| Color | Distribution | Observed Frequency O | Expected Frequency E | Difference O - E | Square of Difference $(O - E)^2$ | Chi-square subtotal $(O - E)^2/E$ |
|---|---|---|---|---|---|---|
| Brown | 0.30 | 152 | 152.7 | -0.3 | 0.49 | 0.003 |
| Yellow | 0.20 | 114 | 101.8 | 12.2 | 148.84 | 1.462 |
| Red | 0.20 | 106 | 101.8 | 4.2 | 17.64 | 0.173 |
| Orange | 0.10 | 51 | 50.9 | 0.1 | 0.01 | 0.000 |
| Green | 0.10 | 43 | 50.9 | -7.9 | 62.41 | 1.226 |
| Blue | 0.10 | 43 | 50.9 | -7.9 | 62.41 | 1.226 |
| | | 509 | | | | 4.091 |

Step 1:   $H_0$:   The color distribution of M&Ms is the same as that reported by M&M/Mars consumer affairs.

   $H_a$:   The color distribution of M&Ms is different from that reported by M&M/Mars consumer affairs.

Step 2:   Expected frequencies are presented in column 4 of the table.

Step 3: Assumptions 1 and 2 are satisfied since all of the expected frequencies are at least 5.
Step 4: $\alpha$ = 0.05
Step 5: $\chi^2$ = 4.091 (See column 7 of the table.)
Step 6: df = 5; From Table VII, the critical value = 11.070
Step 7: Since 4.091 < 11.070, do not reject $H_0$.
Step 8: There is not sufficient evidence to conclude that ehe color distribution of M&Ms is different from that reported by M&M/Mars consumer affairs.

For the P-value approach, $P(\chi^2 > 4.091) > 0.10$. Since the P-value is larger than the significance level, do not reject $H_0$.

**13.31**

| Number | Distribution | Observed Frequency O | Expected Frequency E | Difference O - E | Square of Difference $(O - E)^2$ | Chi-square subtotal $(O - E)^2/E$ |
|---|---|---|---|---|---|---|
| 1 | 0.167 | 23 | 25 | -2 | 4 | 0.160 |
| 2 | 0.167 | 26 | 25 | 1 | 1 | 0.040 |
| 3 | 0.167 | 23 | 25 | -2 | 4 | 0.160 |
| 4 | 0.167 | 21 | 25 | -4 | 16 | 0.640 |
| 5 | 0.167 | 31 | 25 | 6 | 36 | 1.440 |
| 6 | 0.167 | 26 | 25 | 1 | 1 | 0.040 |
| | | 150 | | | | 2.480 |

Step 1: $H_0$: The die is not loaded.
$H_a$: The die is loaded.
Step 2: Expected frequencies are presented in column 4 of the table.
Step 3: Assumptions 1 and 2 are satisfied since all expected frequencies are at least 5.
Step 4: $\alpha$ = 0.05
Step 5: $\chi^2$ = 2.480 (See column 7 of the table.)
Step 6: df = 5; From Table VII, the critical value = 11.071
Step 7: Since 2.480 < 11.071, do not reject $H_0$.
Step 8: The data do not provide sufficient evidence to conclude that the die is loaded.

For the P-value approach, $P(\chi^2 > 2.480) > 0.10$. Since the P-value is larger than the significance level, do not reject $H_0$.

**13.33** (a) Using Minitab, choose **Stat/Tables/Chi-Square Goodness-of-Fit Test**, select the **Observed Counts** option button, enter O in the **Observed counts** text box, enter GAMES in the **Category names** text box, select the **Specific proportions** option button from the **Test** list, enter P in the **Specific proportions** text box, and click **OK**. The results are

**Chi-Square Goodness-of-Fit Test for Observed Counts in Variable: O**

Using category names in GAMES

| Category | Observed | Test Proportion | Expected | Contribution to Chi-Sq |
|---|---|---|---|---|
| 4 | 17 | 0.1250 | 12.1250 | 1.96005 |
| 5 | 23 | 0.2500 | 24.2500 | 0.06443 |
| 6 | 22 | 0.3125 | 30.3125 | 2.27951 |
| 7 | 35 | 0.3125 | 30.3125 | 0.72487 |

| N | DF | Chi-Sq | P-Value |
|---|---|---|---|
| 97 | 3 | 5.02887 | 0.170 |

The P-value = 0.170, which is larger than the significance level 0.05, so do not reject $H_0$. The data do not provide sufficient evidence to conclude that world Series teams are not evenly matched.

(b)  The usual assumption is that the data comprise a random sample from some population about which we want to make some inference. In this case, the data include the entire population of World Series records through 2004. We can think of these records as a sample from the population of all World Series that have been played and have yet to be played, but they are not a random sample. Nevertheless, each World Series is independent of the others, and the procedure does allow us to gain some insight into whether the overall records of the number of games played in each series follow a theoretical model based on the teams being evenly matched.

**13.35** Using Minitab, choose **Stat/Tables/Chi-Square Goodness-of-Fit Test**, select the **Observed Counts** option button, enter O in the **Observed counts** text box, enter REASON in the **Category names** text box, select the **Specific proportions** option button from the **Test** list, enter P in the **Specific proportions** text box, and click **OK**. The results are

**Chi-Square Goodness-of-Fit Test for Observed Counts in Variable: O**

Using category names in REASON

| Category | Observed | Test Proportion | Expected | Contribution to Chi-Sq |
|---|---|---|---|---|
| Help from friends/relatives | 43 | 0.062 | 31.0 | 4.6452 |
| Industry/business | 108 | 0.178 | 89.0 | 4.0562 |
| Job assignment | 23 | 0.072 | 36.0 | 4.6944 |
| Job transfer | 20 | 0.048 | 24.0 | 0.6667 |
| Joining family | 45 | 0.068 | 34.0 | 3.5588 |
| Marriage | 205 | 0.368 | 184.0 | 2.3967 |
| Other | 9 | 0.035 | 17.5 | 4.1286 |
| Study/training | 47 | 0.169 | 84.5 | 16.6420 |

| N | DF | Chi-Sq | P-Value |
|---|---|---|---|
| 500 | 7 | 40.7886 | 0.000 |

The P-value = 0.000, which is smaller than the significance level 0.01, so reject $H_0$. The data provide sufficient evidence to conclude that distribution of reasons for migration between provinces in China differs from that for migration within provinces.

**13.37** (a)  When k =2, the chi-square goodness-of-fit test is equivalent to the one-sample z-test for one proportion (Procedure 12.2) since when there are only two categories in the chi-square test, one of them can be thought of as a success and the other as a failure. The values of p in the chi-square test correspond to p and (1 - p) in the z-test. The two observed frequencies in the chi-square test correspond to x and n - x in the z-test. In other words, knowing one of the probabilities in the chi-square test when k = 2 allows us to know the other one. The same is true for the observed frequencies and for the expected frequencies. Finally, we compare the z and chi-square statistics when k = 2 and $p = p_0$.

$$\chi^2 = \sum \frac{(O-E)^2}{E} = \frac{(x-np)^2}{np} + \frac{[(n-x)-n(1-p)]^2}{n(1-p)}$$

$$= \frac{(x-np)^2}{np} + \frac{(np-x)^2}{n(1-p)} = (x-np)^2\left[\frac{1}{np} + \frac{1}{n(1-p)}\right]$$

$$= (x-np)^2\left[\frac{1}{np(1-p)}\right] = \frac{(x-np)^2}{np(1-p)}$$

Note that since $z = \dfrac{\frac{x}{n}-p}{\sqrt{\frac{p(1-p)}{n}}} = \dfrac{x-np}{\sqrt{np(1-p)}}$, $z^2 = \dfrac{(x-np)^2}{np(1-p)} = \chi^2$.

Thus the computed value of the $\chi^2$ statistic is the same as the square of the z-test statistic. Since the $\chi^2$ statistic is always positive, either a positive or a negative z-value will lead to the same $\chi^2$ value, making the two tests equivalent when the alternative hypothesis for the z-test is two tailed.

(b)   The alternative hypothesis is $H_a: p \neq p_0$. We then follow steps for performing Procedure 12.2 of the text.

(c)   The null hypothesis is that the distribution of the two-category population is given by:

| Category | Probability |
|---|---|
| Has attribute | $p_0$ |
| Doesn't have attribute | $1 - p_0$ |

The alternative hypothesis is that the distribution of the two-category population is different from the one specified in the null hypothesis. We then follow the steps for performing Procedure 13.1 of the text.

**Exercises 13.3**

**13.39** Cells

**13.41** To obtain the total number of observations of bivariate data in a contingency table, one can sum the individual cell frequencies, sum the row subtotals, or sum the column subtotals.

**13.43** Yes. If there were no association between the gender of the physician and specialty of the physician, then the same percentage of male and female physicians would choose family practice. Since different percentages of male and female physicians chose family practice, there is an association between the variables "gender" and "specialty." In other words, knowing the gender of a physician imparts information about the likelihood that the physician specialized in family practice.

**13.45** (a)

|  | M | F | Total |
|---|---|---|---|
| BUS | 2 | 7 | 9 |
| ENG | 10 | 2 | 12 |
| LIB | 3 | 1 | 4 |
| Total | 15 | 10 | 25 |

(b)

|   | BUS | ENG | LIB | Total |
|---|---|---|---|---|
| M | 0.222 | 0.833 | 0.750 | 0.600 |
| F | 0.778 | 0.167 | 0.250 | 0.400 |
| Total | 1.000 | 1.000 | 1.000 | 1.000 |

(c)

|   | M | F | Total |
|---|---|---|---|
| BUS | 0.133 | 0.700 | 0.360 |
| ENG | 0.667 | 0.200 | 0.480 |
| LIB | 0.200 | 0.100 | 0.160 |
| Total | 1.000 | 1.000 | 1.000 |

(d)   Yes.  The conditional distributions of gender are different within each college.

**13.47** (a)

|   | Fresh | Soph | Jun | Sen | Total |
|---|---|---|---|---|---|
| Dem | 2 | 6 | 8 | 4 | 20 |
| Rep | 3 | 9 | 12 | 6 | 30 |
| Other | 1 | 3 | 4 | 2 | 10 |
| Total | 6 | 18 | 24 | 12 | 60 |

(b)

|   | Fresh | Soph | Jun | Sen |
|---|---|---|---|---|
| Dem | 0.333 | 0.333 | 0.333 | 0.333 |
| Rep | 0.500 | 0.500 | 0.500 | 0.500 |
| Other | 0.167 | 0.167 | 0.167 | 0.167 |
| Total | 1.000 | 1.000 | 1.000 | 1.000 |

(c)   There is no association between party affiliation and class level. All of the conditional distributions of political party within class level are identical.

(d)   The marginal distribution of party affiliation will be identical to each of the conditional distributions, that is

| Party | Frequency |
|---|---|
| Dem | 0.333 |
| Rep | 0.500 |
| Other | 0.167 |
| Total | 1.000 |

(e)   True.  Since party affiliation and class level are not associated, all of the conditional distributions of class level within political party will be identical and will be the same as the marginal distribution of class level.

**13.49** (a)   8

(b)   First complete the first row, then the first column total, then the third row total, then the grand total.

|  | Male | Female | Total |
|---|---|---|---|
| White | 10,118 | 1,860 | 11,978 |
| Black | 13,398 | 7,395 | 20,793 |
| Hispanic | 6,041 | 1,643 | 7,684 |
| Other | 520 | 156 | 676 |
| Total | 30,077 | 11,054 | 41,131 |

      (c)   41,131     (d)    7,684     (e)   30,077    (f)   1,860

**13.51** (a)   1,056,500   (b)   48,600   (c)   10,600

        (d)   88,800 + 226,500 − 10,600 = 304,700

        (e)   51,900     (f)   51,900     (g)   1,648,900 − 88,800 = 1,560,100

**13.53** (a)  Complete the first column, then the second column, the third row, and then the remaining totals.

|  | Full Owner | Part Owner | Tenant | Total |
|---|---|---|---|---|
| Under 50 | **639** | 64 | 41 | **744** |
| 50 ← 180 | 487 | 131 | 41 | 659 |
| 180 ← 500 | 203 | **153** | **33** | 389 |
| 500 ← 1000 | 54 | 91 | 17 | 162 |
| 1000 & Over | 46 | 112 | 18 | 176 |
| Total | 1429 | 551 | **150** | **2130** |

      (b)  15  (c)  744,000   (d)   150,000   (e)   91,000   (f)   701,000

      (g)  68,000

**13.55** The table from Exercise 13.49 was first transposed so that the columns became the rows and vice versa.  Then each cell entry was divided by the column total to produce the following table.

|  | White | Black | Hispanic | Other | Total |
|---|---|---|---|---|---|
| Male | 0.845 | 0.644 | 0.786 | 0.769 | 0.731 |
| Female | 0.155 | 0.356 | 0.214 | 0.231 | 0.269 |
| Total | 1.000 | 1.000 | 1.000 | 1.000 | 1.000 |

    (a)  The conditional distributions of gender within each race are given in columns 2, 3, 4, and 5 of the above table.  The conditional distributions of gender by race indicate that in all populations, the largest percentage of AIDS cases is male, but there are differences in the percentages.

    (b)  The marginal distribution of gender is given in the last column of the above table.

    (c)  Yes.  The conditional distributions for gender are different for the four race categories.

    (d)  26.9% of the AIDS cases were females.

    (e)  15.5% of the AIDS cases among whites were females.

    (f)  True.  Since there is an association between gender and race category, the conditional distributions of race by gender cannot be identical (If they were identical, there would be no association between the two variables.).

    (g)  Directly from the table in Exercise 13.49, divide each cell entry by the column total below it to obtain the following table.

|         | Male  | Female | Total |
|---------|-------|--------|-------|
| White   | 0.336 | 0.168  | 0.291 |
| Black   | 0.445 | 0.669  | 0.506 |
| Hispanic| 0.201 | 0.149  | 0.187 |
| Other   | 0.017 | 0.014  | 0.016 |
| Total   | 1.000 | 1.000  | 1.000 |

The conditional distributions of race by gender are given in columns 2 and 3 of the table. The marginal distribution of race is given in the last column of the table. The conditional distributions of race by gender indicate that among both males and females with AIDS, the largest percentage of cases is black. Other interpretations are possible as well.

**13.57**   (a)

|                      | State | Federal | Local | Total |
|----------------------|-------|---------|-------|-------|
| $8^{th}$ grade or less | 0.142 | 0.119   | 0.131 | 0.137 |
| Some high school     | 0.255 | 0.145   | 0.334 | 0.273 |
| GED                  | 0.285 | 0.226   | 0.141 | 0.238 |
| High school diploma  | 0.205 | 0.270   | 0.259 | 0.225 |
| Post secondary       | 0.090 | 0.158   | 0.103 | 0.098 |
| College grad or more | 0.024 | 0.081   | 0.032 | 0.029 |
| Total                | 1.001 | 0.999   | 1.000 | 1.000 |

(b) Yes. The conditional distributions of educational attainment within the facility categories (columns 2, 3, and 4 of the table in part a) are not identical.

(c) The marginal distribution of educational attainment is given in the last column of the table in part (a).

(d)

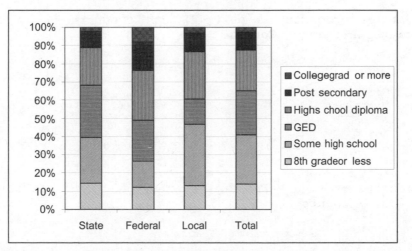

If the conditional distributions of educational attainment within the facility categories (first three bars) had been identical, each of the bars would have been segmented identically and would be identical to the fourth bar, the marginal distribution of educational attainment. The fact that they are not segmented identically means that there is an association between educational attainment and type of facility.

(e) False. Since educational attainment and type of facility are associated, the conditional distributions of facility type within educational attainment will be different.

(f) We interchanged the rows and columns of the table in Exercise 13.51 and then divided each cell entry by its associated column total to obtain the following table.

|  | 8$^{th}$ grade or less | Some high school | GED | High school diploma | Post secondary | College grad or more | Total |
|---|---|---|---|---|---|---|---|
| State | 0.662 | 0.598 | 0.768 | 0.584 | 0.590 | 0.521 | 0.641 |
| Federal | 0.047 | 0.029 | 0.051 | 0.065 | 0.087 | 0.148 | 0.054 |
| Local | 0.291 | 0.374 | 0.181 | 0.352 | 0.323 | 0.331 | 0.305 |
| Total | 1.000 | 1.001 | 1.000 | 1.001 | 1.000 | 1.000 | 1.000 |

The conditional distributions of facility type within educational attainment are given in columns 2, 3, 4, 5, 6, and 7. The marginal distribution of facility type is given by the last column.

(g) From the table in part (f), 5.4% of the prisoners are in federal facilities.

(h) From the table in part (f), 4.7% of the prisoners with at most an eighth grade education are in federal facilities.

(i) From the table in part (a), 11.9% of the prisoners in federal facilities have at most an eighth grade education.

**13.59** Using Minitab, choose **Stat ▶ Tables ▶ Cross Tabulation and Chi-Square,** enter REGION in the **For rows** text box, enter PARTY in the **For columns** text box, check the **Display** box for **Counts** and click **OK**. The following table is produced.

```
                  Democrat  Republican  All

        Midwest          5           7   12
        Northeast        4           5    9
        South            7           9   16
        West             6           7   13
        All             22          28   50
```

(b) Repeat the procedure in part (a), but check the box for **Column percents** instead of **Counts**.

```
                  Democrat  Republican     All

        Midwest      22.73       25.00   24.00
        Northeast    18.18       17.86   18.00
        South        31.82       32.14   32.00
        West         27.27       25.00   26.00
        All         100.00      100.00  100.00


        Cell Contents:      % of Column
```

(c) Repeat the procedure in part (a), but check the box for **Row percents** instead of **Counts**.

```
                  Democrat  Republican     All

        Midwest      41.67       58.33  100.00
        Northeast    44.44       55.56  100.00
        South        43.75       56.25  100.00
        West         46.15       53.85  100.00
        All          44.00       56.00  100.00


        Cell Contents:      % of Row
```

(d) Yes. The conditional distributions of party within each region are not identical.

**13.61** Using Minitab, choose **Stat ▶ Tables ▶ Cross Tabulation and Chi-Square,** enter PARTY in the **For rows** text box, enter CLASS in the **For columns**

text box, check the **Display** box for **Counts** and click **OK**. The following table is produced.

```
            Rows: PARTY    Columns: CLASS

                 I   II  III  All

Democrat        15   14   15   44
Independent      0    1    0    1
Republican      15   22   18   55
All             30   37   33  100

Cell Contents:       Count
```

(b)   Repeat the procedure in part (a), but check the box for **Column percents** instead of **Counts**.

```
            Rows: PARTY    Columns: CLASS

                 I        II       III      All

Democrat        50.00    37.84    45.45    44.00
Independent      0.00     2.70     0.00     1.00
Republican      50.00    59.46    54.55    55.00
All            100.00   100.00   100.00   100.00

Cell Contents:       % of Column
```

(c)   Repeat the procedure in part (a), but check the box for **Row percents** instead of **Counts**.

```
            Rows: PARTY    Columns: CLASS

                 I        II       III      All

Democrat        34.09    31.82    34.09   100.00
Independent      0.00   100.00     0.00   100.00
Republican      27.27    40.00    32.73   100.00
All             30.00    37.00    33.00   100.00

Cell Contents:       % of Row
```

(d)   Yes.  The conditional distributions of party within each class are not identical.

**13.63** (a)   If there were no association between age group and gender, the conditional distribution of age group within gender would be the same as the marginal distribution of age group.  Therefore 7.2% of the resident population would be in the age group 20-24 years.

(b)   If there were no association between age group and gender, the conditional distribution of age group within gender would be the same for each gender.  Therefore 7.2% of the female residents would be in the age group 20-24 years.

(c)   7.2% of 145.0 million, i.e., 10,440,000, females would be in the age group 20-24 years.

(d)   The number of female residents in the age group 20-24 years is not what we would expect IF there were NO association between age group and gender.  Therefore there must be an association between age group and gender.

**Exercises 13.4**

**13.65** $H_0$: The two variables under consideration are statistically independent.
$H_a$: The two variables under consideration are statistically dependent.

**13.67** The degrees of freedom equal $(r - 1)(c - 1) = (6 - 1)(4 - 1) = 15$.

**13.69** If two variables are not associated, then their frequencies do not rise and fall together, whether by causation or as the result of some third variable. If the variables did have a causal relationship, then they would be associated.

**13.71** Step 1:   $H_0$: Siskel's ratings and Ebert's ratings of movies are not associated.

  $H_a$: Siskel's ratings and Ebert's ratings of movies are associated.

Step 2:   Calculate the expected frequencies using the formula $E = RC/n$ where R = row total, C = column total, and n = sample size. The results are shown in the following table.

Ebert's rating

|  |  | Thumbs down | Mixed | Thumbs up | Total |
|---|---|---|---|---|---|
|  | Thumbs down | 11.8 | 8.4 | 24.8 | **45.0** |
| Siskel's | Mixed | 8.4 | 6.0 | 17.6 | **32.0** |
| Rating | Thumbs up | 21.8 | 15.6 | 45.6 | **83.0** |
|  | **Total** | **42.0** | **30.0** | **88.0** | **160.0** |

Step 3:   All of the expected frequencies are greater than 5, so the assumptions for the chi-square test are met.

Step 4:   $\alpha = 0.01$

Step 5:   Compute the value of the test statistic , $\chi^2 = \sum \dfrac{(O-E)^2}{E}$

where O and E represent the observed and expected frequencies respectively.  We show the contributions to this sum from each of the cells of the contingency table in the following table.

|  | Thumbs down | Mixed | Thumbs up |
|---|---|---|---|
| **Thumbs down** | 12.574 | 0.023 | 5.578 |
| **Mixed** | 0.019 | 8.167 | 2.475 |
| **Thumbs up** | 6.377 | 2.767 | 7.376 |

The total of the 9 table entries above is the value of the chi-square statistic, that is,  $\chi^2 = 45.357$. Depending on rounding, your answer could differ slightly.

Step 6:   The degrees of freedom are (3 - 1)((3 - 1) = 4, so the critical value from Table VII is $\chi^2_{0.01} = 13.277$.

Step 7:   Since 45.357 > 13.277, we reject the null hypothesis.

Step 8:   We conclude that there is an association between Siskel's ratings and Ebert's ratings of movies.

**13.73** (a)  The expected frequencies are shown below the observed frequencies in the following contingency table.

| Social Class | A few | Some | Lots | Total |
|---|---|---|---|---|
| Middle | 4<br>4.36 | 13<br>11.64 | 15<br>16.00 | 32 |
| Working | 5<br>4.64 | 11<br>12.36 | 18<br>17.00 | 34 |
| Total | 9 | 24 | 33 | 66 |

None of the expected values is less than one, but two of the six (33%) are less than 5.  Thus Assumptions 1 and 2 are not both satisfied.  The chi-square independence test should not be used.

(b)

| Social Class | Never | Sometimes | Often | Total |
|---|---|---|---|---|
| Middle | 2 6.303 | 8 8.727 | 22 16.970 | 32 |
| Working | 11 6.697 | 10 9.273 | 13 18.030 | 34 |
| Total | 13 | 18 | 35 | 66 |

None of the expected values is less than one, and none of the six are less than 5. Thus Assumptions 1 and 2 are both satisfied. The chi-square independence test may be used.

The following table gives the contributions from each cell to the chi-square statistic.

| Row, Column | O | E | $(O - E)^2/E$ |
|---|---|---|---|
| 1,1 | 2 | 6.303 | 2.9376 |
| 1,2 | 8 | 8.727 | 0.0606 |
| 1,3 | 22 | 16.970 | 1.4909 |
| 2,1 | 11 | 6.697 | 2.7648 |
| 2,2 | 10 | 9.273 | 0.0570 |
| 2,3 | 13 | 18.030 | 1.4033 |
| | 66 | | 8.7142 |

Step 1: $H_0$: The frequency with which parents play "I Spy" games with their children is independent of their economic class.

$H_a$: The frequency with which parents play "I Spy" games with their children is dependent of their economic class.

Step 2: Observed and expected frequencies are presented in the frequency contingency table. Each expected frequency is placed below its corresponding observed frequency. Expected frequencies are calculated using the formula $E = (R \times C)/n$.

The same information about the Os and Es is presented in the table below the contingency table. Column 4 of this table is useful for Step 5.

Step 3: Assumptions 1 and 2 are satisfied since all expected frequencies are at least 5.

Step 4: $\alpha = 0.05$

Step 5: $\chi^2 = 9.070$ (See column 4 of the table below the contingency table.)

Step 6: Critical value = 5.991

Step 7: Since 8.7142 > 5.991, reject $H_0$.

Step 8: The data do provide sufficient evidence to conclude that the frequency with which parents play "I Spy" with their children is dependent on economic class.

For the P-value approach, $0.01 < P(\chi^2 > 8.7142) < 0.025$. Since the P-value is less than the significance level, reject $H_0$.

**13.75**

Size of City

| Status in Practice | Less than 250,000 | 250,000-499,999 | 500,000 or more | Total |
|---|---|---|---|---|
| Government | 12 <br> 15.7 | 4 <br> 4.2 | 14 <br> 10.1 | 30 |
| Judicial | 8 <br> 5.8 | 1 <br> 1.5 | 2 <br> 3.7 | 11 |
| Private Practice | 122 <br> 116.4 | 31 <br> 31.1 | 69 <br> 74.5 | 222 |
| Salaried | 19 <br> 23.1 | 7 <br> 6.2 | 18 <br> 14.8 | 44 |
| Total | 161 | 43 | 103 | 307 |

Step 1:  $H_0$: Size of city and status in practice for lawyers are independent.

$H_a$: Size of city and status in practice for lawyers are dependent.

Step 2:  Observed and expected frequencies are presented in the frequency contingency table.  Each expected frequency is placed below its corresponding observed frequency.  Expected frequencies are calculated using the formula $E = (R \times C)/n$.

Step 3:  Assumption 2 is violated since 25% (3/12) of the expected frequencies are less than 5.  Thus, we do not proceed with the test.

**13.77** The expected frequencies are shown below the observed frequencies in the following contingency table.

| | Depressed | Not Depressed | Total |
|---|---|---|---|
| Osteoporitic | 3 <br> 3.21 | 35 <br> 34.79 | 38 |
| Low BMD | 69 <br> 50.89 | 533 <br> 551.11 | 602 |
| Normal | 97 <br> 114.89 | 1262 <br> 1244.11 | 1359 |
| Total | 169 | 1830 | 1999 |

The table below gives the contributions from each cell to the chi-square statistic.

| Row, Column | O | E | $(O - E)^2/E$ |
|---|---|---|---|
| 1,1 | 3 | 3.21 | 0.014 |
| 1,2 | 35 | 34.79 | 0.001 |
| 2,1 | 69 | 50.89 | 6.441 |
| 2,2 | 533 | 551.11 | 0.595 |
| 3,1 | 97 | 114.89 | 2.787 |
| 3,2 | 1262 | 1244.11 | 0.257 |
| | 1999 | 1999.00 | 10.095 |

Step 1:  $H_0$: Depression and one mineral density in elderly Asian men are statistically independent.

$H_a$: Depression and one mineral density in elderly Asian men are statistically dependent.

Step 2:  Observed and expected frequencies are presented in the frequency contingency table.  Each expected frequency is placed below its corresponding observed frequency.  Expected frequencies are calculated using the formula $E = (R \cdot C)/n$.

The same information about the Os and Es is presented in the table below the contingency table.  Column 4 of this table is useful for Step 6.

Step 3:  Assumptions 1 and 2 are satisfied since only 1 (17%) expected frequency is less than 5.

Step 4:  $\alpha$ = 0.01

Step 5:  $\chi^2$ = 10.095   (See column 4 of the last table above.)
   Note:  Answers may vary slightly due to rounding.

Step 6:  Critical value = 9.210

Step 7:  Since 10.095 > 9.210, reject $H_0$.

Step 8:  We conclude that depression and one mineral density in elderly Asian men are statistically dependent.

For the P-value approach, $0.005 < P(\chi^2 > 91.258) < 0.01$.  Since the P-value is smaller than the significance level, reject $H_0$.

**13.79** Using Minitab, the data are grouped in columns in order of the seven pairs of variables, with the responses "on the Job", "Off the Job", and "Don't Know" in rows 1, 2, and 3.  The procedure is the same for each part.  Choose

**Stat ▶ Tables ▶ Chi-Square Test (Table in Worksheet)**, enter the necessary columns for each part of the exercise [e.g., MALE FEMALE for part (a)] in the **Columns containing the table:** text box, and click **OK**. The results are shown below using the P-value approach.

(a)     Expected counts are printed below observed counts
        Chi-Square contributions are printed below expected counts

|   | Male | Female | Total |
|---|------|--------|-------|
| 1 | 77 | 77 | 154 |
|   | 82.49 | 71.51 | |
|   | 0.366 | 0.422 | |
| 2 | 263 | 215 | 478 |
|   | 256.05 | 221.95 | |
|   | 0.189 | 0.218 | |
| 3 | 28 | 27 | 55 |
|   | 29.46 | 25.54 | |
|   | 0.072 | 0.084 | |
| Total | 368 | 319 | 687 |

Chi-Sq = 1.350, DF = 2, P-Value = 0.509

Since the P-value = 0.509, which is greater than the significance level 0.05, do not reject $H_0$, the data do not provide evidence that gender and response are associated.

(b)     Expected counts are printed below observed counts
        Chi-Square contributions are printed below expected counts

|   | 18-29 years | 30-49 years | 50-64 years | 65 and older | Total |
|---|-------------|-------------|-------------|--------------|-------|
| 1 | 33 | 77 | 35 | 9 | 154 |
|   | 39.12 | 84.53 | 24.73 | 5.62 | |
|   | 0.957 | 0.671 | 4.265 | 2.032 | |
| 2 | 136 | 274 | 56 | 11 | 477 |
|   | 121.16 | 261.83 | 76.60 | 17.41 | |
|   | 1.816 | 0.566 | 5.539 | 2.359 | |
| 3 | 5 | 25 | 19 | 5 | 54 |
|   | 13.72 | 29.64 | 8.67 | 1.97 | |
|   | 5.539 | 0.727 | 12.302 | 4.656 | |
| Total | 174 | 376 | 110 | 25 | 685 |

Chi-Sq = 41.430, DF = 6, P-Value = 0.000
1 cells with expected counts less than 5.

Since the P-value = 0.000, which is less than the significance level 0.05, reject $H_0$, the data provide evidence that age and response are associated.

(c)      Expected counts are printed below observed counts
         Chi-Square contributions are printed below expected counts

|   | Urban | Suburban | Rural | Total |
|---|-------|----------|-------|-------|
| 1 | 62    | 47       | 42    | 151   |
|   | 60.44 | 53.80    | 36.75 |       |
|   | 0.040 | 0.860    | 0.749 |       |
| 2 | 197   | 171      | 109   | 477   |
|   | 190.94| 169.96   | 116.10|       |
|   | 0.192 | 0.006    | 0.435 |       |
| 3 | 14    | 25       | 15    | 54    |
|   | 21.62 | 19.24    | 13.14 |       |
|   | 2.683 | 1.724    | 0.262 |       |
| Total | 273 | 243    | 166   | 682   |

Chi-Sq = 6.952, DF = 4, P-Value = 0.138

Since the P-value = 0.138, which is greater than the significance level 0.05, do not reject $H_0$, the data do not provide evidence that type of community and response are associated.

(d)      Expected counts are printed below observed counts
         Chi-Square contributions are printed below expected counts

|   | Postgraduate | College graduate | Some college | No college | Total |
|---|--------------|------------------|--------------|------------|-------|
| 1 | 23           | 41               | 20           | 68         | 152   |
|   | 18.69        | 41.17            | 29.60        | 62.54      |       |
|   | 0.992        | 0.001            | 3.113        | 0.477      |       |
| 2 | 51           | 126              | 108          | 192        | 477   |
|   | 58.66        | 129.20           | 92.89        | 196.25     |       |
|   | 1.001        | 0.079            | 2.459        | 0.092      |       |
| 3 | 10           | 18               | 5            | 21         | 54    |
|   | 6.64         | 14.63            | 10.52        | 22.22      |       |
|   | 1.699        | 0.778            | 2.893        | 0.067      |       |
| Total | 84       | 185              | 133          | 281        | 683   |

Chi-Sq = 13.651, DF = 6, P-Value = 0.034

Since the P-value = 0.034, which is less than the significance level 0.05, reject $H_0$, the data provide evidence that educational attainment and response are associated.

(e)      Expected counts are printed below observed counts
         Chi-Square contributions are printed below expected counts

|   | Under $20,000 | $20,000-$29,999 | $30,000-$49,999 | $50,000 and over | Total |
|---|---------------|------------------|------------------|------------------|-------|
| 1 | 40            | 32               | 41               | 34               | 147   |
|   | 29.18         | 34.52            | 40.09            | 43.21            |       |
|   | 4.014         | 0.184            | 0.021            | 1.963            |       |
| 2 | 80            | 116              | 131              | 138              | 465   |
|   | 92.30         | 109.20           | 126.82           | 136.68           |       |
|   | 1.638         | 0.423            | 0.138            | 0.013            |       |
| 3 | 11            | 7                | 8                | 22               | 48    |
|   | 9.53          | 11.27            | 13.09            | 14.11            |       |
|   | 0.228         | 1.620            | 1.980            | 4.413            |       |
| Total | 131       | 155              | 180              | 194              | 660   |

Chi-Sq = 16.634, DF = 6, P-Value = 0.011

Since the P-value = 0.011, which is less than the significance level 0.05, reject $H_0$, the data provide evidence that income and response are associated.

(f)       Expected counts are printed below observed counts
Chi-Square contributions are printed below expected counts

|   | Private | Government | Self | Total |
|---|---|---|---|---|
| 1 | 67 | 23 | 61 | 151 |
|   | 93.26 | 25.76 | 31.98 | |
|   | 7.397 | 0.295 | 26.343 | |
| 2 | 326 | 82 | 69 | 477 |
|   | 294.62 | 81.37 | 101.01 | |
|   | 3.343 | 0.005 | 10.145 | |
| 3 | 27 | 11 | 14 | 52 |
|   | 32.12 | 8.87 | 11.01 | |
|   | 0.815 | 0.511 | 0.811 | |
| Total | 420 | 116 | 144 | 680 |

Chi-Sq = 49.665, DF = 4, P-Value = 0.000

Since the P-value = 0.000, which is less than the significance level 0.05, reject $H_0$, the data provide evidence that type of employer and response are associated.

**13.81** The expected frequencies are shown below the observed frequencies in the following contingency table.

|   | PhD | MA/Other | Total |   | PhD | MA | Chi-Square |
|---|---|---|---|---|---|---|---|
| Mail | 65 | 21 | 86 |   | 0.620 | 1.352 | |
|   | 58.95 | 27.05 | | | | | |
| Email | 166 | 73 | 239 |   | 0.029 | 0.062 | |
|   | 163.84 | 75.16 | | | | | |
| Both | 84 | 28 | 112 |   | 0.679 | 1.481 | |
|   | 76.78 | 35.22 | | | | | |
| N/A | 73 | 56 | 129 |   | 2.693 | 5.869 | |
|   | 88.43 | 40.57 | | | | | |
| Total | 388 | 178 | 566 |   | | | 12.785 |

Step 1:   $H_0$:  Degree and preference are non-associated.
          $H_a$:  Degree and preference are associated.
Step 2:   Observed and expected frequencies are presented in the frequency contingency table (left hand table above).  Each expected frequency is placed below its corresponding observed frequency.  Expected frequencies are calculated using the formula $E = (R \cdot C)/n$.
          The right hand table contains each cell's contribution to the Chi-Square value shown in the lower right hand corner.  This total will be used in Step 5.
Step 3:   Assumptions 1 and 2 are satisfied since all of the expected frequencies are at least 5.
Step 4:   $\alpha$ = 0.05
Step 5:   $\chi^2$ = 12.785
Step 6:   Critical value = 7.815
Step 7:   Since 12.785 > 7.815, we reject the null hypothesis.
Step 8:   There is sufficient evidence to conclude that degree and preference are associated.

**13.83** Using Minitab, choose **Stat ▶ Tables ▶ Chi-Square Test (Table in Worksheet)**, enter <u>WHITE BLACK OTHER</u> in the **Columns containing the table:** text box, and click **OK**. The results are shown below using the P-value approach.

```
Expected counts are printed below observed counts
Chi-Square contributions are printed below expected counts

        White   Black   Other   Total
  1       93      14       6      113
        92.47   14.50    6.03
        0.003   0.017   0.000

  2      118      14       4      136
       111.29   17.45    7.25
        0.404   0.683   1.459

  3      167      42       7      216
       176.76   27.72   11.52
        0.539   7.356   1.773

  4      113       7      15      135
       110.48   17.33    7.20
        0.058   6.153   8.450

Total    491      77      32      600

Chi-Sq = 26.897, DF = 6, P-Value = 0.000
```

We are testing the null hypothesis that the racial distribution is the same in each of the four U.S. regions. Since the P-value = 0.000, which is less than the significance level of 0.01, we reject that hypothesis. The data provide sufficient evidence that the racial distribution is not the same in each of the four regions of the U.S.

## Review Problems For Chapter 13

1.   The distributions and curves are distinguished by their numbers of degrees of freedom.
2.   (a) zero   (b)   skewed right   (c)   normal curve
3.   (a)  No.  The degrees of freedom for the $\chi$-square goodness-of-fit test depends on the number of categories, not the number of observations.
     (b)  No.  The degrees of freedom for the $\chi$-square independence test is $(r-1)(c-1)$ where r and c are the number of rows and columns, respectively, in the contingency table.
4.   Values of the test statistic near zero arise when the observed and expected frequencies are in close agreement.  It is only when these frequencies differ enough to produce large values of the test statistic that the null hypothesis is rejected.  These values are in the right tail of the $\chi$-square distribution.
5.   The value would be zero since O - E = 0 for each cell.
6.   (a)  All expected frequencies are 1 or greater, and at most 20% of the expected frequencies are less than 5.
     (b)  Very important.  If either of these assumptions are not met, the test should not be carried out by these procedures.
7.   (a)  7.8%
     (b)  7.8% of 24.7 million, or 1.93 million
     (c)  Since the observed number of Midwest Hispanic households is not equal to the number expected if there were no association between race of the householder and area of residence, we conclude that there *is* an association between the race of the householder and area of residence.
8.   (a)  Compare the conditional distributions of one of the variables within categories of the other variable.  If all of the conditional distributions are identical, there is no association between the variables; if not, there is an association.
     (b)  No.  Since the data are for an entire population, we are not making an

inference from a sample to the population. The association (or non-association) is a fact.

9. (a) Perform a chi-square test of independence. If the null hypothesis (of non-association) is rejected, we conclude that there is an association between the variables.

(b) Yes. It is possible (with probability $\alpha$) that we could reject the null hypothesis when it is, in fact, true. It is also possible, that, even though there is actually an association between the variables, the evidence is not strong enough to draw that conclusion. Either one of these types of errors is due to randomness in selecting a sample which does not exactly reflect the characteristics of the population.

10. For df = 17:

(a) $\chi^2_{0.99} = 6.408$  (b) $\chi^2_{0.01} = 33.409$  (c) $\chi^2_{0.05} = 27.587$

(d) $\chi^2_{0.95} = 8.672$  (e) $\chi^2_{0.975} = 7.564$ and $\chi^2_{0.025} = 30.191$

11.

| Highest level | O | p | E = np | $(O-E)^2/E$ |
|---|---|---|---|---|
| Not HS graduate | 84 | 0.158 | 79.0 | 0.316 |
| HS graduate | 160 | 0.332 | 166.0 | 0.217 |
| Some college | 88 | 0.176 | 88.0 | 0.000 |
| Associate's degree | 32 | 0.078 | 39.0 | 1.256 |
| Bachelor's degree | 87 | 0.170 | 85.0 | 0.047 |
| Advanced degree | 49 | 0.086 | 43.0 | 0.837 |
| Total | 500 | | 500.0 | 2.673 |

Step 1: $H_0$: The educational attainment distribution for adults 25 years old and over this year is the same as in 2000.

$H_a$: The educational attainment distribution for adults 25 years old and over this year differs from that of 2000.

Step 2: Expected frequencies are presented in column 4 of the table.

Step 3: Assumptions 1 and 2 are satisfied since all of the expected frequencies are at least 5.

Step 4: $\alpha = 0.05$

Step 5: $\chi^2 = 2.673$ (See column 5 of the table.)

Step 6: Critical value = 11.070

Step 7: Since 2.673 < 11.070, do not reject $H_0$.

Step 8: It appears that the educational attainment distribution for adults 25 years old and over this year does not differ from that of 2000.

(b) For the P-value approach, $0.10 < P(\chi^2 > 2.673) < 0.90$ (It's closer to 0.90 than to 0.10.) Since the P-value is larger than the significance level, do not reject $H_0$. The evidence against the null hypothesis is very weak or non-existent.

12. (a) The population consists of U.S. presidents.
(b) The two variables are the presidents' region of birth and their political party.
(c)

| | D | DR | F | R | U | W | TOTAL |
|---|---|---|---|---|---|---|---|
| NE | 7 | 1 | 1 | 5 | 0 | 1 | 15 |
| MW | 1 | 0 | 0 | 10 | 0 | 0 | 11 |
| SO | 6 | 3 | 1 | 2 | 1 | 3 | 16 |
| WE | 0 | 0 | 0 | 1 | 0 | 0 | 1 |
| TOTAL | 14 | 4 | 2 | 18 | 1 | 4 | 43 |

13. (a) Dividing each cell entry by the column total in the table in part (a), we obtain the conditional distributions of birth region for each party and in the right hand column, the marginal distribution of birth region.

|      | D     | DR    | F     | R     | U     | W     | TOTAL |
|------|-------|-------|-------|-------|-------|-------|-------|
| NE   | 0.500 | 0.250 | 0.500 | 0.277 | 0.000 | 0.250 | 0.349 |
| MW   | 0.071 | 0.000 | 0.000 | 0.556 | 0.000 | 0.000 | 0.256 |
| SO   | 0.429 | 0.750 | 0.500 | 0.111 | 1.000 | 0.750 | 0.372 |
| WE   | 0.000 | 0.000 | 0.000 | 0.056 | 0.000 | 0.000 | 0.023 |
| TOTAL | 1.000 | 1.000 | 1.000 | 1.000 | 1.000 | 1.000 | 1.000 |

(b) Dividing each cell entry by the row total in the table in part (a), we obtain the conditional distributions of party for each birth region and in the bottom row, the marginal distribution of party.

|       | D     | DR    | F     | R     | U     | W     | TOTAL |
|-------|-------|-------|-------|-------|-------|-------|-------|
| NE    | 0.467 | 0.067 | 0.067 | 0.333 | 0.000 | 0.067 | 1.000 |
| MW    | 0.091 | 0.000 | 0.000 | 0.909 | 0.000 | 0.000 | 1.000 |
| SO    | 0.375 | 0.188 | 0.063 | 0.125 | 0.063 | 0.188 | 1.000 |
| WE    | 0.000 | 0.000 | 0.000 | 1.000 | 0.000 | 0.000 | 1.000 |
| TOTAL | 0.326 | 0.093 | 0.047 | 0.419 | 0.023 | 0.093 | 1.000 |

(c) Yes. The conditional distributions of birth region by party are not all the same.

(d) 41.9% of Presidents are Republicans.

(e) 41.9%

(f) 12.5% of Presidents born in the South are Republicans.

(g) 37.2% of presidents were born in the South.

(h) 37.2%

(i) 11.1% of Republican presidents were born in the South.

14. (a)　2046　(b)　737　(c)　266　(d)　3046　(e)　6580 - 1167 = 5413

　　(f)　5403 + 1167 - 660 = 5910

15. (a)

|       | General | Psychiatric | Chronic | Tuberculosis | Other | Total |
|-------|---------|-------------|---------|--------------|-------|-------|
| GOV   | 0.314   | 0.361       | 0.808   | 0.750        | 0.144 | 0.311 |
| PROP  | 0.122   | 0.486       | 0.038   | 0.000        | 0.361 | 0.177 |
| NP    | 0.564   | 0.153       | 0.154   | 0.250        | 0.495 | 0.512 |
| Total | 1.000   | 1.000       | 1.000   | 1.000        | 1.000 | 1.000 |

The conditional distributions of control type with facility type are given in columns 2, 3, 4, 5, and 6 of the table.

(b) Yes. The conditional distributions of control type within facility type are not all identical.

(c) The marginal distribution of control type is given by the last column of the table above.

(d)

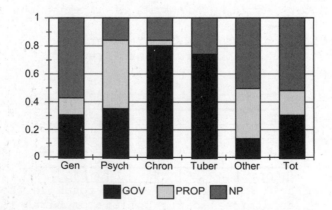

The conditional distributions of control type within facility type are shown be the first five bars of the graph. The marginal distribution of control type is given by the last bar.  Since the bars are not identically shaded, control type and facility type are associated variables.

(e) False.  Since we have established that facility type and control type are associated, the conditional distributions of facility type within control types will not be identical.

(f) After interchanging the rows and columns of the table given in Problem 13, we divided each cell entry by its associated column total to obtain the table below.

|  | GOV | PROP | NP | Total |
|---|---|---|---|---|
| General | 0.829 | 0.566 | 0.905 | 0.821 |
| Psychiatric | 0.130 | 0.307 | 0.034 | 0.112 |
| Chronic | 0.010 | 0.001 | 0.001 | 0.004 |
| Tuberculosis | 0.001 | 0.000 | 0.000 | 0.001 |
| Other | 0.029 | 0.127 | 0.060 | 0.062 |
| Total | 1.000 | 1.000 | 1.000 | 1.000 |

The conditional distributions of facility type within control type are given in columns 2, 3, and 4 of the table.  The marginal distribution of facility type is given by the last column of the table.

(g) From the table in part (a), 17.7% of the hospitals are under proprietary control.

(h) From the table in part (a), 48.6% of the psychiatric hospitals are under proprietary control.

(i) From the table in part (f), 30.7% of the hospitals under proprietary control are psychiatric hospitals.

16.

|  | Positive | Partial | None | Total |
|---|---|---|---|---|
| Lymphocyte depletion | 18 | 10 | 44 | 72 |
| Lymphocyte dominance | 74 | 18 | 12 | 104 |
| Mixed cellularity | 154 | 54 | 58 | 266 |
| Nodular sclerosis | 68 | 16 | 12 | 96 |
| Total | 314 | 98 | 126 | 538 |

| Row, Column | O | E | $(O - E)^2/E$ |
|---|---|---|---|
| 1,1 | 18 | 42.02 | 13.732 |
| 1,2 | 10 | 13.12 | 0.740 |
| 1,3 | 44 | 16.86 | 43.674 |
| 2,1 | 74 | 60.70 | 2.915 |
| 2,2 | 18 | 18.94 | 0.047 |
| 2,3 | 12 | 24.36 | 6.269 |
| 3,1 | 154 | 155.25 | 0.010 |
| 3,2 | 54 | 48.45 | 0.635 |
| 3,3 | 58 | 62.30 | 0.296 |
| 4,1 | 68 | 56.03 | 2.557 |
| 4,2 | 16 | 17.49 | 0.126 |
| 4,3 | 12 | 22.48 | 4.888 |
|  | 538 | 538.00 | 75.889 |

Step 1:   $H_0$:   Treatment response and histological type are statistically independent.

$H_a$:   Treatment response and histological type are statistically dependent.

Step 2:   Observed frequencies are presented in the frequency contingency table.  Expected frequencies are calculated using the formula $E = (R \cdot C)/n$ and are included in the third column of the second table. Column 4 of this table is useful for Step 6.

Step 3:   Assumptions 1 and 2 are satisfied since all expected frequencies are at least 5.

Step 4:   $\alpha = 0.01$

Step 5:   $\chi^2 = 75.889$   (See column 4 of the table below the contingency table.)
Note:  Answers may vary slightly due to rounding.

Step 6:   Critical value = 16.812

Step 7:   Since $75.889 > 16.812$, reject $H_0$.

Step 8:   The data provide sufficient evidence to conclude that treatment response and histological type are associated.

For the P-value approach, $P(\chi^2 > 77.693) < 0.005$.  Since the P-value is smaller than the significance level, reject $H_0$.

17.   (a)   Using Minitab, **Stat ▶ Tables ▶ Cross Tabulation and Chi-Square**, enter REGION in the **For rows:** text box and AGE in the **For columns:** text box, click on the **Counts** box under **Display** to check it and click **OK**.  The results are

Rows: REGION    Columns: AGE

|           | Forties | Fifties | Sixties | All |
|-----------|---------|---------|---------|-----|
| Northeast | 4       | 8       | 3       | 15  |
| Midwest   | 2       | 6       | 3       | 11  |
| South     | 2       | 10      | 4       | 16  |
| West      | 0       | 1       | 0       | 1   |
| All       | 8       | 25      | 10      | 43  |

Cell Contents:     Count

(b)   Repeat the procedure in part (a), checking the box for **Column percents** under **Display**.  The results are

Rows: REGION    Columns: AGE

|           | Forties | Fifties | Sixties | All    |
|-----------|---------|---------|---------|--------|
| Northeast | 50.00   | 32.00   | 30.00   | 34.88  |
| Midwest   | 25.00   | 24.00   | 30.00   | 25.58  |
| South     | 25.00   | 40.00   | 40.00   | 37.21  |
| West      | 0.00    | 4.00    | 0.00    | 2.33   |
| All       | 100.00  | 100.00  | 100.00  | 100.00 |

Cell Contents:     % of Column

(c)   Repeat the procedure in part (a), checking the box for **Row percents** under **Display**.  The results are

Rows: REGION    Columns: AGE

|           | Forties | Fifties | Sixties | All    |
|-----------|---------|---------|---------|--------|
| Northeast | 26.67   | 53.33   | 20.00   | 100.00 |
| Midwest   | 18.18   | 54.55   | 27.27   | 100.00 |
| South     | 12.50   | 62.50   | 25.00   | 100.00 |
| West      | 0.00    | 100.00  | 0.00    | 100.00 |
| All       | 18.60   | 58.14   | 23.26   | 100.00 |

Cell Contents:     % of Row

(d)   Yes.  The conditional distributions of AGE for each REGION are not the same.

18.    Using Minitab, **Stat ▶ Tables ▶ Cross Tabulation and Chi-Square**, enter
       REGION in the **For rows:** text box and AGE in the **For columns:** text box, click
       on the **Counts** box under **Display** to check it. Click on the Chi-square button,
       check the box for Chi-Square analysis and click **OK** twice.   The results are

Rows: OPINION    Columns: EDUCATION

|  | College grad | Some college | HS grad | Not HS grad | All |
|---|---|---|---|---|---|
| Favor | 264 | 205 | 461 | 290 | 1220 |
| Oppose | 17 | 26 | 81 | 81 | 205 |
| No opinion | 6 | 7 | 34 | 56 | 103 |
| All | 287 | 238 | 576 | 427 | 1528 |

Cell Contents:        Count

Pearson Chi-Square = 77.837, DF = 6, P-Value = 0.000
Likelihood Ratio Chi-Square = 79.327, DF = 6, P-Value = 0.000

Since the Pearson Chi-Square P-value = 0.000, which is less than the
significance level 0.01, reject the null hypothesis.  The data provide
sufficient evidence to conclude that opinion on this issue and educational
level are associated.

**Exercises 14.1**

**14.1** (a) $y = b_0 + b_1x$      (b) Constants are $b_0$, $b_1$; variables are $x, y$.
(c) The independent variable is $x$ and the dependent variable is $y$.

**14.3** (a) $b_0$ is the y-intercept; it is the value of $y$ where the line crosses the y-axis.

(b) $b_1$ is the slope; it indicates the change in the value of $y$ for every 1 unit increase in the value of $x$.

**14.5** (a) $y = 68.22 + 0.25x$      (b) $b_0 = 68.22$, $b_1 = 0.25$
(c)                                                    (d)

| Miles x | Cost ($) Y |
|---|---|
| 50 | $68.22 + 0.25(50) = 80.72$ |
| 100 | $68.22 + 0.25(100) = 93.22$ |
| 250 | $68.22 + 0.25(250) = 130.72$ |

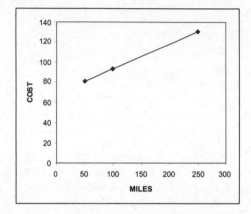

(e) The visual cost estimate of driving the car 150 miles is a little more than $100. The exact cost is
$y = 68.22 + 0.25(150) = 105.72$

**14.7** (a) $b_0 = 32$, $b_1 = 1.8$
(b)                                                    (c)

| X (°C) | y (°F) |
|---|---|
| -40 | $32 + 1.8(-40) = -40$ |
| 0 | $32 + 1.8(0) = 32$ |
| 20 | $32 + 1.8(20) = 68$ |
| 100 | $32 + 1.8(100) = 212$ |

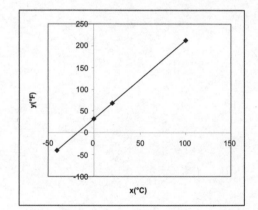

(d) The visual Fahrenheit temperature estimate corresponding to a Celsius temperature of 28° is about 80°F. The exact temperature is $y = 32 + 1.8(28) = 82.4°F$.

**14.9** (a) $b_0 = 68.22$, $b_1 = 0.25$
(b) The y-intercept $b_0 = 68.22$ gives the y-value at which the straight line $y = 68.22 + 0.25x$ intersects the y-axis. The slope $b_1 = 0.25$ indicates that the y-value increases by 0.25 units for every increase in $x$ of one unit.
(c) The y-intercept $b_0 = 68.22$ is the cost (in dollars) for driving the car zero miles. The slope $b_1 = 0.25$ represents the fact that the cost per mile is $0.25; it is the amount the total cost increases for each additional mile driven.

**14.11** (a)  $b_0 = 32$, $b_1 = 1.8$

(b)  The y-intercept $b_0 = 32$ gives the y-value where the line $y = 32 + 1.8x$ intersects the y-axis.  The slope $b_1 = 1.8$ indicates that the y-value increases by 1.8 units for every increase in $x$ of one unit.

(c)  The y-intercept $b_0 = 32$ is the Fahrenheit temperature corresponding to 0℃.  The slope $b_1 = 1.8$ represents the fact that Fahrenheit temperature increases by 1.8° for every increase of the Celsius temperature of 1°.

**14.13** (a)  $b_0 = 3$, $b_1 = 4$

(b)  slopes upward

(c)

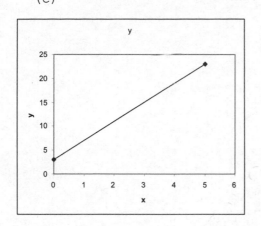

**14.15** (a)  $b_0 = 6$, $b_1 = -7$

(b)  slopes downward

(c)

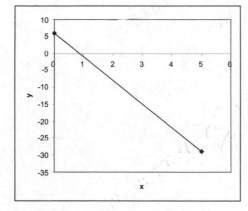

**14.17** (a)  $b_0 = -2$, $b_1 = 0.5$

(b)  slopes upward

(c)

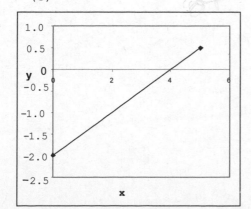

**14.19** (a)   $b_0 = 2$,  $b_1 = 0$
(b)   horizontal
(c)

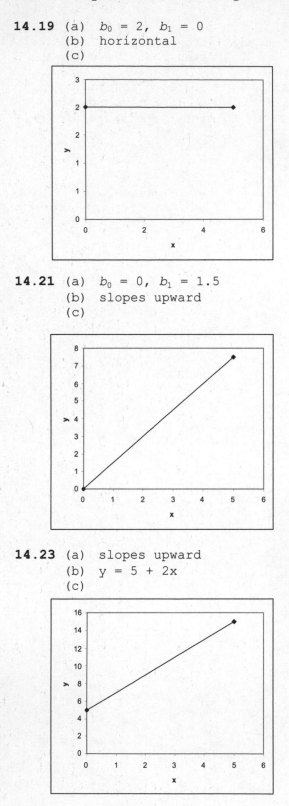

**14.21** (a)   $b_0 = 0$,  $b_1 = 1.5$
(b)   slopes upward
(c)

**14.23** (a)   slopes upward
(b)   $y = 5 + 2x$
(c)

**14.25** (a)  slopes downward
     (b)  y = -2 - 3x
     (c)

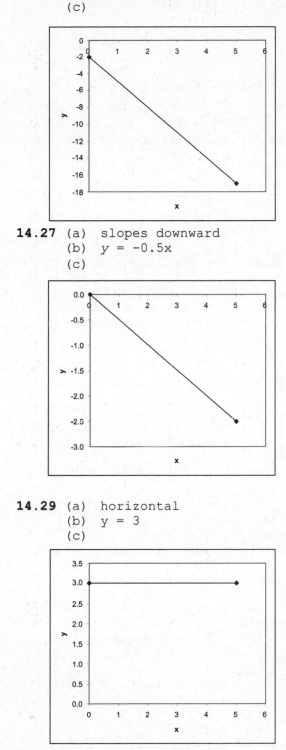

**14.27** (a)  slopes downward
     (b)  $y = -0.5x$
     (c)

**14.29** (a)  horizontal
     (b)  y = 3
     (c)

**14.31** The line defined by the equation must pass through the two points $(x, F)$ — $(2, 32$ and $(3, 16)$. We need to solve for $k$ and $x_0$. We have

$$\begin{cases} 32 = -k(2 - x_0) \\ 16 = -k(3 - x_0) \end{cases} \quad or \quad \begin{cases} 32 = -2k + kx_0 \\ 16 = -3k + kx_0 \end{cases}$$

Subtracting the second equation from the first, we obtain $16 = k$. Substituting 16 for $k$ in the first equation, we find

$$32 = -2(16) + 16x_0 \quad or \quad 64 = 16\, x_0.,$$ from which we find $x_0 = 4$.

(a) Thus, the linear equation that relates the force exerted to the length compressed is $F = -16(x - 4)$.
(b) The spring constant $k = 16$.
(c) The natural length of the spring is $x_0 = 4$ feet.

**14.33** (a) If we can express a line in the form $y = b_0 + b_1 x$, then there will be one and only one y-value corresponding to each $x$-value. However, that is not the case for a vertical line; one value of $x$ results in an infinite number of y-values.
(b) The form of the equation of a vertical line is $x = x_0$, where $x_0$ is the $x$-coordinate of the vertical line.
(c) For a linear equation, the slope indicates how much the y-value on the line increases (or decreases) when the $x$-value increases by one unit. We cannot apply this concept to a vertical line, since the $x$-value is *not* permitted to change. Thus, a vertical line has no slope.

**Exercises 14.2**

**14.35** (a) The criterion used to decide on the line that best fits a set of data points is called the least squares criterion.
(b) The criterion is that the line that best fits a set of data points is the one that has the smallest possible sum of the squares of the errors (errors are the differences between and actual and predicted y values).

**14.37** (a) The dependent variable is called the response variable.
(b) The independent variable is called the predictor variable or the explanatory variable.

**14.39** (a) In the context of regression, an <u>outlier</u> is a data point that lies far from the regression line, relative to the other data points.
(b) In regression analysis, an <u>influential observation</u> is a data point whose removal causes the regression equation to change considerably.

**14.41** (a) Line A: $y = 3 - 0.6x$          Line B: $y = 4 - x$

A  _____
B  - - - - - - - - -
Data points 0

(b)

| Line A: $y = 3 - 0.6x$ | | | | | Line B: $y = 4 - x$ | | | | |
|---|---|---|---|---|---|---|---|---|---|
| x | v | $\hat{v}$ | e | $e^2$ | x | v | $\hat{v}$ | e | $e^2$ |
| 0 | 4 | 3.0 | 1.0 | 1.00 | 0 | 4 | 4 | 0 | 0 |
| 2 | 2 | 1.8 | 0.2 | 0.04 | 2 | 2 | 2 | 0 | 0 |
| 2 | 0 | 1.8 | -1.8 | 3.24 | 2 | 0 | 2 | -2 | 4 |
| 5 | -2 | 0.0 | -2.0 | 4.00 | 5 | -2 | -1 | -1 | 1 |
| 6 | 1 | -0.6 | 1.6 | 2.56 | 6 | 1 | -2 | 3 | 9 |
| | | | | 10.84 | | | | | 14 |

(c)  According to the least-squares criterion, Line A fits the set of data points better than Line B.  This is because the sum of squared errors, i.e., $\sum e^2$, is smaller for Line A than for Line B.

**14.43** (a)

| x | y | xy | $x^2$ |
|---|---|---|---|
| 2 | 1 | 2 | 4 |
| 4 | 3 | 12 | 16 |
| 6 | 4 | 14 | 20 |

Using Formula 14.1,

$$S_{xx} = 20 - 6^2/2 = 2, \ S_{xy} = 14 - (6)(4)/2 = 2, \ b_1 = S_{xy}/S_{xx} = 2/2 = 1$$

$$b_0 = \bar{y} - b_1 \bar{x} = 2 - 1(3) = -1$$

Thus, the regression line is $\hat{y} = -1 + x$.

Without using Formula 14.1, both points must be on the line $y = b_0 + b_1 x$,

so $\begin{cases} 1 = b_0 + b_1 2 \\ 3 = b_0 + b_1 4 \end{cases}$.  Subtracting the first equation from the second, we

get $2 = b_1 2$, or $b_1 = 1$.  Substituting 1 back in the first equation for b₁, $1 = b_0 + 1(2)$, so $b_0 = -1$, and the equation of the line is y = -1 + x.

(b)

| x | y | xy | $x^2$ |
|---|---|---|---|
| 1 | 3 | 3 | 1 |
| 5 | -3 | -15 | 25 |
| 6 | 0 | -12 | 26 |

Using Formula 14.1,

$$S_{xx} = 20 - 6^2/2 = 2, \ S_{xy} = -12 - (6)(0)/2 = -12, \ b_1 = S_{xy}/S_{xx} = -12/8 = -1.5$$

$$b_0 = \bar{y} - b_1 \bar{x} = 0 - (-1.5)(3) = 4.5$$

Thus, the regression line is $\hat{y} = 4.5 - 1.5 \ x$.

Without using Formula 14.1, both points must be on the line $y = b_0 + b_1 x$,

so $\begin{cases} 3 = b_0 + b_1 1 \\ -3 = b_0 + b_1 5 \end{cases}$.  Subtracting the first equation from the second, we

get $-6 = 4 b_1$, or $b_1 = -1.5$.  Substituting -1.5 back in the first equation for $b_1$, $3 = b_0 + (-1.5)(1)$ so $b_0 = 4.5$, and the equation of the line is y = 4.5 - 1.5x.

**14.45**

| x | y | xy | $x^2$ |
|---|---|---|---|
| 3 | -4 | -12 | 9 |
| 1 | 0 | 0 | 1 |
| 2 | -5 | -10 | 4 |
| 6 | -9 | -22 | 14 |

(a)

$$S_{xx} = \sum x^2 - \left(\sum x\right)^2 / n = 14 - 6^2/3 = 2;$$

$$S_{xy} = \sum xy - \left(\sum x\right)\left(\sum y\right)/n = -22 - (6)(-9)/3 = -4$$

$$b_1 = S_{xy}/S_{xx} = -4/2 = -2; \quad b_0 = \bar{y} - b_1\bar{x} = -3 - (-2)(2) = 1$$

$$\hat{y} = 1 - 2x$$

(b) Applying the regression equation for x = 1 and x = 3,

x  $\hat{y}$

1  -1

3  -5

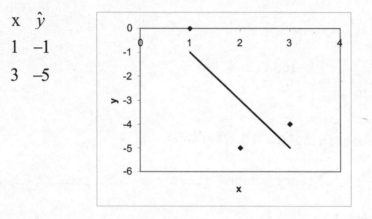

**14.47**

| x | y | xy | $x^2$ |
|---|---|---|---|
| 3 | 4 | 12 | 9 |
| 4 | 5 | 20 | 16 |
| 1 | 0 | 0 | 1 |
| 2 | -1 | -2 | 4 |
| 10 | 8 | 30 | 30 |

(a)

$$S_{xx} = \sum x^2 - \left(\sum x\right)^2 / n = 30 - 10^2/4 = 5;$$

$$S_{xy} = \sum xy - \left(\sum x\right)\left(\sum y\right)/n = 30 - (10)(8)/4 = 10$$

$$b_1 = S_{xy}/S_{xx} = 10/5 = 2; \quad b_0 = \bar{y} - b_1\bar{x} = 2 - (2)(2.5) = -3$$

$$\hat{y} = -3 + 2x$$

(b) Applying the regression equation for x = 1 and x = 4

x  $\hat{y}$

1  −1

4   5

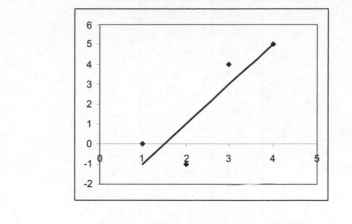

**14.49**

| x | y | xy | x2 |
|---|---|---|---|
| 0 | 4 | 0 | 0 |
| 2 | 2 | 4 | 4 |
| 2 | 0 | 0 | 4 |
| 5 | −2 | −10 | 25 |
| 6 | 1 | 6 | 36 |
| 15 | 5 | 0 | 69 |

(a)

$$S_{xx} = \sum x^2 - \left(\sum x\right)^2 / n = 69 - 15^2 / 5 = 24;$$

$$S_{xy} = \sum xy - \left(\sum x\right)\left(\sum y\right) / n = 0 - (15)(5)/5 = -15$$

$$b_1 = S_{xy} / S_{xx} = -15/24 = -0.625; \quad b_0 = \bar{y} - b_1 \bar{x} = 1 - (-0.625)(3) = 2.875$$

$$\hat{y} = 2.875 - 0.625x$$

(b) Applying the regression equation for x = 0 and x = 6

x  $\hat{y}$

0   2.875

6  −0.875

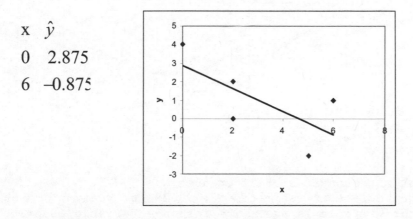

**14.51** (a) Formulas for the slope ($b_1$) and intercept ($b_0$) of the regression equation are, respectively:

$$b_1 = \frac{\sum xy - \sum x \sum y / n}{\sum x^2 - (\sum x)^2 / n} \qquad b_0 = \frac{1}{n}\left(\sum y - b_1 \sum x\right)$$

To compute $b_0$ and $b_1$, construct a table of values for x, y, xy, $x^2$, and their sums:

| x | y | xy | $x^2$ |
|---|---|---|---|
| 6 | 270 | 1620 | 36 |
| 6 | 260 | 1560 | 36 |
| 6 | 275 | 1650 | 36 |
| 2 | 405 | 810 | 4 |
| 2 | 364 | 728 | 4 |
| 5 | 295 | 1475 | 25 |
| 4 | 335 | 1340 | 16 |
| 5 | 308 | 1540 | 25 |
| 1 | 405 | 405 | 1 |
| 4 | 305 | 1220 | 16 |
| 41 | 3222 | 12348 | 199 |

$$\text{Thus}: b_1 = \frac{12348 - 41(3222)/10}{199 - 41^2/10} = -27.9029 \quad b_0 = \frac{1}{10}(3222 - (-27.9029)41) = 436.6019$$

and the regression equation is $\hat{y} = 436.6019 - 27.9029x$

(b) Begin by selecting two x-values within the range of the x-data. For the x-values 1 and 6, the calculated values for y (in $hundreds) are, respectively:

$$\hat{y} = 436.6019 - 27.9029(1) = 408.6990$$

$$\hat{y} = 436.6019 - 27.9029(6) = 269.1845$$

The regression equation can be graphed by plotting the pairs (1, 408.699) and (6, 269.185) and connecting these points with a line. This equation, the original set of 10 data points, and all of the predicted values are shown in the following graph:

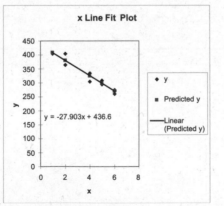

(c)  Price tends to decrease as age increases.
(d)  Corvettes depreciate an estimated $2,790.29 per year.
(e)  The predictor variable is age.  The response variable is price.
(f)  There are no outliers or potential influential observations.

(g)  Price of two-year old Corvette:  $\hat{y} = 436.6019 - 27.9029(2) = 380.7961$

or $38,080.

Price of three-year old Corvette:  $\hat{y} = 436.6019 - 27.9029(3) = 352.8932$

or $35,289.

**14.53** (a)   Formulas for the slope ($b_1$) and intercept ($b_0$) of the regression equation are, respectively:

$$b_1 = \frac{\sum xy - \sum x \sum y / n}{\sum x^2 - (\sum x)^2 / n} \qquad b_0 = \frac{1}{n}\left(\sum y - b_1 \sum x\right)$$

To compute $b_0$ and $b_1$, construct a table of values for x, y, xy, $x^2$, and their sums:

| x | y | xy | $x^2$ |
|---|---|---|---|
| 57 | 8.0 | 456.0 | 3249 |
| 85 | 22.0 | 1870.0 | 7225 |
| 57 | 10.5 | 598.5 | 3249 |
| 65 | 22.5 | 1462.5 | 4225 |
| 52 | 12.0 | 624.0 | 2704 |
| 67 | 11.5 | 770.5 | 4489 |
| 62 | 7.5 | 465.0 | 3844 |
| 80 | 13.0 | 1040.0 | 6400 |
| 77 | 16.5 | 1270.5 | 5929 |
| 53 | 21.0 | 1113.0 | 2809 |
| 68 | 12.0 | 816.0 | 4624 |
| 723 | 156.5 | 10486.0 | 48747 |

$$b_1 = \frac{10486 - 723(1565)/11}{48747 - 723^2/11} = 0.16285 \qquad b_0 = \frac{1}{11}(1565 - 0.16285(723)) = 3.52369$$

and the regression equation is $\hat{y} = 3.52369 + 0.16285x$.

(b)   Begin by selecting two x-values within the range of the x-data. For the x-values 52 and 85, the calculated values for y are, respectively:

$$\hat{y} = 3.52369 + 0.16285(52) = 11.992$$

$$\hat{y} = 3.52369 + 0.16285(85) = 17.366$$

The regression equation can be graphed by plotting the pairs (52,12.000) and (85, 17.379) and connecting these points with a line. This equation and the original set of 11 data points are presented as follows:

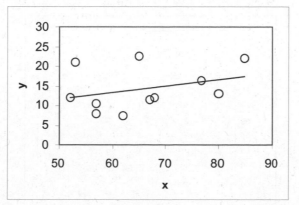

(c)   The quantity of volatile compounds emitted tends to decrease as plant weight increases.

(d)   The quantity of volatile compounds emitted increases by an estimated 0.163 hundred nanograms for each increase in plant weight of one gram.

(e)   The predictor variable is plant weight. The response variable is hundreds of nanograms of volatile compounds emitted by the plant.

(f)   There are no outliers or potential influential observations.

(g)   The quantity of volatile compounds emitted by a 75 gram plant:

$$\hat{y} = 3.52369 + 0.16285(75) = 15.737 \text{ hundred nanograms or 1574 nanograms.}$$

**14.55** (a)   Formulas for the slope ($b_1$) and intercept ($b_0$) of the regression equation are, respectively:

$$b_1 = \frac{\sum xy - \sum x \sum y / n}{\sum x^2 - (\sum x)^2 / n} \qquad b_0 = \frac{1}{n}\left(\sum y - b_1 \sum x\right)$$

To compute $b_0$ and $b_1$, construct a table of values for x, y, xy, $x^2$, and their sums:

| x | y | xy | $x^2$ |
|---|---|---|---|
| 10 | 92 | 920 | 100 |
| 15 | 81 | 1215 | 225 |
| 12 | 84 | 1008 | 144 |
| 20 | 74 | 1480 | 400 |
| 8 | 85 | 680 | 64 |
| 16 | 80 | 1280 | 256 |
| 14 | 84 | 1176 | 196 |
| 22 | 80 | 1760 | 484 |
| 117 | 660 | 9519 | 1869 |

$$b_1 = \frac{9519 - 117(660)/8}{1869 - 117^2/8} = -0.84561 \qquad b_0 = \frac{1}{8}(660 - (-0.84561)(117)) = 94.86698$$

and the regression equation is $\hat{y} = 94.86698 - 0.84561x$.

(b)   Begin by selecting two x-values within the range of the x-data. For the x-values 10 and 28, the calculated values for y are, respectively:

$$\hat{y} = 94.86698 - 0.84561(10) = 86.41$$

$$\hat{y} = 94.86698 - 0.84561(22) = 76.26$$

The regression equation can be graphed by plotting the pairs (10,68.40) and (28, 278.16) and connecting these points with a line. This equation and the original set of 10 data points are presented as follows:

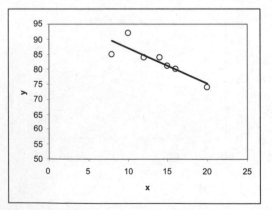

(c)   The exam scores tend to decrease as the number of study hours increases.

(d)   The exam score decreases, on average, about 0.85 points for each hour of study time.

(e)  The predictor variable is study time, in hours.  The response variable is the exam score.

(f)  There may be two outliers, (8,85) and (10,92).  There are no potential influential observations.

(g)  For a student who studies 15 hours the predicted exam score is

$$\hat{y} = 94.86698 - 0.84561(15) = 82.18$$

**14.57** The idea behind finding a regression line is based on the assumption that the data points are actually scattered about a line.  Only the second data set appears to be scattered about a line.  Thus, it is reasonable to determine a regression line only for the second set of data.

**14.59** (a)  It is acceptable to use the regression equation to predict the price of a four-year-old Corvette since that age lies within the range of the ages in the sample data.  It is not acceptable (and would be extrapolation) to use the regression equation to predict the price of a 10-year-old Corvette since that age falls outside the range of the ages in the sample data.

(b)  It is reasonable to use the regression equation to predict price for ages between one and six years old, inclusive.

**14.61** One possibility is that there are students who understand the material well after only a short period of study.  These students do not need to study as long as students who have a more difficult time comprehending the material and possibly never reach the same level of understanding.  There are other possibilities.  Perhaps some students are more organized and do their studying when they are wide awake while others put it off until they are tired and less efficient.

**14.63** (a)

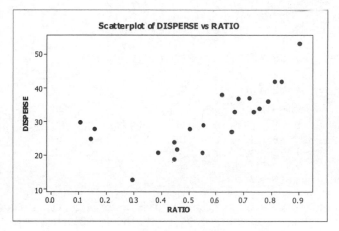

(b)  No.  The data obviously do not fall in a line.

**14.65** (a)  Using Minitab, with the data in columns named BIRDIES and SCORE, choose

**Graph ▶ Scatterplot...**, select the **Simple** version and click **OK**.  Enter DEATH in Row 1 of the **Y variables** column and INAUGURATION in the **X variables** column, and click **OK**.  The result is

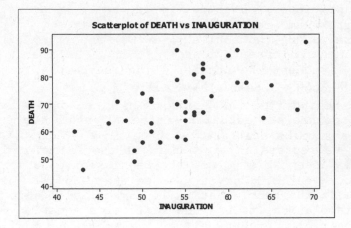

(b)  A line appears to be reasonable, although there may be an outlier or two and an influential observation.  There does not appear to be any curvature in the data.

(c)  Choose **Stat ▶ Regression ▶ Regression**, enter <u>DEATH</u> in the **Response** text box, and <u>INAUGURATION</u> in the **Predictors** text box.  Click **OK**.  The results are
```
The regression equation is
DEATH = 8.7 + 1.12 INAUGURATION


Predictor        Coef  SE Coef     T      P
Constant         8.66    13.84  0.63  0.535
INAUGURATION   1.1172   0.2512  4.45  0.000


S = 9.47690   R-Sq = 35.5%   R-Sq(adj) = 33.7%

Analysis of Variance

Source         DF      SS      MS      F      P
Regression      1  1776.5  1776.5  19.78  0.000
Residual Error  36  3233.2    89.8
Total           37  5009.7

Unusual Observations

Obs  INAUGURATION  DEATH   Fit  SE Fit  Residual  St Resid
 31          54.0  90.00  68.99    1.55     21.01     2.25R
 38          69.0  93.00  85.75    3.90      7.25     0.84 X


R denotes an observation with a large standardized residual.
X denotes an observation whose X value gives it large influence.
```

The regression equation (bold-faced in the output) indicates that the age at death is 8.7 years plus an increase of 1.12 years for every year later that inauguration occurs.

(d)  There is one outlier, the point (54,90).

(e)  After removing the outlier, the regression equation becomes
**DEATH = 7.5 + 1.13 INAUGURATION**
The effect of removing the outlier is to reduce the constant and increase the slope, both slightly.

(f)  After replacing the outlier and removing the potential influential observation (69,93), the regression equation becomes
**DEATH = 13.2 + 1.03 INAUGURATION**
Removing the influential observation increases the constant and reduces the slope, both significantly.  In addition, part of the output is

Unusual Observations

| Obs | INAUGURATION | DEATH | Fit | SE Fit | Residual | St Resid |
|-----|--------------|-------|-------|--------|----------|----------|
| 9 | 68.0 | 68.00 | 83.24 | 4.04 | -15.24 | -1.77 X |
| 37 | 54.0 | 90.00 | 68.83 | 1.57 | 21.17 | 2.26R |

This indicates that the point (54,90) is still an outlier, and there is now a new potential influential observation at (68,68).

**14.67** (a) Using Minitab, with the data in columns named LOT SIZE and VALUE

EXPECTANCY, choose **Graph ▶ Scatterplot...**, select the **Simple** version and click **OK**. Enter <u>VALUE</u> in Row 1 of the **Y variables** column and '<u>LOT SIZE</u>' in the **X variables** column, and click **OK**. The result is

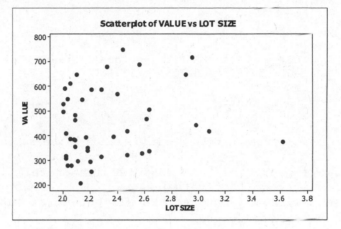

(b) The data points do not have a linear pattern, there is at least one influential observation, and there are likely some outliers. Therefore, find a regression line for these data is not reasonable.

**14.69** (a) Using Minitab, with the data in columns named HIGH and LOW, choose

**Graph ▶ Scatterplot...**, select the **Simple** version and click **OK**. Enter <u>LOW</u> in Row 1 of the **Y variables** column and <u>HIGH</u> in the **X variables** column, and click **OK**. The result is

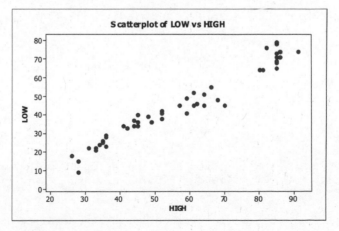

(b) It is reasonable to find a regression line for these data. There is a linear pattern without influential observations.

(c) Choose **Stat ▶ Regression ▶ Regression**, enter <u>LOW</u> in the **Response** text box, and <u>HIGH</u> in the **Predictors** text box. Click **OK**. The results are

The regression equation is

```
LOW = - 7.57 + 0.917 HIGH

Predictor      Coef   SE Coef      T       P
Constant     -7.572     1.786   -4.24   0.000
HIGH        0.91685   0.02967   30.91   0.000

S = 4.15111   R-Sq = 95.2%   R-Sq(adj) = 95.1%

Analysis of Variance

Source           DF      SS      MS        F       P
Regression        1   16459   16459   955.17   0.000
Residual Error   48     827      17
Total            49   17286

Unusual Observations

Obs   HIGH     LOW      Fit   SE Fit   Residual   St Resid
 15   70.0   45.000   56.607   0.705    -11.607     -2.84R
 17   82.0   76.000   67.610   0.949      8.390      2.08R
 30   85.0   79.000   70.360   1.021      8.640      2.15R
 45   28.0    9.000   18.100   1.038     -9.100     -2.26R
```

R denotes an observation with a large standardized residual.

As one would expect, there is positive linear relationship between the high and low average temperatures.

(d) Observations 15, 17, 30 and 45 are all potential outliers.  There are no influential observations.

(e) Delete the four data points mentioned in part (d), starting with number 45 and ending with number 15.  Repeat the regression procedure.  The result is

```
The regression equation is
LOW = - 5.66 + 0.884 HIGH

Predictor      Coef   SE Coef      T       P
Constant     -5.655     1.442   -3.92   0.000
HIGH        0.88407   0.02433   36.34   0.000

S = 3.18323   R-Sq = 96.8%   R-Sq(adj) = 96.7%

Analysis of Variance

Source           DF      SS      MS         F       P
Regression        1   13381   13381   1320.50   0.000
Residual Error   44     446      10
Total            45   13826

Unusual Observations

Obs   HIGH     LOW      Fit   SE Fit   Residual   St Resid
 32   68.0   48.000   54.462   0.552     -6.462     -2.06R
 37   85.0   78.000   69.491   0.847      8.509      2.77R
```

R denotes an observation with a large standardized residual.

Deleting the outliers resulted in an increase in the intercept and a slight decrease in the slope of the regression line.  It also resulted in two new potential outliers.

(f) Not applicable

**14.71** (a)   Using Minitab, with the data in columns named INCOME and BEER, choose

**Graph ▶ Scatterplot...**, select the **Simple** version and click **OK**.

Enter <u>BEER</u> in Row 1 of the **Y variables** column and <u>INCOME</u> in the **X variables** column, and click **OK**.  The result is

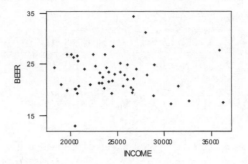

    (b)     There is no linear pattern in these data, so it is inappropriate to carry out a linear regression.  Parts (c)-(f) are omitted.

**14.73** (a)    Using Minitab, with the data in columns named MANAGERS and ASSETS,

choose **Graph ▶ Scatterplot...**, select the **Simple** version and click **OK**.  Enter <u>MANAGERS</u> in Row 1 of the **Y variables** column and <u>ASSETS</u> in the **X variables** column, and click **OK**.  When you see the scatterplot, you might wonder just where the regression line will go.  To see, put the mouse cursor on the graph and right click.  Select Add ▶ Regression fit…, select the button for Linear and check the box for Fit intercept.  Click OK twice. The result is (You could reverse the roles of ASSETS and MANAGERS, but the basic conclusion will remain the same.)

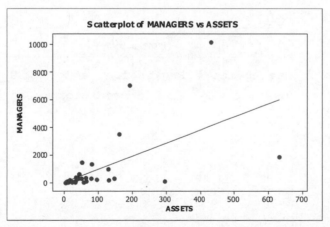

    (b)     Clearly, there will be outliers and influential data points if we fit a line to these data.  A linear regression fit will be of little use, so we omit parts (c-f).  [You can do them if you wish just to confirm this opinion.]

**14.75** (a) From Exercise 14.47,

$$n = 4; \quad \sum x_i = 10; \quad \sum y_i = 8; \quad \sum xy_i = 30; \quad \sum x_i^2 = 30$$

Therefore, $s_{xy} = \dfrac{\sum x_i y_i - \left(\sum x_i\right)\left(\sum y_i\right)/n}{n-1} = \dfrac{30 - (10)(8)/4}{3} = 10/3 =$ Covariance

$$s_x^2 = \frac{\sum x_i^2 - \left(\sum x_i\right)^2 / n}{n-1} = \frac{30 - (10)^2/4}{3} = 5/3$$

(b)

$$b_1 = s_{xy} / s_x^2 = (10/3)/(5/3) = 2; \quad b_0 = \bar{y} - b_1 \bar{x} = 2 - 2(10/4) = -3$$

$$\hat{y} = -3 + 2x$$

This is the same result as was obtained in Exercise 14.47.

**14.77** (a) Using Minitab, with the data in columns named YEAR and POPULATION, choose **Graph ▶ Scatterplot...**, select the **Simple** version and click **OK**. Enter POPULATION in Row 1 of the **Y variables** column and YEAR in the **X variables** column, and click **OK**. The result is

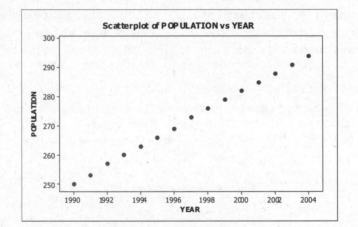

(b) Choose **Stat ▶ Regression ▶ Regression**, enter POPULATION in the **Response** text box, and YEAR in the **Predictors** text box. Click **OK**. The results are

```
The regression equation is
POPULATION = - 6011 + 3.15 YEAR


Predictor      Coef    SE Coef        T       P
Constant   -6011.02      41.82  -143.73   0.000
YEAR        3.14643    0.02094   150.24   0.000


S = 0.350431    R-Sq = 99.9%    R-Sq(adj) = 99.9%


Analysis of Variance


Source          DF       SS       MS         F       P
Regression       1   2772.0   2772.0  22572.91   0.000
Residual Error  13      1.6      0.1
Total           14   2773.6
```

The regression equation above indicates that the population is increasing 3.15 million per year.

(c) The population forecast for 2005 is −6011 + 3.15(2005)  304.75 million
The population forecast for 2006 is −6011 + 3.15(2006)  307.90 million

**Exercises 14.3**

**14.79** (a) The coefficient of determination is a descriptive measure of the utility of the regression equation for making predictions. The symbol for the coefficient of determination is $r^2$.

(b) Two interpretations of the coefficient of determination are
(i) It represents the percentage reduction obtained in the total squared error by using the regression equation to predict the observed values, instead of simply using the mean of the y-values.
(ii) It represents the percentage of variation in the observed $y$-values that is explained by the regression.

**14.81** (a) The coefficient of determination is $r^2$ = SSR/SST = 7626.6/8291.0 = 0.9199. 91.99% of the variation in the observed values of the response variable is explained by the regression. The fact that $r^2$ is near 1 indicates that the regression equation is extremely useful for making predictions.

(b) SSE = SST - SSR = 8291.0 - 7626.6 = 664.4 **14.82** To use the defining formulas, we begin with the following table.

| x | Y | $(y-\bar{y})$ | $(y-\bar{y})^2$ | $\hat{y}$ | $(\hat{y}-\bar{y})$ | $(\hat{y}-\bar{y})^2$ | $(y-\hat{y})$ | $(y-\hat{y})^2$ |
|---|---|---|---|---|---|---|---|---|
| 2 | 3 | -2.0 | 4 | 4 | -1.0 | 1.00 | -1 | 1 |
| 4 | 5 | 0.0 | 0 | 6 | 1.0 | 1.00 | -1 | 1 |
| 3 | 7 | 2.0 | 4 | 5 | 0.0 | 0.00 | 2 | 4 |
| 9 | 15 | 0 | 8 | | 0 | 2.00 | 0 | 6 |

Each $\hat{y}$ value is obtained by substituting the respective value of $x$ into the regression equation $\hat{y} = 2 + x$. (This equation was derived in Exercise 14.44).

(a) $SST = \sum(y-\bar{y})^2 = 8$    $SSR = \sum(\hat{y}-\bar{y})^2 = 2$    $SSE = \sum(y-\hat{y})^2 = 6$

(b) $8 = 2 + 6$

(c) $r^2 = 1 - \dfrac{SSE}{SST} = 1 - \dfrac{6}{8} = \dfrac{1}{4} = 0.25$ $\quad or \quad$ $r^2 = \dfrac{SSR}{SST} = \dfrac{2}{8} = 0.25$

(d) The percentage of variation in the observed $y$-values that is explained by the regression is 25%.

(e) Based on the answer in part (d), the regression equation does not appear to be very useful for making predictions.

**14.83** To use the defining formulas, we begin with the following table.

| x | y | $(y-\bar{y})$ | $(y-\bar{y})^2$ | $\hat{y}$ | $(\hat{y}-\bar{y})$ | $(\hat{y}-\bar{y})^2$ | $(y-\hat{y})$ | $(y-\hat{y})^2$ |
|---|---|---|---|---|---|---|---|---|
| 3 | -4 | -1.0 | 1 | -5 | -2.0 | 4.00 | 1 | 1 |
| 1 | 0 | 3.0 | 9 | -1 | 2.0 | 4.00 | 1 | 1 |
| 2 | -5 | -2.0 | 4 | -3 | 0.0 | 0.00 | -2 | 4 |
| 6 | -9 | 0.0 | 14 | | 0.0 | 8.00 | 0 | 6 |

Each $\hat{y}$ value is obtained by substituting the respective value of $x$ into the regression equation $\hat{y} = 1 - 2x$. (This equation was derived in Exercise 14.45).

(a) $SST = \sum(y - \bar{y})^2 = 14$   $SSR = \sum(\hat{y} - \bar{y})^2 = 8$   $SSE = \sum(y - \hat{y})^2 = 6$

(b) $14 = 8 + 6$

(c) $r^2 = 1 - \dfrac{SSE}{SST} = 1 - \dfrac{6}{14} = \dfrac{4}{7} = 0.571$   or   $r^2 = \dfrac{SSR}{SST} = \dfrac{8}{14} = 0.571$

(d) The percentage of variation in the observed $y$-values that is explained by the regression is 57.1%.

(e) Based on the answer in part (d), the regression equation is somewhat useful for making predictions.

**14.85** To use the defining formulas, we begin with the following table.

| x | y | $(y - \bar{y})$ | $(y - \bar{y})^2$ | $\hat{y}$ | $(\hat{y} - \bar{y})$ | $(\hat{y} - \bar{y})^2$ | $(y - \hat{y})$ | $(y - \hat{y})^2$ |
|---|---|---|---|---|---|---|---|---|
| 3 | 4 | 2.0 | 4 | 3 | 1.0 | 1.00 | 1 | 1 |
| 4 | 5 | 3.0 | 9 | 5 | 3.0 | 9.00 | 0 | 0 |
| 1 | 0 | -2.0 | 4 | -1 | -3.0 | 9.00 | 1 | 1 |
| 2 | -1 | -3.0 | 9 | 1 | -1.0 | 1.00 | -2 | 4 |
| 10 | 8 | 0.0 | 26 | | 0.0 | 20.00 | 0 | 6 |

Each $\hat{y}$ value is obtained by substituting the respective value of $x$ into the regression equation $\hat{y} = -3 + 2x$. (This equation was derived in Exercise 14.47).

(a) $SST = \sum(y - \bar{y})^2 = 26$   $SSR = \sum(\hat{y} - \bar{y})^2 = 20$   $SSE = \sum(y - \hat{y})^2 = 6$

(b) $46 = 40 + 6$

(c) $r^2 = 1 - \dfrac{SSE}{SST} = 1 - \dfrac{6}{26} = \dfrac{10}{13} = 0.769$   or   $r^2 = \dfrac{SSR}{SST} = \dfrac{20}{26} = 0.769$

(d) The percentage of variation in the observed $y$-values that is explained by the regression is 76.9%.

(e) Based on the answer in part (d), the regression equation is useful for making predictions.

**14.87** To use the defining formulas, we begin with the following table.

| x | Y | $(y - \bar{y})$ | $(y - \bar{y})^2$ | $\hat{y}$ | $(\hat{y} - \bar{y})$ | $(\hat{y} - \bar{y})^2$ | $(y - \hat{y})$ | $(y - \hat{y})^2$ |
|---|---|---|---|---|---|---|---|---|
| 0 | 4 | 3 | 9 | 2.875 | 1.875 | 3.51563 | 1.125 | 1.26563 |
| 2 | 2 | 1 | 1 | 1.625 | 0.625 | 0.39063 | 0.375 | 0.14063 |
| 2 | 0 | -1 | 1 | 1.625 | 0.625 | 0.39063 | -1.625 | 2.64062 |
| 5 | -2 | -3 | 9 | -0.250 | -1.250 | 1.56250 | -1.750 | 3.06250 |
| 6 | 1 | 0 | 0 | -0.875 | -1.875 | 3.51563 | 1.875 | 3.51563 |
| 15 | 5 | 0 | 20 | | 0 | 9.375 | 0 | 10.625 |

Each $y$ value is obtained by substituting the respective value of $x$ into the regression equation $\hat{y} = 2.875 - 0.625x$. (This equation was derived in Exercise 14.49).

(a) $SST = \sum(y - \bar{y})^2 = 20 \quad SSR = \sum(\hat{y} - \bar{y})^2 = 9.375 \quad SSE = \sum(y - \hat{y})^2 = 10.625$

(b) $20 = 9.365 + 10.625$

(c) $r^2 = 1 - \dfrac{SSE}{SST} = 1 - \dfrac{10.625}{20} = 0.46875 \quad or \quad r^2 = \dfrac{SSR}{SST} = \dfrac{6.375}{20} = 0.46875$

(d) The percentage of variation in the observed $y$-values that is explained by the regression is 46.875%.

(e) Based on the answer in part (c), the regression equation appears to be moderately useful for making predictions.

**14.89** To use the computing formulas, we begin with the following table. Notice that columns 1-4 are presented in the solution to part (a) of Exercise 14.51, so that only the column for $y^2$ needs to be developed.

| x | y | xy | $x^2$ | $y^2$ |
|---|---|---|---|---|
| 6 | 270 | 1620 | 36 | 72900 |
| 6 | 260 | 1560 | 36 | 67600 |
| 6 | 275 | 1650 | 36 | 75625 |
| 2 | 405 | 810 | 4 | 164025 |
| 2 | 364 | 728 | 4 | 132496 |
| 5 | 295 | 1475 | 25 | 87025 |
| 4 | 335 | 1340 | 16 | 112225 |
| 5 | 308 | 1540 | 25 | 94864 |
| 1 | 405 | 405 | 1 | 164025 |
| 4 | 305 | 1220 | 16 | 93025 |
| 41 | 3222 | 12348 | 199 | 1063810 |

(a) Using the last row of the table and Formula 14.2 of the text, we obtain the three sums of squares as follows.

$$SST = S_{yy} = \sum y^2 - \left(\sum y\right)^2 / n = 1063810 - 3222^2 / 10 = 25681.6$$

$$SSR = \frac{S_{xy}^2}{S_{xx}} = \frac{\left[\sum xy - \left(\sum x\right)\left(\sum y\right)/n\right]^2}{\sum x^2 - \left(\sum x\right)^2/n} = \frac{[12348 - (41)(3222)/10]^2}{199 - 41^2/10} = 24057.9$$

$$SSE = S_{yy} - \frac{S_{xy}^2}{S_{xx}} = 25681.6 - 24057.9 = 1623.7$$

(b) $r^2 = \dfrac{SSR}{SST} = \dfrac{24057.9}{25681.6} = 0.9368$

(c) The percentage of variation in the observed $y$-values that is explained by the regression is 93.68% In words, 93.68% of the variation in the tax efficiency data is explained by the percentage of the mutual funds investments that are in energy securities.

(d) Based on the answers to parts (b) and (c), the regression equation appears to be very useful for making predictions.

**14.91** To use the computing formulas, we begin with the following table.  Notice that columns 1-4 are presented in the solution to part (a) of Exercise 14.53, so that only the column for $y^2$ needs to be developed.

| x | y | xy | $x^2$ | $y^2$ |
|---|---|---|---|---|
| 57 | 8.0 | 456.0 | 3249 | 64.00 |
| 85 | 22.0 | 1870.0 | 7225 | 484.00 |
| 57 | 10.5 | 598.5 | 3249 | 110.25 |
| 65 | 22.5 | 1462.5 | 4225 | 506.25 |
| 52 | 12.0 | 624.0 | 2704 | 144.00 |
| 67 | 11.5 | 770.5 | 4489 | 132.25 |
| 62 | 7.5 | 465.0 | 3844 | 56.25 |
| 80 | 13.0 | 1040.0 | 6400 | 169.00 |
| 77 | 16.5 | 1270.5 | 5929 | 272.25 |
| 53 | 21.0 | 1113.0 | 2809 | 441.00 |
| 68 | 12.0 | 816.0 | 4624 | 144.00 |
| 723 | 156.5 | 10486.0 | 48747 | 2523.25 |

(a)  Using the last row of the table and Formula 14.2 of the text, we obtain the three sums of squares as follows.

$$SST = S_{yy} = \sum y^2 - (\sum y)^2 / n = 2523.25 - 156.5^2 / 11 = 296.68$$

$$SSR = \frac{S_{xy}^2}{S_{xx}} = \frac{[\sum xy - (\sum x)(\sum y)/n]^2}{\sum x^2 - (\sum x)^2 / n} = \frac{[10486 - (723)(1565)/11]^2}{48747 - 723^2 / 11} = 32.52$$

$$SSE = S_{yy} - \frac{S_{xy}^2}{S_{xx}} = 296.68 - 32.52 = 264.16$$

(b)  $r^2 = \dfrac{SSR}{SST} = \dfrac{32.52}{7296.68} = 0.1096$

(c)  The percentage of variation in the observed y-values that is explained by the regression is 10.96%.  In words, only 10.96% of the variation in the price data is explained by size.

(d)  Based on the answers to parts (b) and (c), the regression equation appears to be useless for making predictions.

**14.93** To use the computing formulas, we begin with the following table.  Notice that columns 1-4 are presented in the solution to part (a) of Exercise 14.55, so that only the column for $y^2$ needs to be developed.

| x | y | xy | $x^2$ | $y^2$ |
|---|---|---|---|---|
| 10 | 92 | 920 | 100 | 8464 |
| 15 | 81 | 1215 | 225 | 6561 |
| 12 | 84 | 1008 | 144 | 7056 |
| 20 | 74 | 1480 | 400 | 5476 |
| 8 | 85 | 680 | 64 | 7225 |
| 16 | 80 | 1280 | 256 | 6400 |
| 14 | 84 | 1176 | 196 | 7056 |
| 22 | 80 | 1760 | 484 | 6400 |
| 117 | 660 | 9519 | 1869 | 54638 |

(a)  Using the last row of the table and Formula 14.2 of the text, we obtain the three sums of squares as follows.

$$SST = S_{yy} = \sum y^2 - (\sum y)^2 / n = 54638 - 660^2 / 8 = 188$$

$$SSR = \frac{S_{xy}^2}{S_{xx}} = \frac{[\sum xy - (\sum x)(\sum y)/n]^2}{\sum x^2 - (\sum x)^2 / n} = \frac{[9519 - (117)(660)/8]^2}{1869 - 117^2 / 8} = 112.89$$

$$SSE = S_{yy} - \frac{S_{xy}^2}{S_{xx}} = 188 - 112.89 = 75.11$$

(b)    $r^2 = \dfrac{SSR}{SST} = \dfrac{112.89}{188} = 0.6005$

(c)    The percentage of variation in the observed y-values that is explained by the regression is 60.05%. In words, 60.05% of the variation in the score data is explained by study-time.

(d)    Based on the answers to parts (b) and (c), the regression equation appears to be somewhat useful for making predictions.

**14.95** (a)    In Exercise 14.65, we saw from the scatterplot that finding a regression line for these data was reasonable.

(b)    From the regression output in that exercise, the coefficient of determination (shown as R-Sq) was found to be 0.355.

(c)    The percentage of variation in the observed scores explained by the regression line is 35.5%; i.e., 35.5% of the variation in the death ages was explained by inauguration ages.

(d)    Although the pattern is linear, the variation around the fitted line is so great that the regression equation is only marginally useful for making predictions.

**14.97** (a)    In Exercise 14.67, we saw from the scatterplot that finding a regression line for these data was not reasonable. Parts (b-d) are omitted.

**14.99** (a)    In Exercise 14.69, we saw from the scatterplot that finding a regression line for these data was very reasonable.

(b)    From the regression output in that exercise, the coefficient of determination (shown as R-Sq) was found to be 0.952.

(c)    The percentage of variation in the observed values explained by the regression line is 95.2%; i.e., 95.2% of the variation in the average low temperatures was explained by the average high temperatures.

(d)    The regression equation is very useful for making predictions.

**14.101** (a)    Using Minitab, with the data in columns named INCOME and BEER, we

choose **Graph ▶ Scatterplot...**, select the **Simple** version and click **OK**.

Enter <u>BEER</u> in the first row of the **Y variables** column and <u>INCOME</u> in the first row of the **X variables** column. Click **OK**. The result is

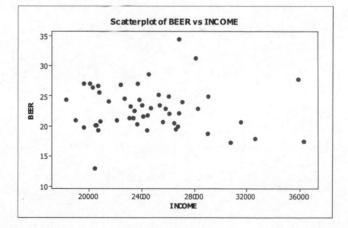

There does not appear to be a linear relationship between beer consumption and income. It is not reasonable to find a regression line. Parts (b-d) are omitted.

**14.103** (a)    In Exercise 14.74, we saw from the scatterplot that finding a regression line for these data was not reasonable because the data exhibited an increasing concave upward curved patternn. Parts (b-d) are omitted.

**14.105**(a)   Using Minitab, with the data in columns named ESTRIOL and WEIGHT, we
choose **Graph ▶ Scatterplot...**, select the **Simple** version and click **OK**.
Enter <u>WEIGHT</u> in the first row of the **Y variables** column and <u>ESTRIOL</u> in
the first row of the **X variables** column.  Click **OK**.  The result is

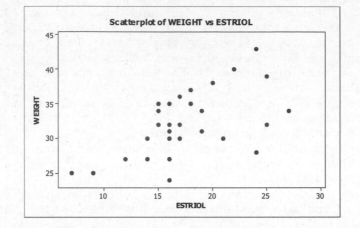

It appears that there is a linear relationship between Estriol and
Weight.  It is reasonable to find a linear regression equation.

(b)   Using Minitab, with the data in columns named ESTRIOL and WEIGHT, we
choose **Stat ▶ Regression ▶ Regression...**, select WEIGHT in the
**Response** text box, select ESTRIOL in the **Predictors** text box, and click
**OK**.  The results are

The regression equation is
WEIGHT = 21.5 + 0.608 ESTRIOL

| Predictor | Coef | StDev | T | P |
|---|---|---|---|---|
| Constant | 21.523 | 2.620 | 8.21 | 0.000 |
| ESTRIOL | 0.6082 | 0.1468 | 4.14 | 0.000 |

S = 3.821     **R-Sq = 37.2%**     R-Sq(adj) = 35.0%

Analysis of Variance

| Source | DF | SS | MS | F | P |
|---|---|---|---|---|---|
| Regression | 1 | 250.57 | 250.57 | 17.16 | 0.000 |
| Residual Error | 29 | 423.43 | 14.60 | | |
| Total | 30 | 674.00 | | | |

Thus the coefficient of determination is $r^2 = 0.372$.

(c)   Only 37.2% of the variation in the observed values of birth weight is
explained by a linear relationship with the predictor variable,
estriol level.

(d)   Based solely on the coefficient of determination, the regression
equation is not very useful for making predictions.

**14.107** (a) $r^2 = 1 - SSE/SST = (SST - SSE)/SST$. If the mean were used to predict the observed values of the response variable, the total squared error would be SST. By using the regression line to predict the observed values of the response variable, the sum of squares of the differences between the predicted values and the mean is SSR = SST - SSE. This is a reduction in the error sum of squares. Dividing this quantity by SST (and converting to a percentage) gives us the percentage reduction in the squared error when we use the regression equation instead of the mean to predict the observed values of the response variable.

(b) From Exercise 14.65, $r^2 = 0.9368$, so the percentage reduction obtained in the total squared error by using the regression equation instead of the mean of observed prices to predict the observed prices is 93.68%.

**Exercises 14.4**

**14.109** Pearson product moment correlation coefficient

**14.111** (a) $+1$ (b) not very useful

**14.113** False. It is possible that both variables are effects from a third variable. Correlation does not imply causation.

**14.115** The sign of the slope and of r are always the same. This can be seen

from $r = \dfrac{S_{xy}}{\sqrt{S_{xx}S_{yy}}} = \dfrac{S_{xy}}{S_{xx}}\dfrac{\sqrt{S_{xx}}}{\sqrt{S_{yy}}} = b_1\dfrac{\sqrt{S_{xx}}}{\sqrt{S_{yy}}}$ .

Since the ratio of square roots in the last term is always positive, r will have the same sign as $b_1$. Since the slope of the regression line is -3.58, r will also be negative. Therefore, $r = -\sqrt{r^2} = -\sqrt{0.709} = -0.842$

**14.117** To compute the linear correlation coefficient using the defining formula, begin with the following table.

| $x$ | $y$ | $x-\bar{x}$ | $y-\bar{y}$ | $(x-\bar{x})^2$ | $(y-\bar{y})^2$ | $(x-\bar{x})(y-\bar{y})$ |
|---|---|---|---|---|---|---|
| 3 | -4 | 1 | -1 | 1 | 1 | -1 |
| 1 | 0 | -1 | 3 | 1 | 9 | -3 |
| 2 | -5 | 0 | -2 | 0 | 4 | 0 |
| 6 | -9 | | | 2 | 14 | -4 |

The columns after the first two have used $\bar{x} = 2$ and $\bar{y} = -3$ .

We will need $s_x$ and $s_y$. These are given by

$$s_x = \sqrt{\frac{\sum(x_i - \bar{x})^2}{n-1}} = \sqrt{\frac{2}{2}} = 1 \ and \ s_y = \sqrt{\frac{\sum(y_i - \bar{y})^2}{n-1}} = \sqrt{\frac{14}{2}} = 2.646$$

$$r = \frac{\frac{1}{n-1}\sum(x_i - \bar{x})(y_i - \bar{y})}{s_x s_y} = \frac{\frac{1}{2}(-4)}{(1)(2)} = \frac{-2}{2.646} = -0.76$$

**14.119** To compute the linear correlation coefficient using the defining formula, begin with the following table.

| $x$ | $y$ | $x-\bar{x}$ | $y-\bar{y}$ | $(x-\bar{x})^2$ | $(y-\bar{y})^2$ | $(x-\bar{x})(y-\bar{y})$ |
|---|---|---|---|---|---|---|
| 3 | 4 | 0.5 | 2 | 0.25 | 4 | 1.0 |
| 4 | 5 | 1.5 | 3 | 2.25 | 9 | 4.5 |
| 1 | 0 | -1.5 | -2 | 2.25 | 4 | 3.0 |
| 2 | -1 | -0.5 | -3 | 0.25 | 9 | 1.5 |
| 10 | 8 | | | 5.00 | 26 | 10.0 |

The columns after the first two have used $\bar{x} = 2.5$ and $\bar{y} = 2$ .

We will need $s_x$ and $s_y$. These are given by

$$s_x = \sqrt{\frac{\sum(x_i - \bar{x})^2}{n-1}} = \sqrt{\frac{5}{3}} = 1.291 \ and \ s_y = \sqrt{\frac{\sum(y_i - \bar{y})^2}{n-1}} = \sqrt{\frac{26}{3}} = 2.944$$

$$r = \frac{\frac{1}{n-1}\sum(x_i - \bar{x})(y_i - \bar{y})}{s_x s_y} = \frac{\frac{1}{3}(10)}{(1.291)(2.944)} = \frac{10/3}{3.801} = 0.877$$

**14.121**   To compute the linear correlation coefficient using the defining formula, begin with the following table.

| $x$ | $y$ | $x - \bar{x}$ | $y - \bar{y}$ | $(x-\bar{x})^2$ | $(y-\bar{y})^2$ | $(x-\bar{x})(y-\bar{y})$ |
|---|---|---|---|---|---|---|
| 0 | 4 | -3 | 3 | 9 | 9 | -9 |
| 2 | 2 | -1 | 1 | 1 | 1 | -1 |
| 2 | 0 | -1 | -1 | 1 | 1 | 1 |
| 5 | -2 | 2 | -3 | 4 | 9 | -6 |
| 6 | 1 | 3 | 0 | 9 | 0 | 0 |
| 15 | 5 | | | 24 | 20 | -15 |

The columns after the first two have used $\bar{x} = 3$ and $\bar{y} = 1$.
We will need $s_x$ and $s_y$. These are given by

$$s_x = \sqrt{\frac{\sum(x_i - \bar{x})^2}{n-1}} = \sqrt{\frac{24}{4}} = 2.449 \ and \ s_y = \sqrt{\frac{\sum(y_i - \bar{y})^2}{n-1}} = \sqrt{\frac{20}{4}} = 2.236$$

$$r = \frac{\frac{1}{n-1}\sum(x_i - \bar{x})(y_i - \bar{y})}{s_x s_y} = \frac{\frac{1}{4}(-15)}{(2.449)(2.236)} = \frac{-3.75}{5.476} = -0.685$$

**14.123**   To compute the linear correlation coefficient, return to the table presented at the beginning of the solution to Exercise 14.89. Use the last row of this table to perform the calculations in part (a).

(a)   The linear correlation coefficient r is computed using the formula in Definition 14.6 of the text:

$$r = \frac{S_{xy}}{\sqrt{S_{xx}S_{xy}}} = \frac{\sum xy - (\sum x)(\sum y)/n}{\sqrt{[\sum x^2 - (\sum x)^2/n][\sum y^2 - (\sum y)^2/n]}}$$

$$= \frac{12348 - (41)(3222)/10}{\sqrt{[199 - 41^2/10][1063810 - 3222^2/10]}} = -0.967872$$

(b)   The value of r in part (a) suggests a strong negative linear relationship between age and price of Corvettes.

(c)   Data points are clustered closely about the regression line.

(d)   $r^2 = (-0.967872)^2 = 0.9368$. This matches the coefficient of determination that was calculated in part (b) of Exercise 14.89.

**14.125**   To compute the linear correlation coefficient, return to the table presented at the beginning of the solution to Exercise 14.91. Use the last row of this table to perform the calculations in part (a).

(a)   The linear correlation coefficient r is computed using the formula in Definition 14.6 of the text:

$$r = \frac{S_{xy}}{\sqrt{S_{xx}S_{xy}}} = \frac{\sum xy - (\sum x)(\sum y)/n}{\sqrt{[\sum x^2 - (\sum x)^2/n][\sum y^2 - (\sum y)^2/n]}}$$

$$= \frac{10486 - (723)(156.5)/11}{\sqrt{[48747 - 723^2/11][2523.252 - 156.5^2/11]}} = 0.331067$$

(b)    The value of r in part (a) suggests a weak positive linear relationship between potato plant weight and quantity of volatile emissions.

(c)    Data points are clustered very loosely about the regression line.

(d)    $r^2 = (0.331037)^2 = 0.1096$.  This matches the coefficient of determination that was calculated in part (b) of Exercise 14.91.

**14.127** To compute the linear correlation coefficient, return to the table presented at the beginning of the solution to Exercise 14.93.  Use the last row of this table to perform the calculations in part (a).

(a)    The linear correlation coefficient r is computed using the formula in Definition 14.6 of the text:

$$r = \frac{S_{xy}}{\sqrt{S_{xx}S_{xy}}} = \frac{\sum xy - (\sum x)(\sum y)/n}{\sqrt{[\sum x^2 - (\sum x)^2/n][\sum y^2 - (\sum y)^2/n]}}$$

$$= \frac{9519 - (117)(660)/8}{\sqrt{[1869 - 117^2/8][54638 - 660^2/8]}} = -0.7749$$

(b)    The value of r in part (a) suggests a fairly strong negative linear relationship between study-time and score for calculus students.

(c)    Data points are clustered fairly closely about the regression line.

(d)    $r^2 = (-0.7749)^2 = 0.6005$.  This matches the coefficient of determination that was calculated in part (b) of Exercise 14.93.

**14.129**    To compute the linear correlation coefficient, begin with the following table.

| x | y | xy | $x^2$ | $y^2$ |
|---|---|----|-----|-----|
| -3 | 9 | -27 | 9 | 81 |
| -2 | 4 | -8 | 4 | 16 |
| -1 | 1 | -1 | 1 | 1 |
| 0 | 0 | 0 | 0 | 0 |
| 1 | 1 | 1 | 1 | 1 |
| 2 | 4 | 8 | 4 | 16 |
| 3 | 9 | 27 | 9 | 81 |
| 0 | 28 | 0 | 28 | 196 |

(a)  The linear correlation coefficient r is computed using the formula in Definition 14.6 of the text:

$$r = \frac{S_{xy}}{\sqrt{S_{xx}S_{xy}}} = \frac{\sum xy - (\sum x)(\sum y)/n}{\sqrt{[\sum x^2 - (\sum x)^2/n][\sum y^2 - (\sum y)^2/n]}}$$

$$= \frac{0 - (0)(282)/7}{\sqrt{[28 - 0^2/7][196 - 28^2/7]}} = 0.0$$

(b)    We cannot conclude from the result in part (a) that x and y are unrelated.  We can conclude only that there is no *linear* relationship between x and y.

(c)                                          Graph for part (e)

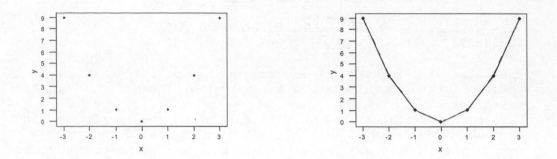

(d)   It is not appropriate to use the linear correlation coefficient as a
      descriptive measure for the data because the data points are not
      scattered about a line.

(e)   For each data point $(x, y)$, we have $y = x^2$.  (See columns 1 and 2 of
      the previous table.)  See the graph above at the right.

**14.131** (a)  r is zero.     (b)   r is negative.     (c)   r is positive.

**14.133** (a)  In Exercise 14.65, we saw that a linear relationship between the age at
      inauguration and the age at death was reasonable.  The scatterplot
      showed that the line relating the two variables had a positive slope.

(b)   From the output in Exercise 14.65, the coefficient of determination was
      0.355.  Since this is $r^2$ and r is positive because of the positive slope

      of the line, $r = \sqrt{0.355} = 0.596$.  You can also obtain the correlation

      coefficient in Minitab by choosing **Stat ▶ Basic Statistics ▶**
      **Correlation...**, select INAUGURATION and DEATH in the **Variables** text
      box, and click **OK**. The result is
           Pearson correlation of INAUGURATION and DEATH = 0.595

(c)   The correlation coefficient is positive, indicating that the linear
      relationship has a positive slope.  The coefficient is not very close
      to 1, indicating that the linear relationship is not very strong.

**14.135** (a)  In Exercise 14.67, we saw that a linear relationship between LOT SIZE
      and VALUE was not reasonable.  Parts (b-c) are omitted.

**14.137** (a)  In Exercise 14.69, we saw that a linear relationship between HIGH and
      LOW was reasonable.

(b)   In Minitab, choose **Stat ▶ Basic Statistics ▶ Correlation...**, select
      HIGH and LOW in the **Variables** text box, and click **OK**. The result is
           Pearson correlation of HIGH and LOW = 0.976

(c)   The correlation coefficient is positive, indicating that the linear
      relationship has a positive slope.  The coefficient is very close to 1,
      indicating that the linear relationship is very useful.

**14.139** (a)  In Exercise 14.71, we saw that a linear relationship between INCOME and
      BEER was not reasonable as there was no linear pattern in the data.
      Parts (b-c) are omitted.

**14.141** (a)  In Exercise 14.74, we saw that there was an increasing relationship
      between diameter and volume, but the relationship was concave upward,
      not linear.  Parts (b-c) are omitted.

**14.143** (a)  In Exercise 14.105, we saw that there was an increasing relationship
      between estriol levels of pregnant women and birth weights of their
      children.  Use of the correlation coefficient is appropriate as a
      descriptive measure of the strength of the linearity of that
      relationship.

(b) Using Minitab, with the data in columns named ESTRIOL and WEIGHT, we choose **Stat ▶ Basic Statistics ▶ Correlation...**, select ESTRIOL and WEIGHT in the **Variables** text box, and click **OK**. The result is

Pearson correlation of ESTRIOL and WEIGHT = 0.610, P-Value = 0.000

(c) There is a moderately positive linear relationship between estriol concentration and birth weight.  The data points will be moderately clustered about the regression line.

**14.145** (a) No.  We only know that the linear correlation coefficient is the square root of the coefficient of determination or it is the negative of that square root.

(b) No.  The slope is positive if r is positive and negative if r is negative, but we can not determine the sign of r.

(c) Yes.  $r = -\sqrt{0.716} = -0.846$

(d) Yes.  $r = \sqrt{0.716} = 0.846$

**14.147** (a) Using Minitab, choose **Graph ▶ Scatterplot**, select the **Simple** version, enter SCORE in the first row of the **Y variables** column and TIME in the first row of the **X variables** column, and click **OK**.  The result is

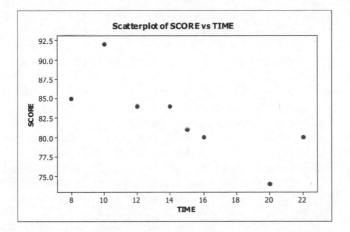

(b) Score decreases as time increases, so the using the rank correlation coefficient is reasonable.

(c) Score decreases linearly as time increases, so the using the linear correlation coefficient is reasonable.

(d) To find the rank correlation coefficient using Minitab, we must first find the ranks for each of the data values in the SCORE and TIME columns.  This is done for TIME by choosing **Data ▶ Rank**, entering TIME in the **Rank data in** text box and RANKTIME in the **Store ranks in** text box and clicking **OK**.  Then choose **Data ▶ Rank**, enter SCORE in the **Rank data in** text box and RANKSCORE in the **Store ranks in** text box and click **OK**.  Now choose **Stat ▶ Basic Statistics ▶ Correlation...**, select RANKTIME and RANKSCORE in the **Variables** text box, and click **OK**. The result is

Pearson correlation of RANKTIME and RANKSCORE = -0.928
P-Value = 0.001

The rank correlation coefficient is $-0.928$, indicating that the ranks are strongly negatively correlated and that SCORE decreases as TIME increases.

### Review Problems for Chapter 14

1.  (a) x  (b)  y  (c) $b_1$  (d)  $b_0$
2.  (a) It intersects the y-axis when $x = 0$.  Therefore $y = 4$.
    (b) It intersects the x-axis when $y = 0$.  Therefore $x = 4/3$.
    (c) slope = $-3$
    (d) The y-value decreases by 3 units when x increases by 1 unit.
    (e) The y-value increases by 6 units when x decreases by 2 units.
3.  (a) True.  The y-intercept is determined by $b_0$ and that value is independent of $b_1$, the slope.
    (b) False.  A horizontal line has a slope of zero.
    (c) True.  If the slope is positive, the x-values and y-values increase and decrease together.
4.  Scatterplot or scatter diagram
5.  A regression equation can be used to make predictions of the response variable for specific values of the predictor variable within the range of the observed values of the predictor variable.
6.  (a) predictor variable or explanatory variable
    (b) response variable
7.  (a) Based on the least-squares criterion, the line that best fits a set of data points is the one have the smallest possible sum of squared errors.
    (b) The line that best fits a set of data points according to the least-squares criterion is called the regression line.
    (c) Using a regression equation to make predictions for values of the predictor variable outside the range of the observed values of the predictor variable is called extrapolation.
8.  (a) An outlier is a data point that lies far from the regression line relative to the other data points.
    (b) An influential observation is a data point whose removal causes the regression equation to change considerably.  Often this is a data point for which the x-value is considerably to the left or right of the rest of the data points.
9.  The coefficient of determination is the percentage of the total variation in the y-values that is explained by the regression equation.
10.  (a) SST is the total sum of squares and measures the variation in the observed values of the response variable.
    (b) SSR is the regression sum of squares and measures the variation in the observed values of the response variable that is explained by the regression.  It can also be thought of as the variation in the predicted values of the response variable corresponding to the x-values in the data points.
    (c) SSE is the error sum of squares and measures the variation in the observed values of the response variable that is not explained by the regression.
11.  (a) One use of the linear correlation coefficient is as a descriptive measure of the strength of the linear relationship between two variables.
    (b) A positive linear relationship between two variables means that one variable tends to increase linearly as the other increases.
    (c) A value of r close to $-1$ suggests a strong negative linear relationship between the variables.

(d) A value of r close to <u>zero</u> suggests at most a weak linear relationship between the variables.

**12.** True.  It is quite possible that both variables are strongly affected by one (or more) other variables (called lurking variables).

**13.** (a) $y = 72 - 12x$        (b)      $b_0 = 72$, $b_1 = -12$
(c) The line slopes downward since $b_1 < 0$.
(d) After two years:  $y = 72 - 12(2) = \$48$ hundred $= \$4800$
After five years:  $y = 72 - 12(5) = \$12$ hundred $= \$1200$.

(e)

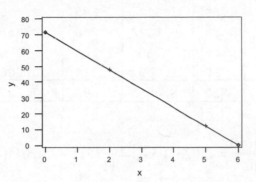

(f) From the graph, we estimate the value to be about $2500 after 4 years. The actual value is $y = 7200 - 1200(4) = \$2400$.

**14.** (a)

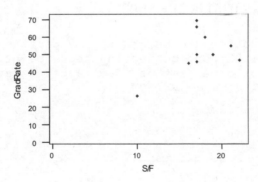

(b) It is moderately reasonable to find a regression line for the data because the data points appear to be scattered about a line.  This perception is, however, heavily influenced by the single data point at $x = 10$.

(c) The regression equation can be determined by calculating its slope and intercept.  Formulas for the slope ($b_1$) and intercept ($b_0$) of the regression equation are, respectively:

$$b_1 = \frac{\sum xy - \sum x \sum y / n}{\sum x^2 - (\sum x)^2 / n} \qquad b_0 = \frac{1}{n}\left(\sum y - b_1 \sum x\right)$$

To compute $b_0$ and $b_1$, construct a table of values for x, y, xy, $x^2$, and their sums. (Note:  A column for $y^2$ is also presented.  This is used in Problems 15 and 16.)

| x | y | xy | $x^2$ | $y^2$ |
|---|---|---|---|---|
| 16 | 45 | 720 | 256 | 2,025 |
| 20 | 55 | 1,100 | 400 | 3,025 |
| 17 | 70 | 1,190 | 289 | 4,900 |
| 19 | 50 | 950 | 361 | 2,500 |
| 22 | 47 | 1,034 | 484 | 2,209 |
| 17 | 46 | 782 | 289 | 2,116 |
| 17 | 50 | 850 | 289 | 2,500 |
| 17 | 66 | 1,122 | 289 | 4,356 |
| 10 | 26 | 260 | 100 | 676 |
| 18 | 60 | 1,080 | 324 | 3,600 |
| 173 | 515 | 9,088 | 3,081 | 27,907 |

Thus,

$$b_1 = \frac{9088 - 173(515)/10}{3081 - 173^2/10} = 2.02611 \quad b_0 = \frac{1}{11}(515 - 2.02611(173)) = 16.448$$

and the regression equation is: $\hat{y} = 16.448 + 2.02611x$ .

To graph the regression equation, begin by selecting two x-values within the range of the x-data. For the x-values 10 and 22, the calculated values for y are, respectively:

$$\hat{y} = 16.448 + 2.02611(10) = 36.71$$

$$\hat{y} = 16.448 + 2.02611(22) = 61.02$$

The regression equation can be graphed by plotting the pairs (10, 36.71) and (22, 61.02) and connecting these points with a line. This equation and the original set of 10 data points are presented as follows:

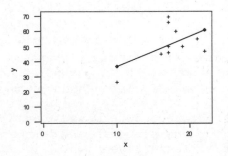

(d) Graduation rate tends to increase as the student-to-faculty ratio increases.

(e) Graduation rate increases an estimated 2.026% for each additional 1 unit increase in the student-to-faculty ratio.

(f) The predicted graduation rate for a university with a student-to-faculty ratio of 17 is 16.448 + 2.0261(17) = 50.89%.

(g) There is one potential influential observation, (10, 26) and no outliers. Looking at the scatterplot in part (c), we suspect that there may be little relationship between the student-to-faculty ratio and the graduation rate if the influential observation is removed from the data.

15. (a) To compute SST, SSR, and SSE using the computing formulas, begin with the table presented in Problem 14(c). Using the last row of this table and the formula in Definition 14.6 of the text, we obtain the three sums of squares as follows.

$$SST = S_{yy} = \sum y^2 - (\sum y)^2 / n = 27907 - 515/10 = 1384.5$$

$$SSR = \frac{S_{xy}^2}{S_{xx}} = \frac{[\sum xy - (\sum x)(\sum y)/n]^2}{\sum x^2 - (\sum x)^2 / n} = \frac{[9088 - (173)(515)/10]^2}{3081 - 173^2/10} = 361.66$$

$$SSE = S_{yy} - \frac{S_{xy}^2}{S_{xx}} = 1384.50 - 361.66 = 1022.846$$

$$r^2 = \frac{SSR}{SST} = \frac{361.66}{1384.50} = 0.261$$

(b) The percentage reduction obtained in the total squared error by using the regression equation, instead of the sample mean y, to predict the observed costs is 26.1%.

(c) The percentage of the variation in the graduation rate that is explained by the student-to-faculty ratio is 26.1%.

(d) The regression equation appears to be not very useful for making predictions.

16. (a) To compute the linear correlation coefficient, begin with the table presented in Problem 14(c). Using the last row of this table and the formula in Definition 14.6 of the text, we get

$$r = \frac{S_{xy}}{\sqrt{S_{xx}S_{xy}}} = \frac{\sum xy - (\sum x)(\sum y)/n}{\sqrt{[\sum x^2 - (\sum x)^2 / n][\sum y^2 - (\sum y)^2 / n]}}$$

$$= \frac{9088 - (173)(515)/10}{\sqrt{[3081 - 173^2/10][27907 - 515^2/10]}} = 0.511$$

(b) The value of r suggests a weak to moderate positive linear correlation.

(c) Data points are clustered about the regression line, but not very closely.

(d) $r^2 = (0.511)^2 = 0.261$

17. Using Minitab, with the data in three columns named POPULATION, AREA, and PLANTS, choose **Stat ▶ Basic statistics ▶ Correlation**, **enter** POPULATION, AREA, and PLANTS in the **Variables** text **box**. Make sure that the **Display P-values** box is checked. **Click OK.** The output is

|       | POPULATION | AREA   |
|-------|------------|--------|
| AREA  | 0.109      |        |
|       | 0.453      |        |
|       |            |        |
| PLANTS| 0.721      | -0.309 |
|       | 0.000      | 0.029  |

Cell Contents: Pearson correlation
                P-Value

(a) The correlation coefficient between population and area is 0.109.

(b) The correlation coefficient between population and number of exotic plants is 0.721.

(c) The correlation coefficient between area and number of exotic plants is -0.309.

(d) Any linear association between population and area is very weak since the correlation coefficient is near zero. There is a moderately strong positive linear relationship between population and number of exotic plants. There is a fairly weak negative linear association between area and number of exotic plants. One possible explanation of the stronger positive relationship between population and number of exotic plants might be that the more people there are, the greater the possibility that exotic plants are introduced into the state by those people, intentionally or unintentionally.

18. (a) Using Minitab, with the data in columns named FAT and DEATHRATE, choose

**Graph ▶ Scatterplot...**, select the **Simple** version, enter 'LIFE EXPECTANCY' in row 1 of the **Y variables** column and IMR in the **X variables** column and click **OK**.

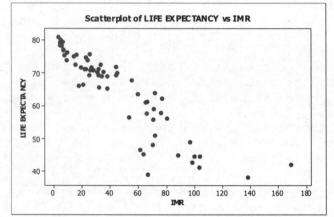

Life expectancy appears to decrease linearly as IMR increases.

(b) It appears to be reasonable to find a regression line for the data, although we expect that there will be both outliers and influential observations.

(c) Choose **Stat ▶ Regression ▶ Regression**, enter 'LIFE EXPECTANCY' in the **Response box** and IMR in the **Predictor box**. Click on the Options button and enter 30 in the Prediction intervals for new observations text box. Click **OK** twice. The results are

The regression equation is
**LIFE EXPECTANCY = 78.8 - 0.312 IMR**

| Predictor | Coef | SE Coef | T | P |
|---|---|---|---|---|
| Constant | 78.811 | 1.031 | 76.43 | 0.000 |
| IMR | -0.31173 | 0.01840 | -16.94 | 0.000 |

S = 5.24948   R-Sq = 81.8%   R-Sq(adj) = 81.5%

Analysis of Variance

| Source | DF | SS | MS | F | P |
|---|---|---|---|---|---|
| Regression | 1 | 7909.7 | 7909.7 | 287.03 | 0.000 |
| Residual Error | 64 | 1763.6 | 27.6 | | |
| Total | 65 | 9673.4 | | | |

```
Unusual Observations
                LIFE
Obs   IMR   EXPECTANCY      Fit   SE Fit   Residual   St Resid
  2   169       42.000   26.161    2.393     15.839      3.39RX
 32    63       45.200   59.048    0.741    -13.848     -2.66R
 39   138       38.200   35.856    1.849      2.344      0.48 X
 50    61       46.600   59.859    0.719    -13.259     -2.55R
 66    67       39.000   58.082    0.771    -19.082     -3.67R
```

R denotes an observation with a large standardized residual.
X denotes an observation whose X value gives it large influence.

```
Predicted Values for New Observations
New
Obs     Fit   SE Fit        95% CI               95% PI
  1   69.460   0.693   (68.074, 70.845)   (58.882, 80.038)
```

```
Values of Predictors for New Observations
New
Obs   IMR
  1   30.0
```

The regression equation, **LIFE EXPECTANCY = 78.8 - 0.312 IMR**, indicates that life expectancy decreases as the infant mortality rate increases.

(d)  For a country with an IMR = 30, the predicted life expectancy is 78.8 - 0.312(30) = 69.44 years.  Note:  The output above shows 69.460 years, using more decimal places in the calculation.

(e)  Choose **Stat ▶ Basic statistics ▶ Correlation**, enter `LIFE EXPECTANCY`

and IMR in the **Variables** text box.  Click **OK**.  The results are
```
Pearson correlation of IMR and LIFE EXPECTANCY = -0.904
P-Value = 0.000
```
Life expectancy and IMR are highly negatively correlated. Thus, there is strong negative linear association between the two variables.

(f)  The output shown in part (c) indicates that observations 2, 32, 50, and 66 are potential outliers and observations 2 and 39 are influential observations.

**19.** (a)  Using Minitab, with the data in columns named IMR and LIFE EXPECTANCY,

choose **Graph ▶ Scatterplot…**, select the **Simple** version, enter `LIFE

EXPECTANCY` in row 1 of the **Y variables** column and IMR in the **X variables** column and click **OK**.

The graph at the right does not show any relationship between average July precipitation and average July high temperature.

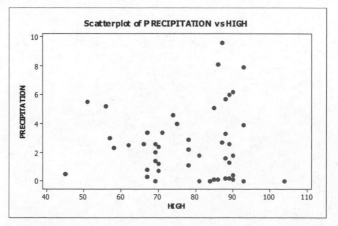

(b)  The graph in part (a) does not indicate that there is any linear relationship between average precipitation and average high temperature in July for 48 cities.  Parts (c-f) are omitted.

20.    (a)  Using Minitab, with the data in columns named FAT and DEATHRATE, choose
            **Graph ▶ Scatterplot...**, select the **Simple** version, enter DEATHRATE in row
            1 of the **Y variables** column and FAT in the **X variables** column and click
            OK.

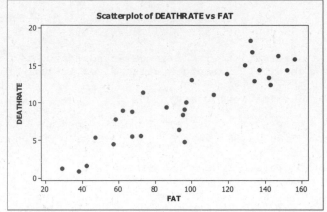

            The scatter diagram shows that the death rate from prostate cancer
            tends to increase linearly with an increase in fat consumption.
       (b)  Obtaining a linear regression equation for the data appears
            reasonable.  There is no noticeable curvature in the data, and there
            is an increasing trend along a line.

       (c)  Choose **Stat ▶ Regression ▶ Regression**, enter DEATHRATE in the
            **Response box** and FAT in the **Predictor box**.  Click on the Options
            button and enter in the **Prediction intervals for new observations** text
            box.  Click **OK** twice.   The results are

            The regression equation is
            **DEATHRATE = - 1.06 + 0.113 FAT**

            Predictor      Coef   SE Coef       T       P
            Constant     -1.063     1.170   -0.91   0.371
            FAT         0.11336   0.01126   10.06   0.000

            S = 2.29453   R-Sq = 78.3%   R-Sq(adj) = 77.6%

            Analysis of Variance

            Source           DF       SS        MS       F       P
            Regression        1   533.31    533.31  101.30   0.000
            Residual Error   28   147.42      5.26
            Total            29   680.73

            Unusual Observations

            Obs   FAT   DEATHRATE      Fit   SE Fit   Residual   St Resid
              5    96       4.800    9.820    0.419     -5.020     -2.23R
             30   132      18.400   13.901    0.575      4.499      2.03R

            R denotes an observation with a large standardized residual.

Predicted Values for New Observations

```
New
Obs    Fit  SE Fit       95% CI          95% PI
  1  9.366   0.423  (8.500, 10.232)  (4.587, 14.145)
```

Values of Predictors for New Observations

```
New
Obs    FAT
  1   92.0
```

The slope of the regression line is 0.11336.  This means that for each 1 unit increase in fat consumption, the death rate goes up by 0.11336. Note that this is an observational study, and therefore this relationship cannot be construed as a cause and effect relationship. There may be other lurking variables.  For example, the countries with the highest fat consumption and death rates tend to be those in which the life expectancy is greatest.  Since prostate cancer is typically a disease that strikes older men, some of the countries with lower death rates from prostate cancer may be experiencing higher rates of other diseases that cause death before a man can experience prostate cancer.

(d) For a country with a per capita fat consumption of 92 grams per day, the predicted prostate cancer death rate, from the output above, is 9.366 per 100.000 males.  This can also be obtained by substituting 92 for FAT in the regression equation.

(e) Choose **Stat ▶ Basic statistics ▶ Correlation, enter** <u>DEATHRATE</u> and <u>FAT</u> in the **Variables** text **box.**  Click **OK.**  The results are
Pearson correlation of FAT and DEATHRATE = 0.885 P-Value = 0.000

The correlation coefficient is quite high, indicating a fairly strong linear association between fat consumption and prostate cancer death rate.

(f) From the regression output in part (c), we see that there are two potential outliers, observations 5 and 30, which arise from the countries of Greece and Sweden.  The data point for Greece lies below the fitted line and the one for Sweden lies above the line.  No observations are identified as potential influential observations.

**21.** (a) Using Minitab, with the data in columns named PAYROLL and WINNING,
choose **Graph ▶ Scatterplot…,** select the **Simple** version, enter <u>WINNING</u> in row 1 of the **Y variables** column and <u>PAYROLL</u> in the **X variables** column and click **OK.**

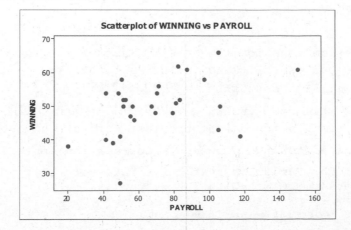

There is a moderate increasing linear trend shown in the scatterplot.

(b)  Finding a regression line is reasonable, although there may be potential outliers or influential observations.

(c)  Choose **Stat ▶ Regression ▶ Regression, enter** WINNING in the **Response** box and PAYROLL in the **Predictor** box. Enter 50 and 100 in column C4 of the worksheet. Click on the **Options** button and enter <u>C4</u> in the **Prediction intervals for new observations** text box. Click **OK** twice. The results are

```
The regression equation is
WINNING = 40.9 + 0.128 PAYROLL

Predictor     Coef   SE Coef      T      P
Constant    40.949    3.879   10.56  0.000
PAYROLL     0.12839  0.05134    2.50  0.019

S = 7.65251   R-Sq = 18.3%   R-Sq(adj) = 15.3%

Analysis of Variance

Source          DF       SS      MS      F      P
Regression       1   366.29  366.29   6.25  0.019
Residual Error  28  1639.71   58.56
Total           29  2006.00

Unusual Observations

Obs  PAYROLL  WINNING    Fit  SE Fit  Residual  St Resid
  3      150    61.00  60.21    4.31      0.79      0.13 X
 25      117    41.00  55.97    2.77    -14.97     -2.10R
 30       49    27.00  47.24    1.78    -20.24     -2.72R

R denotes an observation with a large standardized residual.
X denotes an observation whose X value gives it large influence.

Predicted Values for New Observations

New
Obs    Fit  SE Fit      95% CI           95% PI
  1  47.37    1.75  (43.79, 50.95)  (31.29, 63.45)
  2  53.79    2.06  (49.57, 58.01)  (37.55, 70.02)

Values of Predictors for New Observations

New
Obs  PAYROLL
  1       50
  2      100
```

The regression equation is $\hat{y} = 40.9 + 0.128x$. In general, the winning percentage and payroll tend to increase together.

(d)  For a $50 million payroll, $\hat{y} = 40.9 + 0.128(50) = 47.3$.

For a $100 million payroll, $\hat{y} = 40.9 + 0.128(100) = 53.7$.

These can also be found from the last part of the output in part (c).

(e)  Choose **Stat ▶ Basic statistics ▶ Correlation, enter** <u>PAYROLL</u> and <u>WINNING</u> in the **Variables** text box. Make sure that the **Display P-values** box is checked. Click **OK**. The output is

```
Pearson correlation of PAYROLL and WINNING = 0.427,    P-Value = 0.019
```

There is a moderate correlation between payroll and winning percentage. From the scatterplot, we see that there are several data points that would fall well away from a fitted line.  A line will not do very well as a model for these data.

(f)  The output in part (c) indicates that observations 25 and 30 (Mets and Tigers) are potential outliers and observation 3 (Yankees) is an influential observation.

**Exercises 15.1**

**15.1** Conditional distribution, conditional mean, and conditional standard deviation

**15.3** (a) population regression line        (b)   $\sigma$

(c) normal, $\beta_0 + 6\,\beta_1$, $\sigma$

**15.5** The sample regression line is the best estimate of the population regression line.

**15.7** Residual

**15.9** The plot of the residuals against the values of the predictor variable provides the same information as a scatter diagram of the data points. However, it has the advantage of making it easier to spot patterns such as curvature and non-constant standard deviation.

**15.11** (a) We first create the table below.

| x | y | $\hat{y}$ | $e = y - \hat{y}$ | $e^2$ |
|---|---|---|---|---|
| 3 | -4 | -5 | 1 | 1 |
| 1 | 0 | -1 | 1 | 1 |
| 2 | -5 | -3 | -2 | 4 |
|  |  |  |  | 6 |

$$s_e = \sqrt{\frac{SSE}{n-2}} = \sqrt{\frac{6}{1}} = 2.449$$

(b-c)     For the residual plot, we plot e against x in the graph below at the left. We can use Excel for the normal probability plot. With the values of e in a column named 'e', highlight the data with the mouse, choose **DDXL**, then select **Charts and Plots**, click on the **Function Type** drop down box, and click on **Normal Probability Plot**. Drag the 'e' from the **Names and Columns** box into the **Quantitative Variable** box and click **OK**. The graph is shown below at the right.

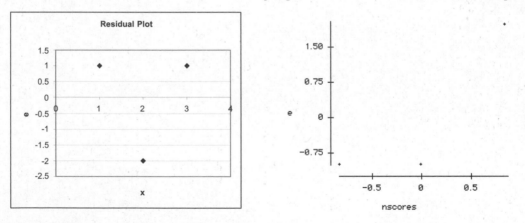

**15.13** (a) We first create the table below.

| x | y | $\hat{y}$ | $e = y - \hat{y}$ | $e^2$ |
|---|---|---|---|---|
| 3 | 4 | 3 | 1 | 1 |
| 4 | 5 | 5 | 0 | 0 |
| 1 | 0 | -1 | 1 | 1 |
| 2 | -1 | 1 | -2 | 4 |
|  |  |  |  | 6 |

$$s_e = \sqrt{\frac{SSE}{n-2}} = \sqrt{\frac{6}{2}} = 1.732$$

(b-c)     For the residual plot, we plot e against x in the graph following

at the left.  We can use Excel for the normal probability plot.
With the values of e in a column named 'e', highlight the data with
the mouse, choose **DDXL**, then select **Charts and Plots**, click on the
**Function Type** drop down box, and click on **Normal Probability Plot**.
Drag the 'e' from the **Names and Columns** box into the **Quantitative
Variable** box and click **OK**.  The graph is shown following at the
right.

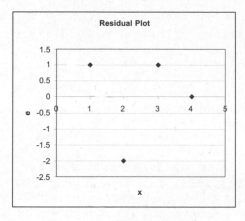

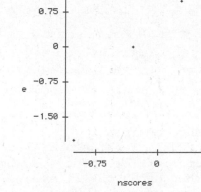

**15.15** (a)  We first create the table below.

| x | y | $\hat{y}$ | $e = y - \hat{y}$ | $e^2$ |
|---|---|---|---|---|
| 0 | 4 | 2.875 | 1.125 | 1.265625 |
| 2 | 2 | 1.625 | 0.375 | 0.140625 |
| 2 | 0 | 1.625 | −1.625 | 2.640625 |
| 5 | −2 | −0.250 | −1.750 | 3.062500 |
| 6 | 1 | −0.875 | 1.875 | 3.515625 |
| | | | | 10.625000 |

$$s_e = \sqrt{\frac{SSE}{n-2}} = \sqrt{\frac{10.625}{3}} = 1.882$$

(b-c)  For the residual plot, we plot e against x in the graph following at
the left.  We can use Excel for the normal probability plot.  With
the values of e in a column named 'e', highlight the data with the
mouse, choose **DDXL**, then select **Charts and Plots**, click on the
**Function Type** drop down box, and click on **Normal Probability Plot**.
Drag the 'e' from the **Names and Columns** box into the **Quantitative
Variable** box and click **OK**.  The graph is shown following at the
right.

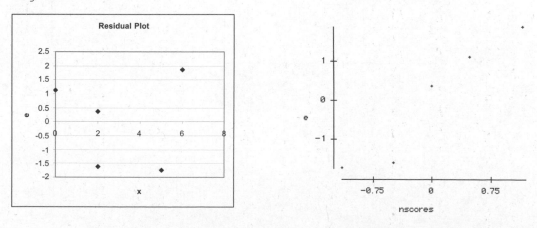

**15.17** If the assumptions for regression inferences are satisfied for a model relating a Corvette's age to its price, this means that there are constants $\beta_0$, $\beta_1$, and $\sigma$ such that, for each age $x$, the prices for Corvettes of that age are normally distributed with mean $\beta_0 + \beta_1 x$ and standard deviation $\sigma$.

**15.19** If the assumptions for regression inferences are satisfied for a model relating the volume of plant emissions of volatile compounds to the weight of the plant, this means that there are constants $\beta_0$, $\beta_1$, and $\sigma$ such that, for each weight $x$, the volumes of emissions $y$ are normally distributed with mean $\beta_0 + \beta_1 x$ and standard deviation $\sigma$.

**15.21** If the assumptions for regression inferences are satisfied for a model relating the test scores of calculus students to their study-times, this means that there are constants $\beta_0$, $\beta_1$, and $\sigma$ such that, for each study-time $x$, the test scores $y$ are normally distributed with mean $\beta_0 + \beta_1 x$ and standard deviation $\sigma$.

**15.23** To compute the standard error of the estimate, first retrieve the computation for the error sum of squares (SSE) in part (a) of Exercise 14.89 and then apply the formula for $s_e$ in Definition 15.1 of the text.

(a)  In part (a) of Exercise 14.89, $n = 10$ and SSE was computed as 1,623.7. Applying Definition 15.1, the standard error of the estimate is

$$s_e = \sqrt{\frac{SSE}{n-2}} = \sqrt{\frac{1623.7}{10-2}} = 14.2465$$

(b)  Presuming that the variables age ($x$) and price ($y$) for Corvettes satisfy Assumptions (1) - (3) for regression inferences, the standard error of the estimate $s_e = 14.2465$ provides an estimate for the common population standard deviation $\sigma$ of prices for all Corvettes of any given age.

(c)  Obtain the necessary normal scores from Table III in the text.

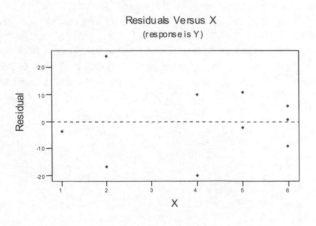

Residuals Versus X
(response is Y)

| Age x | Residual e |
|---|---|
| 6 | 0.82 |
| 6 | -9.18 |
| 6 | 5.82 |
| 2 | 24.20 |
| 2 | -16.80 |
| 5 | -2.09 |
| 4 | 10.01 |
| 5 | 10.91 |
| 1 | -3.70 |
| 4 | -19.99 |

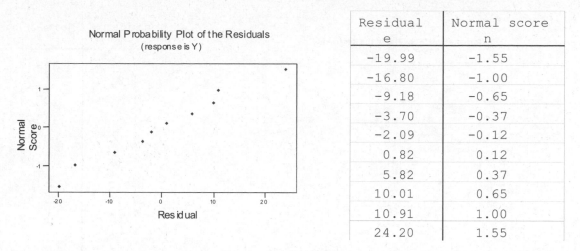

Normal Probability Plot of the Residuals
(response is Y)

| Residual e | Normal score n |
|---|---|
| -19.99 | -1.55 |
| -16.80 | -1.00 |
| -9.18 | -0.65 |
| -3.70 | -0.37 |
| -2.09 | -0.12 |
| 0.82 | 0.12 |
| 5.82 | 0.37 |
| 10.01 | 0.65 |
| 10.91 | 1.00 |
| 24.20 | 1.55 |

(d)  Taking into account the small sample size, we can say that the residuals fall roughly in a horizontal band centered and symmetric about the x-axis.  We can also say that the normal probability plot for residuals is very roughly linear.  Therefore, it appears reasonable to consider the assumptions for regression inferences for the variables age and price of Corvettes to be met.

15.25 To compute the standard error of the estimate, first retrieve the computation for the error sum of squares (SSE) in part (a) of Exercise 14.91 and then apply the formula for $s_e$ in Definition 15.1 of the text.

(a)  In part (a) of Exercise 14.91, n = 10 and SSE was computed as 88.5. Applying Definition 15.1, the standard error of the estimate is

$$s_e = \sqrt{\frac{SSE}{n-2}} = \sqrt{\frac{264.16}{11-2}} = 5.4177 \text{ hundred nanograms}$$

(b)  Presuming that the weight (x) of the potato plants Solanum tubersom and volume of volatile emissions (y) from the plants satisfy Assumptions (1) - (3) for regression inferences, the standard error of the estimate $s_e$ = 5.4177 hundred nanograms (541.77 nanograms) provides an estimate for the common population standard deviation $\sigma$ of emission volumes for all plants of any particular weight.

(c)  Obtain the necessary normal scores from Table III in the text.

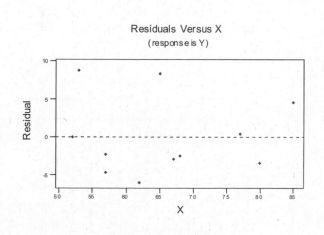

Residuals Versus X
(response is Y)

| Weight x | Residual e |
|---|---|
| 57 | -4.81 |
| 85 | 4.63 |
| 57 | -2.31 |
| 65 | 8.39 |
| 52 | 0.01 |
| 67 | -2.93 |
| 62 | -6.12 |
| 80 | -3.55 |
| 77 | 0.44 |
| 53 | 8.85 |
| 68 | -2.60 |

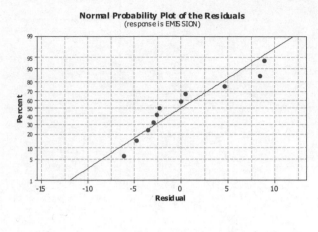

Normal Probability Plot of the Residuals
(response is EMISSION)

| Residual e | Normal score n |
|---|---|
| −6.12 | −1.59 |
| −4.81 | −1.06 |
| −3.55 | −0.73 |
| −2.93 | −0.46 |
| −2.60 | −0.22 |
| −2.31 | 0.00 |
| 0.01 | 0.22 |
| 0.44 | 0.46 |
| 4.63 | 0.73 |
| 8.39 | 1.06 |
| 8.85 | 1.59 |

(d)  Taking into account the small sample size, we can say that the residuals fall roughly in a horizontal band centered and symmetric about the x-axis.  We can also say that the normal probability plot for residuals is approximately linear.

Therefore, based on the sample data, there are no obvious violations of the assumptions for regression inferences for the variables plant weight and volume of volatile emissions.

**15.27** To compute the standard error of the estimate, first retrieve the computation for the error sum of squares (SSE) in part (a) of Exercise 14.93 and then apply the formula for $s_e$ in Definition 15.1 of the text.

(a)  In part (a) of Exercise 14.93, n = 8 and SSE was computed as 75.11. Applying Definition 15.1, the standard error of the estimate is

$$s_e = \sqrt{\frac{SSE}{n-2}} = \sqrt{\frac{75.11}{8-2}} = 3.5382$$

(b)  Presuming that the study-time (x) of the calculus students and their test scores (y) satisfy Assumptions (1) - (3) for regression inferences, the standard error of the estimate $s_e$ = 3.5382 provides an estimate for the common population standard deviation $\sigma$ of test scores for all calculus students with any particular study-time.

(c)  Obtain the necessary normal scores from Table III in the text.

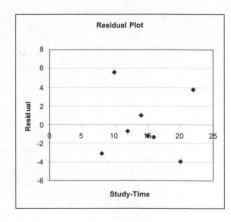

Residual Plot

| Study Time x | Residual e |
|---|---|
| 10 | 5.59 |
| 15 | −1.18 |
| 12 | −0.72 |
| 20 | −3.95 |
| 8 | −3.10 |
| 16 | −1.34 |
| 14 | 0.97 |
| 22 | 3.74 |

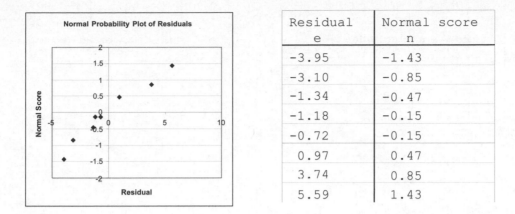

| Residual e | Normal score n |
|---|---|
| -3.95 | -1.43 |
| -3.10 | -0.85 |
| -1.34 | -0.47 |
| -1.18 | -0.15 |
| -0.72 | -0.15 |
| 0.97 | 0.47 |
| 3.74 | 0.85 |
| 5.59 | 1.43 |

(d) Taking into account the small sample size, we can say that the residuals fall roughly in a horizontal band centered and symmetric about the x-axis.  We can also say that the normal probability plot for residuals is approximately linear.  Therefore, based on the sample data, there are no obvious violations of the assumptions for regression inferences for the variables study-time and test scores of calculus students.

**15.29** (a) The assumption of linearity (Assumption 1) may be violated since the band is not horizontal, as may be the assumption of equal standard deviations (Assumption 2) since there is more variation in the residuals for small x than for large x.

(b) It appears that the standard deviation does not remain constant; thus, Assumption 2 is violated.

(c) The graph does not suggest violation of one or more of the assumptions for regression inferences.

(d) The normal probability plot appears to be more curved than linear; thus, the assumption of normality is violated.

**15.31** Using Minitab, with the data in columns named INAUGURATION and DEATH, we choose **Stat ▶ Regression ▶ Regression...**, select <u>DEATH</u> in the **Response** text box, and select <u>INAUGURATION</u> in the **Predictors** text box.  Click the **Graphs** button, select the **Regular** option button from the **Residuals for Plots** list, select the **Normal plot of residuals** check box from the **Residual Plots** list, click in the **Residuals versus the variables** text box and specify <u>INAUGURATION</u>, click **OK**, and click **OK**.  The results are

```
The regression equation is
DEATH = 8.7 + 1.12 INAUGURATION

Predictor       Coef   SE Coef     T      P
Constant        8.66    13.84    0.63   0.535
INAUGURATION   1.1172    0.2512  4.45   0.000

S = 9.47690    R-Sq = 35.5%    R-Sq(adj) = 33.7%

Analysis of Variance

Source          DF     SS       MS      F      P
Regression       1   1776.5   1776.5  19.78  0.000
Residual Error  36   3233.2     89.8
Total           37   5009.7
```

(a) The standard error of the estimate is 9.47690.  Roughly speaking, the predicted death ages differ, on average, from the observed death ages by 9.47690 years.

(b)

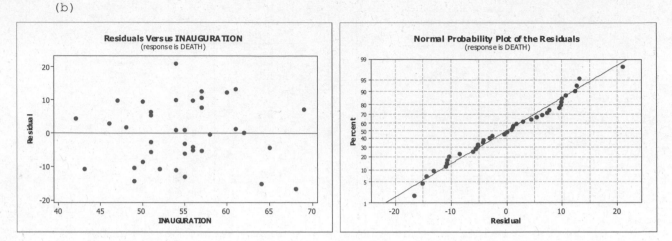

(c)   The residuals fall roughly in a horizontal band centered and symmetric
      about the x-axis.  We can also say that the normal probability plot for
      residuals is approximately linear.  Therefore, based on the sample
      data, there are no obvious violations of the assumptions for regression
      inferences for the variables inauguration age and death age of U.S.
      Presidents.

**15.33** Using Minitab, with the data in columns named LOT SIZE and VALUE, we choose

**Stat ▶ Regression ▶ Regression...,** select <u>VALUE</u> in the **Response** text box,

and select <u>'LOT SIZE'</u> in the **Predictors** text box.   Click the **Graphs** button,
select the **Regular** option button from the **Residuals for Plots** list, select
the **Normal plot of residuals** check box from the **Residual Plots** list, click
in the **Residuals versus the variables** text box and specify <u>'LOT SIZE'</u>, click
**OK,** and click **OK.**   The results are

```
The regression equation is
VALUE = 292 + 67.1 LOT SIZE

Predictor    Coef   SE Coef     T       P
Constant     292.0    140.3   2.08   0.043
LOT SIZE     67.11     59.87  1.12   0.269

S = 139.012   R-Sq = 2.9%   R-Sq(adj) = 0.6%

Analysis of Variance
Source          DF       SS     MS      F      P
Regression       1     24286  24286   1.26  0.269
Residual Error  42    811619  19324
Total           43    835905
```

(a)   The standard error of the estimate is 139.012 thousand dollars.
      Roughly speaking, the predicted values differ, on average, from the
      observed values by 139.012 thousand dollars.

(b)

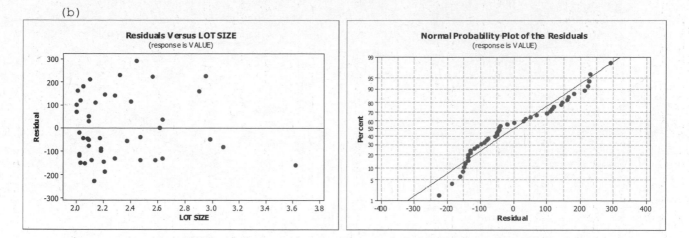

(c)  The residuals appear to fall roughly in a band centered and symmetric
     about the x-axis.  We can also say that the normal probability plot for
     residuals deviates from a linear pattern, but it is difficult to say
     whether the deviation is enough to rule out normality.  Therefore, for
     the time being, based on the sample data, the assumptions for
     regression inferences seem to be reasonable for the variables lot size
     and value.  More on these data later.

15.35 Using Minitab, with the data in columns named HIGH and LOW, we choose **Stat**

▶ **Regression** ▶ **Regression...**, select <u>LOW</u> in the **Response** text box, and
select <u>HIGH</u> in the **Predictors** text box.  Click the **Graphs** button, select the
**Regular** option button from the **Residuals for Plots** list, select the **Normal
plot of residuals** check box from the **Residual Plots** list, click in the
**Residuals versus the variables** text box and specify <u>HIGH</u>, click **OK**, and
click **OK**.   The results are

```
The regression equation is
LOW = - 7.57 + 0.917 HIGH


Predictor      Coef  SE Coef      T      P
Constant     -7.572    1.786  -4.24  0.000
HIGH        0.91685  0.02967  30.91  0.000


S = 4.15111   R-Sq = 95.2%   R-Sq(adj) = 95.1%


Analysis of Variance


Source          DF     SS     MS       F      P
Regression       1  16459  16459  955.17  0.000
Residual Error  48    827     17
Total           49  17286
```

(a)  The standard error of the estimate is 4.15111 degrees.  Roughly
     speaking, the predicted low temperatures differ, on average, from the
     observed low temperatures by 4.15111 degrees.

(b)

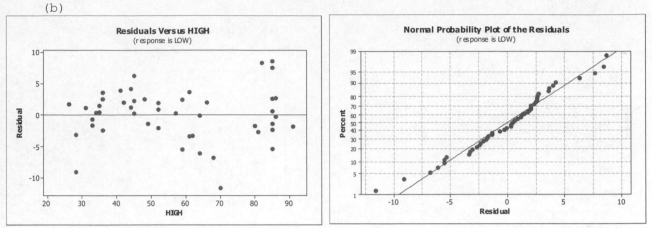

(c)  The residuals appear to fall roughly in a band centered and symmetric
about the x-axis.  We can also say that the normal probability plot for
residuals deviates only slightly from a linear pattern.  Therefore,
based on the sample data, the assumptions for regression inferences
seem to be reasonable for the variables average high January
temperature and average low January temperature.

**15.37** Using Minitab, with the data in columns named DISP and MPG, we choose **Stat**

▶ **Regression** ▶ **Regression...**, select <u>DISP</u> in the **Response** text box, select

<u>MPG</u> in the **Predictors** text box, select **Residuals** from the **Storage** check-box
list.  Click the **Graphs...** button, select the **Regular** option button from the
**Residuals for Plots** list, select the **Normal plot of residuals** check box from
the **Residual Plots** list, click in the **Residuals versus the variables** text
box and specify <u>DISP</u>, click **OK**, and click **OK**.  The results are

(a)      The regression equation is
         MPG = 32.3 - 3.58 DISP

         Predictor       Coef        StDev          T        P
         Constant     32.2732       0.6704      48.14    0.000
         DISP         -3.5762       0.2099     -17.03    0.000


         **S = 2.248**      R-Sq = 70.9%     R-Sq(adj) = 70.7%

         Analysis of Variance

         Source         DF          SS          MS         F         P
         Regression      1        1465.8      1465.8    290.18    0.000
         Residual Error 119         601.1         5.1
         Total         120        2067.0

The standard error of the estimate is the first entry in the sixth
line of the computer output.  It is reported as s = 2.248.  Presuming
that the variables DISP and MPG satisfy the assumptions for regression
inferences, the standard error of the estimate, 2.248, provides an
estimate for the common population standard deviation, $\sigma$, of miles
per gallon for all cars with a particular engine displacement.

(b)    The two graphs that result follow.

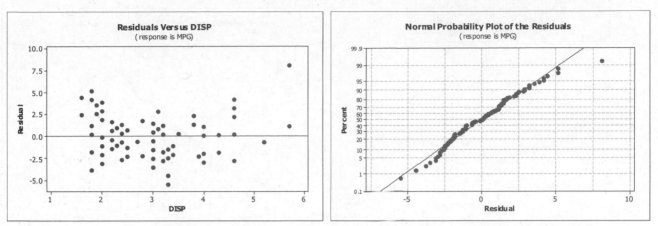

(c)    The residuals plotted against engine displacement display a concave
       upward curve, indicating that the linear model is not appropriate for
       the data. Thus, Assumption 1 is violated and the statement made in
       part (a) interpreting the meaning of the standard error does not hold.

**15.39** Using Minitab, with the data in columns named DIAMETER and VOLUME, we choose

**Stat ▶ Regression ▶ Regression...**, select <u>VOLUME</u> in the **Response** text box,
select <u>DIAMETER</u> in the **Predictors** text box, select **Residuals** from the
**Storage** check-box list.  Click the **Graphs...** button, select the **Regular**
option button from the **Residuals for Plots** list, select the **Normal plot of
residuals** check box from the **Residual Plots** list, click in the **Residuals
versus the variables** text box and specify <u>DIAMETER</u>, click **OK**, and click **OK**.
The results are

(a)          The regression equation is
             VOLUME = - 41.6 + 6.84 DIAMETER


             Predictor      Coef       StDev          T         P
             Constant     -41.568      3.427      -12.13     0.000
             DIAMETER      6.8367      0.2877       23.77     0.000


             S = 9.875        R-Sq = 89.3%      R-Sq(adj) = 89.1%


             Analysis of Variance

             Source          DF         SS          MS         F         P
             Regression       1       55083       55083     564.88     0.000
             Residual Error   68       6631          98
             Total            69      61714

The standard error of the estimate is the first entry in the sixth line
of the computer output.  It is reported as s = 9.875.  Presuming that
the variables DIAMETER and VOLUME satisfy the assumptions for
regression inferences, the standard error of the estimate, 9.875,
provides an estimate for the common population standard deviation, $\sigma$ ,
of volumes for all trees having a particular diameter at breast height.

(b)   The two graphs that result follow.

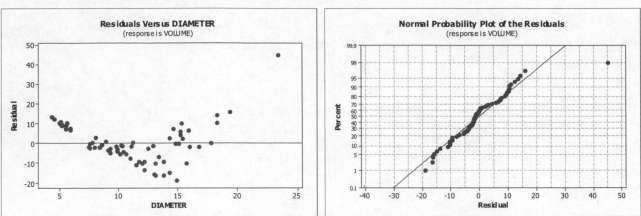

(c)   The residuals plotted against diameter are displayed in a curved
      concave upward pattern, indicating that the linear model is not
      appropriate.  This is a violation of Assumption 1. In addition, both
      plots indicate the presence of an outlier.  Finally, it appears that
      there is more variability in the center of the first graph than at the
      left side, so Assumption 2 is also violated.  Removal of the outlier
      will not change the curved pattern of the other residuals or the
      unequal variation in the residuals, so this data should not be
      analyzed using a linear model.

### Exercises 15.2

**15.41** normal distribution with mean $\beta_1$ = -3.5

**15.43** We can also use the coefficient of determination, $r^2$, and the linear
correlation coefficient, r, as a basis for a test to decide whether a
regression equation is useful for prediction.

**15.45** From Exercise 15.11, we have $s_e = 2.44949$ .

Also, $\sum x_i = 6; \sum x_i^2 = 14; n=3; \quad S_{xx} = \sum x_i^2 - \left(\sum x_i\right)^2 / n = 14 - 6^2/3 = 2$

(a)   $t = \dfrac{b_1}{s_e / \sqrt{S_{xx}}} = \dfrac{-2}{2.44949/\sqrt{2}} = -1.155$ ; the critical values with 1 df are

±6.314.  Since -6.314 < t < 6.314, we cannot reject $\beta_1 = 0$ and conclude
that x is not useful for predicting y.

(b)   The 90% confidence interval for the slope of the regression line is

$$b_1 \pm t_{\alpha/2} \frac{s_e}{\sqrt{S_{xx}}} = -2 \pm 6.314 \frac{2.44949}{\sqrt{2}} = (-12.936, 8.936)$$

**15.47** From Exercise 15.13, we have $s_e = 1.73205$ .

Also, $\sum x_i = 10; \sum x_i^2 = 30; n=4; \quad S_{xx} = \sum x_i^2 - \left(\sum x_i\right)^2 / n = 30 - 10^2/4 = 5$

(a)   $t = \dfrac{b_1}{s_e / \sqrt{S_{xx}}} = \dfrac{2}{1.73205/\sqrt{5}} = 2.582$ ; the critical values with 2 df are

±2.920.  Since  -2.920 < t < 2.920, we cannot reject $\beta_1 = 0$ and conclude
that x is not useful for predicting y.

(b)   The 90% confidence interval for the slope of the regression line is

$$b_1 \pm t_{\alpha/2}\frac{s_e}{\sqrt{S_{xx}}} = 2 \pm 2.920\frac{1.73205}{\sqrt{5}} = (-0.260, 4.600)$$

**15.49** From Exercise 15.15, we have $s_e = 1.88193$.

Also, $\sum x_i = 15; \sum x_i^2 = 69; n=5; \; S_{xx} = \sum x_i^2 - \left(\sum x_i\right)^2/n = 69 - 15^2/5 = 24$

(a) $t = \dfrac{b_1}{s_e/\sqrt{S_{xx}}} = \dfrac{-0.625}{1.88193/\sqrt{24}} = -1.627$; the critical values with 3 df are

$\pm 2.353$. Since $-2.353 < t < 2.353$, we cannot reject $\beta_1 = 0$ and conclude that x is not useful for predicting y.

(b) The 90% confidence interval for the slope of the regression line is

$$b_1 \pm t_{\alpha/2}\frac{s_e}{\sqrt{S_{xx}}} = -0.625 \pm 2.353\frac{1.88193}{\sqrt{24}} = (-1.512, 0.282)$$

**15.51** From Exercise 14.51, $\sum x = 41$, $\sum x^2 = 199$, and $b_1 = -27.9029$. From Exercise 15.23, $s_e = 14.2464$.

Step 1:  $H_0$: $\beta_1 = 0$,    $H_a$: $\beta_1 \neq 0$

Step 2:  $\alpha = 0.10$

Step 3:  $t = \dfrac{b_1}{s_e/\sqrt{\sum x^2 - (\sum x)^2/n}} = \dfrac{-27.9029}{14.2465/\sqrt{199 - 41^2/10}} = -10.887$

Step 4:  df = n - 2 = 8;    critical values = $\pm 1.860$

Step 5:  Since $-10.887 < -1.860$, reject $H_0$.

Step 6:  The data provide sufficient evidence to conclude that the slope of the population regression line is not zero and, hence, that age is useful as a predictor of price for Corvettes.

For the p-value approach, p < 0.01.  Therefore, because the p-value is less than the significance level of 0.10, we can reject $H_0$.

**15.53** From Exercise 14.53, $\sum x = 723$, $\sum x^2 = 48747$, and $b_1 = 0.16285$.  From Exercise 15.25, $s_e = 5.418$.

Step 1:  $H_0$: $\beta_1 = 0$,    $H_a$: $\beta_1 \neq 0$

Step 2:  $\alpha = 0.05$

Step 3:  $t = \dfrac{b_1}{s_e/\sqrt{\sum x^2 - (\sum x)^2/n}} = \dfrac{0.16285}{5.418/\sqrt{48747 - 723^2/11}} = 1.053$

Step 4:  df = n - 2 = 9;    critical values = $\pm 2.262$

Step 5:  Since $1.053 < 2.262$, do not reject $H_0$.

Step 6:  Evidently, plant weight is not useful as a predictor of volume of volatile emissions.

For the p-value approach, p > 0.20.  Therefore, because the p-value is greater than the significance level of 0.05, we cannot reject $H_0$.

**15.55** From Exercise 14.55, $\sum x = 117$, $\sum x^2 = 1869$, and $b_1 = -0.846$.  From Exercise 15.27, $s_e = 3.538$.

Step 1:  $H_0$: $\beta_1 = 0$,    $H_a$: $\beta_1 \neq 0$

Step 2:  $\alpha = 0.01$

Step 3:  $t = \dfrac{b_1}{s_e/\sqrt{\sum x^2 - (\sum x)^2/n}} = \dfrac{-0.846}{3.538/\sqrt{1869 - 117^2/8}} = -3.004$

Step 4:  df = n - 2 = 6;    critical values = $\pm 3.707$

Step 5:  Since $-3.707 < -3.004 < 3.707$, do not reject $H_0$.

Step 6:  Evidently, study-time is not useful as a predictor of test scores of calculus students.

For the p-value approach, 0.02 <p < 0.05.  Therefore, because the p-value is less than the significance level of 0.01, we cannot reject $H_0$.

**15.57** From Exercise 14.51, $\sum x = 41$, $\sum x^2 = 199$, and $b_1 = -27.9029$.  From Exercise 15.23, $s_e = 14.2465$.

(a)  Step 1:    For a 90% confidence interval, $\alpha = 0.10$.  With df = n - 2 = 8, $t_{\alpha/2} = t_{0.05} = 1.860$.

Step 2:    The endpoints of the confidence interval for $\beta_1$ are

$$b_1 \pm t_{\alpha/2} \cdot s_e / \sqrt{\sum x^2 - (\sum x)^2 / n}$$

$$-27.9029 \pm 1.860 \cdot 14.2465 / \sqrt{199 - 41^2 / 10}$$

$$-27.9029 \pm 4.7670$$

$$-32.6699 \ to \ -23.1359$$

(b)  We can be 90% confident that the yearly decrease in mean price for Corvettes is somewhere between $2314 and $3267.

**15.59** From Exercise 14.53, $\sum x = 723$, $\sum x^2 = 48747$, and $b_1 = 0.16285$.  From Exercise 15.25, $s_e = 5.4177$.

(a)  Step 1:  For a 95% confidence interval, $\alpha = 0.05$.  With df = n - 2 = 9, $t_{\alpha/2} = t_{0.025} = 2.262$.

Step 2:  The endpoints of the confidence interval for $\beta_1$ are

$$b_1 \pm t_{\alpha/2} \cdot s_e / \sqrt{\sum x^2 - (\sum x)^2 / n}$$

$$0.16285 \pm 2.262 \cdot 5.4177 / \sqrt{48747 - 723^2 / 11}$$

$$0.16285 \pm 0.34997$$

$$-0.187 \ to \ 0.513$$

(b)  We can be 95% confident that the increase in mean volatile plant emissions per one gram increase in weight is somewhere between -0.187 and 0.513 hundred nanograms.

**15.61** From Exercise 14.55, $\sum x = 117$, $\sum x^2 = 1869$, and $b_1 = -0.846$.  From Exercise 15.27, $s_e = 3.53816$.

(a)  Step 1:  For a 99% confidence interval, $\alpha = 0.01$. With df = n - 2 = 6, $t_{\alpha/2} = t_{0.005} = 3.707$.

Step 2:  The endpoints of the confidence interval for $\beta_1$ are

$$b_1 \pm t_{\alpha/2} \cdot s_e / \sqrt{\sum x^2 - (\sum x)^2 / n}$$

$$-0.846 \pm 3.707 \cdot 3.53816 / \sqrt{1869 - 117^2 / 8}$$

$$-0.846 \pm 1.044$$

$$-1.890 \ to \ 0.198$$

(b)    We can be 99% confident that the change in test score per one hour increase in study-time for calculus students is somewhere between -1.890 and 0.198.

**15.63** (a)  In Exercise 15.31, part (c), we determined that it was reasonable to consider Assumptions 1-3 for regression inferences met by these variables.

(b)  Part of the Minitab output from Exercise 15.31 is reproduced below.

```
Predictor          Coef   SE Coef      T      P
Constant           8.66     13.84   0.63  0.535
INAUGURATION     1.1172    0.2512   4.45  0.000
```

The t value and the P-value for the regression t-test are 4.45 and 0.000, respectively. Since 0.000 < 0.05, we conclude that the inauguration ages are useful for predicting the presidents' death ages.

(c) The P-value of the regression t-test is 0.000. Therefore the evidence in favor of the utility of the regression equation for making predictions is very strong.

**15.65** (a) In Exercise 15.33, part (c), we determined that it was temporarily reasonable to consider Assumptions 1-3 for regression inferences met by these variables, although Assumption 3 was questionable. Note: In Exercise 14.67, we concluded that the observations did not follow a linear pattern. This exercise will help to resolve the conflict.

(b) Part of the Minitab output from Exercise 15.33 is reproduced below.
```
Predictor   Coef   SE Coef     T      P
Constant   292.0     140.3  2.08  0.043
LOT SIZE   67.11     59.87  1.12  0.269
```
The t value and the P-value for the regression t-test are 1.12 and 0.269, respectively. Since 0.269 > 0.05, we conclude that the lot size is not useful for predicting the home values.

(c) The P-value of the regression t-test is 0.269. Therefore the evidence in favor of the utility of the regression equation for making predictions is weak or none. We conclude that our initial impressions in Exercise 14.67 were correct and that it is not reasonable to use these data for making regression inferences.

**15.67** (a) In Exercise 15.35, part (c), we determined that it was reasonable to consider Assumptions 1-3 for regression inferences met by these variables.

(b) Part of the Minitab output from Exercise 15.35 is reproduced below.
```
Predictor      Coef   SE Coef       T      P
Constant     -7.572     1.786   -4.24  0.000
HIGH        0.91685   0.02967   30.91  0.000
```
The t value and the P-value for the regression t-test are 30.91 and 0.000, respectively. Since 0.000 < 0.05, we conclude that average high temperatures are useful for predicting average low temperatures in January.

(c) The P-value of the regression t-test is 0.000. Therefore the evidence in favor of the utility of the regression equation for making predictions is very strong.

**15.69** (a) In Exercise 15.37, part (c), we determined that it was not reasonable to consider Assumption 1 for regression inferences met by these variables. The residual plot was convex upward. Therefore, parts (b) and (c) are omitted.

**15.71** (a) In Exercise 15.39, part (c), we determined that it was not reasonable to consider Assumptions 1 and 2 for regression inferences met by these variables. Therefore, parts (b) and (c) are omitted.

### Exercises 15.3

**15.73** $11,443

**15.75** From Exercise 15.11, $\hat{y} = 1 - 2x$. From Exercise 15.45, $s_e = 2.44949$ $and$ $S_{xx} = 2$.

(a) The point estimate for the conditional mean of y at x = 2 is
$$\hat{y}_p = 1 - 2x = 1 - 2(2) = -3.$$

(b) The 95% confidence interval for the conditional mean at x = 2 is

$$\hat{y}_p \pm t_{\alpha/2}\, s_e \sqrt{\frac{1}{n} + \frac{\left(x_p - \sum x_i / n\right)^2}{S_{xx}}} = -3 \pm 12.706(2.44949)\sqrt{\frac{1}{3} + \frac{(2 - 6/3)^2}{2}} = -3 \pm 17.97$$

$$= (-20.97, 14.97)$$

(c)  The predicted value of y at x = 2 is $\hat{y}_p = 1 - 2x = 1 - 2(2) = -3$ .

(d)  The 95% prediction interval for y at x = 2 is

$$\hat{y}_p \pm t_{\alpha/2}\, s_e \sqrt{1 + \frac{1}{n} + \frac{\left(x_p - \sum x_i / n\right)^2}{S_{xx}}} = -3 \pm 12.706(2.44949)\sqrt{1 + \frac{1}{3} + \frac{(2 - 6/3)^2}{2}}$$

$$= -3 \pm 35.94 = (-38.94, 32.94)$$

**15.77**  From Exercise 15.13, $\hat{y} = -3 + 2x$ .  From Exercise 15.47,

$s_e = 1.41421$ *and* $S_{xx} = 5$ .

(a)  The point estimate for the conditional mean of y at x = 4 is
$\hat{y}_p = -3 + 2x = -3 + 2(4) = 5$ .

(b)  The 95% confidence interval for the conditional mean at x = 4 is

$$\hat{y}_p \pm t_{\alpha/2}\, s_e \sqrt{\frac{1}{n} + \frac{\left(x_p - \sum x_i / n\right)^2}{S_{xx}}} = 5 \pm 4.303(1.73205)\sqrt{\frac{1}{4} + \frac{(4 - 10/4)^2}{5}} = 5 \pm 6.24$$

$$= (-1.24, 11.24)$$

(c)  The predicted value of y at x = 4 is $\hat{y}_p = -3 + 2x = -3 + 2(4) = 5$ .

(d)  The 95% prediction interval for y at x = 4 is

$$\hat{y}_p \pm t_{\alpha/2}\, s_e \sqrt{1 + \frac{1}{n} + \frac{\left(x_p - \sum x_i / n\right)^2}{S_{xx}}} = 5 \pm 4.303(1.73205)\sqrt{1 + \frac{1}{4} + \frac{(4 - 10/4)^2}{5}}$$

$$= 5 \pm 9.72 = (-4.72, 14.72)$$

**15.79**  From Exercise 15.15, $\hat{y} = 2.875 + 0.625x$ .  From Exercise 15.49,

$s_e = 1.88193$ *and* $S_{xx} = 24$ .

(a)  The point estimate for the conditional mean of y at x = 3 is
$\hat{y}_p = 2.875 + 0.625x = 2.875 + 0.625(3) = 4.75$ .

(b)  The 95% confidence interval for the conditional mean at x = 3 is

$$\hat{y}_p \pm t_{\alpha/2}\, s_e \sqrt{\frac{1}{n} + \frac{\left(x_p - \sum x_i / n\right)^2}{S_{xx}}} = 4.75 \pm 3.182(1.88193)\sqrt{\frac{1}{5} + \frac{(3 - 15/5)^2}{24}} = 4.75 \pm 2.68$$

$$= (2.07, 7.43)$$

(c)  The predicted value of y at x = 3 is
$\hat{y}_p = 2.875 + 0.625x = 2.875 + 0.625(3) = 4.75$ .

(d)  The 95% prediction interval for y at x = 3 is

$$\hat{y}_p \pm t_{\alpha/2}\, s_e \sqrt{1 + \frac{1}{n} + \frac{\left(x_p - \sum x_i / n\right)^2}{S_{xx}}} = 4.75 \pm 3.182(1.88193)\sqrt{1 + \frac{1}{5} + \frac{(3 - 15/5)^2}{24}}$$

$$= 4.75 \pm 6.56 = (-1.81, 11.31)$$

**15.81**  From Exercise 14.51, $\hat{y} = 436.602 - 27.9029x$, $\sum x = 41$, and $\sum x^2 = 199$.  From Exercise 15.23, $s_e = 14.2465$.

(a)   $\hat{y}_p = 436.602 - 27.9029(4) = 324.9904 = \$32,499$.

(b)  Step 1:  For a 90% confidence interval, $\alpha = 0.10$.

With df $= n - 2 = 8$, $t_{\alpha/2} = t_{0.05} = 1.860$.

Step 2:  $\hat{y}_p = 324.9904$

Step 3:  The endpoints of the confidence interval are

$$\hat{y}_p \pm t_{\alpha/2} \cdot s_e \cdot \sqrt{\frac{1}{n} + \frac{(x_p - \sum x/n)^2}{\sum x^2 - (\sum x)^2/n}}$$

$$= 324.9904 \pm 1.860 \cdot 14.2465 \sqrt{\frac{1}{10} + \frac{(4 - 41/10)^2}{199 - 41^2/10}}$$

$$= 324.9904 \pm 8.3931 = (316.60, 333.38)$$

The interpretation of this interval is as follows:  We can be 90% confident that the mean price of all four-year-old Corvettes is somewhere between $31,660 and $33,338.

(c)  This is the same as the answer in part (a):  $\hat{y}_p = 324.9904$.

(d)  For a 90% prediction interval, Steps 1 and 2 are the same as Steps 1 and 2, respectively, in part (b).  Thus, they are not repeated here. Only Step 3 is presented.

Step 3:  The endpoints of the prediction interval are

$$\hat{y}_p \pm t_{\alpha/2} \cdot s_e \cdot \sqrt{1 + \frac{1}{n} + \frac{(x_p - \sum x/n)^2}{\sum x^2 - (\sum x)^2/n}}$$

$$= 324.9904 \pm 1.860 \cdot 14.265 \sqrt{1 + \frac{1}{10} + \frac{(4 - 41/10)^2}{199 - 41^2/10}}$$

$$= 324.9904 \pm 27.7959 = (297.20, 352.78)$$

The interpretation of this interval is as follows:  We can be 90% certain that the price of a randomly selected four-year-old Corvette will be somewhere between $29,720 and $35,278.

(e)

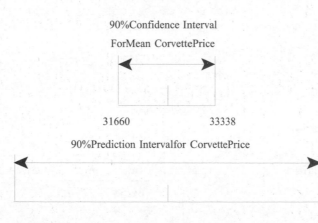

90%Confidence Interval

ForMean CorvettePrice

31660          33338

90%Prediction Intervalfor CorvettePrice

29720                              35278

(f)  The error in the estimate of the mean price of four-year-old Corvettes is due only to the fact that the population regression line is being estimated by a sample regression line; whereas, the error in the prediction of the price of a randomly selected four-year-old Corvette is due to that fact plus the variation in prices for four-year-old Corvettes.

**15.83** From Exercise 14.53, $\hat{y} = 3.523688 + 0.162848x$, $\sum x = 726$, and $\sum x^2 = 48747$. From Exercise 15.25, $s_e = 5.4177$.

(a)    $\hat{y}_p = 3.523688 + 0.162848(60) = 13.2946$

(b)    Step 1:    For a 95% confidence interval, $\alpha = 0.05$.

With df = n - 2 = 9, $t_{\alpha/2} = t_{0.025} = 2.262$.

Step 2:    $\hat{y}_p = 13.2946$

Step 3:    The endpoints of the confidence interval are

$$\hat{y}_p \pm t_{\alpha/2} \cdot s_e \cdot \sqrt{\frac{1}{n} + \frac{(x_p - \sum x/n)^2}{\sum x^2 - (\sum x)^2/n}}$$

$$= 13.2946 \pm 2.262 \cdot 5.4177 \cdot \sqrt{\frac{1}{11} + \frac{(60 - 723/11)^2}{48747 - 723^2/11}}$$

$$= 813.2946 \pm 4.2036 = (9.0910, 17.4962)$$

The interpretation of this interval is as follows:  We can be 95% confident that the mean quantity of volatile emissions of all plants that weigh 60 grams is somewhere between 9.0910 and 17.4982 hundred nanograms.

(c)    This is the same as the answer in part (a):  $\hat{y}_p = 13.2946$.

(d)    For a 95% prediction interval, Steps 1 and 2 are the same as Steps 1 and 2, respectively, in part (b).  Thus, they are not repeated here. Only Step 3 is presented.

Step 3:    The endpoints of the prediction interval are

$$\hat{y}_p \pm t_{\alpha/2} \cdot s_e \cdot \sqrt{1 + \frac{1}{n} + \frac{(x_p - \sum x/n)^2}{\sum x^2 - (\sum x)^2/n}}$$

$$= 13.2946 \pm 2.262 \cdot 5.4177 \cdot \sqrt{1 + \frac{1}{11} + \frac{(60 - 723/11)^2}{48747 - 723^2/11}}$$

$$= 813.2946 \pm 12.9557 = (0.3389, 26.2503)$$

The interpretation of this interval is as follows:  We can be 95% certain that the quantity of volatile emissions of a randomly selected plant that weighs 60 grams is somewhere between 0.3389 and 26.2503 hundred nanograms.

**15.85** From Exercise 14.55, $\hat{y} = 94.86698 - 0.84561x$, $\sum x = 117$, and $\sum x^2 = 1869$. From Exercise 15.27, $s_e = 3.5382$.

(a)    $\hat{y}_p = 94.86698 - 0.84561(15) = 82.1828$

(b)    Step 1:    For a 99% confidence interval, $\alpha = 0.01$.

With df = n - 2 = 6, $t_{\alpha/2} = t_{0.005} = 3.707$.

Step 2:    $\hat{y}_p = 82.1828$

Step 3:    The endpoints of the confidence interval are

$$\hat{y}_p \pm t_{\alpha/2} \cdot s_e \cdot \sqrt{\frac{1}{n} + \frac{(x_p - \sum x/n)^2}{\sum x^2 - (\sum x)^2/n}}$$

$$= 82.1828 \pm 3.707 \cdot (3.5382) \cdot \sqrt{\frac{1}{8} + \frac{(15 - 117/8)^2}{1869 - 117^2/8}}$$

$$= 82.1828 \pm 4.6537 = (77.5291, 86.8365)$$

The interpretation of this interval is as follows:  We can be 99% confident that the mean calculus test score of students who study 15 hours is somewhere between 77.53 and 86.84 mm.

(c)    This is the same as the answer in part (a):  $\hat{y}_p = 82.1828$.

(d)   For a 99% prediction interval, Steps 1 and 2 are the same as Steps 1 and 2, respectively, in part (b). Thus, they are not repeated here. Only Step 3 is presented.

   Step 3:      The endpoints of the prediction interval are

$$\hat{y}_p \pm t_{\alpha/2} \cdot s_e \cdot \sqrt{1 + \frac{1}{n} + \frac{(x_p - \sum x/n)^2}{\sum x^2 - (\sum x)^2/n}}$$

$$= 82.1828 \pm 3.707 \cdot (3.5382) \cdot \sqrt{1 + \frac{1}{8} + \frac{(15 - 117/8)^2}{1869 - 117^2/8}}$$

$$= 82.1828 \pm 13.9172 = (68.2656, 96.0100)$$

The interpretation of this interval is as follows:  We can be 99% certain that a randomly selected calculus student who studies 125 hours will have a test score somewhere between 68.27 and 96.01 mm.

**15.87** (a)  In Exercise 15.31, part (c), we determined that it was reasonable to consider Assumptions 1-3 for regression inferences met by these variables.

(b)  Using Minitab, with the data in columns named INAUGURATION and DEATH,

   we choose **Stat ▶ Regression ▶ Regression...**, select <u>DEATH</u> in the

   **Response** text box, and select <u>INAUGURATION</u> in the **Predictors** text box. Click the **Options** button, select the **Regular** option button from the **Residuals for Plots** list, enter <u>53</u> in the **Prediction intervals for new observations** box and <u>95</u> in the **Confidence level** box, click **OK**, and click **OK**.  Since much of the output was shown in Exercise 15.31, we show below only that part of the output pertinent to this exercise.

   Predicted Values for New Observations

```
New
Obs    Fit   SE Fit      95% CI          95% PI
  1   67.88   1.60   (64.63, 71.12)  (48.38, 87.37)
```

The point estimate for the conditional mean of DEATH for inauguration age 53 is found under the word 'Fit' in the output as 67.88.  This is the same as the predicted score for inauguration at age 53.

(c)  The 95% confidence interval for the conditional mean death age for presidents inaugurated at age 53 is (64.63, 71.12).  We can be 95% confident that the mean death age of U.S. Presidents inaugurated at age 53 lies somewhere between 64.63 and 71.12 years.

(d)  The predicted death age for Presidents inaugurated at age 53 is found under the word 'Fit' in the output as 67.88.

(e)  The 95% prediction interval for the death age of a President inaugurated at age 53 is (48.38, 87.37).  We can be 95% certain that the death age of a president inaugurated at age 53 will fall somewhere between 48.38 and 87.37.  Note:  The lower end of this prediction interval can be safely increased to 53 since a president cannot die prior to being inaugurated.

(f)  The confidence interval and the prediction interval are both centered at the predicted value for an inauguration age of 53.  The prediction interval is longer because there is variation associated with the additional observation that is not present in the confidence interval for the mean.

**15.89** (a)  In Exercise 15.33, part (c), we determined that it was reasonable to consider Assumptions 1-3 for regression inferences met by these variables.  Note:  In Exercise 14.67, we concluded that the pattern of the observations was NOT linear, meaning that Assumption 1 for regressions inferences was not met by these data.  In Exercise 15.65, we found that the evidence for a linear pattern was weak or none.  Thus we will not continue with parts (b)-(f).

**15.91** (a)  In Exercise 15.35, part (c), we determined that it was reasonable to consider Assumptions 1-3 for regression inferences met by these variables.

(b)  Using Minitab, with the data in columns named HIGH and LOW, we choose

**Stat ▶ Regression ▶ Regression...**, select LOW in the **Response** text box, and select HIGH in the **Predictors** text box.  Click the **Options** button, select the **Regular** option button from the **Residuals for Plots** list, enter 55 in the **Prediction intervals for new observations** box and 95 in the **Confidence level** box, click **OK**, and click **OK**.  Since much of the output was shown in Exercise 15.35, we show below only that part of the output pertinent to this exercise.

Predicted Values for New Observations

```
New
Obs    Fit   SE Fit      95% CI              95% PI
  1  42.855  0.590  (41.669, 44.040)  (34.425, 51.285)
```

The point estimate for the conditional mean January low temperature for an average high temperature of 55 degrees is found under the word 'Fit' in the output as 42.855 degrees.  This is the same as the predicted value for an average high temperature of 55 degrees.

(c)  The 95% confidence interval for the conditional mean low January temperature for cities with mean high temperature of 55 degrees is (41.669, 44.040).  We can be 95% confident that the mean low January temperature for cities with a mean January high temperature of 55 degrees lies somewhere between 41.669 and 44.040 degrees.

(d)  The predicted mean low January temperature for a city with a mean high January temperature of 55 degrees is found under the word 'Fit' in the output as 42.855 degrees.

(e)  The 95% prediction interval for the mean January low temperature for a city with a mean January high temperature is (34.425, 51.285).  We can be 95% certain that the mean low January temperature for a city with a mean January high temperature of 55 degrees will lie somewhere between 34.425 and 51.285 degrees.

(f)  The confidence interval and the prediction interval are both centered at the predicted value for a city with a mean January high temperature of 55 degrees.  The prediction interval is longer because there is variation associated with the additional observation that is not present in the confidence interval for the mean.

**15.93** (a)  In Exercise 15.37, part (c), we determined that it was not reasonable to consider Assumption 1 for regression inferences met by these variables.  The residual plot was convex upward.  Therefore, parts (b)-(f) are omitted.

**15.95** (a)  In Exercise 15.39, part (c), we determined that it was not reasonable to consider Assumptions 1 and 2 for regression inferences met by these variables.  Therefore, parts (b)-(f) are omitted.

**15.97** (a)-(b)  In Example 15.8 of the text, the 95% confidence interval for the mean price of all Orions of a given age is given by

$$134.69 \pm 2.262 \cdot 12.58 \cdot \sqrt{\frac{1}{11} + \frac{(x_p - 58/11)^2}{326 - 58^2/11}}$$

Substituting the values 2 through 7 into this formula for $x_p$, we obtain the following confidence intervals and a plot showing the lower and upper confidence limits for the six different values of $x_p$.  The margin of error (half of the difference between the upper and lower confidence limits) is shown in the last column of the table and is plotted against $x_p$ in the second graph.

| $x_p$ | Lower Confidence Limit | Upper Confidence Limit | Margin of Error |
|---|---|---|---|
| 2 | 112.25 | 157.13 | 22.44 |
| 3 | 117.93 | 151.45 | 16.76 |
| 4 | 122.92 | 146.46 | 11.77 |
| 5 | 125.94 | 143.44 | 8.75 |
| 6 | 124.95 | 144.43 | 9.74 |
| 7 | 120.79 | 148.59 | 13.90 |

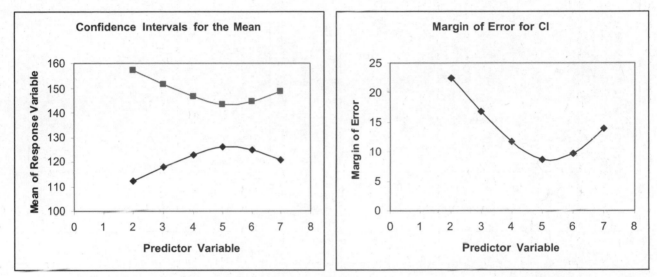

We see from the table and the first plot that the confidence intervals get wider as the predictor variable gets farther from the mean of the predictor variable. This is also seen in the second plot, which shows that the margin of error first decreases as $x_p$ increases to the mean 5.27 of the predictor variable and then increases again as $x_p$ moves away from 5.27.

(c)  In Example 15.9 of the text, the 95% prediction interval for an Orion of a given age is given by

$$= 134.69 \pm 2.262 \cdot 12.58 \cdot \sqrt{1 + \frac{1}{11} + \frac{(x_p - 58/11)^2}{326 - 58^2/11}}$$

Substituting the values 2 through 7 into this formula for $x_p$, we obtain the following confidence intervals and a plot showing the lower and upper confidence limits for the six different values of $x_p$. The margin of error (half of the difference between the upper and lower confidence limits) is shown in the last column of the table and is plotted against $x_p$ in the second graph.

| $x_p$ | Lower Prediction Limit | Upper Prediction Limit | Margin of Error |
|---|---|---|---|
| 2 | 98.45 | 170.93 | 36.24 |
| 3 | 101.67 | 167.71 | 33.02 |
| 4 | 103.89 | 165.49 | 30.80 |
| 5 | 104.92 | 164.46 | 29.77 |
| 6 | 104.61 | 164.77 | 30.08 |
| 7 | 103.02 | 166.36 | 31.67 |

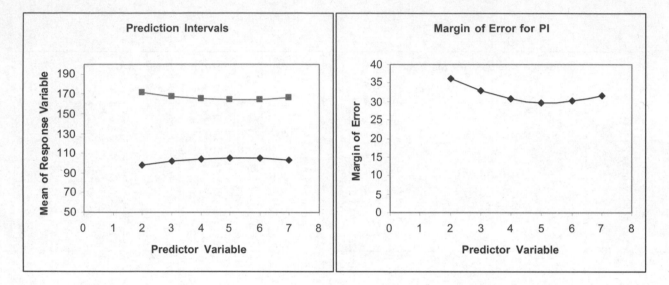

We see from the table and the first plot that the prediction intervals get wider as the predictor variable gets farther from the mean of the predictor variable.  This is also seen in the second plot which shows that the margin of error first decreases as $x_p$ increases to the mean 5.27 of the predictor variable and then increases again as $x_p$ moves away from 5.27.

## Exercises 15.4

**15.99**      r, the sample linear correlation coefficient

**15.101** (a)   uncorrelated

(b)   increases

(c)   negatively

**15.103**      Step 1  $H_0: \rho = 0$,  $H_a: \rho < 0$

Step 2  $\alpha = 0.10$

Step 3  From Exercise 14.117, $r = -0.76$, $n = 3$

$$t = \frac{r}{\sqrt{\dfrac{1-r^2}{n-2}}} = \frac{-0.76}{\sqrt{\dfrac{1-(-0.76)^2}{1}}} = -1.169$$

Step 4  The critical value is $-3.0784$

Step 5  Since $-1.169 > -3.078$, do not reject $H_0$.

**15.105**      Step 1  $H_0: \rho = 0$,  $H_a: \rho > 0$

Step 2  $\alpha = 0.10$

Step 3  From Exercise 14.119, $r = 0.877$, $n = 4$

$$t = \frac{r}{\sqrt{\dfrac{1-r^2}{n-2}}} = \frac{0.877}{\sqrt{\dfrac{1-0.877^2}{2}}} = 2.581$$

Step 4  The critical value is 1.886

Step 5  Since 2.581 > 1.886, reject $H_0$.

**15.107**  Step 1  $H_0: \rho = 0$, $H_a: \rho \neq 0$

Step 2  $\alpha = 0.10$

Step 3  From Exercise 14.121, $r = -0.685$, $n = 5$

$$t = \frac{r}{\sqrt{\dfrac{1-r^2}{n-2}}} = \frac{-0.685}{\sqrt{\dfrac{1-(-0.685)^2}{3}}} = -1.629$$

Step 4  The critical values are $\pm 2.353$

Step 5  Since $-2.353 < -1.629 < 2.353$, do not reject $H_0$.

**15.109**  From Exercise 14.123, $r = -0.96787$ and $n = 10$.

Step 1: $H_0: \rho = 0$   $H_a: \rho < 0$

Step 2: $\alpha = 0.05$

Step 3: $t = \dfrac{r}{\sqrt{\dfrac{1-r^2}{n-2}}} = \dfrac{-0.96787}{\sqrt{\dfrac{1-(-0.96787)^2}{10-2}}} = -10.8870$

Step 4: The critical value for $n - 2 = 8$ df is $-1.860$.

Step 5: Since $-10.8870 < -1.860$, reject $H_0$ and conclude that $\rho < 0$.

For the p-value approach, $p < 0.005$.  Since $p < 0.05$, reject $H_0$ and conclude that $\rho < 0$.

**15.111**  From Exercise 14.125, $r = 0.33107$ and $n = 11$.

Step 1: $H_0: \rho = 0$   $H_a: \rho \neq 0$

Step 2: $\alpha = 0.05$

Step 3: $t = \dfrac{r}{\sqrt{\dfrac{1-r^2}{n-2}}} = \dfrac{0.33107}{\sqrt{\dfrac{1-(0.33107)^2}{11-2}}} = 1.0526$

Step 4: The critical values for $n - 2 = 9$ df are $\pm 2.262$.

Step 5: Since $-2.262 < 1.0526 < 2.262$, do not reject $H_0$ and conclude that $\rho = 0$ is reasonable.

For the p-value approach, $p > 0.200$.  Since $p > 0.05$, do not reject $H_0$ and conclude that $\rho = 0$ is reasonable.

**15.113**  From Exercise 14.127, $r = -0.7749$ and $n = 8$.

Step 1: $H_0: \rho = 0$   $H_a: \rho < 0$

Step 2: $\alpha = 0.01$

Step 3: $t = \dfrac{r}{\sqrt{\dfrac{1-r^2}{n-2}}} = \dfrac{-0.7749}{\sqrt{\dfrac{1-(-0.7749)^2}{8-2}}} = -3.003$

Step 4: The critical value for $n - 2 = 6$ df is $-3.143$.

Step 5: Since $-3.003 > 3.143$, do not reject $H_0$.  There is not sufficient evidence to conclude that $\rho \neq 0$.

For the p-value approach, $0.010 < p < 0.025$.  Since $p > 0.01$, do not reject $H_0$.

**15.115**  The population linear correlation coefficient $\rho$ is a parameter (constant) that measures the linear correlation between the population of all data points.  The sample linear correlation coefficient $r$ is a statistic (random variable); it measures the linear correlation between a sample of data points, and so its value depends on chance, namely, on which data points are obtained from sampling.

**15.117**  In Exercise 15.31, part (c), we determined that it was reasonable to consider Assumptions 1-3 for regression inferences met by these variables.

**Using Minitab,** choose **Stat ▶ Basic Statistics ▶ Correlation...**, select INAUGURATION and DEATH in the **Variables** text box, and click **OK**. The result is
    Pearson correlation of INAUGURATION and DEATH = 0.595
    P-Value = 0.000
Note:  The p-value shown is for a two-tailed test.  The p-value for the right-tail test would be even smaller.  Since the p-value < 0.05, reject $H_0$ and conclude that the inauguration ages and death ages of U.S. Presidents are positively linearly correlated.

**15.119**  In Exercise 15.65, part (a), we determined that it was not reasonable to consider Assumptions 1 for regression inferences met by these variables.  The correlation t-test should not be used.  Over several exercises, there was some controversy regarding this decision.  If you choose to continue with the correlation t-test, the results from Minitab are
  Pearson correlation of LOT SIZE and VALUE = 0.170
  P-Value = 0.269
The high P-value confirms our decision not to reject $\rho = 0$.

**15.121**  In Exercise 15.35, part (c), we determined that it was reasonable to consider Assumptions 1-3 for regression inferences met by these variables.

**Using Minitab,** choose **Stat ▶ Basic Statistics ▶ Correlation...**, select HIGH and LOW in the **Variables** text box, and click **OK**. The result is
  Pearson correlation of HIGH and LOW = 0.976
  P-Value = 0.000
Since the P-value < 0.05, reject $\rho = o$.

**15.123**  In Exercise 15.37, part (c), we determined that it was not reasonable to consider Assumption 1 for regression inferences met by these variables.  The correlation t-test should not be used.

**15.125**  In Exercise 15.39, part (c), we determined that it was not reasonable to consider Assumptions 1 and 2 for regression inferences met by these variables.  The correlation t-test should not be used.

**Exercises 15.5**
**15.127**(a)  A normal probability plot is a plot of normal scores against sample data.
  (b)  An important use of such plots is to help assess the normality of a variable.
  (c)  If the sample data come from a normal distribution, the plot should be roughly linear.  Therefore, if the plot is roughly linear, we accept as reasonable that the variable is normally distributed; if the plot shows systematic deviations from a line, then conclude that the variable is probably not normally distributed.
  (d)  The method in (c) is subjective because what constitutes "roughly linear" is a matter of opinion.

**15.129**  If the population under consideration is normally distributed, then the correlation between the sample data and its normal scores should be near 1.  Note that since the normal scores increase as the data values

increase, the correlation between the sample data and their normal
scores cannot be negative.  Thus, the correlation test for normality is
always left-tailed because if the correlation (the test statistic of
interest) is significantly smaller than one, the null hypothesis that
the population is normally distributed is rejected.

**15.131**   Step 1: $H_0$: The population is normally distributed.
        $H_a$: The population is not normally distributed.
Step 2: $\alpha = 0.05$
Step 3:

| Exam score | Normal score | xw | $x^2$ | $w^2$ |
|---|---|---|---|---|
| 34 | −1.87 | −63.58 | 1156 | 3.4969 |
| 39 | −1.40 | −54.60 | 1521 | 1.9600 |
| 63 | −1.13 | −71.19 | 3969 | 1.2769 |
| 64 | −0.92 | −58.88 | 4096 | 0.8464 |
| 67 | −0.74 | −49.58 | 4489 | 0.5476 |
| 70 | −0.59 | −41.30 | 4900 | 0.3481 |
| 75 | −0.45 | −33.75 | 5625 | 0.2025 |
| 76 | −0.31 | −23.56 | 5776 | 0.0961 |
| 81 | −0.19 | −15.39 | 6561 | 0.0361 |
| 82 | −0.06 | −4.92 | 6724 | 0.0036 |
| 84 | 0.06 | 5.04 | 7056 | 0.0036 |
| 85 | 0.19 | 16.15 | 7225 | 0.0361 |
| 86 | 0.31 | 26.66 | 7396 | 0.0961 |
| 88 | 0.45 | 39.60 | 7744 | 0.2025 |
| 89 | 0.59 | 52.51 | 7921 | 0.3481 |
| 90 | 0.74 | 66.60 | 8100 | 0.5476 |
| 1555 | 0.00 | 302.48 | 126791 | 17.6284 |

$$R_p = \frac{\sum xw}{\sqrt{[\sum x^2 - (\sum x)^2/n][\sum w^2]}} = \frac{302.48}{\sqrt{[126791 - (1555)^2/20][17.6284]}} = 0.939$$

Step 4:   From Table IX, the critical value is = 0.951.
Step 5:   From Step 4 the value of the test statistic is $R_p = 0.939$,
        which is less than 0.952 and therefore falls in the rejection
        region.  Hence we reject $H_0$.
        The test results are statistically significant at the 5%
        level, that is, at the 5% significance level, the data provide
        sufficient evidence to conclude that final exam scores are not
        normally distributed.

**15.133**   Step 1:   $H_0$: The population is normally distributed.
        $H_a$: The population is not normally distributed.
Step 2: $\alpha = 0.10$

Step 3:

| Times x | Normal score w | xw | $x^2$ | $w^2$ |
|---|---|---|---|---|
| 93.37 | -1.71 | -159.6627 | 8717.9569 | 2.9241 |
| 94.15 | -1.21 | -113.9215 | 8864.2225 | 1.4641 |
| 95.10 | -0.90 | -85.5900 | 9044.0100 | 0.8100 |
| 95.57 | -0.66 | -63.0762 | 9133.6249 | 0.4356 |
| 96.63 | -0.45 | -43.4835 | 9337.3569 | 0.2025 |
| 97.19 | -0.27 | -26.2413 | 9445.8961 | 0.0729 |
| 97.47 | -0.09 | -8.7723 | 9500.4009 | 0.0081 |
| 97.73 | 0.09 | 8.7957 | 9551.1529 | 0.0081 |
| 97.91 | 0.27 | 26.4357 | 9586.3681 | 0.0729 |
| 98.44 | 0.45 | 44.2980 | 9690.4336 | 0.2025 |
| 99.38 | 0.66 | 65.5908 | 9876.3844 | 0.4356 |
| 101.05 | 0.90 | 90.9450 | 10211.1025 | 0.8100 |
| 101.09 | 1.21 | 101.0900 | 10219.1881 | 1.4641 |
| 103.02 | 1.71 | 176.1642 | 10613.1204 | 2.9241 |
| 1368.10 | 0.00 | 12.5719 | 133791.2182 | 11.3706 |

$$R_p = \frac{\sum xw}{\sqrt{[\sum x^2 - (\sum x)^2 / n][\sum w^2]}} = \frac{12.5719}{\sqrt{[133791.2182 - (1368.10)^2 / 14][11.3706]}} = 0.990$$

Step 4: From Table IX, the critical value is = 0.948.

Step 5: From Step 4 the value of the test statistic is $R_p = 0.990$, which does not fall in the rejection region.  Hence we do not reject $H_0$.

Step 6: The test results are not statistically significant at the 10% level; that is, at the 10% significance level, the data do not provide sufficient evidence to conclude that the finishing times of 1-mile thoroughbred horse races are not normally distributed.

**15.135**    Step 1:   $H_0$: The population is normally distributed.

   $H_a$: The population is not normally distributed.

Step 2: $\alpha = 0.10$

Step 3:

| Time x | Normal Score w | xw | $x^2$ | $w^2$ |
|---|---|---|---|---|
| 5.8 | -1.74 | -10.092 | 33.64 | 3.0276 |
| 8.0 | -1.24 | -9.92 | 64.00 | 1.5376 |
| 8.1 | -0.94 | -7.614 | 65.61 | 0.8836 |
| 8.3 | -0.71 | -5.893 | 68.89 | 0.5041 |
| 8.4 | -0.51 | -4.284 | 70.56 | 0.2601 |
| 9.0 | -0.33 | -2.970 | 81.00 | 0.1089 |
| 9.1 | -0.16 | -1.456 | 82.81 | 0.0256 |
| 11.1 | 0.00 | 0.000 | 123.21 | 0.0000 |
| 13.0 | 0.16 | 2.080 | 169.00 | 0.0256 |
| 13.3 | 0.33 | 4.389 | 176.89 | 0.1089 |
| 13.5 | 0.51 | 6.885 | 182.25 | 0.2601 |
| 15.1 | 0.71 | 10.721 | 228.01 | 0.5041 |
| 15.6 | 0.94 | 14.664 | 243.36 | 0.8836 |
| 16.3 | 1.24 | 20.212 | 265.69 | 1.5376 |
| 17.1 | 1.74 | 29.754 | 292.41 | 3.0276 |
| 171.7 | 0.00 | 46.476 | 2147.33 | 12.6950 |

$$R_p = \frac{\sum xw}{\sqrt{[\sum x^2 - (\sum x)^2 / n][\sum w^2]}} = \frac{46.476}{\sqrt{[2147.33 - (171.7)^2 / 15][12.6950]}} = 0.967$$

Step 4: The critical value is = 0.951.

Step 5: From Step 4 the value of the test statistic is $R_p$ = 0.967, which is not in the rejection region. Hence we do not reject $H_0$. The test results are not statistically significant at the 5% level, the data do not provide sufficient evidence to conclude that the time spent per user per month on shoe and apparel web sites is not normally distributed.

**15.137**    Step 1: $H_0$: The population is normally distributed.

               $H_a$: The population is not normally distributed.

Step 2: $\alpha$ = 0.01

Step 3:

| DOU x | w | xw | $x^2$ | $w^2$ |
|---|---|---|---|---|
| 0.7 | -1.91 | -1.3370 | 0.49 | 3.6481 |
| 1.0 | -1.45 | -1.4500 | 1.00 | 2.1025 |
| 1.1 | -1.18 | -1.2980 | 1.21 | 1.3924 |
| 1.1 | -0.98 | -1.0780 | 1.21 | 0.9604 |
| 1.2 | -0.81 | -0.9720 | 1.44 | 0.6561 |
| 1.5 | -0.66 | -0.9900 | 2.25 | 0.4356 |
| 1.8 | -0.53 | -0.9540 | 3.24 | 0.2809 |
| 1.8 | -0.40 | -0.7200 | 3.24 | 0.1600 |
| 1.8 | -0.28 | -0.5040 | 3.24 | 0.0784 |
| 1.8 | -0.17 | -0.3060 | 3.24 | 0.0289 |
| 1.9 | -0.06 | -0.1140 | 3.61 | 0.0036 |
| 2.0 | 0.06 | 0.1200 | 4.00 | 0.0036 |
| 2.0 | 0.17 | 0.3400 | 4.00 | 0.0289 |
| 2.0 | 0.28 | 0.5600 | 4.00 | 0.0784 |
| 2.3 | 0.40 | 0.9200 | 5.29 | 0.1600 |
| 2.7 | 0.53 | 1.4310 | 7.29 | 0.2809 |
| 3.3 | 0.66 | 2.1780 | 10.89 | 0.4356 |
| 3.4 | 0.81 | 2.7540 | 11.56 | 0.6561 |
| 3.6 | 0.98 | 3.5280 | 12.96 | 0.9604 |
| 3.8 | 1.18 | 4.4840 | 14.44 | 1.3924 |
| 6.7 | 1.45 | 9.7150 | 44.89 | 2.1025 |
| 7.6 | 1.91 | 14.5160 | 57.76 | 3.6481 |
| 55.1 | 0.00 | 30.8230 | 201.25 | 19.4938 |

$$R_p = \frac{\sum xw}{\sqrt{[\sum x^2 - (\sum x)^2 / n][\sum w^2]}} = \frac{30.8230}{\sqrt{[201.25 - (55.1)^2 / 22][19.4938]}} = 0.877805$$

Step 4: The critical value is = 0.933.

Step 5: From Step 4 the value of the test statistic is $R_p$ = 0.878, which is in the rejection region. Hence we reject $H_0$. The test results are statistically significant at the 5% level, the data provide sufficient evidence to conclude that the DOU is not normally distributed.

**15.139**    If a variable is normally distributed, a normal probability plot of the sample data should be close to linear with a positive slope. If the plot is roughly linear, we assume that the variable is approximately normally distributed, and, if the plot is not roughly linear, we assume that the variable is not approximately normally distributed. If the variable is normally distributed, the data values and their normal scores should have a correlation coefficient near 1 because a plot of those points should be roughly linear. If the correlation coefficient is too much smaller than 1, this implies that the plot is not linear, and the data values are not approximately normally distributed. We will reject the null hypothesis of a normal distribution if the correlation coefficient is too much smaller than 1, i.e., if it is less than established critical values of the correlation coefficient.

**15.141**  In Exercise 15.23, we obtained the residuals and their normal scores. From the table in that exercise, we compute the following sums:

$$\sum ew_i = 112.4596; \quad \sum e_i^2 = 1623.58; \quad \sum w_i^2 = 7.9526. \quad \text{Then}$$

$$R_p = \frac{\sum ew_i}{\sqrt{[\sum e_i^2 - (\sum e_i)^2 / n][\sum w_i^2]}} = \frac{112.4596}{\sqrt{[1623.58 - (0)^2 / 10][7.9526]}} = 0.990494$$

The critical value for n = 10 and $\alpha$ = 0.05 is 0.917. Since 0.990 > 0.917, do not reject $H_0$. The data do not provide sufficient evidence to conclude that the normality assumption for regression inferences is violated.

**15.143**  In Exercise 15.25, we obtained the residuals and their normal scores. From the table in that exercise, we compute the following sums:

$$\sum ew_i = 45.8901; \quad \sum e_i^2 = 264.2192; \quad \sum w_i^2 = 8.8892. \quad \text{Then}$$

$$R_p = \frac{\sum ew_i}{\sqrt{[\sum e_i^2 - (\sum e_i)^2 / n][\sum w_i^2]}} = \frac{45.8901}{\sqrt{[264.2192 - (0)^2 / 11][8.8892]}} = 0.946903$$

The critical value for n = 11 and $\alpha$ = 0.05 is 0.923. Since 0.947 > 0.923, do not reject $H_0$. The data do not provide sufficient evidence to conclude that the normality assumption for regression inferences is violated.

**15.145**  In Exercise 15.27, we obtained the residuals and their normal scores. From the table in that exercise, we compute the following sums:

$$\sum ew_i = 20.8369; \quad \sum e_i^2 = 79.0955; \quad \sum w_i^2 = 6.0216. \quad \text{Then}$$

$$R_p = \frac{\sum ew_i}{\sqrt{[\sum e_i^2 - (\sum e_i)^2 / n][\sum w_i^2]}} = \frac{20.8269}{\sqrt{[79.0955 - (0)^2 / 8][6.0216]}} = 0.980338$$

The critical value for n = 8 and $\alpha$ = 0.05 is 0.905. Since 0.980 > 0.905, do not reject $H_0$. The data do not provide sufficient evidence to conclude that the normality assumption for regression inferences is violated.

**15.147**  Using Minitab with the data in a column named TEMPERATURE, we choose

**Stat ▶ Basic statistics ▶ Normality test...**, select <u>TEMPERATURE</u> in the **Variable** text box, select the **Ryan-Joiner** option button from the **Tests for Normality** field, and click **OK**. The result is shown below. The p-value shown in the upper right is > 0.100. Since this is larger than the 0.05 significance level, we do not reject the null hypothesis of normality.

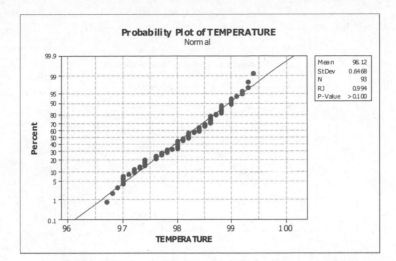

**15.149**   Using Minitab with the data in a column named CHIPS, we choose **Stat ▶**

**Basic statistics ▶ Normality test...**, select <u>CHIPS</u> in the **Variable** text box, select the **Ryan-Joiner** option button from the **Tests for Normality** field, and click **OK**.  The result is shown below.  The p-value shown in the upper right is 0.058.

(a)   Since this is larger than the 0.05 significance level, we do not reject the null hypothesis of normality.

(b)   Since this is smaller than the 0.10 significance level, we reject the null hypothesis of normality.

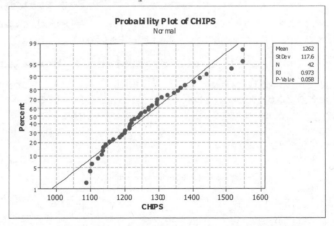

**15.151** (a)   In Exercise 15.31, part (c), we determined that it was reasonable to consider Assumptions 1-3 for regression inferences met by these variables.

(b)   Using Minitab, in order to do the correlation test for normality of the residuals, we must perform the regression while saving the residuals to an additional column of the worksheet.  That column will be named RESI1 by Minitab.  Choose **Stat ▶ Regression ▶ Regression...**, select <u>DEATH</u> in the **Response:** text box, select <u>INAUGURATION</u> in the **Predictors:** text box, click on the **Storage...** button, check the **Residuals** box, click **OK**, and click **OK**.  Then choose **Stat ▶ Basic statistics ▶ Normality test...**, select <u>RESI1</u> in the **Variable** text box, select the **Ryan-Joiner** option button from the **Tests for Normality** field, and click **OK**.  The result is shown below.  The p-value shown in the upper right is > 0.100.  Since this is larger than the 0.05 significance level, we do not reject the null hypothesis of normality.

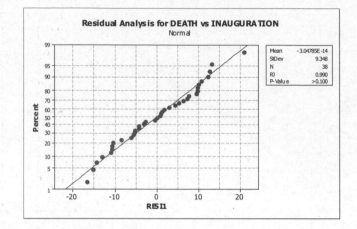

**15.153** (a) In Exercise 15.33, part (c), and exercise 15.65, we determined that it was not reasonable to consider Assumption 1 for regression inferences met by these variables.  Therefore, parts (b) and (c) are omitted.

**15.155** (a) In Exercise 15.35, part (c), we determined that it was reasonable to consider Assumptions 1-3 for regression inferences met by these variables.

(b) Using Minitab, in order to do the correlation test for normality of the residuals, we must perform the regression while saving the residuals to an additional column of the worksheet.  That column will be named RESI1 by Minitab.  Choose **Stat ▶ Regression ▶ Regression...**, select <u>LOW</u> in the **Response:** text box, select <u>HIGH</u> in the **Predictors:** text box, click on the **Storage...** button, check the **Residuals** box, click **OK**, and click **OK**.  Then choose **Stat ▶ Basic statistics ▶ Normality test...**, select <u>RESI1</u> in the **Variable** text box, select the **Ryan-Joiner** option button from the **Tests for Normality** field, and click **OK**.  The result is shown below.  The p-value shown in the upper right is         > 0.100.  Since this is larger than the 0.05 significance level, we do not reject the null hypothesis of normality.

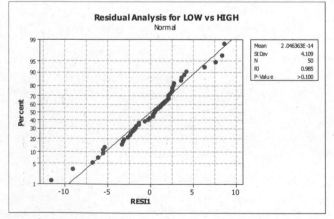

**15.157** (a) In Exercise 15.37, part (c), we determined that it was not reasonable to consider Assumption 1 for regression inferences met by these variables.  The residual plot was convex upward.  Therefore, parts (b) and (c) are omitted.

**15.159**(a) In Exercise 15.39, part (c), we determined that it was not reasonable to consider Assumptions 1 and 3 for regression inferences met by these variables. Therefore, parts (b) and (c) are omitted.

**15.161**Minitab does not have the capability of performing a normality test using grouped data. However, it does have an easy way to enter repeated data values so that every value occurs in the column as often as its frequency dictates. First, name one of the columns LENGTH. Then, in the Session Window, following the MTB > prompt, enter the command <u>SET 'LENGTH'</u>. A new prompt, DATA> will appear. Following this prompt, enter each data value in parentheses preceded by its frequency. The number 9.8 with its frequency of 4 is entered as 4(9.8). Put a space or a comma before entering the next number. You may hit the enter key at any time and continue the data on the next line. When you have entered the last data values, type <u>END</u> at an MTB> prompt. Once you have named one column LENGTH, the entire data entry process will look like this:

```
MTB > SET 'LENGTH'
DATA> 9.5 4(9.8) 24(10.1) 67(10.4) 193(10.7) 417(11.0) 575(11.3)
DATA> 691(11.6) 509(11.9) 306(12.2) 131(12.5) 63(12.8) 16(13.1) 3(13.4)
DATA> END
```

Now, we choose **Stat ▶ Basic Statistics ▶ Normality test...**, select <u>LENGTH</u> in the **Variable** text box, select the **Ryan-Joiner** option button from the **Tests for Normality** field, and click **OK**. The result is shown following. The value of $R_p$ is 1.000 and the P-value shown in the upper right is > 0.100. Since this is larger than the 0.05 significance level, we do not reject the null hypothesis of normality.

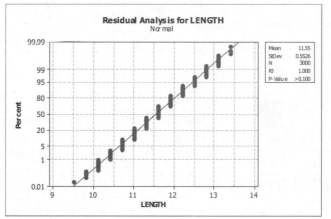

**15.163** (a) In Minitab, name four columns T1, T2, T3, and T4. Then choose **Calc ▶ Random data ▶ Exponential...**, type <u>75</u> in the **Generate rows of data** text box, select T1, T2, T3, and T4 in the **Store in column(s):** text box, type <u>8.7</u> in the **Mean** text box, and click **OK**.

(b) To perform the correlation test on T1, choose **Stat ▶ Basic statistics ▶ Normality test...**, select T1 in the **Variable** text box, click on the **Ryan-Joiner** button and click **OK**. Repeat this process for T2, T3, and T4. The results are shown in the four following graphs.

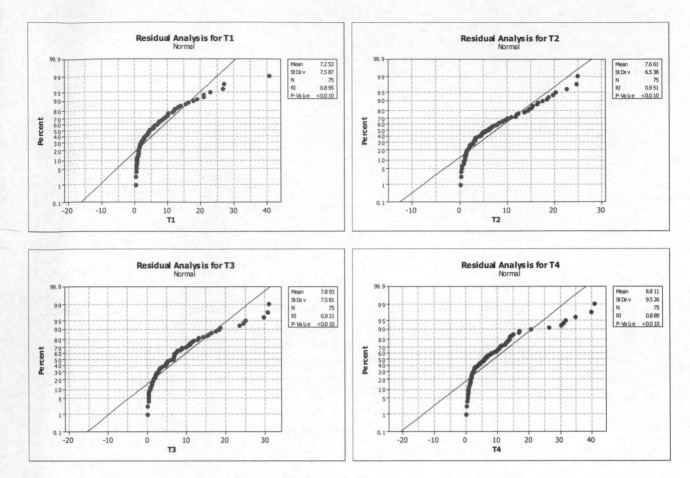

The four values of $R_p$ are 0.895, 0.951, 0.921, and 0.889.  Their P-values are all less than 0.0100.  Thus all of them are significant at the 0.05 significance level.

(c)   Since the samples were generated from a non-normal (exponential) distribution and we expect that samples will reflect the properties of the population, the results in (b) are what we should have expected.

## Review Problems for Chapter 15

1.    (a)  conditional

      (b)  The four assumptions for regression inferences are:

        (1)  *Population regression line:*  There is a line $y = \beta_0 + \beta_1 x$ such that, for each $x$-value, the mean of the corresponding population of $y$-values lies on that line.

        (2)  *Equal standard deviations:*  The standard deviation $\sigma$ of the population of $y$-values corresponding to a particular $x$-value is the same, regardless of the $x$-value.

        (3)  *Normality:*  For each $x$-value, the corresponding population of $y$-values is normally distributed.

        (4)  *Independence:* The observations of the response variable are independent of one another.

2.    (a)  The slope of the sample regression line, $b_1$

      (b)  The y-intercept of the sample regression line, $b_0$

      (c)  $s_e$

3.    We used a plot of the residuals against the values of the predictor variable x and a normal probability plot of the residuals.  In the first plot, the residuals should lie in a horizontal band centered on and symmetric about

the x-axis. In the second plot, the points should lie roughly in a line.

4.  (a) A residual plot showing curvature indicates that the first assumption, that of linearity, is probably not valid.

(b) This type of plot indicates that the second assumption, that of constant standard deviation, is probably not valid.

(c) A normal probability plot with extreme curvature indicates that the third assumption, that the conditional distribution of the response variable is normally distributed, is not valid.

(d) A normal probability plot that is roughly linear, but shows outliers may be indicating that: the linear model is appropriate for most values of x, but not for all; or, that the standard deviation is not constant for all values of x, allowing for a few y values to lie far from the regression line; or, that a few data values are 'faulty', that is, the experimental conditions represented by the value of the x variable were not as they should have been or that the y value was in error.

5.  If we reject the null hypothesis, we are claiming that $\beta_1$ is not zero. This means that different values of x will lead to different values of y. Hence, the regression equation is useful for making predictions.

6.  t (for $b_1$), r, $r^2$

7.  No. The best estimate of conditional mean of the response variable and the best prediction of a single future value of y are the same.

8.  A confidence interval estimates the value of a parameter; a prediction interval is used to predict a future value of a random variable.

9.  $\rho$

10. (a) If $\rho > 0$, the variables are positively correlated, that is, there is a tendency for one variable to increase linearly as the other one increases.

(b) If we know only that $\rho \neq 0$, the two variables are linearly correlated, that is, there is a linear relationship between them.

(c) If $\rho < 0$, the variables are negatively correlated, that is, there is a tendency for one variable to decrease linearly as the other one increases.

11. If the regression assumptions are satisfied, then there is a linear relationship between the student/faculty ratio and the graduation rate, the conditional standard deviation of the graduation rate is the same for all values of the student/faculty ratio, the conditional distribution of the graduation rate is a normal distribution for all values of the student/faculty ratio, and the values of the graduation rate are independent of each other.

12. (a) To determine $b_0$ and $b_1$ for the regression line, construct a table of values for x, y, $x^2$, xy and their sums. A column for $y^2$ is also presented for later use.

| x | y | $x^2$ | Xy | $y^2$ |
|---|---|---|---|---|
| 16 | 45 | 256 | 720 | 2025 |
| 20 | 55 | 400 | 1100 | 3025 |
| 17 | 70 | 289 | 1190 | 4900 |
| 19 | 50 | 361 | 950 | 2500 |
| 22 | 47 | 484 | 1034 | 2209 |
| 17 | 46 | 289 | 782 | 2116 |
| 17 | 50 | 289 | 850 | 2500 |
| 17 | 66 | 289 | 1122 | 4356 |
| 10 | 26 | 100 | 260 | 676 |
| 18 | 60 | 324 | 1080 | 3600 |
| 173 | 515 | 3081 | 9088 | 27907 |

$S_{xx}$ = 3081 - $173^2$/10 = 88.10

$S_{xy}$ = 9088 - (173)(515)/10 = 178.50

$b_1$ = 178.50/88.10 = 2.0261

$b_0$ = 515/10 - 2.0261(173/10) = 16.4484

$\hat{y}_p$ = 16.448 + 2.026x is the equation of the regression line.

(b)  $S_{yy}$ = 27907 - $515^2$/10 = 1384.50

SSE = $S_{yy}$ - $S^2_{xy}$/$S_{xx}$ = 1384.50 - $178.50^2$/88.10 = 1022.8400

$$s_e = \sqrt{\frac{SSE}{n-2}} = \sqrt{\frac{1922.8400}{10-2}} = 11.3073$$

This value of $s_e$ indicates that, roughly speaking, the predicted values of the graduation rate differ from the observed values of the graduation rate by about 11.31.

(c)  Presuming that the variables Student/Faculty ratio and Graduation rate satisfy the assumptions for regression inferences, the standard error of the estimate, $s_e$ = 11.31, provides an estimate for the common population standard deviation, $\sigma$, of graduation rates of entering freshmen at universities with any particular student/faculty ratio.

13.   For each value of x, we compute $e = y - \hat{y}$ in the following table and plot e against x.

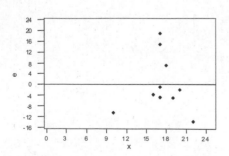

| x | e |
|---|---|
| 16 | -3.87 |
| 20 | -1.97 |
| 17 | 19.11 |
| 19 | -4.94 |
| 22 | -14.02 |
| 17 | -4.89 |
| 17 | -0.89 |
| 17 | 15.11 |
| 10 | -10.71 |
| 18 | 7.08 |

| Residual | Normal score |
|----------|--------------|
| -14.02 | -1.55 |
| -10.71 | -1.00 |
| -4.94 | -0.65 |
| -4.89 | -0.37 |
| -3.87 | -0.12 |
| -1.97 | 0.12 |
| -0.89 | 0.37 |
| 7.08 | 0.65 |
| 15.11 | 1.00 |
| 19.11 | 1.55 |

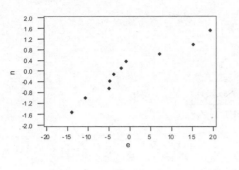

The small sample size makes it difficult to evaluate the plot of e against x.  While there are no obvious patterns in the first plot, there are two points (in the lower left and right corners of the plot) that are cause for concern.  The normal probability plot is not far from linear.

**14.**  (a)  From Review Problem 11, $\sum x = 173$, $\sum x^2 = 3{,}081$, and $b_1 = 2.0261$.  From Review Problem 12, $s_e = 11.3073$.

Step 1: $H_0$: $\beta_1 = 0$

$H_a$: $\beta_1 \neq 0$

Step 2: $\alpha = 0.05$

Step 3:  $t = \dfrac{b_1}{s_e / \sqrt{\sum x^2 - (\sum x)^2 / n}} = \dfrac{2.0261}{11.3073 / \sqrt{3081 - 173^2 / 10}} = 1.68$

Step 4:  df = n - 2 = 8; critical values = ±2.306
Step 5:  Since -2.306 < 1.68 < 2.306, do not reject the null hypothesis at the 5% significance level.
Step 6:  The data do not provide sufficient evidence to conclude that the student-to-faculty ratio is useful as a predictor of graduation rate.  For the P-value approach, note that 0.10 < p < 0.20.  Since p > $\alpha$, do not reject the null hypothesis.

(b)  Step 1:  For a 95% confidence interval, $\alpha = 0.05$.  With df = n - 2 = 8, $t_{\alpha/2} = t_{0.025} = 2.306$.

Step 2:  The endpoints of the confidence interval for $\beta_1$ are

$$b_1 \pm t_{\alpha/2} \cdot s_e / \sqrt{\sum x^2 - (\sum x)^2 / n}$$

$$2.0261 \pm 2.306 \cdot 11.3073 / \sqrt{3081 - 173^2 / 10}$$

$$2.0261 \pm 2.7780$$

$$-0.7519 \; to \; 4.8041$$

We can be 95% confident that the change in graduation rate for universities per 1 unit increase in the student/faculty ratio is somewhere between -0.7519 and 4.8041 percent.

**15.**  From Review Exercise 11, $\hat{y} = 16.4484 + 2.0261x$, $\sum x = 173$, and $\sum x^2 = 3{,}081$.  From Review Exercise 12, $s_e = 11.3073$.

(a)  $\hat{y}_p = 16.4484 + 2.0261(17) = 50.89$
(b)  Step 1:  For a 95% confidence interval, $\alpha = 0.05$.

With df = n - 2 = 8, $t_{\alpha/2} = t_{0.025} = 2.306$.

Step 2:  $\hat{y}_p = 50.89$

Step 3:  The endpoints of the confidence interval are:

$$\hat{y}_p \pm t_{\alpha/2} \cdot s_e \cdot \sqrt{\frac{1}{n} + \frac{(x_p - \sum x/n)^2}{\sum x^2 - (\sum x)^2/n}}$$

$$= 50.89 \pm 2.306 \cdot 11.3073 \cdot \sqrt{\frac{1}{10} + \frac{(17 - 173/10)^2}{3081 - 173^2/10}}$$

$$= 50.89 \pm 8.29 = (42.60, 59.18)$$

The interpretation of this interval is as follows:  We can be 95% confident that the graduation rate of universities with a student/faculty ratio of 17 is somewhere between 42.60% and 59.18%.

(c)  This is the same as the answer in part (a): $\hat{y}_p = 50.89$.

(d)  For a 95% prediction interval, Steps 1 and 2 are the same as Steps 1 and 2, respectively, in part (b).  Thus, they are not repeated here. Only Step 3 is presented.

Step 3: The endpoints of the prediction interval are:

$$\hat{y}_p \pm t_{\alpha/2} \cdot s_e \cdot \sqrt{1 + \frac{1}{n} + \frac{(x_p - \sum x/n)^2}{\sum x^2 - (\sum x)^2 n}}$$

$$= 50.89 \pm 2.306 \cdot 11.3073 \cdot \sqrt{1 + \frac{1}{10} + \frac{(17 - 173/10)^2}{3081 - 173^2/10}}$$

$$= 50.89 \pm 27.36 = (23.53, 78.25)$$

The interpretation of this interval is as follows:  We can be 95% certain that the graduation rate of a randomly selected university with a student/faculty ratio of 17 will be somewhere between 23.53% and 78.25%.

(e)  The error in the estimate of the mean graduation rate for universities with a student/faculty ratio of 17 is due only to the fact that the population regression line is being estimated by a sample regression line.  The error in the prediction of the graduation rate of a randomly chosen university with a student/faculty ratio is due to the estimation error mentioned above plus the variation in graduation rates of universities with a student/faculty ratio of 17.

16.  From Review Problem 11,

$S_{xx} = 88.10$

$S_{xy} = 178.50$

$S_{yy} = 1384.50$

$$r = \frac{S_{xy}}{\sqrt{S_{xx}S_{yy}}} = \frac{178.50}{\sqrt{(88.10)(1384.50)}} = 0.5111$$

Step 1:  $H_0$: $\rho = 0$, $H_a$: $\rho > 0$

Step 2:  $\alpha = 0.025$

Step 3:

$$t = \frac{r}{\sqrt{\frac{1 - r^2}{n - 2}}} = \frac{0.5111}{\sqrt{\frac{1 - (0.5111)^2}{10 - 2}}} = 1.682$$

Step 4:  df = n - 2 = 8; critical value = 2.306

> Step 5:  Since 1.682 < 2.306, do not reject $H_0$.
>
> Step 6:  The data do not provide sufficient evidence to conclude that graduation rate and student /faculty ratio are positively linearly correlated.
>
> For the P-value approach, 0.05 < p < 0.10.  Therefore, because the p-value is larger than the significance level of 0.025, we cannot reject $H_0$.

**17.**  normal scores

**18.**  Step 1:  $H_0$: The population is normally distributed.
          $H_a$: The population is not normally distributed.

   Step 2:  $\alpha$ = 0.05

   Step 3:

| Mileage x | Normal score w | xw | $x^2$ | $w^2$ |
|---|---|---|---|---|
| 25.9 | −1.74 | −45.066 | 670.81 | 3.0276 |
| 27.3 | −1.24 | −33.852 | 745.29 | 1.5376 |
| 27.3 | −0.94 | −25.662 | 745.29 | 0.8836 |
| 27.6 | −0.71 | −19.596 | 761.76 | 0.5041 |
| 27.8 | −0.51 | −14.178 | 772.84 | 0.2601 |
| 27.8 | −0.33 | −9.174 | 772.84 | 0.1089 |
| 28.5 | −0.16 | −4.560 | 812.25 | 0.0256 |
| 28.6 | 0.00 | 0.000 | 817.96 | 0.0000 |
| 28.8 | 0.16 | 4.608 | 829.44 | 0.0256 |
| 28.9 | 0.33 | 9.537 | 835.21 | 0.1089 |
| 29.4 | 0.51 | 14.994 | 864.36 | 0.2601 |
| 29.7 | 0.71 | 21.087 | 882.09 | 0.5041 |
| 30.9 | 0.94 | 29.046 | 954.81 | 0.8836 |
| 31.2 | 1.24 | 38.688 | 973.44 | 1.5376 |
| 31.6 | 1.74 | 54.984 | 998.56 | 3.0276 |
| 431.3 | 0.00 | 20.856 | 12,436.95 | 12.6950 |

$$R_p = \frac{\sum xw}{\sqrt{[\sum x^2 - (\sum x)^2 / n][\sum w^2]}} = \frac{20.856}{\sqrt{[12436.95 - (431.3)^2 / 15][12.6950]}} = 0.981$$

   Step 4:  The critical value is = 0.938.

   Step 5:  From Step 4 the value of the test statistic is $R_p$ = 0.981, which does not fall in the rejection region.  Hence we do not reject $H_0$.

   Step 6:  The test results are not statistically significant at the 5% level, that is, at the 5% significance level, the data do not provide sufficient evidence to conclude that the gas mileages for this model are not normally distributed.  Note:  If Minitab is used to compute the normal scores and $R_p$, you will get 0.983.  The conclusion remains the same.

**19.**  From Problem 13, we have the residuals and their normal scores.  We construct the table below to get the sums needed for computing $R_p$.

| e | w | ew | $e^2$ | $w^2$ | $w^2$ |
|---|---|---|---|---|---|
| -14.02 | -1.55 | 21.7310 | 196.5604 | 2.4025 | 2.4025 |
| -10.71 | -1.00 | 10.7100 | 114.7041 | 1.0000 | 1.0000 |
| -4.94 | -0.65 | 3.2110 | 24.4036 | 0.4225 | 0.4225 |
| -4.89 | -0.37 | 1.8093 | 23.9121 | 0.1369 | 0.1369 |
| -3.87 | -0.12 | 0.4644 | 14.9769 | 0.0144 | 0.0144 |
| -1.97 | 0.12 | -0.2364 | 3.8809 | 0.0144 | 0.0144 |
| -0.89 | 0.37 | -0.3293 | 0.7921 | 0.1369 | 0.1369 |
| 7.08 | 0.65 | 4.6020 | 50.1264 | 0.4225 | 0.4225 |
| 15.11 | 1.00 | 15.1100 | 28.3121 | 1.0000 | 1.0000 |
| 19.11 | 1.55 | 29.6205 | 365.1921 | 2.4025 | 2.4025 |
| 0.00 | 0.00 | 86.6900 | 1022.8600 | 7.9500 | 7.9500 |

$$R_p = \frac{\sum ew}{\sqrt{[\sum e^2 - (\sum e)^2 / n][\sum w^2]}} = \frac{86.69}{\sqrt{[1022.86 - (0)^2/10][7.95]}} = 0.961211$$

At the 5% significance level, the critical value from Table IX is 0.905. Since $R_p > 0.905$, do not reject the null hypothesis of normality.

20. Using Minitab, with the data in columns named IMR and LIFE EXPECTANCY, we choose **Stat ▶ Regression ▶ Regression...**, select 'LIFE EXPECTANCY' in the **Response:** text box, select IMR in the **Predictors:** text box, click on the **Graphs** button, click in the **Residuals versus the variables** text box and select IMR, and click **OK**. Looking ahead to parts (b-h) and (j), click on the **Options...** button, enter 30 in the **Prediction intervals for new observations:** text box, enter 95 in the **Confidence level:** text box, and click **OK**. Click on the **Storage...** button, check the **Residuals** box, click **OK**, and click **OK**. Then choose **State ▶ Basic statistics ▶ Normality test...**, select RESI1 in the **Variable** text box, select the **Ryan-Joiner** option button from the **Tests for Normality** field, and click **OK**. We will first show all of the written output and then answer the questions. The graphical output will be shown in part (k).

```
The regression equation is
LIFE EXPECTANCY = 78.8 - 0.312 IMR

Predictor      Coef   SE Coef       T       P
Constant     78.811     1.031   76.43   0.000
IMR         -0.31173   0.01840  -16.94   0.000

S = 5.24948   R-Sq = 81.8%   R-Sq(adj) = 81.5%

Analysis of Variance

Source          DF      SS      MS       F       P
Regression       1  7909.7  7909.7  287.03   0.000
Residual Error  64  1763.6    27.6
Total           65  9673.4
```

Unusual Observations

| Obs | IMR | LIFE EXPECTANCY | Fit | SE Fit | Residual | St Resid |
|-----|-----|-----------------|--------|--------|----------|----------|
| 2 | 169 | 42.000 | 26.161 | 2.393 | 15.839 | 3.39RX |
| 32 | 63 | 45.200 | 59.048 | 0.741 | -13.848 | -2.66R |
| 39 | 138 | 38.200 | 35.856 | 1.849 | 2.344 | 0.48 X |
| 50 | 61 | 46.600 | 59.859 | 0.719 | -13.259 | -2.55R |
| 66 | 67 | 39.000 | 58.082 | 0.771 | -19.082 | -3.67R |

R denotes an observation with a large standardized residual.
X denotes an observation whose X value gives it large influence.

Predicted Values for New Observations

| New Obs | Fit | SE Fit | 95% CI | 95% PI |
|---------|--------|--------|------------------|------------------|
| 1 | 69.460 | 0.693 | (68.074, 70.845) | (58.882, 80.038) |

Values of Predictors for New Observations

| New Obs | IMR |
|---------|------|
| 1 | 30.0 |

(a) LIFE EXPECTANCY = 78.8 - 0.312 IMR

(b) The standard error of the estimate is 5.24948. Roughly speaking, this indicates how much, on average, the predicted life expectancies differ from the observed values of life expectancies.

(c) The coefficient of determination is 0.818, indicating that 81.8% of the variation in life expectancy is explained by a linear relationship with the infant mortality rate. This is a fairly strong relationship and indicates that the IMR may be useful for predicting life expectancy if the assumptions for making such inferences are valid.

(d) The point estimate for the conditional mean of life expectancy for IMR = 30 is 69.460 years. This is the point on the sample regression line that best estimates the corresponding point on the population regression line.

(e) The 95% confidence interval for the conditional mean of life expectancy for IMR = 30 is 68.074 to 70.845 years. We are 95% confident that the mean life expectancy is somewhere between these limits when the IMR is 30.

(f) The predicted value of life expectancy for IMR = 30 is 69.460 years. This is our best estimate of life expectancy for a country with an IMR equal to 30.

(g) The 95% prediction interval for the life expectancy of a country with IMR = 30 is 58.882 to 80.038 years. We are 95% certain that the life expectancy is somewhere between these limits when the IMR is 30.

(h) The error in the estimate of the mean life expectancy of countries with IMR = 30 is due only to the fact that the population regression line is being estimated by a sample regression line; whereas, the error in the prediction of the life expectancy of a single country with an IMR = 30 is due to that fact plus the variation in life expectancies for countries with IMR = 30.

(j) Choose **Stat ▶ Basic Statistics ▶ Correlation**, enter <u>IMR</u> and "<u>LIFE EXPECTANCY</u>' in the **Variables** text box, check the box to **Display P-values**, and click **OK**. The result is
    Pearson correlation of IMR and LIFE EXPECTANCY = -0.904
    P-Value = 0.000

The P-value for a left-tail test is half of that shown (still 0.000). Since the P-value is less than 0.05, we reject the null hypothesis that $\rho = 0$ and since r is negative, the data provide evidence that there is a negative correlation between IMR and life expectancy assuming that the assumptions for making regression inferences are reasonable.

(k) For the residual analysis, we look at the graph produced by the procedure at the beginning of this solution and also the results of a Ryan-Joiner normality test on the residuals, which were stored earlier.

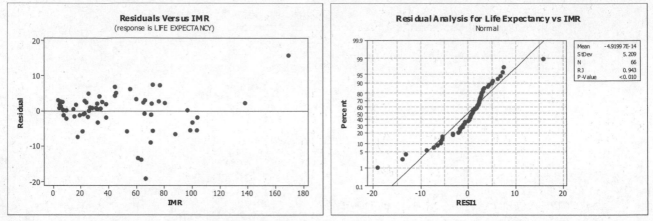

The first graph shows several residuals somewhat farther from the horizontal axis than the rest of the points and there are two residuals at the far right that indicate potential influential observations. This graph does not show a band of residuals centered on and symmetric about the horizontal axis. The normal probability plot shows several points well off the line and the Ryan-Joiner test has a P-value that is less than 0.01. Thus, Assumption 3 for making regression inferences appears to be violated. In addition, the concern about several outliers and potential influential observations indicates that the tests, confidence interval and prediction interval produced earlier may not be valid.

21. Using Minitab, with the data in columns named HIGH and PRECIPITATION, we choose **Stat ▶ Regression ▶ Regression...**, select PRECIPITATION in the **Response:** text box, select HIGH in the **Predictors:** text box, click on the **Graphs** button, click in the **Residuals versus the variables** text box and select HIGH, and click **OK**. Looking ahead to parts (b-h) and (j), click on the **Options...** button, enter 83 in the **Prediction intervals for new observations:** text box, enter 95 in the **Confidence level:** text box, and click **OK**. Click on the **Storage...** button, check the **Residuals** box, click **OK**, and click **OK**. Then choose **State ▶ Basic statistics ▶ Normality test...**, select RESI1 in the **Variable** text box, select the **Ryan-Joiner** option button from the **Tests for Normality** field, and click **OK**. We will first show all of the written output and then answer the questions. The graphical output will be shown in part (k).

```
The regression equation is
PRECIPITATION = 2.05 + 0.0067 HIGH

Predictor      Coef   SE Coef      T      P
Constant      2.048     2.170   0.94  0.350
HIGH        0.00667   0.02742   0.24  0.809

S = 2.41336   R-Sq = 0.1%   R-Sq(adj) = 0.0%
```

```
Analysis of Variance

Source              DF      SS      MS      F      P
Regression           1   0.344   0.344   0.06   0.809
Residual Error      46 267.919   5.824
Total               47 268.263

Unusual Observations

Obs  HIGH  PRECIPITATION    Fit  SE Fit  Residual  St Resid
  6    86          8.100  2.621   0.410     5.479     2.30R
 10    87          9.600  2.628   0.425     6.972     2.93R
 22    45          0.500  2.348   0.972    -1.848    -0.84 X
 45    93          7.900  2.668   0.537     5.232     2.22R


R denotes an observation with a large standardized residual.
X denotes an observation whose X value gives it large influence.

Predicted Values for New Observations

New
Obs    Fit   SE Fit      95% CI            95% PI
  1  2.601    0.373  (1.850, 3.353)  (-2.314, 7.517)

Values of Predictors for New Observations

New
Obs  HIGH
  1  83.0
```

(a)  PRECIPITATION = 2.05 + 0.0067 HIGH

(b)  The standard error of the estimate is 2.41336.  Roughly speaking, this indicates how much, on average, the predicted average July precipitations differ from the observed values of precipitations.

(c)  The coefficient of determination is 0.001, indicating that 0.1% of the variation in average July precipitation is explained by a linear relationship with the average July high temperature.  This is a very weak relationship and indicates that the average July high temperature is useless for predicting average July precipitation if the assumptions for making such inferences are valid.

(d)  The point estimate for the conditional mean of average July precipitation for an average July high temperature of 83 degrees is 0.373 inches.  This is the point on the sample regression line that best estimates the corresponding point on the population regression line.

(e)  The 95% confidence interval for the conditional mean of average July precipitation for an average July high temperature of i3 degrees is 1.850 to 3.353 inches.  We are 95% confident that the mean average July precipitation is somewhere between these limits when the average July high temperature is 83 degrees.

(f)  The predicted value of average July precipitation for a city with an average July high temperature of 83 degrees is 0.373 inches.  This is our best estimate of average July precipitation for a city with an average July high temperature of 83 degrees.

(g)  The 95% prediction interval for the average July precipitation of a city with an average high July temperature of 83 degrees is -2.314 to 7.517 inches.  The lower prediction limit can be changed to 0.00 since negative precipitation amounts are not possible.  We are 95% certain that the average July precipitation is somewhere between these limits for a city with an average July high temperature of 83 degrees.

(h)  The error in the estimate of the mean average July precipitation of cities with average July high temperature of 83 degrees is due only to the fact that the population regression line is being estimated by a sample regression line; whereas, the error in the prediction of the average July precipitation of a single city with an average July high temperature of 83 degrees is due to that fact plus the variation in average July precipitation for cities with average high July temperatures of 83 degrees.

(j) Choose **Stat ▶ Basic Statistics ▶ Correlation**, enter HIGH and PRECIPITATION in the **Variables** text box, check the box to **Display P-values**, and click **OK**. The result is

   Pearson correlation of HIGH and PRECIPITATION = 0.036
   P-Value = 0.809

Since the P-value is greater than 0.05, we do not reject the null hypothesis that $\rho = 0$.

(k) For the residual analysis, we look at the graph produced by the procedure at the beginning of this solution and also the results of a Ryan-Joiner normality test on the residuals, which were stored earlier.

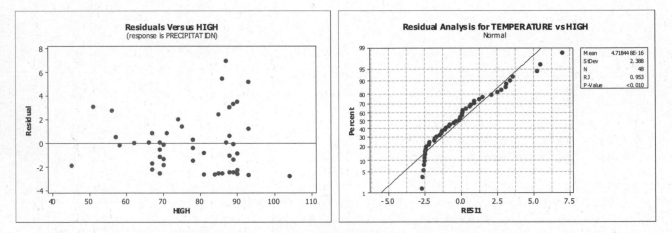

The first graph does not show a band of residuals centered on and symmetric about the horizontal axis. They may even be somewhat U-shaped. The normal probability plot shows several points well off the line and the Ryan-Joiner test has a P-value that is less than 0.01. Thus, Assumption 3 for making regression inferences appears to be violated. In addition, since a linear model accounts for only 0.1% of the variation in average July precipitation, it appears that Assumption 1 (of a linear model) is also violated. The regression inferences above are very unlikely to be valid and are pretty useless.

22. Using Minitab, with the data in columns named FAT CONSUMPTION and DEATH RATE, we choose **Stat ▶ Regression ▶ Regression...**, select 'DEATH RATE' in the **Response:** text box, select 'FAT CONSUMPTION' in the **Predictors:** text box, click on the **Graphs** button, click in the **Residuals versus the variables** text box and select 'FAT CONSUMPTION', and click **OK**. Looking ahead to parts (b-h) and (j), click on the **Options...** button, enter 92 in the **Prediction intervals for new observations:** text box, enter 95 in the **Confidence level:** text box, and click **OK**. Click on the **Storage...** button, check the **Residuals** box, click **OK**, and click **OK**. Then choose **State ▶ Basic statistics ▶ Normality test...**, select RESI1 in the **Variable** text box, select the **Ryan-Joiner** option button from the **Tests for Normality** field, and click **OK**. We will first show all of the written output and then answer the questions. The graphical output will be shown in part (k).

```
The regression equation is
DEATH RATE = - 1.06 + 0.113 FAT CONSUMPTION

Predictor            Coef   SE Coef      T      P
Constant           -1.063     1.170  -0.91  0.371
FAT CONSUMPTION   0.11336   0.01126  10.06  0.000

S = 2.29453   R-Sq = 78.3%   R-Sq(adj) = 77.6%

Analysis of Variance

Source           DF      SS      MS       F      P
Regression        1  533.31  533.31  101.30  0.000
Residual Error   28  147.42    5.26
Total            29  680.73

Unusual Observations

Obs  FAT CONSUMPTION  DEATH RATE    Fit  SE Fit  Residual  St Resid
  5               96       4.800  9.820   0.419    -5.020     -2.23R
 30              132      18.400 13.901   0.575     4.499      2.03R

R denotes an observation with a large standardized residual.

Predicted Values for New Observations

New
Obs    Fit  SE Fit       95% CI           95% PI
  1  9.366   0.423  (8.500, 10.232)  (4.587, 14.145)

Values of Predictors for New Observations

New
Obs  FAT CONSUMPTION
  1             92.0
```

(a) The regression equation is DEATH RATE = - 1.06 + 0.113 FAT CONSUMPTION
(b) The standard error of the estimate is 2.29453.  Roughly speaking, this indicates how much, on average, the predicted death rates differ from the observed values of death rates.
(c) The coefficient of determination is 0.783, indicating that 78.3% of the variation in death rates is explained by a linear relationship with the fat consumption.  The t-test for the slope results in t = 10.06 with a P-value of 0.00.  Thus, we would reject the null hypothesis that $\beta_1 = 0$. This is a fairly strong relationship and indicates that fat consumption is quite useful for predicting death rates if the assumptions for making such inferences are reasonable.
(d) The point estimate for the conditional mean death rate for a country with a fat consumption of 92 grams per day is 9.366 per 100,000 males. This is the point on the sample regression line that best estimates the corresponding point on the population regression line.
(e) The 95% confidence interval for the conditional mean of death rates for countries with a fat consumption of 92 grams per day is 8.500 to 10.232 per 100,000 males.  We are 95% confident that the mean death rate is somewhere between these limits when the average fat consumption is 92 grams per day.
(f) The predicted death rate for a country with an average fat consumption of 92 grams per day is 9.366 per 100,000 males.  This is our best estimate of the death rate for a country with an average fat consumption of 92 grams per day.
(g) The 95% prediction interval for the death rate of a country with an average fat consumption of 92 grams per day is 4.587 to 14.145 per

100,000 males. We are 95% certain that the death rate is somewhere between these limits for a country with an average fat consumption of 92 grams per day.

(h) The error in the estimate of the mean death rate of countries with an average fat consumption of 92 grams per day is due only to the fact that the population regression line is being estimated by a sample regression line; whereas, the error in the prediction of the death rates of a single country with an average fat consumption of 92 grams per day is due to that fact plus the variation in death rates for countries with an average fat consumption of 92 grams per day.

(j) Choose **Stat ▶ Basic Statistics ▶ Correlation**, enter 'DEATH RATE' and 'FAT CONSUMPTION' in the **Variables** text box, check the box to **Display P-values**, and click **OK**. The result is
    Pearson correlation of DEATH RATE and FAT CONSUMPTION = 0.885
    P-Value = 0.000

The P-value for a right-tailed test is half of that shown (still 0.000) if r is positive. Since the P-value is less than 0.05, we reject the null hypothesis that $\rho = 0$.

(k) For the residual analysis, we look at the graph produced by the procedure at the beginning of this solution and also the results of a Ryan-Joiner normality test on the residuals, which were stored earlier.

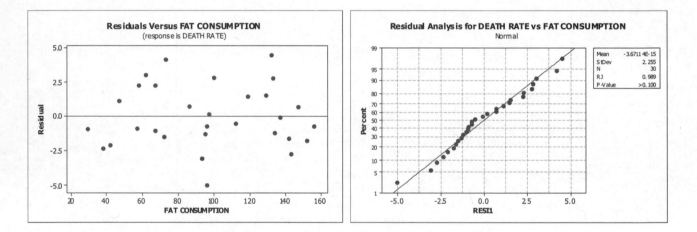

In the left-hand graph, the residuals are located in a horizontal band centered on and somewhat symmetrical about the horizontal axis. The P-value for the Ryan-Joiner test is > 0.100 indicating that normality of the residuals is a reasonable assumption. Since the linear model also appears to be appropriate, the regression inferences made earlier in this problem are likely to be reasonable.

23. Using Minitab, with the data in columns named PAYROLL and PERCENTAGE, we choose **Stat ▶ Regression ▶ Regression...**, select PERCENTAGE in the **Response:** text box, select PAYROLL in the **Predictors:** text box, click on the **Graphs** button, click in the **Residuals versus the variables** text box and select FAT PAYROLL, and click **OK**. Looking ahead to parts (b-h) and (j), click on the **Options...** button, enter 50 in the **Prediction intervals for new observations:** text box, enter 95 in the **Confidence level:** text box, and click **OK**. Click on the **Storage...** button, check the **Residuals** box, click OK, and click **OK**. Then choose **State ▶ Basic statistics ▶ Normality test...**, select RESI1 in the **Variable** text box, select the **Ryan-Joiner**

option button from the **Tests for Normality** field, and click **OK**. We will
first show all of the written output and then answer the questions. The
graphical output will be shown in part (k).

```
     The regression equation is
     PERCENTAGE = 40.9 + 0.128 PAYROLL

     Predictor      Coef    SE Coef      T       P
     Constant      40.949     3.879    10.56   0.000
     PAYROLL       0.12839   0.05134    2.50   0.019

     S = 7.65251    R-Sq = 18.3%    R-Sq(adj) = 15.3%

     Analysis of Variance

     Source           DF       SS        MS      F       P
     Regression        1     366.29    366.29   6.25   0.019
     Residual Error   28    1639.71     58.56
     Total            29    2006.00

     Unusual Observations

     Obs  PAYROLL  PERCENTAGE    Fit   SE Fit  Residual  St Resid
       3     150        61.00   60.21    4.31      0.79      0.13 X
      25     117        41.00   55.97    2.77    -14.97     -2.10R
      30      49        27.00   47.24    1.78    -20.24     -2.72R

     R denotes an observation with a large standardized residual.
     X denotes an observation whose X value gives it large influence.

     Predicted Values for New Observations

     New
     Obs    Fit   SE Fit       95% CI            95% PI
       1   47.37    1.75   (43.79, 50.95)   (31.29, 63.45)

     Values of Predictors for New Observations

     New
     Obs  PAYROLL
       1     50.0
```

(a)  The regression equation is PERCENTAGE = 40.9 + 0.128 PAYROLL
(b)  The standard error of the estimate is 7.65251. Roughly speaking, this
     indicates how much, on average, the predicted percentages differ from
     the observed values of percentages.
(c)  The coefficient of determination is 0.183, indicating that 18.3% of the
     variation in percentages is explained by a linear relationship with the
     average July high temperature. The t-test for the slope results in $t =$
     2.50 with a P-value of 0.019. Thus, we would reject the null
     hypothesis that $\beta_1 = 0$. Even though the P-value is small, the low
     coefficient of determination indicates that this is a fairly weak
     relationship and that payroll is only a little useful for predicting
     percentages if the assumptions for making such inferences are valid.
(d)  The point estimate for the conditional mean percentage for a team with
     a payroll of $50 million is 47.37%. This is the point on the sample
     regression line that best estimates the corresponding point on the
     population regression line.
(e)  The 95% confidence interval for the conditional mean percentage for
     teams with a payroll of $50 million is 43.79 to 50.95 percent. We are
     95% confident that the mean percentage is somewhere between these
     limits for teams with a payroll of $50 million.
(f)  The predicted percentage for a team with a payroll of $50 million is
     47.37%. This is our best estimate of the percentage for a team with a

payroll of $50 million.

(g) The 95% prediction interval for the percentage of a team with a payroll of $50 million is 31.29 to 63.45 percent.  We are 95% certain that the percentages is somewhere between these limits for a team with a payroll of $50 million.

(h) The error in the estimate of the mean percentage of teams with a payroll of $50 million is due only to the fact that the population regression line is being estimated by a sample regression line; whereas, the error in the prediction of the percentage of a single team with a payroll of $50 million is due to that fact plus the variation in percentages for teams with payrolls of $50 million.

(j) Choose **Stat ▶ Basic Statistics ▶ Correlation**,  enter <u>PERCENTAGE</u> and <u>PAYROLL</u> in the **Variables** text box, check the box to **Display P-values**, and click **OK**.  The result is

    Pearson correlation of PAYROLL and PERCENTAGE = 0.427
    P-Value = 0.019

The P-value for a right-tailed test is half of that shown (0.019/2 = 0.0095) if r is positive.  Since the P-value is less than 0.05, we reject the null hypothesis that $\rho = 0$.

(k) For the residual analysis, we look at the graph produced by the procedure at the beginning of this solution and also the results of a Ryan-Joiner normality test on the residuals, which were stored earlier.

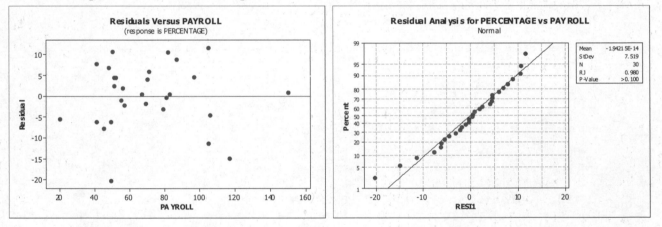

In the left-hand graph, the residuals are located in a horizontal band centered on and somewhat symmetrical about the horizontal axis except for the two most negative residuals for the Mets and the Tigers.  The residual at the far right has been labeled in the output as an influential observation (Yankees).  The P-value for the Ryan-Joiner test is > 0.100 indicating that normality of the residuals is a reasonable assumption.  Since the linear model also appears to be appropriate, although not very useful, the regression inferences made earlier in this problem are likely to be reasonable.

## CHAPTER 16 ANSWERS

### Exercises 16.1

**16.1** We state the two numbers of degrees of freedom.

**16.3** $F_{0.05}$, $F_{0.025}$, $F_\alpha$

**16.5** (a) The first number in parentheses—12—is the number of degrees of freedom for the numerator.

(b) The second number in parentheses—7—is the number of degrees of freedom for the denominator.

**16.7** (a) $F_{0.05} = 1.89$ (b) $F_{0.01} = 2.47$ (c) $F_{0.025} = 2.14$

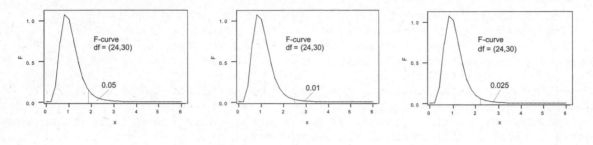

**16.9** (a) $F_{0.01} = 2.88$ (b) $F_{0.05} = 2.10$ (c) $F_{0.10} = 1.78$

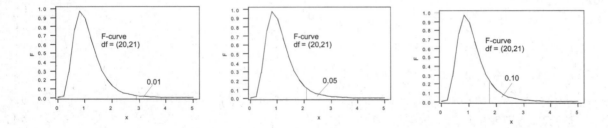

**16.11** A straightforward method for finding $F_{0.05}$ for df = (25, 20) using Table VIII is by interpolation. Specifically, proceed along the top row of the table until locating the numbers 24 and 30, between which is 25, the desired number of degrees of freedom for the numerator. Notice that 25 is not reported in this row but is 1/6 of the way in moving from 24 to 30; i.e., (25 - 24)/(30 - 24) = 1/6.
The outside columns of the table give the degrees of freedom for the denominator, which is 20 in this case. Go down either of the outside columns to the row labeled "20." Then go across that row until you are under the columns headed "24" and "30." The numbers in the body of the table intersecting these row and column positions are the F-values 2.08 and 2.04. Thus, 24 is matched with 2.08, and 30 is matched with 2.04.
Now, proceed with interpolation. With 25 known to be 1/6 of the distance between 24 and 30, we likewise want to find the F-value that is 1/6 of the distance between 2.08 and 2.04. We do this by taking (1/6) of (2.04 - 2.08) = -0.0066. Adding -0.0066 to 2.08 gives us the desired F-value, which is 2.0734, or 2.07.
Thus, an approximation to $F_{0.05}$ for df = (25, 20) using Table VIII is 2.07. The limitation of this approximation is that it uses a linear technique on values that are related in a nonlinear fashion. However, because the difference between 2.08 and 2.04 is so small in the first place, the approximation errors will likewise be small.

**16.13** The pooled-t procedure (Procedure 10.1) is a method for comparing the means of two populations. One-way ANOVA is a procedure for comparing the means of several populations. Thus, one-way ANOVA is a generalization of the pooled-t procedure.

**16.15** The reason for the word "variance" in the phrase "analysis of variance" is because the analysis-of-variance procedure for comparing means involves analyzing the *variation* in the sample data.

**16.17** (a) MSTR (or SSTR) is a statistic that measures the variation among the sample means for a one-way ANOVA.

(b) MSE (or SSE) is a statistic that measures the variation within the samples for a one-way ANOVA.

(c) F = MSTR/MSE is a statistic that compares the variation among the sample means to the variation within the samples.

**16.19** The term *one-way* is used because the type of ANOVA compares the means of a variable for populations that result from a classification by *one* other variable, call the factor.

**16.21** No, because the variation among the sample means is not large relative to the variation within the samples.

**16.23** In a one-way ANOVA, the residual of an observation is the difference between an observation and the mean of the sample containing it.

**16.25**

$$\bar{x}_1 = 18/3 = 6; \ \bar{x}_2 = 6/3 = 2; \ \bar{x}_3 = 16/4 = 4$$

$$\bar{x} = \sum x_i / n = (18 + 6 + 16)/(3 + 3 + 4) = 40/10 = 4$$

$$(n_1 - 1)s_1^2 = (8-6)^2 + (4-6)^2 + (6-6)^2 = 8$$

$$(n_2 - 1)s_2^2 = (2-2)^2 + (1-2)^2 + (3-2)^2 = 2$$

$$(n_3 - 1)s_3^2 = (4-4)^2 + (3-4)^2 + (6-4)^2 + (3-4)^2 = 6$$

(a) $SSTR = n_1(\bar{x}_1 - \bar{x})^2 + n_2(\bar{x}_2 - \bar{x})^2 + n_3(\bar{x}_3 - \bar{x})^2 = 3(6-4)^2 + 3(2-4)^2 + 4(4-4)^2 = 24$

(b) $MSTR = SSTR/(k-1) = 24/(3-1) = 12$

(c-d)

$$SSE = (n_1 - 1)s_1^2 + (n_2 - 1)s_2^2 + (n_3 - 1)s_3^2 = 8 + 2 + 6 = 16$$

$$MSE = SSE/(n-k) = 16/(10-3) = 2.286$$

(e) $F = MSTR/MSE = 12/2.286 = 5.25$

**16.27**

$$\bar{x}_1 = 20/4 = 5; \ \bar{x}_2 = 18/3 = 6; \ \bar{x}_3 = 30/5 = 6; \ \bar{x}_4 = 21/5 = 5; \ \bar{x}_5 = 27/3 = 9$$

$$\bar{x} = \sum x_i / n = (20 + 18 + 30 + 21 + 27)/(4 + 3 + 5 + 5 + 3) = 120/20 = 6$$

$$(n_1 - 1)s_1^2 = (7-5)^2 + (4-5)^2 + (5-5)^2 + (4-5)^2 = 6$$

$$(n_2 - 1)s_2^2 = (5-6)^2 + (9-6)^2 + (4-6)^2 = 14$$

$$(n_3 - 1)s_3^2 = (6-6)^2 + (7-6)^2 + (5-6)^2 + (4-6)^2 + (8-6)^2 = 10$$

$$(n_4 - 1)s_4^2 = (3-5)^2 + (7-5)^2 + (7-5)^2 + (4-5)^2 + (4-5)^2 = 14$$

$$(n_5 - 1)s_5^2 = (7-9)^2 + (9-9)^2 + (11-9)^2 = 8$$

(a) $SSTR = n_1(\bar{x}_1 - \bar{x})^2 + n_2(\bar{x}_2 - \bar{x})^2 + n_3(\bar{x}_3 - \bar{x})^2 + n_4(\bar{x}_4 - \bar{x})^2 + n_5(\bar{x}_5 - \bar{x})^2$

$$= 4(5-6)^2 + 3(6-6)^2 + 5(6-6)^2 + 5(5-6)^2 + 3(9-6)^2 = 36$$

(b) $MSTR = SSTR/(k-1) = 36/(5-1) = 9$

(c-d)

$$SSE = (n_1 - 1)s_1^2 + (n_2 - 1)s_2^2 + (n_3 - 1)s_3^2 + (n_4 - 1)s_4^2 + (n_5 - 1)s_5^2$$

$$= 6 + 14 + 10 + 14 + 8 = 52$$

$$MSE = SSE/(n - k) = 52/(20 - 5) = 3.467$$

(e)   $F = MSTR/MSE = 9/3.467 = 2.600$

**16.29** If the hypothesis test situation abides by the characteristics outlined in this exercise, we can use the pooled-t test or we can use the one-way ANOVA test discussed in this section using k = 2.

**Exercises 16.3**

**16.31** A small value of F results when MSTR is small compared to MSE, i.e., when the variation between sample means is small compared to the variation within samples. This describes what should happen when the null hypothesis is true; thus it does not comprise evidence that the null hypothesis is false. Only when the variation between sample means is large compared to the variation within samples, i.e., when F is large, do we have evidence that the null hypothesis is not true.

**16.33** SST = SSTR + SSE. This means that the total variation can be partitioned into a component representing variation among the sample means and another component representing variation within the samples.

**16.35** (a)  One-way ANOVA
(b)  Two-way ANOVA

**16.37** Since 0.708 = 2.124/df(Treatment), we have df(Treatment) = 3
It follows that df (Total) = 3 + 20 = 23.
Since 0.75 = 0.708/MSE, we have MSE = 0.708/0.75 = 0.944.
Then since 0.944 = SSE/20, we have SSE = 0.944(20) = 18.88.
Finally SST = SSTR + SSE = 2.124 + 18.880 = 21.004
Thus, the completed table is

| Source | df | SS | MS=SS/df | F-statistic |
|---|---|---|---|---|
| Treatment | 3 | 2.124 | 0.708 | 0.75 |
| Error | 20 | 18.880 | 0.944 | |
| Total | 23 | 21.004 | | |

**16.39** Since the Total df is 16, the df for Treatment is 14 - 12 = 2.
Since MSTR = SSTR/2, SSTR = 2(MSTR) = 2(1.4) = 2.8
Similarly, SSE = 12(MSE) = 12(0.9) = 10.8
Now SST = SSTR + SSE = 2.8 + 10.8 = 13.6
Finally, F = MSTR/MSE = 1.4/0.9 = 1.556
Thus, the completed table is

| Source | df | SS | MS=SS/df | F-statistic |
|---|---|---|---|---|
| Treatment | 2 | 2.8 | 1.4 | 1.556 |
| Error | 12 | 10.8 | 0.9 | |
| Total | 14 | 13.6 | | |

**16.41** We have the following:
k = 3
$n_1 = 3$        $n_2 = 3$,        $n_3 = 4$
$T_1 = 18$       $T_2 = 6$        $T_3 = 16$
$n = \sum n_j = 3 + 3 + 4 = 10$
$\sum x_i = \sum T_j = 18 + 6 + 16 = 40$

$$\sum x_i^2 = (8)^2 + (4)^2 + (6)^2 + ... + (6)^2 + (3)^2 = 200$$

(a) Consequently,

$$SST = \sum x_i^2 - \left(\sum x_i\right)^2 / n = 200 - 40^2/10 = 200 - 160 = 40$$

$$SSTR = \sum (T_j^2/n_j) - \left(\sum x_i\right)^2 / n = (18)^2/3 + (6)^2/3 + (16)^2/4 - (40)^2/10 = 184 - 160 = 24$$

$$SSE = SST - SSTR = 40 - 24 = 16$$

(b) The results are the same as in Exercise 16.25.

(c)

| Source | df | SS | MS=SS/df | F-statistic |
|--------|----|----|----------|-------------|
| Treatment | 2 | 24 | 12.00 | 5.25 |
| Error | 7 | 16 | 2.29 | |
| Total | 9 | 40 | | |

(d) The df are (2,7). The critical value is 4.74. Since 5.25 > 4.74, we reject the null hypothesis that the population means are equal.

**16.43** We have the following:

k = 5

| | | | | |
|--|--|--|--|--|
| $n_1 = 4$ | $n_2 = 3$ | $n_3 = 5$ | $n_4 = 5$ | $n_5 = 3$ |
| $T_1 = 20$ | $T_2 = 18$ | $T_3 = 30$ | $T_4 = 25$ | $T_5 = 27$ |

$$n = \sum n_j = 4 + 3 + 5 + 5 + 3 = 20$$

$$\sum x_i = \sum T_j = 20 + 18 + 30 + 25 + 27 = 120$$

$$\sum x_i^2 = (7)^2 + (4)^2 + (5)^2 + ... + (9)^2 + (11)^2 = 808$$

(a) Consequently,

$$SST = \sum x_i^2 - \left(\sum x_i\right)^2 / n = 808 - 120^2/20 = 808 - 720 = 88$$

$$SSTR = \sum (T_j^2/n_j) - \left(\sum x_i\right)^2 / n$$

$$= (20)^2/4 + (18)^2/3 + (30)^2/5 + (25)^2/5 + (27)^2/3 - (120)^2/20 = 756 - 720 = 36$$

$$SSE = SST - SSTR = 88 - 36 = 52$$

(b) The results are the same as in Exercise 16.26.

(c)

| Source | df | SS | MS=SS/df | F-statistic |
|--------|----|----|----------|-------------|
| Treatment | 4 | 36 | 9.00 | 2.60 |
| Error | 15 | 52 | 3.47 | |
| Total | 19 | 88 | | |

(d) The df are (4,15). The critical value is 3.06. Since 2.60 > 3.06, we do not reject the null hypothesis that the population means are equal.

**16.45** The total number of populations being sampled is k = 3. Let the subscripts 1, 2, and 3 refer to diatoms, bacteria, and macroalgae, respectively. The total number of observations is n = 12. Also, $n_1 = 4$, $n_2 = 4$, and $n_3 = 4$.

Step 1: $H_0: \mu_1 = \mu_2 = \mu_3$ (population means are equal)

$H_a$: population means are different

Step 2: $\alpha = 0.05$

Step 3: Let $T_1$, $T_2$ and $T_3$ refer to the sum of the data values in each of the three samples, respectively. Thus,

$T_1 = 1828$            $T_2 = 1225$            $T_3 = 1175$.

Also, the sum of all the data values is $\sum x_i = 4228$, and their sum of squares is $\sum x_i^2 = 1,561,154$. Then,

$$SST = \sum x_i^2 - \left(\sum x_i\right)^2 / n = 1561.154 - 4228^2/12 = 71488.67$$

$$SSTR = \frac{T_1^2}{n_1} + \frac{T_2^2}{n_2} + \frac{T_3^2}{n_3} - \frac{\left(\sum x_i\right)^2}{n} = \frac{1828^2}{4} + \frac{12255^2}{4} + \frac{1175^2}{4} - \frac{4228^2}{12} = 66043.17$$

and SSE = SST - SSTR = 71488.67 - 66043.17 = 5445.50
Step 4:   MSTR = SSTR/(k - 1) = 66043.17/(3 - 1) = 33021.585
          MSE = SSE/(n - k) = 5445.50/(12 - 3) = 605.056
          F = MSTR/MSE = 33021.585/605.056 = 54.98

The one-way ANOVA table is

| Source | Df | SS | MS = SS/df | F-statistic |
|---|---|---|---|---|
| Treatment | 2 | 66043.17 | 33021.585 | 54.58 |
| Error | 9 | 5445.50 | 605.056 | |
| Total | 11 | 71488.67 | | |

Step 5:   df = (k - 1, n - k) = (2, 9); critical value = 4.26
Step 6:   Since 54.58 > 4.26, we reject the null hypothesis and conclude that there is a difference in the mean number of copepods among the three different diets. For the p-value approach, P(F > 54.58) < 0.005. Since p < 0.05, we reject the null hypothesis.

**16.47** The number of populations being sampled is k = 5. Let the subscripts 1-5 refer to strains A-E, respectively, The total number observations is n = 25. Also, $n_1 = n_2 = n_3 = n_4 = n_5 = 5$.

Step 1:   $H_0$:   $\mu_1 = \mu_2 = \mu_3 = \mu_4 = \mu_5$   (population means are equal)
          $H_a$:   Not all of the population means are equal
Step 2:   $\alpha$ = 0.05
Step 3:   Let $T_1 - T_5$ refer to the sum of the data values in each of the five samples, respectively. Thus,
          $T_1 = 104$  $T_2 = 129$  $T_3 = 185$  $T_4 = 98$  $T_5 = 194$.

Also, the sum of all the data values is $\sum x_i = 710$, and their sum of squares is $\sum x_i^2 = 25424$.

Consequently,  $SST = \sum x_i^2 - \left(\sum x_i\right)^2 / n = 25424 - 710^2/25 = 5260$

$$SSTR = \frac{T_1^2}{n_1} + \frac{T_2^2}{n_2} + \frac{T_3^2}{n_3} + \frac{T_4^2}{n_4} + \frac{T_5^2}{n_5} - \frac{\left(\sum x_i\right)^2}{n}$$

$$= \frac{104^2}{5} + \frac{129^2}{5} + \frac{185^2}{5} + \frac{98^2}{5} + \frac{194^2}{5} - \frac{710^2}{25} = 1620$$

and SSE = SST - SSTR = 5260 - 1620 = 3640.
Step 4:   MSTR = SSTR/(k - 1) = 1620/(5 - 1) = 405
          MSE = SSE/(n - k) = 3640/(25 - 5) = 182
          F = MSTR/MSE = 405/182 = 2.23

The one-way ANOVA table is:

| Source | df | SS | MS = SS/df | F-statistic |
|---|---|---|---|---|
| Treatment | 4 | 1620 | 405 | 2.23 |
| Error | 20 | 3640 | 182 | |
| Total | 24 | 5260 | | |

Step 5: df = (k - 1, n - k) = (4, 20); critical value = 2.87
Step 6: From Step 4, F = 2.23.  From Step 5, $F_{0.05}$ = 2.87.  Since 2.23 < 2.87, do not reject $H_0$.
Step 7:   At the 5% significance level, the data do not provide sufficient evidence to conclude that the population means are different.
For the P-value approach, P(F > 2.23) > 0.10.  Therefore, since the P-value is larger than the significance level, do not reject $H_0$.

**16.49** The total number of populations being sampled is k = 3.  Let the subscripts 1, 2, and 3 refer to 2000, 2001, and 2002, respectively.  The total number of observations is n = 34.  Also, $n_1$ = 11 $n_2$ = 13, $n_3$ = 10.

Step 1:   $H_0$: $\mu_1 = \mu_2 = \mu_3 = \mu_4$ (mean calibration constants are equal)

   $H_a$: Not all the means are equal.

Step 2:   $\alpha$ = 0.05

Step 3:   Let $T_1$, $T_2$, and $T_3$, refer to the sum of the data values in each of the four samples, respectively.  Thus,
   $T_1$ = 55.0    $T_2$ = 84.2    $T_3$ = 46.0.

   Also, the sum of all the data values is $\sum x_i$ = 185.2, and their sum of squares is $\sum x_i^2$ = 1108.48.

   Consequently, $SST = \sum x_i^2 - (\sum x_i)^2 / n = 1108.48 - 185.2^2 / 34 = 99.685$

$$SSTR = \frac{T_1^2}{n_1} + \frac{T_2^2}{n_2} + \frac{T_3^2}{n_3} + \frac{T_4^2}{n_4} - \frac{(\sum x_i)^2}{n}$$

$$= \frac{55^2}{11} + \frac{84.2^2}{13} + \frac{46^2}{10} - \frac{185.2^2}{34} = 23.162$$

and SSE = SST - SSTR = 99.685 - 23.162 = 76.523

Step 4:   MSTR = SSTR/(k - 1) = 23.162/(3 - 1) = 11.581
   MSE = SSE/(n - k) = 76.523/(34 - 3) = 2.468
   F = MSTR/MSE = 11.581/2.468 = 4.692

The one-way ANOVA table is:

| Source | df | SS | MS = SS/df | F-statistic |
|---|---|---|---|---|
| Treatment | 2 | 23.162 | 11.581 | 4.692 |
| Error | 31 | 76.523 | 2.468 | |
| Total | 33 | 99.685 | | |

Step 5:   df = (k - 1, n - k) = (2, 31); critical value = 3.32
Step 6:   From Step 4, F = 4.692.  From Step 5, $F_{0.05}$ = 3.32.  Since 4.692 > 3.32, reject $H_0$.
Step 7:   At the 5% significance level, the data provide sufficient evidence to conclude that a difference exists in mean resistance to *Stagonospora nororum* among the three years of wheat harvests.
For the p-value approach, 0.005 < P(F > 4.692) < 0.001.  Therefore, since the p-value is smaller than the significance level, reject $H_0$.

**16.51** (a)  To carry out the Analysis of variance and residual analysis, we will use the data in a single column, which is named RENT.  In the next column, which is named REGION, are the names of each of the regions corresponding to each data value in RENT.  The residual analysis consists of plotting the residuals against the means (or fits) and constructing a normal probability plot.  This is done at the same time as the analysis of variance in Minitab.  To do this, we choose **Stat ▶**

**Anova ▶One-way...**, select <u>RENT</u> for the **Response:** text box and <u>REGION</u> in
the **Factor:** text box. Then click on the **Graphs** button and click to
place check marks in the boxes for **Normal plot of residuals** and
**Residuals versus fits,** and click **OK.**  The printed results are

### One-way ANOVA: RENT versus REGION

```
Source  DF      SS      MS     F     P
REGION   3  400513  133504  7.54  0.002
Error   16  283265   17704
Total   19  683778
```

S = 133.1   R-Sq = 58.57%   R-Sq(adj) = 50.81%

```
                              Individual 99% CIs For Mean Based on
                              Pooled StDev
Level       N    Mean   StDev  -+---------+---------+---------+--------
Midwest     6   743.0    92.1  (-------*-------)
Northeast   5  1055.2   150.2                    (--------*-------)
South       4   854.5   168.0    (---------*--------)
West        5  1064.4   128.4                 (-------*--------)
                              -+---------+---------+---------+--------
                              600       800      1000      1200
```

Pooled StDev = 133.1

(b)  The F-value of 7.54 is significant at the 0.05 level (P-Value = 0.002).
If the assumptions of normal populations and equal standard deviations
are met, we reject the null hypothesis of equal population means and
conclude that there is a difference in the mean monthly rents among
newly completed apartments in the four U.S. regions.

(c)  The graphs produced by the procedure in part (a) are shown following.

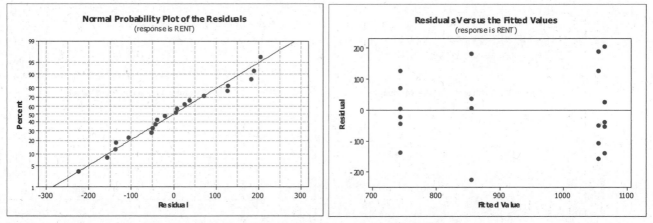

From the printed output, the ratio of the largest standard deviation to
the smallest is 168.0/92.1 = 1.82.  This is less than 2, indicating
that the rule of 2 is not violated, leading us to conclude that the
assumption of equal standard deviations is reasonable.  This can also
be seen in the right-hand graph above.  There are no outliers in the
normal probability plot and the points are linear, indicating that
there are no problems with the normality assumption as well.  [A Ryan-
Joiner normality test of the residuals has a P-value of >0.10, leading
us to conclude that the normality assumption is reasonable.]

**16.53** (a)  To carry out the Analysis of variance and residual analysis, we will
use the data in the single column named RATE.  In the next column,

which is named BREAST, are names of each of the three feather
treatments corresponding to each data value in RATE.  The residual
analysis consists of plotting the residuals against the means (or fits)
and constructing a normal probability plot.  This is done at the same
time as the analysis of variance in Minitab.  To do this, we choose

**Stat ▶ Anova ▶One-way...**, select <u>RATE</u> for the **Response:** text box and

<u>BREAST</u> in the **Factor:** text box. Enter <u>99</u> in the **Confidence level** box.
Then click on the **Graphs** button and click to place check marks in the
boxes for **Normal plot of residuals** and **Residuals versus fits**, and click
**OK.**  The printed results are

### One-way ANOVA: RATE versus BREAST

```
Source   DF      SS      MS     F      P
BREAST    2    960.3   480.1   6.09  0.008
Error    21   1655.3    78.8
Total    23   2615.6

S = 8.878    R-Sq = 36.71%    R-Sq(adj) = 30.69%

                                Individual 99% CIs For Mean Based on
                                Pooled StDev
Level        N     Mean   StDev   ----+---------+---------+---------+-----
Control      8   14.838   6.556              (--------*--------)
Enlarged     8   19.775  13.048                   (--------*--------)
Reduced      8    4.588   4.821   (--------*-------)
                                  ----+---------+---------+---------+-----
                                      0        10        20        30

Pooled StDev = 8.878
```

(b)  The F-value of 6.09 is significant at the 0.01 level (P-Value = 0.008).
     If the assumptions of normal populations and equal standard deviations
     are met, we reject the null hypothesis of equal population means.
     There is sufficient evidence to conclude that there is a difference in
     the mean singing rates among male rock sparrows exposed to the three
     types of breast treatments.
(c)  The graphical results are

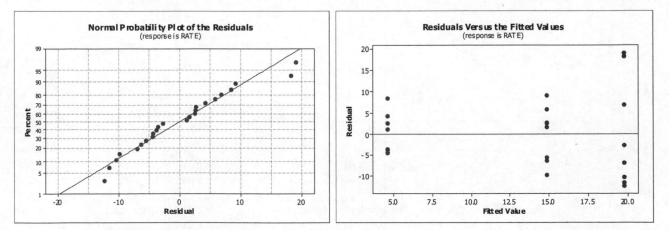

From the printed output, the ratio of the largest standard deviation to
the smallest is 13.048/4.821 = 2.71.  This is greater than 2,
indicating that the rule of 2 is violated and leading us to conclude
that the assumption of equal standard deviations may not be reasonable.
The small sample sizes may be partly responsible for this occurrence,

which can also be seen the right-hand graph above.  The points in the probability plot are linear, indicating no problems with the normality assumption.  [A Ryan-Joiner normality test of the residuals has a P-value of >0.10, leading us to conclude that the normality assumption is reasonable.]

**16.55** (a) To carry out the analysis of variance and residual analysis, we will use the data in the single column named VOLUME.  In the next column, which is named MATERIAL, are names of each of the three feather treatments corresponding to each data value in VOLUME.  The residual analysis consists of plotting the residuals against the means (or fits) and constructing a normal probability plot.  This is done at the same time as the analysis of variance in Minitab.  To do this, we choose

**Stat ▶ Anova ▶One-way...,** select <u>VOLUME</u> for the **Response:** text box and

<u>MATERIAL</u> in the **Factor:** text box. Enter <u>95</u> in the **Confidence level** box. Then click on the **Graphs** button and click to place check marks in the boxes for **Normal plot of residuals** and **Residuals versus fits,** and click **OK.**  The printed results are

### One-way ANOVA: HARDNESS versus MATERIAL

```
Source      DF     SS      MS       F       P
MATERIAL     2  805.31  402.66  114.71   0.000
Error       15   52.65    3.51
Total       17  857.97
```

```
S = 1.874   R-Sq = 93.86%   R-Sq(adj) = 93.04%
```

```
                                Individual 95% CIs For Mean Based on
                                Pooled StDev
Level       N    Mean   StDev  -----+---------+---------+---------+----
Duracross   6  39.633   3.120                                 (--*---)
Duradent    6  24.017   0.615  (--*--)
Endura      6  27.533   0.647        (--*--)
                               -----+---------+---------+---------+----
                               25.0      30.0      35.0      40.0
```

```
Pooled StDev = 1.874
```

(b) The F-value of 114.71 is significant at the 0.05 level (P-Value = 0.000).  If the assumptions of normal populations and equal standard deviations are met, we reject the null hypothesis of equal population means.  There is sufficient evidence to conclude that there is a difference in the mean hardness among the three materials for making artificial teeth.

(c) The graphical results are

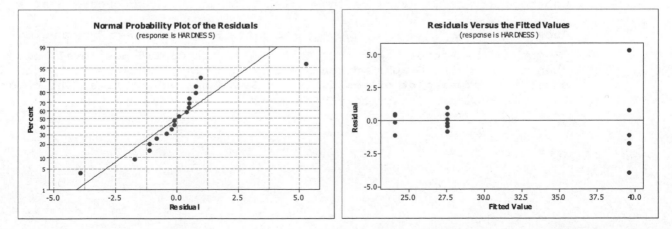

From the printed output, the ratio of the largest standard deviation to the smallest is 3.120/0.615 = 5.07.  This is greater than 2. The rule of 2 is violated, leading us to conclude that the assumption of equal standard deviations may not be reasonable.  The small sample sizes may be partly responsible for this occurrence, which can also be seen the right-hand graph above.  The points in the probability plot are not linear, indicating that there may also be problems with the normality assumption.  [A Ryan-Joiner normality test of the residuals has a P-value of < 0.010, so we conclude that the normality assumption is not reasonable.]  Although the F value is so large that it is unlikely that either violation would have any serious effect on our conclusions, these data should probably be analyzed with a more robust method.

**16.57** (a)  We will obtain the plots and standard deviations using the normal

probability plot in Minitab.  Choose **Graph ▶ Probability plot...**,

select the **Simple** version, enter <u>Some</u>, <u>Associate</u>, and <u>Bachelor</u> in the **Variables** text box, Click on the **Multiple graphs** button and select **In separate panels of the same graph**, and click **OK**.  Repeat this process with Master and Doctorate.  The two sets of graphs are

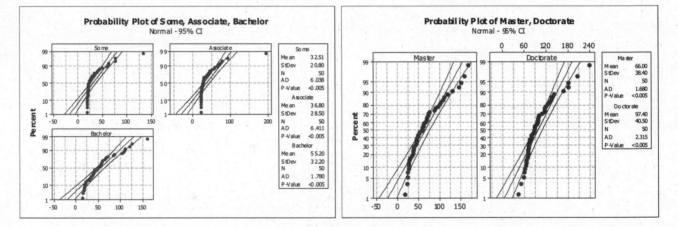

The standard deviations shown at the right of each set of graphs are 20.80, 28.50, 32.20, 38.40, and 40.50.

(b)  To carry out the residual analysis (and possibly the ANOVA), we will use the data in the single column named EARNINGS.  In the next column, which is named DEGREE, are names of each of the five degrees corresponding to each data value in EARNINGS.  The residual analysis consists of plotting the residuals against the means (or fits) and constructing a normal probability plot.  This is done at the same time as the analysis of variance in Minitab.  To do this, we choose **Stat ▶**

**Anova ▶One-way...**, select <u>EARNINGS</u> for the **Response:** text box and

<u>DEGREE</u> in the **Factor:** text box. Enter <u>95</u> in the **Confidence level** box.  Then click on the **Graphs** button and click to place check marks in the boxes for **Normal plot of residuals** and **Residuals versus fits**, and click **OK**.  The residual analysis results are

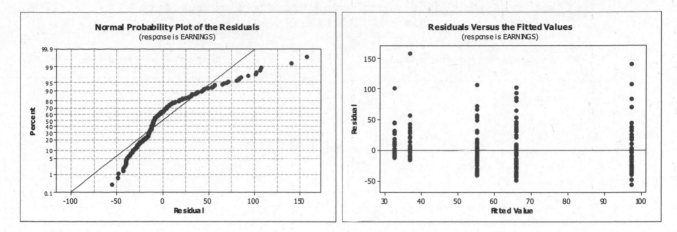

(c) The ratio of the largest to smallest standard deviation is 40.50/20.80
    = 1.95.  The rule of 2 is not violated.  The normal probability plots
    of the individual samples and of the residuals are far from linear and
    have at least one outlier in every plot. Conducting a one-way ANOVA
    test on these data does not appear to be reasonable.  Parts (d)-(e) are
    omitted.

**16.59** (a) We will obtain the plots and standard deviations using the normal

    probability plot in Minitab.  Choose **Graph ▶ Probability plot...**,

    select the **Simple** version, enter <u>Meadow Pipit</u>, <u>Tree Pipit</u>, and <u>Hedge
    Sparrow</u> in the **Variables** text box, Click on the **Multiple graphs** button
    and select **In separate panels of the same graph**, and click **OK**.  Then
    repeat with Robin, Pied Wagtail, and Wren.  The graphs are

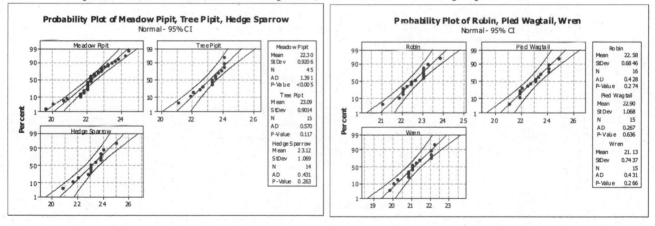

The standard deviations for the six birds are, respectively, 0.9206,
0.9014, 1.069, 0.6846, 1.068, and 0.7437.

(b) To carry out the residual analysis (and possibly the ANOVA), we will
    use the data in the single column named LENGTH.  In the next column,
    which is named SPECIES, are the names of each of the six species
    corresponding to each data value in LENGTH.  The residual analysis
    consists of plotting the residuals against the means (or fits) and
    constructing a normal probability plot.  This is done at the same time

    as the analysis of variance in Minitab.  To do this, we choose **Stat ▶**

    **Anova ▶One-way...**, select <u>LENGTH</u> for the **Response:** text box and <u>SPECIES</u>

    in the **Factor:** text box. Enter <u>95</u> in the **Confidence level** box.  Then
    click on the **Graphs** button and click to place check marks in the boxes

for **Normal plot of residuals** and **Residuals versus fits**, and click **OK**. The residual analysis results are

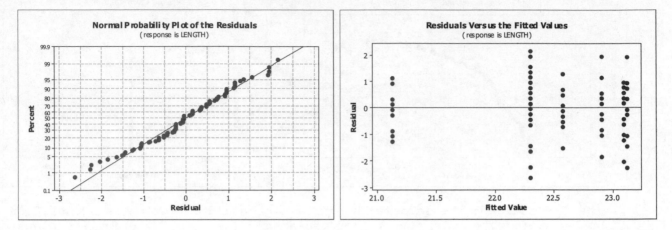

(c) The ratio of the largest to smallest standard deviation is 1.069/0.7437 = 1.44.  The rule of 2 is not violated.  This is also seen in the graph at the right above.  The normal probability plots of the individual samples and of the residuals are all linear except for that of the Meadow Pipit, which appears to have two possible outliers with low values.  Conducting a one-way ANOVA test on these data appears to be reasonable, although a more robust alternative procedure may be preferable.

(d) The printed portion of the output from the procedure in part (b) is

**One-way ANOVA: LENGTH versus SPECIES**

```
Source    DF      SS     MS      F      P
SPECIES    5  42.940  8.588  10.39  0.000
Error    114  94.248  0.827
Total    119  137.188


S = 0.9093   R-Sq = 31.30%   R-Sq(adj) = 28.29%
```

Individual 95% CIs For Mean Based on Pooled StDev

| Level | N | Mean | StDev | --+---------+---------+---------+------- |
|---|---|---|---|---|
| Hedge Sparrow | 14 | 23.121 | 1.069 | (-----*-----) |
| Meadow Pipit | 45 | 22.299 | 0.921 | (---*--) |
| Pied Wagtail | 15 | 22.903 | 1.068 | (-----*-----) |
| Robin | 16 | 22.575 | 0.685 | (----*-----) |
| Tree Pipit | 15 | 23.090 | 0.901 | (-----*----) |
| Wren | 15 | 21.130 | 0.744 | (-----*-----) |

```
                       --+---------+---------+---------+-------
                      20.80    21.60    22.40    23.20
```

Pooled StDev = 0.909

Since the F-value of 10.39 has a P-value of 0.000, we reject the null hypothesis of equal population means.

(e) We conclude that the data provide sufficient evidence that the mean egg lengths for the eggs laid by cuckoos in the nests of the six species are different.

**16.61** (a) We will obtain the plots and standard deviations using the normal probability plot in Minitab.  Choose **Graph ▶ Probability plot...**, select the **Simple** version, enter Law, Science, Medicine, and Technology in the **Variables** text box, click on the **Multiple graphs** button and select **In separate panels of the same graph**, and click **OK**.  The graphs

are

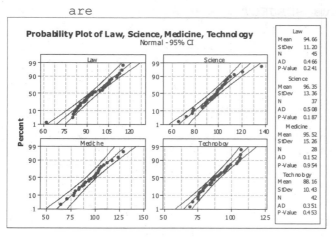

The standard deviations for the four subject areas are, respectively, 11.20, 13.36, 15.26, and 10.43.

(b)  To carry out the residual analysis (and possibly the ANOVA), we will use the data in the single column named PRICE.  In the next column, which is named SUBJECT, are the names of each of the four subject areas corresponding to each data value in PRICE.  The residual analysis consists of plotting the residuals against the means (or fits) and constructing a normal probability plot.  This is done at the same time as the analysis of variance in Minitab.  To do this, we choose **Stat ▶**

**Anova ▶One-way...,** select PRICE for the **Response:** text box and SUBJECT in the **Factor:** text box. Enter 95 in the **Confidence level** box.  Then click on the **Graphs** button and click to place check marks in the boxes for **Normal plot of residuals** and **Residuals versus fits,** and click **OK.** The residual analysis results are

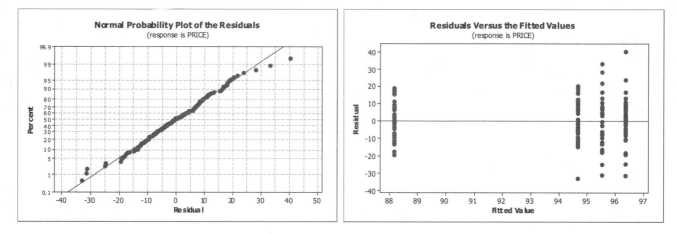

(c)  The ratio of the largest to smallest standard deviation is 15.26/10.43 = 1.46.  The rule of 2 is not violated.  This is also seen in the graph at the right above.  The normal probability plots of the individual samples and of the residuals are all linear.  Conducting a one-way ANOVA test on these data appears to be reasonable.

(d)  The printed portion of the output from the procedure in part (b) is

The endpoints for $\mu_1 - \mu_2$ are

$$5 - 6 \pm \frac{4.41}{\sqrt{2}} \cdot \sqrt{3.467} \cdot \sqrt{\frac{1}{4} + \frac{1}{3}} = -1 \pm 4.43 = -5.43 \quad to \quad 3.43$$

The endpoints for $\mu_1 - \mu_3$ are

$$5 - 6 \pm \frac{4.41}{\sqrt{2}} \cdot \sqrt{3.467} \cdot \sqrt{\frac{1}{4} + \frac{1}{5}} = -1 \pm 3.89 = -4.89 \quad to \quad 2.89$$

The endpoints for $\mu_1 - \mu_4$ are

$$5 - 5 \pm \frac{4.41}{\sqrt{2}} \cdot \sqrt{3.467} \cdot \sqrt{\frac{1}{4} + \frac{1}{5}} = 0 \pm 3.89 = -3.89 \quad to \quad 3.89$$

The endpoints for $\mu_1 - \mu_5$ are

$$5 - 9 \pm \frac{4.41}{\sqrt{2}} \cdot \sqrt{3.467} \cdot \sqrt{\frac{1}{4} + \frac{1}{3}} = -4 \pm 4.43 = -8.43 \quad to \quad 0.43$$

The endpoints for $\mu_2 - \mu_3$ are

$$6 - 6 \pm \frac{4.41}{\sqrt{2}} \cdot \sqrt{3.467} \cdot \sqrt{\frac{1}{3} + \frac{1}{5}} = 0 \pm 4.24 = -4.24 \quad to \quad 4.24$$

The endpoints for $\mu_2 - \mu_4$ are

$$6 - 5 \pm \frac{4.41}{\sqrt{2}} \cdot \sqrt{3.467} \cdot \sqrt{\frac{1}{3} + \frac{1}{5}} = 1 \pm 4.24 = -3.24 \quad to \quad 5.24$$

The endpoints for $\mu_2 - \mu_5$ are

$$6 - 9 \pm \frac{4.41}{\sqrt{2}} \cdot \sqrt{3.467} \cdot \sqrt{\frac{1}{3} + \frac{1}{3}} = -3 \pm 4.74 = -7.74 \quad to \quad 1.74$$

The endpoints for $\mu_3 - \mu_4$ are

$$6 - 5 \pm \frac{4.41}{\sqrt{2}} \cdot \sqrt{3.467} \cdot \sqrt{\frac{1}{5} + \frac{1}{5}} = 1 \pm 3.6 = -2.67 \quad to \quad 4.67$$

The endpoints for $\mu_3 - \mu_5$ are

$$6 - 9 \pm \frac{4.41}{\sqrt{2}} \cdot \sqrt{3.467} \cdot \sqrt{\frac{1}{5} + \frac{1}{3}} = -3 \pm 4.24 = -7.24 \quad to \quad 1.24$$

The endpoints for $\mu_4 - \mu_5$ are

$$5 - 9 \pm \frac{4.41}{\sqrt{2}} \cdot \sqrt{3.467} \cdot \sqrt{\frac{1}{5} + \frac{1}{3}} = -4 \pm 4.24 = -8.24 \quad to \quad 0.24$$

Step 4:   Based on the confidence intervals in Step 3, all of which contain zero, none of the means is different from the others. This is consistent with the result of Exercise 16.43 in which the null hypothesis of equal means was not rejected.

**16.85** Step 1:   Family confidence level = 0.95.

Step 2:   $\kappa = 3$ and $\nu = n - k = 12 - 3 = 9$.  Consulting Table XI, we find that $q_{0.05} = 3.95$.

Step 3:   Before obtaining all possible confidence intervals for $\mu_i - \mu_j$, we construct a table giving the sample means and sample sizes.

| j | 1 | 2 | 3 |
|---|---|---|---|
| $\bar{x}_j$ | 457.00 | 306.25 | 293.75 |
| $n_j$ | 4 | 4 | 4 |

In Exercise 16.45, we found that MSE = 605.056. Now we are ready to obtain the required confidence intervals.

The endpoints for $\mu_1 - \mu_2$ are

$$457.00 - 306.25 \pm \frac{3.95}{\sqrt{2}} \cdot \sqrt{605.056} \cdot \sqrt{\frac{1}{4} + \frac{1}{4}} = 150.75 \pm 48.58 = 102.17 \quad to \quad 199.33$$

The endpoints for $\mu_1 - \mu_3$ are:

$$457.00 - 293.75 \pm \frac{3.95}{\sqrt{2}} \cdot \sqrt{605.056} \cdot \sqrt{\frac{1}{4} + \frac{1}{4}} = 163.25 \pm 48.58 = 114.67 \quad to \quad 211.83$$

The endpoints for $\mu_2 - \mu_3$ are:

$$306.5 - 293.75 \pm \frac{3.95}{\sqrt{2}} \cdot \sqrt{605.056} \cdot \sqrt{\frac{1}{4} + \frac{1}{4}} = 12.50 \pm 48.58 = -36.08 \quad to \quad 61.08$$

Step 4: Based on the confidence intervals in Step 3, $\mu_1$ is different from $\mu_2$, and $\mu_1$ is different from $\mu_3$; $\mu_2$ and $\mu_3$ cannot be declared different.

Step 5: We summarize the results with the following diagram.

| Microalgae | Bacteria | Diatoms |
|---|---|---|
| (3) | (2) | (1) |
| 293.75 | 306.25 | 457.00 |

Interpreting this diagram, we conclude with 95% confidence that $\mu_1$ is different from $\mu_2$, and $\mu_1$ is different from $\mu_3$.

**16.87** (a) From Exercise 16.47, we have MSE = 182, $\kappa$ = 5, and $\nu$ = 20.

Step 1: Family confidence level = 0.95.

Step 2: $\kappa$ = 5 and $\nu$ = n - k = 25 - 5 = 20. Consulting Table XI, we find that $q_{0.05}$ = 4.23.

Step 3: Before obtaining all possible confidence intervals for $\mu_i - \mu_j$, we construct a table giving the sample means and sample sizes.

| j | 1 | 2 | 3 | 4 | 5 |
|---|---|---|---|---|---|
| $\bar{x}_j$ | 20.8 | 25.8 | 37.0 | 24.6 | 38.8 |
| $n_j$ | 5 | 5 | 5 | 5 | 5 |

Now we are ready to obtain the required confidence intervals.

The endpoints for $\mu_1 - \mu_2$ are

$$20.8 - 25.8 \pm \frac{4.23}{\sqrt{2}} \cdot \sqrt{182} \cdot \sqrt{\frac{1}{5} + \frac{1}{5}} = -5.0 \pm 25.5 = -30.5 \quad to \quad 20.5$$

The endpoints for $\mu_1 - \mu_3$ are

$$20.8 - 37.0 \pm \frac{4.23}{\sqrt{2}} \cdot \sqrt{182} \cdot \sqrt{\frac{1}{5} + \frac{1}{5}} = -16.2 \pm 25.5 = -41.7 \quad to \quad 9.3$$

The endpoints for $\mu_1 - \mu_4$ are

$$20.8 - 24.6 \pm \frac{4.23}{\sqrt{2}} \cdot \sqrt{182} \cdot \sqrt{\frac{1}{5} + \frac{1}{5}} = -3.8 \pm 25.5 = -29.3 \ \textit{to} \ 21.7$$

The endpoints for $\mu_1 - \mu_5$ are

$$20.8 - 38.8 \pm \frac{4.23}{\sqrt{2}} \cdot \sqrt{182} \cdot \sqrt{\frac{1}{5} + \frac{1}{5}} = -18.0 \pm 25.5 = -43.8 \ \textit{to} \ 7.5$$

The endpoints for $\mu_2 - \mu_3$ are

$$25.8 - 37.0 \pm \frac{4.23}{\sqrt{2}} \cdot \sqrt{182} \cdot \sqrt{\frac{1}{5} + \frac{1}{5}} = -11.2 \pm 25.5 = -36.7 \ \textit{to} \ 14.3$$

The endpoints for $\mu_2 - \mu_4$ are

$$25.8 - 24.6 \pm \frac{4.23}{\sqrt{2}} \cdot \sqrt{182} \cdot \sqrt{\frac{1}{5} + \frac{1}{5}} = 1.2 \pm 25.5 = -24.3 \ \textit{to} \ 26.7$$

The endpoints for $\mu_2 - \mu_5$ are

$$25.8 - 38.8 \pm \frac{4.23}{\sqrt{2}} \cdot \sqrt{182} \cdot \sqrt{\frac{1}{5} + \frac{1}{5}} = -13.0 \pm 25.5 = -38.5 \ \textit{to} \ 12.5$$

The endpoints for $\mu_3 - \mu_4$ are

$$37.0 - 24.6 \pm \frac{4.23}{\sqrt{2}} \cdot \sqrt{182} \cdot \sqrt{\frac{1}{5} + \frac{1}{5}} = 12.4 \pm 25.5 = -13.1 \ \textit{to} \ 37.9$$

The endpoints for $\mu_3 - \mu_5$ are

$$37.0 - 38.8 \pm \frac{4.23}{\sqrt{2}} \cdot \sqrt{182} \cdot \sqrt{\frac{1}{5} + \frac{1}{5}} = -1.8 \pm 25.5 = -27.3 \ \textit{to} \ 23.7$$

The endpoints for $\mu_4 - \mu_5$ are

$$24.6 - 38.8 \pm \frac{4.23}{\sqrt{2}} \cdot \sqrt{182} \cdot \sqrt{\frac{1}{5} + \frac{1}{5}} = -12.2 \pm 25.5 = -37.7 \ \textit{to} \ 13.2$$

Step 4: Based on the confidence intervals in Step 3, all of which contain zero, none of the means is different from the others.

(b) The data do not provide sufficient evidence at the 5% significance level to conclude that a difference exists in mean bacteria counts among the five strains of *Staphylococcus aureus*. If there were such evidence, then at least one of the intervals found in part (a) would not have contained zero.

**16.89** Step 1: Family confidence level = 0.95.

Step 2: $\kappa = 3$ and $\nu = n - k = 34 - 3 = 31$. Consulting Table XI, we find that $q_{0.05} = 3.49$ for the closest entry with df = (3,30).

Step 3: Before obtaining all possible confidence intervals for $\mu_i - \mu_j$, we construct a table giving the sample means and sample sizes.

| j | 1 | 2 | 3 |
|---|---|---|---|
| $\bar{x}_j$ | 5.00 | 6.48 | 4.60 |
| $n_j$ | 11 | 13 | 10 |

In Exercise 16.49, we found that MSE = 2.468. Now we are ready to obtain the required confidence intervals.

The endpoints for $\mu_1 - \mu_2$ are

$$5.00 - 6.48 \pm \frac{3.49}{\sqrt{2}} \cdot \sqrt{2.468} \cdot \sqrt{\frac{1}{11} + \frac{1}{13}} = -1.48 \pm 1.59 = -3.07 \quad to \quad 0.11$$

The endpoints for $\mu_1 - \mu_3$ are:

$$5.00 - 4.60 \pm \frac{3.49}{\sqrt{2}} \cdot \sqrt{2.468} \cdot \sqrt{\frac{1}{11} + \frac{1}{10}} = 0.40 \pm 1.69 = -1.29 \quad to \quad 2.09$$

The endpoints for $\mu_2 - \mu_3$ are:

$$6.48 - 4.60 \pm \frac{3.49}{\sqrt{2}} \cdot \sqrt{2.468} \cdot \sqrt{\frac{1}{13} + \frac{1}{10}} = 1.88 \pm 1.63 = 0.25 \quad to \quad 3.51$$

Step 4:  Based on the confidence intervals in Step 3, $\mu_2$ is different from $\mu_3$; $\mu_1$ cannot be declared different from $\mu_2$ or $\mu_3$.

Step 5:  We summarize the results with the following diagram.

```
2002          2000          2001
 (3)           (1)           (2)
4.60          5.00          6.48
_____

_____
```

Interpreting this diagram, we conclude with 95% confidence that $\mu_1$ is different from $\mu_3$.

**16.91** Using Minitab, with all of the data in a column named RENT and the regions of the U.S. in a column named REGION, we choose **Stat ▶ ANOVA ▶ Oneway...**, select RENT in the **Response** text box and REGION in the **Factor** text box, click on the **Comparisons** button, click on **Tukey's, family error rate** so that an X shows in its check-box, type 5 in its text box, and click **OK**. The result is [The first part of the output will be identical to the output given in Exercise 16.51.]

```
Tukey 95% Simultaneous Confidence Intervals
All Pairwise Comparisons among Levels of REGION

Individual confidence level = 98.87%
```

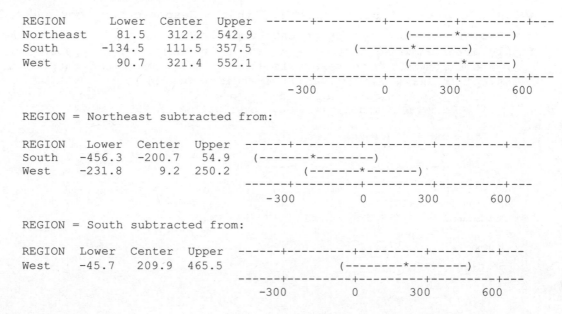
```
REGION = Midwest subtracted from:

REGION      Lower   Center   Upper   ------+---------+---------+---------+---
Northeast    81.5    312.2   542.9                   (------*-------)
South      -134.5    111.5   357.5          (-------*-------)
West         90.7    321.4   552.1                  (-------*------)
                                    ------+---------+---------+---------+---
                                        -300        0       300       600

REGION = Northeast subtracted from:

REGION   Lower   Center   Upper   ------+---------+---------+---------+---
South   -456.3   -200.7    54.9   (-------*--------)
West    -231.8      9.2   250.2        (-------*-------)
                                  ------+---------+---------+---------+---
                                      -300        0       300       600

REGION = South subtracted from:

REGION  Lower   Center   Upper   ------+---------+---------+---------+---
West    -45.7    209.9   465.5                (--------*--------)
                                 ------+---------+---------+---------+---
                                     -300        0       300       600
```

Looking at the intervals that do not include zero, we conclude that the mean monthly rent in the Midwest is different from that in the Northeast and from that in the West.  All other intervals contain zero; therefore, we cannot declare the means for those pairs to be different.

**16.93** Using Minitab, with all of the data in a column named RATE and the patch size manipulations in a column named BREAST, we choose **Stat ▶ ANOVA ▶ Oneway...**, select <u>RATE</u> in the **Response** text box and <u>BREAST</u> in the **Factor** text box, click on the **Comparisons** button, click on **Tukey's, family error rate** so that an X shows in its check-box, type <u>1</u> in its text box, and click **OK**.  The result is [The first part of the output will be identical to the output given in Exercise 16.53.]

```
Tukey 99% Simultaneous Confidence Intervals
All Pairwise Comparisons among Levels of BREAST

Individual confidence level = 99.63%

BREAST = Control subtracted from:

BREAST     Lower     Center    Upper    +---------+---------+---------+---------
Enlarged   -9.533    4.937     19.408                (---------*---------)
Reduced    -24.721   -10.250   4.221          (---------*---------)
                                             +---------+---------+---------+---------
                                           -30       -15        0        15


BREAST = Enlarged subtracted from:

BREAST     Lower     Center    Upper    +---------+---------+---------+---------
Reduced    -29.658   -15.188   -0.717    (----------*---------)
                                         +---------+---------+---------+---------
                                       -30       -15        0        15
```

Looking at the intervals that do not include zero, we conclude that the mean singing time in the enlarged patch-size group is different from that in the reduced patch-size group.  The mean singing time in the control group cannot be declared different from that of either the reduced or enlarged patch-size groups.

**16.95** Using Minitab, with all of the data in a column named HARDNESS and the type of artificial tooth material in a column named MATERIAL, we choose **Stat ▶ ANOVA ▶ Oneway...**, select <u>HARDNESS</u> in the **Response** text box and <u>MATERIAL</u> in the **Factor** text box, click on the **Comparisons** button, click on **Tukey's, family error rate** so that an X shows in its check-box, type <u>5</u> in its text box, and click **OK**.  The result is [The first part of the output will be identical to the output given in Exercise 16.55.]

```
Tukey 99% Simultaneous Confidence Intervals
All Pairwise Comparisons among Levels of MATERIAL

Individual confidence level = 99.62%

MATERIAL = Duracross subtracted from:

MATERIAL   Lower     Center    Upper    --------+---------+---------+---------+-
Duradent   -19.319   -15.617   -11.915  (-----*----)
Endura     -15.802   -12.100   -8.398      (-----*----)
                                         --------+---------+---------+---------+-
                                       -14.0      -7.0      0.0       7.0
```

```
MATERIAL = Duradent subtracted from:

MATERIAL   Lower  Center  Upper  --------+---------+---------+---------+-
Endura     -0.185  3.517  7.219                      (----*----)
                                  --------+---------+---------+---------+-
                                       -14.0     -7.0      0.0      7.0
```

Looking at the intervals that do not include zero, we conclude that the mean hardness of Duracross is different from that of Duradent and of Endura, but we cannot declare that the hardness of Duradent is different from that of Endura.

**16.97** In Exercise 16.57, we determined that conducting a one-way ANOVA with these data was not reasonable. Tukey comparisons will not be carried out.

**16.99** In Exercise 16.59, we determined that conducting a one-way ANOVA with these data was reasonable, although normality was in question for one sample. Using Minitab, with all of the data in a column named LENGTH and the group

in a column named SPECIES, we choose **Stat ▶ ANOVA ▶ Oneway...**, select

LENGTH in the **Response** text box and SPECIES in the **Factor** text box, click on the **Comparisons** button, click on **Tukey's, family error rate** so that an X shows in its check-box, type 5 in its text box, and click **OK**.  The Tukey comparison part of the output is

```
Tukey 95% Simultaneous Confidence Intervals
All Pairwise Comparisons among Levels of SPECIES

Individual confidence level = 99.55%

SPECIES = Hedge Sparrow subtracted from:

SPECIES         Lower    Center    Upper
Meadow Pipit  -1.6292   -0.8225   -0.0158
Pied Wagtail  -1.1977   -0.2181    0.7615
Robin         -1.5111   -0.5464    0.4183
Tree Pipit    -1.0110   -0.0314    0.9482
Wren          -2.9710   -1.9914   -1.0118

SPECIES            +---------+---------+---------+---------
Meadow Pipit             (-----*----)
Pied Wagtail              (------*-----)
Robin                   (-----*------)
Tree Pipit                (------*-----)
Wren              (------*-----)
                   +---------+---------+---------+---------
                  -3.0      -1.5       0.0       1.5

SPECIES = Meadow Pipit subtracted from:

SPECIES         Lower    Center    Upper
Pied Wagtail  -0.1815    0.6044    1.3904
Robin         -0.4912    0.2761    1.0434
Tree Pipit     0.0052    0.7911    1.5770
Wren          -1.9548   -1.1689   -0.3830

SPECIES            +---------+---------+---------+---------
Pied Wagtail                   (----*----)
Robin                        (----*----)
Tree Pipit                     (----*-----)
Wren                     (----*----)
                   +---------+---------+---------+---------
                  -3.0      -1.5       0.0       1.5
```

```
          SPECIES = Pied Wagtail subtracted from:

          SPECIES       Lower    Center    Upper
          Robin        -1.2757   -0.3283   0.6191
          Tree Pipit   -0.7759    0.1867   1.1492
          Wren         -2.7359   -1.7733  -0.8108

          SPECIES         +---------+---------+---------+---------
          Robin                       (------*-----)
          Tree Pipit                     (-----*------)
          Wren                 (-----*------)
                          +---------+---------+---------+---------
                        -3.0      -1.5       0.0       1.5

          SPECIES = Robin subtracted from:

          SPECIES       Lower    Center    Upper
          Tree Pipit   -0.4324    0.5150   1.4624
          Wren         -2.3924   -1.4450  -0.4976

          SPECIES         +---------+---------+---------+---------
          Tree Pipit                     (-----*------)
          Wren                 (-----*------)
                          +---------+---------+---------+---------
                        -3.0      -1.5       0.0       1.5

          SPECIES = Tree Pipit subtracted from:

          SPECIES    Lower   Center   Upper   +---------+---------+---------+--------
          -
          Wren      -2.9225  -1.9600  -0.9975   (-----*-----)
                                                +---------+---------+---------+--------
          -
                                              -3.0      -1.5       0.0       1.5
```

Looking at the intervals that do not contain zero, we conclude that the mean lengths of cuckoo eggs laid in the nests of hedge sparrows were different from those laid in the nests of meadow pipits and wrens; those in nests of meadow pipits differed from those laid in the nests of tree pipits and wrens; and those laid in the nests of pied wagtails, robins, and tree pipits differed from those laid in the nests of wrens.

16.101 In Exercise 16.61, we determined that conducting a one-way ANOVA with these data was reasonable. Using Minitab, with all of the data in a column named PRICE and the type of book in a column named SUBJECT, we choose **Stat ▶**

**ANOVA ▶ Oneway...**, select PRICE in the **Response** text box and SUBJECT in the **Factor** text box, click on the **Comparisons** button, click on **Tukey's, family error rate** so that an X shows in its check-box, type 5 in its text box, and click **OK**. The Tukey comparison part of the output is

```
          Tukey 95% Simultaneous Confidence Intervals
          All Pairwise Comparisons among Levels of SUBJECT

          Individual confidence level = 98.96%

          SUBJECT = Law subtracted from:

          SUBJECT     Lower  Center  Upper  ---------+---------+---------+---------+
          Medicine    -6.87    0.86   8.59          (---------*---------)
          Science     -5.44    1.69   8.82           (--------*--------)
          Technology -13.39   -6.50   0.39    (--------*-------)
                                             ---------+---------+---------+---------+
                                                   -8.0      0.0       8.0      16.0
```

```
SUBJECT = Medicine subtracted from:

SUBJECT      Lower   Center  Upper  ---------+---------+---------+---------+
Science      -7.22   0.83    8.88                 (---------*---------)
Technology  -15.20  -7.36    0.48   (---------*---------)
                                    ---------+---------+---------+---------+
                                         -8.0      0.0       8.0      16.0


SUBJECT = Science subtracted from:

SUBJECT      Lower   Center  Upper  ---------+---------+---------+---------+
Technology  -15.44  -8.19   -0.94   (--------*--------)
                                    ---------+---------+---------+---------+
                                         -8.0      0.0       8.0      16.0
```

Looking at the intervals that do not contain zero, we conclude that the mean price of science books differs from the mean price of technology books.

**16.103** In Exercise 16.63, we determined that conducting a one-way ANOVA with these data was reasonable, although the rule of 2 was violated. Using Minitab, with all of the data in a column named HEMOLEVEL and the type of sickle cell disease in a column named TYPE, we choose **Stat ▶ ANOVA ▶ Oneway...**, select HEMOLEVEL in the **Response** text box and TYPE in the **Factor** text box, click on the **Comparisons** button, click on **Tukey's, family error rate** so that an X shows in its check-box, type 5 in its text box, and click **OK**. The Tukey comparison part of the output is

```
Tukey 95% Simultaneous Confidence Intervals
All Pairwise Comparisons among Levels of TYPE

Individual confidence level = 98.05%

TYPE = HB SC subtracted from:

TYPE    Lower   Center  Upper  -------+---------+---------+---------+--
HB SS  -4.224  -3.048  -1.871  (----*----)
HB ST  -1.405  -0.213   0.978            (----*----)
                               -------+---------+---------+---------+--
                                  -2.5      0.0       2.5       5.0


TYPE = HB SS subtracted from:

TYPE   Lower  Center  Upper  -------+---------+---------+---------+--
HB ST  1.785  2.834   3.883                       (---*----)
                             -------+---------+---------+---------+--
                                -2.5      0.0       2.5       5.0
```

Looking at the intervals that do not contain zero, we conclude that the mean hemoglobin level of patients with type HB SS sickle cell disease differs than that of patients with types HB SC and HB ST.

**16.105** Step 1:  Family confidence level = 0.99.

Step 2:  $K = 3$ and $V = n - k = 81 - 3 = 78$. Consulting Table X, we find that $q_{0.01} = 4.28$.

Step 3:  Before obtaining all possible confidence intervals for $\mu_i - \mu_j$, we construct a table giving the sample means and sample sizes.

| j | 1 | 2 | 3 |
|---|---|---|---|
| $\bar{x}_j$ | 25.8 | 22.1 | 26.6 |
| $n_j$ | 32 | 20 | 29 |

In Exercise 16.39, we found that MSE = 74.64. Now we are ready to obtain the required confidence intervals.

The endpoints for $\mu_1 - \mu_2$ are

$$25.8 - 22.1 \pm \frac{4.28}{\sqrt{2}} \cdot \sqrt{74.64} \cdot \sqrt{\frac{1}{32} + \frac{1}{20}} = -3.8 \; to \; 11.2$$

The endpoints for $\mu_1 - \mu_3$ are

$$25.8 - 26.6 \pm \frac{4.28}{\sqrt{2}} \cdot \sqrt{74.64} \cdot \sqrt{\frac{1}{32} + \frac{1}{29}} = -7.5 \; to \; 5.9$$

The endpoints for $\mu_2 - \mu_3$ are

$$22.1 - 26.6 \pm \frac{4.28}{\sqrt{2}} \cdot \sqrt{74.64} \cdot \sqrt{\frac{1}{20} + \frac{1}{29}} = -12.3 \; to \; 3.1$$

Step 4:   Based on the confidence intervals in Step 3, we declare none of the means different.

Step 5:   We summarize the results with the following diagram.

```
Remitted Chronic     None
   (2)       (1)      (3)
  22.1            25.8   26.6
```

Interpreting this diagram, we conclude with 99% confidence that there is no difference in the mean ages at time of arrest among the three groups of former East Germany political prisoners classified by the current status relative to posttraumatic stress disorder.

**16.107** (a)  If i > j, then we obtain the confidence interval for $\mu_i - \mu_j$, by taking the negative of the confidence interval for $\mu_j - \mu_i$.

(b)  The confidence intervals are the following:

$\mu_2 - \mu_1$ is from -2.43 to 5.43

$\mu_3 - \mu_1$ is from -7.36 to 1.36

$\mu_4 - \mu_1$ is from -7.91 to 0.31

$\mu_3 - \mu_2$ is from -8.69 to -0.31

$\mu_4 - \mu_2$ is from -9.23 to -1.37

$\mu_4 - \mu_3$ is from -5.16 to 3.56.

**Exercises 16.5**

**16.109** The conditions required for using the Kruskal-Wallis test to compare k population means are:
(1)   Simple random samples
(2)   Independent samples
(3)   Same-shape populations
(4)   All sample sizes are 5 or greater.

**16.111**    equal

**16.113** H has approximately a chi-square distribution with k - 1 degrees of freedom, so for five populations, we use critical values from the right hand side of the chi-square distribution with 4 degrees of freedom.

**16.115** The total number of populations being sampled is k = 3.  Let the subscripts 1, 2, and 3 refer to 1980, 1990, and 1997, respectively.  The total number of pieces of sample data is n = 24.  Also, $n_1 = 8$, $n_2 = 7$, and $n_3 = 9$.

Step 1:      H₀: $\mu_1 = \mu_2 = \mu_3$ (mean consumption of low-fat milk are equal)

H_a: Not all the means are equal.

Step 2:      $\alpha = 0.01$

Step 3:

| Sample 1 | Rank | Sample 2 | Rank | Sample 3 | Rank |
|----------|------|----------|------|----------|------|
| 8.3 | 1 | 11.4 | 7 | 11.6 | 8.5 |
| 8.6 | 2 | 14.6 | 14 | 11.8 | 10 |
| 9.2 | 3 | 15.4 | 17 | 13.1 | 11 |
| 9.4 | 4 | 15.9 | 19 | 13.8 | 12 |
| 10.7 | 5 | 16.0 | 20 | 14.3 | 13 |
| 11.1 | 6 | 16.1 | 21 | 15.0 | 15 |
| 11.6 | 8.5 | 17.0 | 22 | 15.6 | 18 |
| 15.1 | 16 | | | 17.5 | 23 |
| | | | | 17.8 | 24 |
| | 45.5 | | 120 | | 134.5 |

Step 4:

$$H = \frac{12}{n(n+1)} \sum \frac{R_j^2}{n_j} - 3(n+1)$$

$$= \frac{12}{24(24+1)} \left( \frac{45.5^2}{8} + \frac{120^2}{7} + \frac{134.5^2}{9} \right) - 3(24+1)$$

$$= 11.519$$

Step 5:  df = k - 1 = 2; critical value = 9.210
Step 6:  Since 11.519 > 9.210, reject $H_0$.
Step 7:  At the 1% significance level, the data provide sufficient evidence to conclude that there is a difference in mean consumption of low-fat milk for 1980, 1990, and 2000.

For the p-value approach, $P(\chi^2 > 11.519) < 0.005$.  Therefore, because the p-value is smaller than the significance level, reject $H_0$.

**16.117** The total number of populations being sampled is k = 4.  Let the subscripts 1, 2, 3, and 4 refer to Northeast, Midwest, South, and West, respectively. The total number of observations is n = 32.  Also, $n_1 = n_2 = n_3 = n_4 = 8$.

Step 1:  $H_0$: $\eta_1 = \eta_2 = \eta_3 = \eta_4$ (median asking rents are equal)

   $H_a$: Not all the medians are equal.
Step 2:  $\alpha = 0.10$
Step 3:

| Sample 1 | Rank | Sample 2 | Rank | Sample 3 | Rank | Sample 4 | Rank |
|----------|------|----------|------|----------|------|----------|------|
| 1293 | 4 | 1605 | 11 | 642 | 1 | 694 | 2 |
| 1581 | 9 | 1639 | 12.5 | 722 | 3 | 1345 | 5 |
| 1781 | 18 | 1655 | 15.5 | 1354 | 6 | 1565 | 8 |
| 2130 | 23 | 1691 | 17 | 1513 | 7 | 1649 | 14 |
| 2149 | 25 | 2058 | 20 | 1591 | 10 | 1655 | 15.5 |
| 2286 | 27 | 2115 | 22 | 1639 | 12.5 | 2068 | 21 |
| 2989 | 30 | 2413 | 28 | 1982 | 19 | 2203 | 26 |
| 3182 | 31 | 3361 | 32 | 2135 | 24 | 2789 | 29 |
| | 167 | | 158 | | 82.5 | | 120.5 |

Step 4:

$$H = \frac{12}{n(n+1)}\sum \frac{R_j^2}{n_j} - 3(n+1)$$

$$= \frac{12}{32(32+1)}(\frac{167^2}{8} + \frac{158^2}{8} + \frac{82.5^2}{8} + \frac{120.5^2}{8}) - 3(32+1) = 6.369$$

Step 5:  df = k - 1 = 3; critical value = 6.251
Step 6:  Since 6.369 > 6.251, reject $H_0$.
Step 7:  At the 10% significance level, the data provide sufficient evidence to conclude that a difference exists among the median asking rents in the four U.S. regions.

For the p-value approach, $0.05 < P(\chi^2 > 6.369) < 0.10$.   Therefore, because the p-value is smaller than the significance level,  reject $H_0$.

**16.119**    The total number of populations being sampled is k = 3.  Let the subscripts 1, 2, and 3 refer to Cuba, Nicaragua, and Costa Rica, respectively.  The total number of observations is n = 19.  Also, $n_1 = 8$, $n_2 = 6$, and $n_3 = 5$.

Step 1:  $H_0$: $\mu_1 = \mu_2 = \mu_3$ (mean earnings are equal)
         $H_a$: Not all the means are equal.
Step 2:  $\alpha = 0.10$
Step 3:

| Cuba | Rank | Nicaragua | Rank | Costa Rica | Rank |
|------|------|-----------|------|-----------|------|
| 5.36 | 11 | 25.66 | 19 | 1.51 | 5 |
| 23.88 | 18 | 2.01 | 6 | 2.32 | 8 |
| 9.70 | 14 | 1.06 | 3 | 14.20 | 16 |
| 11.44 | 15 | 1.02 | 2 | 8.75 | 13 |
| 0.54 | 1 | 5.49 | 12 | 4.75 | 10 |
| 3.64 | 9 | 2.28 | 7 | | |
| 15.36 | 17 | | | | |
| 1.33 | 4 | | | | |
| | 89 | | 49 | | 52 |

Step 4:

$$H = \frac{12}{n(n+1)}\sum \frac{R_j^2}{n_j} - 3(n+1)$$

$$= \frac{12}{19(19+1)}(\frac{89^2}{8} + \frac{49^2}{6} + \frac{52^2}{5}) - 3(19+1) = 0.982$$

Step 5:  df = k - 1 = 2; critical value = 4.605
Step 6:  Since 0.982 < 4.605, do not reject $H_0$.
Step 7:  At the 5% significance level, the data do not provide sufficient evidence to conclude that a difference exists among the mean speeds of the sea turtles found in the waters of the three countries.

For the p-value approach, $P(\chi^2 > 0.982) > 0.10$.   Therefore, because the p-value is larger than the significance level, do not reject $H_0$.

**16.121** (a)  The total number of populations being sampled is k = 4.  Let the subscripts 1, 2, 3, 4, and 5 refer to the five strains.  The total number of observations is n = 25.  Also, $n_1 = 5$, $n_2 = 5$, $n_3 = 5$, $n_4 = 5$ and $n_5 = 5$.

Step 1: $H_0$: $\mu_1 = \mu_2 = \mu_3 = \mu_4 = \mu_5$ (mean earnings are equal)

$H_a$: Not all the means are equal.

Step 2: $\alpha = 0.05$

Step 3:

| Strain A | Rank | Strain B | Rank | Strain C | Rank | Strain D | Rank | Strain E | Rank |
|---|---|---|---|---|---|---|---|---|---|
| 9 | 2 | 3 | 1 | 10 | 3 | 14 | 5 | 33 | 18 |
| 27 | 12 | 32 | 17 | 47 | 22 | 18 | 8 | 43 | 20 |
| 22 | 10 | 37 | 19 | 50 | 23 | 17 | 7 | 28 | 13 |
| 30 | 15 | 45 | 21 | 52 | 24 | 29 | 14 | 59 | 25 |
| 16 | 6 | 12 | 4 | 26 | 11 | 20 | 9 | 31 | 16 |
| | 45 | | 62 | | 83 | | 43 | | 92 |

Step 4:

$$H = \frac{12}{n(n+1)} \sum \frac{R_j^2}{n_j} - 3(n+1)$$

$$= \frac{12}{25(25+1)} \left( \frac{45^2}{5} + \frac{62^2}{5} + \frac{83^2}{5} + \frac{43^2}{5} + \frac{92^2}{5} \right) - 3(25+1) = 7.185$$

Step 5: df = k - 1 = 4; critical value = 9.488

Step 6: Since 7.185 < 9.488, do not reject $H_0$.

Step 7: At the 5% significance level, the data do not provide sufficient evidence to conclude that a difference exists among the mean bacteria counts among the five strains of *Staphylococcus aureus*.

For the p-value approach, $P(\chi^2 > 7.185) > 0.10$. Therefore, because the p-value is larger than the significance level, do not reject $H_0$.

(b) The Kruskal-Wallace test may be used whenever the distributions of the variable have the same shape. If the bacteria counts in all five strains have normal distributions with equal standard deviations, then the five distributions will have the same shape, and the Kruskal-Wallace may be used. If the populations are actually normally distributed with equal standard deviations, then the one-way ANOVA is the better choice since it was designed expressly for this situation and is more powerful than the Kruskal-Wallace test under these circumstances.

**16.123** (a) Using Minitab, with all of the data in one column named RENT and four regions of the U.S. corresponding to each of the rent values in a

column named CLASS, we choose **Stat ▶ Nonparametrics ▶ Kruskal-Wallis...**, select RENT in the **Response** text box, select REGION in the **Factor** text box, and click **OK**. The result is

Kruskal-Wallis Test on RENT

| REGION | N | Median | Ave Rank | Z |
|---|---|---|---|---|
| Midwest | 6 | 734.5 | 4.5 | -2.97 |
| Northeast | 5 | 1005.0 | 14.4 | 1.70 |
| South | 4 | 876.0 | 8.5 | -0.76 |
| West | 5 | 1025.0 | 15.4 | 2.14 |
| Overall | 20 | | 10.5 | |

H = 12.23   DF = 3   P = 0.007

* NOTE * One or more small samples

(b) Since the P-value is 0.007, which is less than the 0.05 significance level, we reject the null hypothesis of equal means; the data provide sufficient evidence to conclude that there are differences between the mean monthly rents in the four regions of the U.S.

**16.125** (a) Using Minitab, with all of the data in one column named RATE and categories of patch-size manipulations corresponding to each of the

singing rates in a column named BREAST, we choose **Stat ▶ Nonparametrics**

**▶ Kruskal-Wallis...**, select RATE in the **Response** text box, select

BREAST in the **Factor** text box, and click **OK**. The result is

Kruskal-Wallis Test on RATE

| BREAST | N | Median | Ave Rank | Z |
|--------|----|--------|----------|-------|
| Control | 8 | 16.900 | 15.1 | 1.29 |
| Enlarged | 8 | 14.900 | 16.3 | 1.84 |
| Reduced | 8 | 3.350 | 6.1 | -3.12 |
| Overall | 24 | | 12.5 | |

H = 9.86  DF = 2  P = 0.007

(b) Since the P-value is 0.007, which is less than the 0.01 significance level, we reject the null hypothesis of equal means; the data provide sufficient evidence to conclude that there are differences between the singing rates for the three patch-size manipulations.

**16.127** (a) Using Minitab, with all of the data in one column named HARDNESS and categories artificial teeth materials corresponding to each of the

volumes worn off in a column named MATERIAL, we choose **Stat ▶**

**Nonparametrics ▶ Kruskal-Wallis...**, select HARDNESS in the **Response**

text box, select MATERIAL in the **Factor** text box, and click **OK**. The result is

Kruskal-Wallis Test on HARDNESS

| MATERIAL | N | Median | Ave Rank | Z |
|----------|----|--------|----------|-------|
| Duracross | 6 | 39.45 | 15.5 | 3.37 |
| Duradent | 6 | 24.15 | 3.5 | -3.37 |
| Endura | 6 | 27.45 | 9.5 | 0.00 |
| Overall | 18 | | 9.5 | |

H = 15.16  DF = 2  P = 0.001
H = 15.20  DF = 2  P = 0.000  (adjusted for ties)

(b) Since the P-value is 0.001, which is less than the 0.05 significance level, we reject the null hypothesis of equal means; the data provide sufficient evidence to conclude that there are differences between the hardnesses for the three artificial teeth materials.

**16.129** (a) Since the populations are non-normal and do not have the same shape, neither test can be performed.

(b) Normal distributions with different shapes have different standard deviations, so the ANOVA test cannot be used. Since the shapes are different, the Kruskal-Wallis test cannot be used either.

**16.131** (a) In Exercise 16.57, the normal probability plots were all non-linear. To check whether the distributions are similar in shape, we will obtain

box plots using Minitab. Choose **Graph ▶ Boxplot**, select the **Multiple**

**Y's Simple** version and click **OK**. Enter Some, Associate, Bachelor, Master, and Doctorate in the **Graph variables** text box and click **OK**. The result is

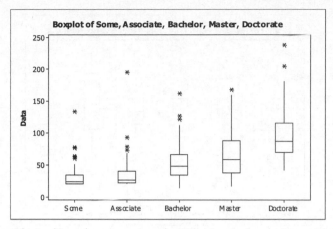

All of the distributions are right-skewed with outliers, so the shapes are similar, if not exactly the same.  We will proceed with the Kruskal-Wallis test.

(b)  Using Minitab, choose **Stat ▶ Nonparametrics ▶ Kruskal-Wallis...,**

select <u>EARNINGS</u> in the **Response** text box, select <u>DEGREE</u> in the **Factor** text box, and click **OK.**  The result is

        Kruskal-Wallis Test on EARNINGS

        DEGREE        N    Median   Ave Rank       Z
        Associate    50    26.60       83.0    -4.65
        Bachelor     50    48.35      127.2     0.18
        Doctorate    50    86.90      200.3     8.17
        Master       50    58.05      148.9     2.55
        Some         50    23.60       68.3    -6.26
        Overall     250              125.5

        H = 107.27  DF = 4  P = 0.000
        H = 107.28  DF = 4  P = 0.000  (adjusted for ties)

(c)  The P-value of the test is 0.00, which is less than the 0.05 significance level.  Reject the null hypothesis of equal means.  The data provide sufficient evidence to conclude that the mean earnings of the five degree groups are not equal.

(d)  No one-way ANOVA was carried out.

**16.133** (a)  In Exercise 16.59, normal probability plots of the lengths of cuckoo eggs laid in the nests of six species of birds were linear with approximately the same standard deviations, with the exception of the Meadow Pipit, which had several outliers.  To check whether the distributions are similar in shape, we will obtain box plots using

Minitab.  Choose **Graph ▶ Boxplot,** select the **Multiple Y's Simple**

version and click **OK.**  Enter <u>Meadow Pipit</u>, <u>Tree Pipit</u>, <u>Hedge Sparrow</u>, <u>Robin</u>, <u>Pied Wagtail</u>, and <u>Wren</u> in the **Graph variables** text box and click **OK.**  The result is

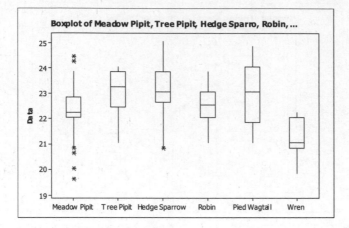

Thus, the six distributions appear to range from symmetric to slightly left skewed.  The Kruskal-Wallis test is reasonable.

(b)  Using Minitab, choose **Stat ▶ Nonparametrics ▶ Kruskal-Wallis...**,

select <u>LENGTH</u> in the **Response** text box, select <u>SPECIES</u> in the **Factor** text box, and click **OK**.  The result is

```
Kruskal-Wallis Test on LENGTH

SPECIES           N   Median   Ave Rank      Z
Hedge Sparrow    14   23.05      80.9      2.33
Meadow Pipit     45   22.25      54.8     -1.40
Pied Wagtail     15   23.05      72.6      1.44
Robin            16   22.55      64.7      0.52
Tree Pipit       15   23.25      82.8      2.65
Wren             15   21.05      19.8     -4.85
Overall         120               60.5

H = 34.80   DF = 5   P = 0.000
H = 35.04   DF = 5   P = 0.000   (adjusted for ties)
```

(c)  The P-value of the test is 0.000, which is less than the 0.05 significance level.  Reject the null hypothesis of equal means.  The data provide sufficient evidence to conclude that the mean egg lengths in the six types of nests are not equal.

(d)  The one-way ANOVA performed in Exercise 16.59 had a P-value of 0.006, very close to this result, and we reached the same conclusion.

**16.135** (a)  In Exercise 16.61, normal probability plots of the book prices for the four subject areas were linear with approximately the same standard deviations.  Thus, the four distributions appear to be the same shape, and the Kruskal-Wallis test is reasonable.

(b)  Using Minitab, choose **Stat ▶ Nonparametrics ▶ Kruskal-Wallis...**,

select <u>PRICE</u> in the **Response** text box, select <u>SUBJECT</u> in the **Factor** text box, and click **OK**.  The result is

```
Kruskal-Wallis Test on PRICE

SUBJECT        N   Median   Ave Rank      Z
Law           45   93.11      82.0      1.00
Medicine      28   96.85      82.2      0.76
Science       37   98.05      87.0      1.66
Technology    42   86.91      57.6     -3.27
Overall      152               76.5

H = 11.02   DF = 3   P = 0.012
H = 11.02   DF = 3   P = 0.012   (adjusted for ties)
```

(c) The P-value of the test is 0.012, which is less than the 0.05 significance level. Reject the null hypothesis of equal means. The data provide sufficient evidence to conclude that the mean price for hardcover books in the four subject areas are not equal.

(d) The one-way ANOVA performed in Exercise 16.61 had a P-value of 0.014, and we reached the same conclusion.

**16.137**(a) In Exercise 16.63, normal probability plots of the hemoglobin levels of patients with three different types of sickle cell disease were linear, but with different standard deviations. To check whether the distributions are similar in shape, we will obtain box plots using

Minitab. Choose **Graph ▶ Boxplot**, select the **Multiple Y's Simple** version and click **OK**. Enter <u>High</u>, <u>Medium</u>, and <u>Low</u> in the **Graph variables** text box and click **OK**. The result is

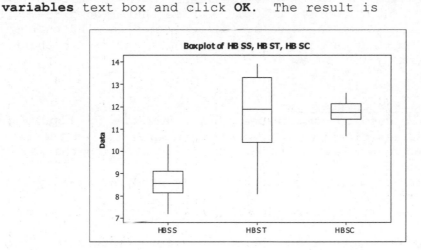

The shapes of the distributions are close to being the same, although with different spreads. The assumptions for Kruskal-Wallis test are satisfied.

(b) Using Minitab, with all of the data in one column named HEMOLEVEL and magazine educational levels in a column named TYPE, we choose **Stat ▶ Nonparametrics ▶ Kruskal-Wallis...**, select <u>HEMOLEVEL</u> in the **Response** text box, select <u>TYPE</u> in the **Factor** text box, and click **OK**. The result is

```
Kruskal-Wallis Test on HEMOLEVEL
```

| TYPE | N | Median | Ave Rank | Z |
|------|------|--------|----------|-------|
| HB SC | 10 | 11.750 | 29.4 | 2.53 |
| HB SS | 16 | 8.550 | 9.6 | -4.88 |
| HB ST | 15 | 11.900 | 27.6 | 2.68 |
| Overall | 41 | | 21.0 | |

```
H = 23.92  DF = 2  P = 0.000
H = 23.94  DF = 2  P = 0.000 (adjusted for ties)
```

(c) The P-value is 0.000, which is less than the significance level of 0.05, so we reject the null hypothesis that the mean hemoglobin levels are the same for the three types of sickle cell disease. We conclude that there are differences in the mean hemoglobin levels for the three types of disease. Without a formal analysis, it would appear that the mean hemoglobin level for the HB SS group is lower than either of the other two groups, and the other two groups have approximately equal means.

(d) The one-way ANOVA performed in Exercise 16.63 also had a P-value of

0.000, and we reached the same conclusion.

**16.139** (a) The defining formula and the computing formula for H are equivalent as long as there are no ties in the data. The computing formula uses the facts that the sum of the first n integers is n(n+1)/2 and that the sum of the squares of the first n integers is n(n+1)(2n+1)/6. Thus it assumes that all of the ranks are distinct integers. The discrepancy between the formulas results, as in this case, when there are ties in the ranks, invalidating the formula for the sum of squares of the first n integers. The formula for the sum of the first n integers remains valid.

(b) No. In this example, both values of H are greater than the critical value of 5.991.

(c) No. Using Table VII, either value leads to the P-value being between 0.005 and 0.01, but close to 0.01. While there would be a difference in P-values anytime the H values differ, keep in mind that the distribution of H is only approximately chi-square, so the P-values aren't exact in any case.

## Review Problems For Chapter 16

1.  One-way ANOVA is used to compare means of a variable for populations that result from a classification by one other variable called the factor.

2.  (i) Simple random samples: Check by carefully studying the way that the sampling was done.
    (ii) Independent samples: Check by carefully studying the way that the sampling was done.
    (iii) Normal populations; Check with normal probability plots, histograms, dotplots.
    (iv) All populations have equal standard deviations; this is a reasonable assumption if the ratio of the largest standard deviation to the smallest one is less than 2.

3.  F distribution

4.  There are n = 17 observations and k = 3 samples. The degrees of freedom are (k - 1, n - k) = (2, 14).

5.  (a) Variation among sample means is measured by the mean square for treatments, MSTR = SSTR/(k - 1).
    (b) The variation within samples is measured by the error mean square, MSE = SSE/(n - k).

6.  (a) SST is the total sum of squares. It measures the total variation among all of the sample data. $SST = \sum (x - \bar{x})^2$

    SSTR is the treatment sum of squares. It measures the variation among the sample means. $SSTR = \sum n_i (\bar{x}_i - \bar{x})^2$

    SSE is the error sum of squares. It measures the variation within the samples. $SSE = \sum (x - \bar{x}_i)^2 = \sum (n_i - 1)s_i^2$

    (b) SST = SSTR + SSE. This means that the total variation in the sample can be broken down into two components, one representing the variation between the sample means and one representing the variation within the samples.

7.  (a) One purpose of a one-way ANOVA table is to organize and summarize the quantities required for ANOVA.

(b)

| Source | df | SS | MS=SS/df | F-statistic |
|--------|-----|-----|--------------------|-------------|
| Treatment | k - 1 | SSTR | MSTR=SSTR/(k - 1) | F=MSTR/MSE |
| Error | n - k | SSE | MSE=SSE/(n - k) | |
| Total | n - 1 | SST | | |

**8.** If the null hypothesis is rejected, a multiple comparison is done to determine which means are different.

**9.** The individual confidence level gives the confidence that we have that any particular confidence interval will contain the population quantity being estimated. The family confidence level gives the confidence that we have that all of the confidence intervals will contain all of the population quantities being estimated. The family confidence level is appropriate for multiple comparisons because we are interested in all of the possible comparisons.

**10.** Tukey's multiple-comparison procedure is based upon the Studentized Range distribution or q-distribution.

**11.** Larger. One has to be less confident about the truth of several statements at once than about the truth of a single statement. For example, if one were 99% confident about a single statement, the confidence that two statements were both true must be smaller since there is no way to have 100% confidence in the second statement. Similarly, each time a statement is added to the list, the overall confidence that all of them are true must decrease.

**12.** The parameters for the q-curve are $K = k = 3$, and $V = n - k = 17 - 3 = 14$. [k = number of samples and n = total number of observations.]

**13.** Kruskal-Wallis test

**14.** Chi-square distribution with k - 1 degree of freedom [k = number of samples]

**15.** If the null hypothesis of equal population means is true, then the means of the ranks of the k samples should be about equal. If the variation in the means of the ranks for the k samples is too large, then we have evidence against the null hypothesis.

**16.** Use the Kruskal-Wallis test. The outliers will have a greater effect on the ANOVA than on the Kruskal-Wallis test since an outlier will be replaced by its rank in the Kruskal-Wallis test and the rank of the most distant outlier is either 1 or n regardless of how large or small the actual data value is. In other words, the Kruskal-Wallis test is more robust to outliers than is the ANOVA.

**17.** (a) 2 (b) 14 (c) 3.74 (d) 6.51 (e) 3.74

**18.** (a) The total number of populations being sampled is k = 3. Let the subscripts 1, 2, and 3 refer to A, B, and C, respectively. The total number of pieces of sample data is n = 12. Also, $n_1 = 3$, $n_2 = 5$, $n_3 = 4$. The following statistics for each sample are: $\bar{x}_1 = 3$, $\bar{x}_2 = 3$, and $\bar{x}_3 = 6$. Also, $s_1 = 2.000$, $s_2 = 2.449$, and $s_3 = 4.243$. Note, the mean of all the sample data is $\bar{x} = 4$. We use this information as follows:

(b) SST $= \sum (x - \bar{x})^2 = (1 - 4)^2 + \ldots + (3 - 4)^2 = 110.0$

SSTR $= n_1(\bar{x}_1 - \bar{x})^2 + n_2(\bar{x}_2 - \bar{x})^2 + n_3(\bar{x}_3 - \bar{x})^2$
$= 3(3 - 4)^2 + 5(3 - 4)^2 + 4(6 - 4)^2 = 24.0$

SSE $= (n_1 - 1)s_1^2 + (n_2 - 1)s_2^2 + (n_3 - 1)s_3^2$
$= (3 - 1)(2.000)^2 + (5 - 1)(2.449)^2 + (4 - 1)(4.243)^2 = 86.0$

SST = SSTR + SSE since 110.0 = 24.0 + 86.0

(c) Let $T_1$, $T_2$, and $T_3$ refer to the sum of the data values in each of the three samples, respectively. Thus: $T_1 = 9$   $T_2 = 15$   $T_3 = 24$.

Also, the sum of all the data values is $\Sigma x = 48$, and their sum of squares is $\sum x^2 = 302$.

Consequently:  $SST = \sum x^2 - \dfrac{\left(\sum x\right)^2}{n} = 302 - \dfrac{48^2}{12} = 110$

$$SSTR = \dfrac{T_1^2}{n_1} + \dfrac{T_2^2}{n_2} + \dfrac{T_3^2}{n_3} - \dfrac{\left(\sum x\right)^2}{n} = \dfrac{9^2}{3} + \dfrac{15^2}{5} + \dfrac{24^2}{4} - \dfrac{48^2}{12} = 24$$

and SSE = SST - SSTR = 110 - 24 = 86 .

(d)

| Source | Df | SS | MS=SS/df | F-statistic |
|--------|----|----|----------|-------------|
| Treatment | 2 | 24 | 12.000 | 1.255 |
| Error | 9 | 86 | 9.556 | |
| Total | 11 | 110 | | |

19. (a) MSTR is a measure of the variation between the sample mean losses for highway robberies, gas station robberies, and convenience store robberies.

(b) MSE is a measure of the variation within the three samples.

(c) The four assumptions for one-way ANOVA, given in Key Fact 16.2: simple random samples, independent samples, normal populations, and equal standard deviations.  Assumptions 1 and 2 on simple random independent samples are absolutely essential to the one-way ANOVA procedure. Assumption 3 on normality is not too critical as long as the populations are not too far from being normally distributed. Assumption 4 on equal standard deviations is also not that important provided the sample sizes are roughly equal.

20. (a) The samples and their normal scores are shown in the table below along with the residuals which are $x_i - \overline{x}_i$ .

| | Hwy | w | Residual | Gas | w | Residual | Conv. St. | w | Residual |
|--|-----|---|----------|-----|---|----------|-----------|---|----------|
| | 797 | -1.18 | -27.8 | 825 | -1.28 | 129.17 | 665 | -1.28 | 82.83 |
| | 841 | -0.50 | 16.2 | 722 | -0.64 | 26.17 | 742 | -0.64 | 159.83 |
| | 684 | 0.00 | -140.8 | 701 | -0.20 | 5.17 | 701 | -0.20 | 118.83 |
| | 933 | 0.50 | 108.2 | 640 | 0.20 | -55.83 | 527 | 0.20 | -55.17 |
| | 869 | 1.18 | 44.2 | 480 | 0.64 | -215.83 | 423 | 0.64 | -159.17 |
| | | | | 807 | 1.28 | 111.17 | 435 | 1.28 | -147.17 |
| Total | 4124 | | | 4175 | | | 3493 | | |
| Mean | 824.8 | | | 695.83 | | | 582.17 | | |
| St.Dev. | 92.90 | | | 126.06 | | | 138.97 | | |

We will plot the normal scores w against the data for each type of store. The graphs have been produced using Minitab, but you can easily do them by hand.  The standard deviations for each type of robbery are shown in the above table.

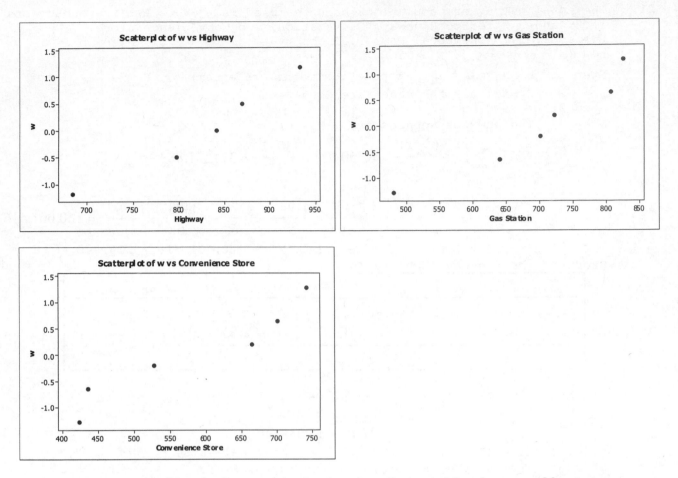

(b)  Now order all of the residuals in the above table from smallest to largest and plot them against the normal scores in Table III for n = 17.  Also plot the residuals against the means for each type of robbery.  The results are

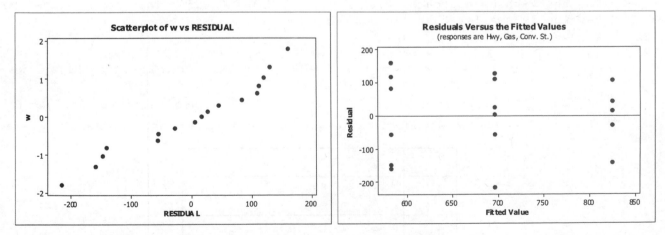

(c)  All three of the normal probability plots are close to linear, and the ratio of the largest to smallest standard deviation is 1.50.  The residual analysis shows that the normal probability plot of the residuals is close to linear and the last graph shows that the spreads of the residuals are about the same for each type of robbery.  Thus the ANOVA assumptions of normality and equal standard deviations appear to be reasonable.

21.   The total number of populations being sampled is k = 3.   Let the subscripts 1, 2, and 3 refer to Highway, Gas station, and Convenience store, respectively.   The total number of observations is n = 17. Also, $n_1$ = 5, $n_2$ = 6, and $n_3$ = 6.

Step 1:   $H_0$: $\mu_1 = \mu_2 = \mu_3$   (population means are equal)

$H_a$: Not all the means are equal.

Step 2:   $\alpha$ = 0.05

Step 3:   The sums of squares are

$$SST = \sum x^2 - \frac{\left(\sum x\right)^2}{n} = 8{,}550{,}628 - \frac{11792^2}{17} = 371{,}141.882$$

$$SSTR = \frac{T_1^2}{n_1} + \frac{T_2^2}{n_2} + \frac{T_3^2}{n_3} - \frac{\left(\sum x\right)^2}{n} = \frac{4124^2}{5} + \frac{4175^2}{6} + \frac{3493^2}{6} - \frac{11792^2}{17} = 160{,}601.416$$

and SSE = SST - SSTR = 371,141.882 - 160,601.416 = 210,540.467

Step 4:   The one-way ANOVA table is:

| Source | df | SS | MS = SS/df | F-statistic |
|--------|----|----|-----------|-------------|
| Treatment | 2 | 160,601.416 | 80,300.708 | 5.34 |
| Error | 14 | 210,540.467 | 15,038.605 | |
| | 16 | 371,141.882 | | |

The F-statistic is defined and calculated as

$$F = \frac{MSTR}{MSE} = \frac{80{,}300.708}{15{,}038.605} = 5.34$$

Step 5:   df = (k - 1, n - k) = (2, 14); critical value = 3.74

Step 6:   From Step 4, F = 5.34.   From Step 5, $F_{0.05}$ = 3.74.   Since 5.34 > 3.74, reject $H_0$.

Step 7:   At the 5% significance level, the data provide sufficient evidence to conclude that a difference in mean losses exists among the three types of robberies.

For the P-value approach, 0.01 < P(F > 5.34) < 0.025.   Therefore, since the P-value is smaller than the significance level, reject $H_0$.

22.   (a)   $q_{0.05}$ = 3.70          (b)      4.89

23.   (a)   Step 1:   Family confidence level = 0.95.

Step 2:   $\kappa$ = 3 and $\nu$ = n - k = 17 - 3 = 14.   From Exercise 21(a), we find that $q_{0.05}$ = 3.70.

Step 3:   Before obtaining all possible confidence intervals for $\mu_i - \mu_j$, we construct a table giving the sample means and sample sizes.

| j | 1 | 2 | 3 |
|---|---|---|---|
| $\bar{x}_j$ | 824.8 | 695.8 | 583.2 |
| $n_j$ | 5 | 6 | 6 |

In Problem 19, we found that MSE = 15,039.   Now we are ready to obtain the required confidence intervals.

The endpoints for $\mu_1 - \mu_2$ are

$$824.8 - 695.8 \pm \frac{3.70}{\sqrt{2}} \cdot \sqrt{15039} \cdot \sqrt{\frac{1}{5} + \frac{1}{6}} = -65.3 \text{ to } 323.3$$

The endpoints for $\mu_1 - \mu_3$ are

$$824.8 - 583.2 \pm \frac{3.70}{\sqrt{2}} \cdot \sqrt{15039} \cdot \sqrt{\frac{1}{5} + \frac{1}{6}} = 48.3 \text{ to } 436.9$$

The endpoints for $\mu_2 - \mu_3$ are

$$695.8 - 583.2 \pm \frac{3.70}{\sqrt{2}} \cdot \sqrt{15039} \cdot \sqrt{\frac{1}{6} + \frac{1}{6}} = -71.6 \text{ to } 298.9$$

Step 4:      Based on the confidence intervals in Step 3, we declare means $\mu_1$ and $\mu_3$ different. All other pairs of means are not declared different.

Step 5:      We summarize the results with the following diagram.

| Convenience store (3) | Gas station (2) | Highway (1) |
|---|---|---|
| 583.2 | 695.8 | 824.8 |

(b) Interpreting this diagram, we conclude with 95% confidence that the mean loss due to convenience-store robberies is less than that due to highway robberies; no other means can be declared different.

**24.**    (a) The total number of populations being sampled is k = 3. Let the subscripts 1, 2, and 3 refer to highway, gas station, and convenience store, respectively. The total number of pieces of sample data is n = 17. Also, $n_1 = 5$, $n_2 = n_3 = 6$.

Step 1:    $H_0$: $\mu_1 = \mu_2 = \mu_3$    (mean losses are equal)

        $H_a$: Not all the means are equal.

Step 2:    $\alpha$ = 0.05

Step 3:

| Sample 1 | Rank | Sample 2 | Rank | Sample 3 | Rank |
|---|---|---|---|---|---|
| 684 | 7 | 480 | 3 | 423 | 1 |
| 797 | 12 | 640 | 5 | 435 | 2 |
| 841 | 15 | 701 | 8.5 | 527 | 4 |
| 869 | 16 | 722 | 10 | 665 | 6 |
| 933 | 17 | 807 | 13 | 701 | 8.5 |
|  |  | 825 | 14 | 742 | 11 |
|  | 67 |  | 53.5 |  | 32.5 |

Step 4:

$$H = \frac{12}{n(n+1)} \sum \frac{R_j^2}{n_j} - 3(n+1) = \frac{12}{17(17+1)} \left( \frac{67^2}{5} + \frac{53.5^2}{6} + \frac{32.5^2}{6} \right) - 3(17+1) = 6.819$$

Step 5:    df = k - 1 = 2; critical value = 5.991

Step 6:    Since 6.819 > 5.991, reject $H_0$.

Step 7:    At the 5% significance level, the data provide sufficient evidence to conclude that a difference exists in mean losses among the three types of robberies.

For the p-value approach, $0.025 < P(\chi^2 > 6.819) < 0.05$.  Therefore, because the p-value is smaller than the significance level, reject $H_0$.

(b)  It is permissible to perform the Kruskal-Wallis test because normal populations having equal standard deviations have the same shape.  It is better to use the one-way ANOVA test when the assumptions for that test are met because it is more powerful than the Kruskal-Wallis test.

(c)  The P-value for the ANOVA test was between 0.01 and 0.025.  The Kruskal-Wallis P-value was between 0.025 and 0.05.  Thus, we arrive at the same conclusion when testing at the 0.05 significance level.

25.  (a)  Using Minitab, choose **Graph ▶ Probability Plot**, select the **Simple** version and click **OK**.  Enter Loss, Stable, and Gain in the **Graph Variables** text box, click on the **Multiple Graphs** button and select **In separate panels of the same graph**, and click **OK**.  The result is

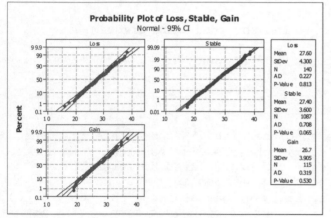

The standard deviations of the three groups are shown in the right hand panel as 4.300, 3.600, and 3.905.

(b)  To carry out the Analysis of variance and residual analysis, we will use the data in a single column, which is named BMI.  In the next column, which is named GROUP, are the names of each of the regions corresponding to each data value in BMI.  The residual analysis consists of plotting the residuals against the means (or fits) and constructing a normal probability plot.  This is done at the same time as the analysis of variance in Minitab.  To do this, we choose **Stat ▶**

**Anova ▶One-way...**, select BMI for the **Response:** text box and GROUP in the **Factor:** text box. Enter 95 in the **Confidence level** box. Then click on the **Graphs** button and click to place check marks in the boxes for **Normal plot of residuals** and **Residuals versus fits**, and click **OK**.  Click on the **Comparisons** button and select **Tukey's**, and enter 5 in the **family error rate** box, and click **OK**.  The graphical results are

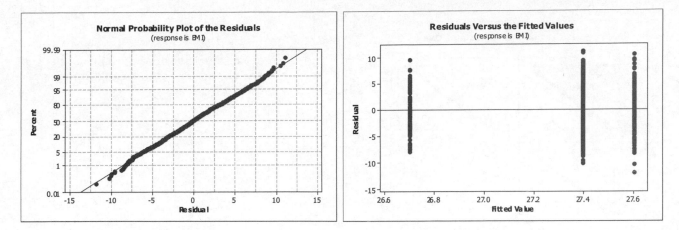

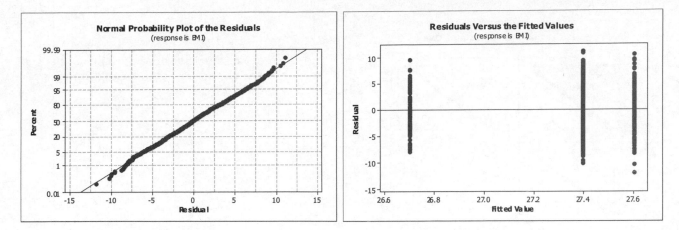

(c) The normal probability plots in part (a) are linear as is the normal probability plot of the residuals in part (b). The plot of the residuals against the fitted values shows approximately equal spreads. The rule of 2 is not violated since the ratio of the largest to smallest standard deviation is 4.3/3.6 = 1.19.  A one-way ANOVA is reasonable.

(d) The one-way ANOVA results from the procedure in part (b) are

**One-way ANOVA: BMI versus GROUP**

```
Source    DF      SS    MS     F      P
GROUP      2    60.0  30.0  2.18  0.113
Error   1339 18379.7  13.7
Total   1341 18439.6

S = 3.705   R-Sq = 0.33%   R-Sq(adj) = 0.18%

                            Individual 95% CIs For Mean Based on
                            Pooled StDev
Level      N    Mean  StDev ------+---------+---------+---------+---
Gain     115  26.700  3.905 (----------*----------)
Loss     140  27.603  4.300                  (---------*---------)
Stable  1087  27.399  3.600                      (---*--)
                            ------+---------+---------+---------+---
                            26.40     27.00     27.60     28.20
```

(e) The P-value of the test is 0.113, which is larger than the 0.05 significance level.  Do not reject the null hypothesis of equal means. The data do not provide sufficient evidence to conclude that there is a difference in body mass index among the three groups of men with different weight losses.

(f) Since the ANOVA test was not significant, it is not necessary to use Tukey's multiple comparison.  None of the means will differ from the others.

26. (a) Using Minitab, choose **Graph ▶ Probability Plot**, select the **Simple** version and click **OK**.  Enter Loss, Stable, and Gain in the **Graph Variables** text box, click on the **Multiple Graphs** button and select **In separate panels of the same graph**, and click **OK**.  The result is

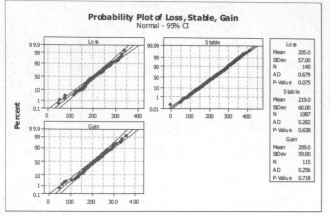

The standard deviations of the three groups are shown in the right hand panel as 57, 60, and 59.

(b)  To carry out the Analysis of variance and residual analysis, we will use the data in a single column, which is named POWER.  In the next column, which is named GROUP, are the names of each of the regions corresponding to each data value in POWER.  The residual analysis consists of plotting the residuals against the means (or fits) and constructing a normal probability plot.  This is done at the same time as the analysis of variance in Minitab.  To do this, we choose **Stat ▶**

**Anova ▶One-way...**, select POWER for the **Response:** text box and GROUP in the **Factor:** text box. Enter 95 in the **Confidence level** box. Then click on the **Graphs** button and click to place check marks in the boxes for **Normal plot of residuals** and **Residuals versus fits**, and click **OK**. Click on the **Comparisons** button and select **Tukey's**, and enter 5 in the **family error rate** box, and click **OK**.  The graphical results are

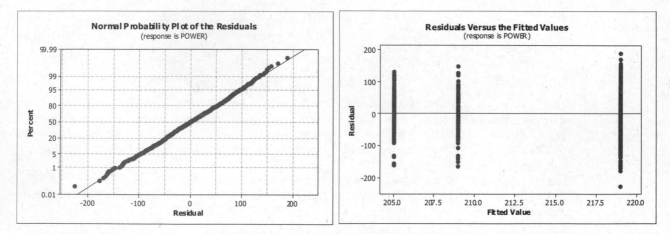

(c)  The normal probability plots in part (a) are linear as is the normal probability plot of the residuals in part (b). The plot of the residuals against the fitted values shows approximately equal spreads. The rule of 2 is not violated since the ratio of the largest to smallest standard deviation is 60/57 = 1.05.  A one-way ANOVA is reasonable.

(d)  The one-way ANOVA results from the procedure in part (b) are

**One-way ANOVA: POWER versus GROUP**

```
Source     DF      SS       MS      F      P
GROUP       2    31742    15871   4.47  0.012
Error    1339  4758204     3554
Total    1341  4789946
```

```
S = 59.61   R-Sq = 0.66%   R-Sq(adj) = 0.51%
```

```
                                Individual 95% CIs For Mean Based on Pooled
                                StDev
Level       N    Mean   StDev   -+---------+---------+---------+--------
Gain      115  209.00   59.00        (--------------*--------------)
Loss      140  205.00   57.00     (-------------*-------------)
Stable   1087  219.00   60.00                                    (----*----)
                                -+---------+---------+---------+--------
                                196.0     203.0     210.0     217.0
```

Pooled StDev = 59.61

(e) The P-value of the test is 0.012, which is smaller than the 0.05 significance level. Reject the null hypothesis of equal means. The data provide sufficient evidence to conclude that there is a difference in leg power among the three groups of men with different weight losses.

(f) The Tukey multiple comparison results produced by the procedure in part (b) are

```
Tukey 95% Simultaneous Confidence Intervals
All Pairwise Comparisons among Levels of GROUP

Individual confidence level = 98.06%

GROUP = Gain subtracted from:

GROUP   Lower  Center  Upper  --------+---------+---------+---------+-
Loss   -21.56   -4.00  13.56     (-----------*-----------)
Stable  -3.68   10.00  23.68                (--------*--------)
                               --------+---------+---------+---------+-
                                     -15        0        15       30

GROUP = Loss subtracted from:

GROUP   Lower  Center  Upper  --------+---------+---------+---------+-
Stable   1.47   14.00  26.53                  (-------*--------)
                               --------+---------+---------+---------+-
                                     -15        0        15       30
```

Looking at the confidence intervals that do not contain zero, we see that there is a difference in mean leg power between the "Loss" group and the "Stable" group, and that there are no other pairs that can be declared different. All of this can be said with 95% confidence.

27. (a) Using Minitab, choose **Graph ▶ Probability Plot**, select the **Simple** version and click **OK**. Enter 25-34, 35-44, 45-54 and 55-64 in the **Graph Variables** text box, click on the **Multiple Graphs** button and select **In separate panels of the same graph**, and click **OK**. The result is

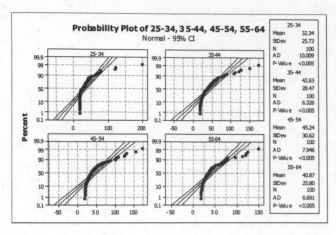

The standard deviations of the incomes in the four age groups are 25.73, 28.47, 30.62, and 25.80.

(b)   To carry out the Analysis of variance and residual analysis, we will use the data in a single column, which is named POWER.  In the next column, which is named GROUP, are the names of each of the regions corresponding to each data value in POWER.  The residual analysis consists of plotting the residuals against the means (or fits) and constructing a normal probability plot.  This is done at the same time as the analysis of variance in Minitab.  To do this, we choose **Stat ▶**

**Anova ▶One-way...**, select <u>POWER</u> for the **Response:** text box and <u>GROUP</u> in the **Factor:** text box. Enter <u>95</u> in the **Confidence level** box. Then click on the **Graphs** button and click to place check marks in the boxes for **Normal plot of residuals** and **Residuals versus fits**, and click **OK**. Click on the **Comparisons** button and select **Tukey's**, and enter <u>5</u> in the **family error rate** box, and click **OK**.  The graphical results are

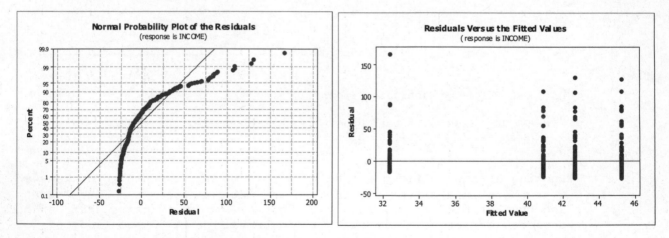

(c)   The normal probability plots for each age group are non-linear, as is the residual normal probability plot in part (b).  The rule of 2 is not violated as the ratio of the largest to smallest standard deviation is $30.62/25.73 = 1.38$.  The normality assumption is not reasonable, so parts (d)-(f) are omitted.

28.   (a)   Since the one-way ANOVA was reasonable, the Kruskal-Wallis test is also reasonable.

(b)   Using Minitab, choose **Stat ▶ Nonparametrics ▶ Kruskal-Wallis...**, select <u>BMI</u> in the **Response** text box, select <u>GROUP</u> in the **Factor** text box, and click **OK**.  The result is

Kruskal-Wallis Test on BMI

| GROUP | N | Median | Ave Rank | Z |
|---|---|---|---|---|
| Gain | 115 | 26.60 | 608.3 | -1.83 |
| Loss | 140 | 27.55 | 696.8 | 0.82 |
| Stable | 1087 | 27.40 | 674.9 | 0.67 |
| Overall | 1342 | | 671.5 | |

H = 3.74  DF = 2  P = 0.154
H = 3.74  DF = 2  P = 0.154  (adjusted for ties)

(c) The P-value for the test is 0.154, which is larger than the 0.05 significance level.  Do not reject the null hypothesis of equal means. There is not sufficient evidence to conclude that a difference exists in the body mass index of the men in the three weight loss groups.

(d) The P-value for the one-way ANOVA test was 0.113.  This is close to the Kruskal-Wallis P-value and results in the same conclusion.

**29.** (a) Since the one-way ANOVA was reasonable, the Kruskal-Wallis test is also reasonable.

(b) Using Minitab, choose **Stat ▶ Nonparametrics ▶ Kruskal-Wallis...**,
select POWER in the **Response** text box, select GROUP in the **Factor** text box, and click **OK**.  The result is

Kruskal-Wallis Test on POWER

| GROUP | N | Median | Ave Rank | Z |
|---|---|---|---|---|
| Gain | 115 | 204.4 | 623.1 | -1.40 |
| Loss | 140 | 202.8 | 594.8 | -2.47 |
| Stable | 1087 | 219.6 | 686.5 | 2.93 |
| Overall | 1342 | | 671.5 | |

H = 8.90  DF = 2  P = 0.012
H = 8.90  DF = 2  P = 0.012  (adjusted for ties)

(c) The P-value for the test is 0.012, which is smaller than the 0.05 significance level.  Reject the null hypothesis of equal means.  There is sufficient evidence to conclude that a difference exists in the leg power of the men in the three weight loss groups.

(d) The P-value for the one-way ANOVA test was 0.012.  This is the same as the Kruskal-Wallis P-value and results in the same conclusion.

**30.** (a) The normal probability plots in Problem 27 were all non-linear.  To check whether the distributions are similar in shape, we will obtain

box plots using Minitab.  Choose **Graph ▶ Boxplot**, select the **Multiple Y's Simple** version and click **OK**.  Enter 25-34, 35-44, 454-54, and 55-64 in the **Graph variables** text box and click **OK**.  The result is

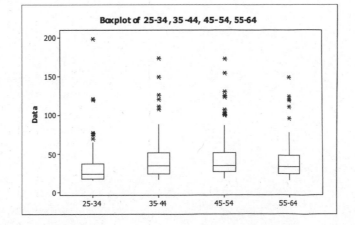

It appears that all four of the distributions have about the same shape, right-skewed, about the same spread.  Thus the assumptions for using the Kruskal-Wallis test are reasonable.

(b)   Choose **Stat ▶ Nonparametrics ▶ Kruskal-Wallis...**, select <u>INCOME</u> in the

**Response** text box, select <u>AGE</u> in the **Factor** text box, and click **OK**.
The result is

Kruskal-Wallis Test on INCOME

| AGE | N | Median | Ave Rank | Z |
|---|---|---|---|---|
| 25-34 | 100 | 24.10 | 146.9 | -5.35 |
| 35-44 | 100 | 34.60 | 216.7 | 1.62 |
| 45-54 | 100 | 34.50 | 227.6 | 2.71 |
| 55-64 | 100 | 33.30 | 210.7 | 1.02 |
| Overall | 400 | | 200.5 | |

H = 29.72   DF = 3   P = 0.000
H = 29.72   DF = 3   P = 0.000   (adjusted for ties)

(c)   The P-value of the test is 0.000, which is smaller than the 0.05 significance level.  Reject the null hypothesis of equal means.  The data provide sufficient evidence to conclude that there is a difference in the mean incomes of the four age groups.

(d)   No one-way ANOVA was performed.